Microseismic Monitoring and Stability Analysis of Large Rock Mass Engineering

Ke Ma · Chunan Tang

Microseismic Monitoring and Stability Analysis of Large Rock Mass Engineering

Science Press
Beijing

Springer

Ke Ma
Department of Civil Engineering
Dalian University of Technology
Dalian, China

Chunan Tang
Department of Civil Engineering
Dalian University of Technology
Dalian, China

ISBN 978-981-96-6482-5 ISBN 978-981-96-6483-2 (eBook)
https://doi.org/10.1007/978-981-96-6483-2

Jointly published with Science Press
The print edition is not for sale in China mainland. Customers from China mainland please order the print book from: Science Press.

This Springer imprint is published by the registered company Springer Nature Singapore Pte Ltd.
The registered company address is: 152 Beach Road, #21-01/04 Gateway East, Singapore 189721, Singapore

Preface

The rapid advancement of China's national economy and technology has significantly improved people's living standards, resulting in a continuous increase in energy demand among the general population. Starting from the 10th Five-Year Plan period in China, the state has proposed and progressively implemented a series of major projects, including Electricity Transmission from West to East China, Natural Gas Transmission from West to East China, China Digs to Earth's Center, and National Petroleum Depot. These initiatives have effectively transformed the resource advantages of western provinces into economic growth drivers, optimized and adjusted traditional energy sectors, and introduced new energy reserve methods. Simultaneously, numerous large-scale rock mass engineering projects have emerged at an accelerating pace. The safety of these large-scale rock mass engineering projects, both during construction and operation, has consistently remained a topic of significant concern. As a three-dimensional "body" monitoring approach, microseismic monitoring has increasingly found widespread application in safety early-warning systems for rock mass engineering projects.

Chapter 1 closely addresses the critical scientific issue of excavation-induced deformation effects and stability evolution mechanisms of highly steep rock slopes in complex geological environments. Taking the right-bank slope of Dagangshan Hydropower Station, located in Southwest China and characterized by a distinctive slope structure, as the research object, a geomechanical model that accurately represents the unloading behavior of slope structures is established. This is achieved by integrating microseismic monitoring with numerical simulation analysis, based on deformation indicators observed during excavation. A systematic evaluation of slope stability is conducted, explaining the mechanisms leading to localized slope failures during excavation, and elucidating the evolution of deformation and failure under excavation-induced effects. Furthermore, numerical models of anti-shear gallery structures for slope stabilization are developed, through which the interaction mechanisms and deformation coordination between landslides and anti-shear galleries during slope failure are analyzed. The appropriateness and effectiveness of the anti-shear gallery for landslide mitigation are validated, thus providing essential references to ensure the long-term safety and stability of slopes.

Chapter 2 introduces large underground water-sealed oil caverns currently being constructed in China, characterized by unique rock-mass structures with high sidewalls, large spans, and no lining. Stability issues associated with excavation-induced disturbances and subsequent operational phases have become increasingly prominent. This chapter takes the large underground water-sealed oil caverns in Jinzhou, Liaoning Province as its research background. Based on the physical and mechanical foundation that microseismic effects are consistent with rock mass damage, the development laws of microfractures in cavern rock masses subjected to excavation unloading are studied through theoretical analysis, microseismic monitoring, and numerical simulation. Meanwhile, leveraging abundant microseismic source information—including the spatial positions of microfracture points, source dimensions, energy, and stress drop, a geological model of the microseismic monitoring area is integrated with a three-dimensional numerical mechanical model, and consequently, a systematic method suitable for evaluating the sealing performance of underground water-sealed oil caverns is proposed.

Chapter 3 takes the double curvature arch dam of Dagangshan Hydropower Station, located in a high seismic region, as the research object. A comprehensive microseismic monitoring system was successfully established at the dam heel and dam galleries, enabling real-time and continuous monitoring of micro-fracture events within the dam structure. This system effectively captured microseismic events initiated during the reservoir impoundment period. Through on-site microseismic monitoring data, numerical simulations, and theoretical analyses, this chapter investigates the initiation, development, propagation, coalescence, and interactive processes of micro-fractures within the dam concrete under construction activities and reservoir impoundment conditions, elucidating their roles in inducing macroscopic instability and failure modes of the dam structure. Detailed seismic source parameters, including the spatial locations of micro-fracture points, seismic source size, energy, and stress drop, were revealed, providing insights into the relationship between dam heel deformation and microseismic activities. Furthermore, based on the spatial-temporal correlation between microseismic activity within dam galleries and the observed cracks at the arch crown of the gallery, a microseismic threshold for cracking in the arch crown of the gallery was established. By comparatively analyzing microseismic monitoring results and three-dimensional numerical simulations, the intrinsic causes of cracking in the arch crown of the gallery were identified, potential hazard zones within the double curvature arch dam during reservoir operation were delineated, and the inherent relationship between microseismic events and dam stress was clarified. The causes of microseismic event initiation at Dagangshan Hydropower Station were also explained. Finally, this chapter proposes a systematic method suitable for stability analysis and evaluation of large concrete double curvature arch dams in hydropower engineering projects in China, thereby providing critical references for the design, construction, and safety management of such dams in Southwest China.

Chapter 4 addresses the issue of coal mining above confined aquifers, taking the 22517-working face of the Dongjiahe Coal Mine of Shaanxi Coal Chenghe Mining Co., Ltd. as the research background. A microseismic monitoring system was established to conduct real-time monitoring of damage and fracture events within the floor rock mass. This system captured extensive data reflecting the progression from gradual fracture initiation to instability, as well as from dispersed micro-fractures to cascading coalescence. The evolution characteristics of micro-fractures within the floor rock mass during initial roof weighting and fault-crossing stages were analyzed. By integrating on-site microseismic monitoring results with the Realistic Failure Process Analysis (RFPA2D), the complete process of initiation, propagation, and development of water-conducting fractured zones in the coal mine floor was accurately reconstructed. The evolution of damage, stress, and seepage fields within coal and rock masses during the incubation period of water inrush channels was systematically studied, emphasizing the precursory response characteristics of micro-fractures and rock damage under mining conditions before water inrush disasters. Furthermore, simplified mechanical models based on actual engineering conditions were developed, and theoretical analyses using the theory of elasticity and structural mechanics were conducted to investigate the stress distribution within the floor rock mass during initial roof weighting and fault-crossing events. The stress components induced by combined effects of mining-induced stress, original rock stress, and water pressure along the coal seam direction were analyzed. Additionally, the stress distribution patterns around faults before and after fault-crossing, as well as the influence of structural deformation modes of the floor rock mass on fault damage characteristics, were explained in detail. Through comprehensive theoretical analysis, numerical simulation, and microseismic monitoring, the chapter elucidates the evolution patterns of micro-fractures and rock damage in different types of coal-floor water inrush hazards and explores the formation mechanisms of water inrush channels.

Chapter 5 centers on the critical scientific issue of rock damage evolution and failure mechanisms during excavation of large-span underground power houses. Taking Huanggou Pumped Storage Power Station in Heilongjiang Province as the research object, a microseismic monitoring system was established to monitor micro-fractures in the surrounding rock during the construction period, capturing real-time microseismic events throughout excavation. The spatiotemporal evolution patterns of microseismic activity were analyzed, together with their correlations with construction dynamics and geological structures. Additionally, using RFPA2D-SRM (Strength Reduction Method), which accurately simulates the gradual rock failure process, the evolution process of rock damage consistent with microseismic monitoring results was reproduced. This enabled the identification of potential instability failure modes and the determination of safety reserve coefficients of the surrounding rock, thereby revealing damage evolution laws and mechanisms under excavation-induced disturbances combined with fault structures and in-situ stresses. Moreover, numerical models based on actual cross-sectional profiles were constructed to investigate how fault locations influence the stability of surrounding rock, defining the influential range of faults. From the perspective of lateral pressure coefficients, the damage evolution patterns and failure modes of underground powerhouse cavern groups under varying deviatoric stresses were systematically studied, establishing relationships between safety reserve coefficients and lateral pressure coefficients. The research outcomes provide a valuable reference for monitoring and evaluating the stability of surrounding rock during excavation of pumped-storage underground power houses.

The research presented in this book received substantial support and recognition from several individuals, including Director Li Quansheng from the 2030 Project Office of China Energy Investment Corporation; Director Yang Zhongwei from the Production Command Center of Dadu River Basin, China Energy Investment Corporation; Secretary Lü Pengfei from the Preparation Office of Laoyingyan Hydropower Station, China Energy Dadu River Corporation; Chairman Lin Dan from China Energy Tibet Electric Power Company; General Manager Li Fangping from China Energy Jinsha River Benzilan Hydropower Co., Ltd.; Professor Wang Sujian from Xi'an University of Science and Technology; General Manager Chen Tong from Shaanxi Coal Fengjing New Energy Technology Co., Ltd.; Researcher Li Shihai, Senior Engineer Feng Chun, Dr. Tang Dehong, and Dr. Wang Lixiang from the Institute of Mechanics, Chinese Academy of Sciences. Here, the authors wish to express sincere gratitude to all of them for their contributions. Additionally, various literature sources were referenced throughout the book, listed individually at the end of each chapter; the authors hereby extend appreciation to the authors of these cited references. Due to the inherent complexity, variability, randomness, and uncertainty of geological conditions in large-scale rock engineering projects, coupled with the fact that microseismic monitoring studies are often carried out on specific aspects or singular engineering cases, inevitable limitations may exist within this book. Therefore, constructive criticism and feedback from experts and professional readers are sincerely welcomed.

Dalian, China

Ke Ma
Chunan Tang

Competing Interests The authors have no competing interests to declare that are relevant to the content of this manuscript.

Contents

Chapter 1
Microseismic Monitoring and Stability Analysis of Highly Steep Rock Slopes

Abstract This chapter presents a comprehensive study on the microseismic monitoring and stability analysis of highly steep rock slopes, with a focus on the right bank slope of the Dagangshan Hydropower Station. It begins by highlighting the research background and significance, followed by a review of domestic and international studies on excavation disturbances, numerical simulations, microseismic monitoring, and reinforcement effects, thereby identifying key challenges as the basis of the chapter. The engineering geological conditions of the right bank slope are then analyzed, and two- and three-dimensional numerical simulations are employed to assess slope stability. Subsequently, the construction and application of a microseismic monitoring system are described, revealing the spatiotemporal distribution of disturbances, the evolution of microseismic energy density, and the primary structural planes influencing slope stability, as well as investigating the causes of fractures during excavation and unloading. Finally, the reinforcement effect of anti-shear galleries is evaluated through microseismic data analysis and RFPA^{2D}-SRM modeling, focusing on energy distribution, stability, and reinforcement performance. By integrating numerical simulations with microseismic monitoring, the chapter offers valuable insights into slope stability and reinforcement strategies, thereby supporting the optimization of engineering practices.

1.1 Background and Significance of the Study

Hydro energy resources are extremely rich in China, with a theoretical reserve of 694 million kW and a technically developable capacity of 542 million kW, both ranking first in the world. For regional distribution, Yunnan, Sichuan and Tibet in Southwest China account for 2/3 of the country's total hydro energy resources, which are mainly distributed in the Yalong, Dadu, Lancang, Jinsha and Nu River basins [1]. In recent years, with the rapid development of the country's economic construction and the continuous improvement of people's living standards, the dependence on clean energy such as hydropower has increased. Further, as the key development field of the Western Development Strategy, there are a large number

K. Ma and C. Tang, *Microseismic Monitoring and Stability Analysis of Large Rock Mass Engineering*, https://doi.org/10.1007/978-981-96-6483-2_1

of large-scale (mega) hydropower resource development projects that have been constructed, are under construction or are proposed to be constructed in Southwest China, which are related to hydropower resources. Specifically, works related to large hydropower hub projects include: Xiaowan, Longyangxia, Baihetan, Manwan, Xiluodu, Diheyan, Xiangjiaba, Wuqiangxi, Zipingpu, Three Gorges, Jinping I Grade, Xiaolangdi, Wudongde, Waterfall Gully, Tianshengqiao, Changheba, Nuozhadu, Ertan, Laxiwa, Dagangshan and other projects. In these large-scale hydropower engineering construction projects in high mountain valley areas with abundant hydro energy resources and complex geological conditions, the slopes of rock-based projects are commonly massive, with their own heights far exceeding that of the dam buildings. For example, the maximum height of the slope at Xiaowan Hydropower Station is 700–800 m, with an excavation height of 670 m; the high and steep slope at Tianshengqiao Hydropower Station is 400 m, with an excavation height of 350 m; and the height of the slope at the shoulder of Zipingpu Hydropower Station is 350 m, with an excavation height of 280 m (Table 1.1). The above mentioned high and steep engineering slopes not only constitute an important regional environment for the dam shoulder buildings of hydropower projects, but also the key to the success of the dam and even the whole hydropower construction project [2–4]. The accident of Meishan continuous arch dam in China and Vajoint dam in Italy [5–7], for example, was caused by the unstable sliding of the mountain caused by expansion and deformation of the rock body after immersion in water, precisely because of the lack of high attention paid to the mountain in the reservoir area during the construction process at that time. Meanwhile, the oversized excavation construction is a huge potential threat to hydraulic structures. Data show that slope destabilization problems occur frequently in excavation of water conservancy and hydropower projects in China, such as high slopes on both sides of the dam area of Ankang Hydropower Station, high slopes on the left bank shoulder of Manwan Hydropower Station, high slopes in the plant area of Tianshengqiao Grade II Hydropower Station, and slopes on the downstream tiger hill slope of Longyangxia Hydropower Station. The management of these landslides has cost a lot which also delayed the construction period. These lessons have led to a sobering realization that the primary prerequisite for ensuring the successful implementation of hydropower energy development projects is the safe and orderly conduct of slope excavation construction on both sides of the dam shoulder, which will be even more urgent and difficult to study in the southwestern region with its unique and complex geological and environmental conditions.

Due to the geological action in the evolutionary process of the river valley, the rock body of the high and steep slopes in the river valley has a strong unloading problem, which is exemplified by the deep fissures in the rock body of the left bank in Jinping first-level hydropower station. Therefore, most of the current studies on the stability of unloading crack slopes are also focused on the Jinping Level 1 Hydropower Station. However, the study of a single project makes this type of slope in the investigation and design stage too much use of uniform specifications, without the other engineering site-specific determination and research, which can easily lead to the lack of evaluation of stability, resulting in catastrophic consequences. For the right bank slope project of Dagangshan Hydropower Station on which this chapter

Table 1.1 High and steep slopes of some constructed, under construction or proposed large (mega) hydropower stations in southwest China

Number	Project name	Natural slope		Major geological formation	Excavation height/m
		Slope height/m	Gradient/(°)		
1	Xiaoan	700–800	47	1 large-scale fault and over 10 general-scale faults	670
2	Longyangxia	–	–	Through tensile cracks and gently dipping cracks	> 200
3	Baihetan	300–600	> 40	Multiple inter- and intra-stratigraphic fault zones and large-scale faults	200–300
4	Manwan	120–160	> 40	Along slope fault and joint development	180
5	Xiluodu	300–350	> 60	Multiple gently dipping inter- and intra-stratigraphic tectonic dislocation zones	300–350
6	Geheyan	190	–	–	190
7	Xiangjiaba	350	> 50	Multiple small-scale faults and inter-stratigraphic dislocation zones	200
8	Wuqiangxi	170	–	Layered structures, creep and slip along rock strata	170
9	Zipingpu	350	> 40	1 large-scale fault and multiple inter-stratigraphic shear zones	280
10	Sanxia	150–200	25–35	Hundreds of small-scale faults	170
11	Bingpingyiji	> 1000	> 55	Multiple faults and inter- stratigraphic fault zones developed	530
12	Xiaolangdi	150–200	30–40	Faults and mudded intercalation	120
13	Wudongde	830–1036	> 43	–	430
14	Pubugou	> 500	35–45	Sub-minor faults are developnig	240

(continued)

Table 1.1 (continued)

Number	Project name	Natural slope		Major geological formation	Excavation height/m
		Slope height/m	Gradient/(°)		
15	Tianshengtiao	400	50	Layered and fissured rock mass	350
16	Changheba	700	45–55	Developed minor faults, compresso-crushed zones and fissured joints	300
17	Nuozadu	800	> 43	4 large-scale faults and over 40 general-scale faults	300–400
18	Ertan	300–400	25–45	16 faults and multiple small-scale tectonic fragmentation zones	200–300
19	Laxiwa	680–700	50–70	Mainly composed of fault structures	700
20	Dagangshan	> 600	> 40	Large unloading fracture zone and faults	380–410

is based, it is the insufficient investigation of the deep unloading fissures, faults and other geological conditions of the slope, resulting in a number of macroscopic cracks and signs of deformation intensification occurring sequentially during the large-scale excavation of the slope, causing the excavation of the slope to be forced out of work for half a year, delaying the entire duration of the project for about two years, and leading to several hundred million RMB of economic losses [8–10].

1.2 Status and Analysis of Domestic and International Research

1.2.1 Research on Excavation Disturbance of Rock Slopes

The stability problem of high and steep slope excavation project is related to the fields of water conservancy and hydropower, open-pit mines, transportation, etc., within the context of the rapid development of a large number of engineering construction. Moreover, slope destabilization accidents caused by excavation disturbance are even more common. For example, the disintegration of nearly 200,000 m^3 of rock mass occurred in 1987 during the excavation of high slopes at the exit of the left bank of the diversion hole of the Geheyan Hydropower Station [11, 12]. The high side slopes of the permanent locks of the Three Gorges Project experienced stress redistribution during the excavation and unloading process. This led to the deterioration of the mechanical properties of the rock body of the septum pier, and induced the primary fissures in the rock body of the slope surface to open up to develop new tensile fissures, which seriously affected the stability of the slope [13–16]. The unloading rebound generated during the excavation of the left bank lock slopes of Wuqiangxi Hydropower Station induced the slopes to undergo dumping deformation failure twice, forcing the excavation to be suspended [17]. The slope instability and collapse that occurred during the excavation of the inlet open channel at the gate head slope of the Tianshengqiao secondary hydropower station killed 48 people who were working in the pit [18]. The left shoulder slope of Manwan Hydropower Station was excavated with a large landslide, which was 100 m high with a total volume of 108,000 m^3, resulting in a delay of more than one year and a direct economic loss of 120 million RMB [19, 20]. In the construction process of Chengkun Railway, 120 slope failures and slides containing red layer soft rock were found, accounting for 65% of the total number of landslides in the whole line [21]. The Yuanmao highway in Yunnan produced many excavation slope disasters during the construction process [22].

Earlier studies on the disaster-causing mechanism of excavated slopes in China were mainly focused on the excavation and unloading project of high slope rock at the locks of the Three Gorges Project [23]. Xia et al. [16] made a detailed description of the excavation loosening zone and traits of the high side slopes of the Three Gorges Project locks. Deng et al. [24] monitored the excavation loosening zone of the Three Gorges Project lock slopes utilizing the sliding deformometer, explored the concept of the excavation loosening zone, concluded that the deformation of the slope rock mass has obvious zoning, and defined the main deformation zone as the slope rock mass loosening zone. Shengqian [25] conducted a systematic study combining the disturbance zone of high slope excavation in Three Gorges Project locks with the mechanical properties of engineered rock body. Xiao et al. [26] concluded through numerical analysis that the slope excavation relaxation zone is a localized failure of the slope body in a limited range caused by stress redistribution and displacement changes after slope excavation. Based on the geomechanical model, Wang et al. [27] pointed out the influencing factors of epigenetic modification on the mechanical mechanism of rock structural features in combination with the slopes of some large hydropower projects.

1.2.2 Numerical Simulation of Rock Slopes

Rock slope failure is a continuous to destructive evolutionary process, containing stages of small deformation, failure evolution, fracture crack formation, crack extension through to the occurrence of bulk movement. With the rapid development of computational technology, the use of numerical methods to simulate the mechanical behavior of rock slopes under complex loading is an effective means to study their damage mechanisms. The deformation and damage of rock slopes are mainly controlled by their internal structure, which is caused by the existence of massive macro–micro fissures and discontinuous structural surfaces, such as joints, inside the rock body of the slopes. The whole catastrophic process goes through four stages of conception, emergence, evolution and development, and it is not possible to apply theoretical analytical methods to study its destructive process and laws. Moreover, since most of the actual slopes locate in complex geological environments, their scale effects and force conditions cannot be solved by indoor tests. Therefore, using numerical simulation methods to study the mechanical behavior of excavated rock slopes is an effective way to analyze their stability.

Zheng et al. [28] analyzed the damage mechanism of rock slopes by numerical simulation using the finite element strength reduction method based on the principle of plastic mechanical damage, which showed that the through-damage process is a progressive failure process occurring between the structural surface and the rock bridge gradually expanding from the local damage to the overall failure. Zhou et al. [29] proposed a slope seismic stability analysis method based on static-dynamic finite elements by taking the displacement mutation of the characteristic point at the moment of penetration of the plastic zone as the limit state, so as to determine whether the slope is locally destabilized or integrally destabilized under seismic loading. Zhou et al. [30] used elasto-plastic macro-deformation finite element to analyze the process of slope destabilization and failure by deriving macro-deformation elasto-plastic finite element equations for the analysis of slope failure process. Cao et al. [31] analyzed the variation rule of the safety factor and the position of the sliding surface of multilayer rock body slopes under different strengths, slope ratios and layer thicknesses by FLAC3D intensity reduction method, which revealed the destructive mechanism of multilayer slopes. Kou et al. [32] applied FLAC3D to the stress-deformation and stability analysis of the excavation process of high slopes in the locks of the Three Gorges Project. Yin [33] studied the deformation and failure characteristics of the mountain body under the influences of gravity, karst and mining activities at the bottom by using FLAC3D simulation with the Jiwei Mountain landslide in Wulong, Chongqing, as an example. Hart et al. [34] used the 3DEC program to simulate the progressive failure process of simple rock slopes. Bhasin et al. [35] used the UDEC software to analyze the static and dynamic stability of a 700 m high rock slope in western Norway, assessing the volume of rock that could potentially slide under both static and dynamic action, and thereby used to assess the height of tsunamis in the Gulf. Cheng et al. [36] reproduced the motion process and state of the large deformation stage after destabilization of the rock body on the high steep slope by numerical simulation of discrete element method based on the high-speed rock landslide dynamics model. Mao et al. [37] used the discrete element method to explore the mechanism of ground shaking-induced landslide start-up drastic movement, and numerically simulated the whole process of deformation and instability of slopes under seismic action by taking the Luozhuoling landslide as an example. Cao et al. [38] simulated the whole process of landslides from deformation accumulation to destructive sliding by using discrete element numerical simulation to study the deformation and destruction process of smooth rock landslides under the action of earthquake, taking the Tangjiashan landslide triggered by Wenchuan earthquake as an example. Sitar et al. [39] analyzed the discontinuous deformation characteristics of slopes containing jointed surfaces and the laws of kinematic damage by using 2D discontinuous deformation analysis (DDA). Chen et al. [40] used DDA to analyze the stability of two slopes located in Hokkaido and Shin-Hu, Japan, by simulating slope failure and toppling failure. Hatzor et al. [41] analyzed the stability of rock slopes with highly discontinuous characteristics under seismic action by using DDA. Wu et al. [42] verified the applicability of the DDA method applied to the stability analysis of rock slopes, and simulated the whole process of the initiation and landslide in Qianshuangping. Sun et al. [43] applied the DDA method to study the instability and deformation mechanism of tipping failure, and pointed out the possible problems of using the G-B method (an analytical method based on the principle of limit equilibrium proposed by Goodman and Bray). ELFEN is a commonly used software package for the finite-discrete element coupling method, which can simulate the crack extension process in brittle rocks by local dissections of the mesh, and ELFEN contains a rich set of fracture, damage, and softening models to characterize different rock features [44]. Zhang et al. [45] used a numerical manifold method to simulate the toppling destruction process of slopes and compared it with the indoor centrifuge test, adding the tracking simulation function of multi-fracture extension, so that it can simulate both the discrete characteristics of the block system and the tensile cracking and shearing of the intact rock mass. Stability analysis of slopes actually obtains the safety reserve deformation characteristics and stress state of the whole slope, which is essential for the study of slope destabilization failure.

1.2.3 Research on Microseismic Monitoring and Its Applications

Engineering practice shows that the stability analysis of slopes under complex geological conditions requires the construction of accurate physical–mechanical models and the use of advanced monitoring technology, which can be accurately assessed by combining the two to complement and compare the analysis with each other [46]. The main monitoring methods used for slope deformation domestically and internationally are: slope surface measurement (rangefinder, total station, level gauge, theodolite, etc.), slope internal measurement (water pressure monitor, borehole inclinometer, and anchor cable dynamometer, etc.), displacement meter, global positioning system (GPS) measurement technology, infrared remote sensing monitoring method, laser micro displacement monitoring technology, synthetic aperture radar interferometric measurement, fiber optic displacement measurement, time-domain reflectance test technology, closed-end and acoustic emission technology. In terms of geotechnical engineering slope monitoring, scholars domestically and internationally have done a lot of research with a series of key results [47–50], especially the successful early warning of landslides such as Lianzi Cliff [51], Xintan, and

Huanglashi. Zhang et al. [52] and Meng et al. [53] have done numerous researches on the monitoring and early warning mechanism of high slopes in Jinping first-level hydropower station. Wu et al. [54] developed a remote monitoring and forecasting system for geologic hazards on roadbed slopes of mountain highways. Zhao et al. [55] evaluated the stability of high slopes in Xiaowan Hydropower Station by studying the monitoring data such as apparent deformation, inclinometer, multipoint displacement and anchor cable loading. Existing slope monitoring techniques are able to accurately monitor the parts where damage has occurred. However, for rock slopes, the surface displacement is not obvious due to the small deformation before macroscopic damage. When the deformation on the surface of the rock slope is detected, the internal damage may have already occurred, making it difficult to effectively monitor the precursor information of slope instability. The traditional monitoring methods cannot provide early warning due to the difficulty of capturing the information of destabilization on rock slopes.

In recent years, as a spatio-temporal dynamic three-dimensional "body" monitoring method, microseismic monitoring technology can capture the information of micro-breakage inside the rock body, reproduce the emergence of cracks in the rock body, development, expansion, evolution, penetration, and up to the generation of macro-slip surface, which can be used as an important monitoring tool for evaluating the stability of the rock body. Microseismic monitoring technology is mainly applied in the fields of mining and petroleum. This technology has now been widely used as an advanced and effective means of ground pressure monitoring in high geostress mines domestically and internationally, making it a fundamental means of research and management of ground pressure in deep mines [56]. Since the establishment of the first seismic observatory in 1908 in Bochum, Ruhr District, Germany, remarkable research results have been achieved in mines [57–59], tunnels [60], underground oil and gas material storage caverns [61], and hot dry rock power generation [62] in countries such as South Africa, Germany, Poland, the United States of America, the United Kingdom, Canada, Japan, and Australia at present. A number of scholars have carried out numerous studies on microseismic monitoring around underground nuclear waste laboratories in Canada, yielding many valuable results [63, 64]. Kaiser et al. [65] systematically investigated the microseismic activity of mines by combining 3D virtual technology and mapped the risk level of microseismic activity in mines, which explored a new research field for mining resources extraction and support. In recent years, with the emergence of rock disaster problems such as impact pressure gas induced by deep mineral resources mining, hydraulic fracturing in the petroleum field, safety of nuclear waste storage, and rock explosion in high-buried tunnels, instability of underground caverns and high slopes in the hydropower industry, domestic scholars have carried out a large number of successful researches in various fields by combining with the microseismic monitoring technology. Li et al. [66] used the microseismic monitoring system of Engineering Seismology Group (ESG) to establish a microseismic monitoring system to monitor the impact ground pressure in Fankou lead–zinc mine. Jiang et al. [67] successfully developed a microseismic monitoring system at the University of Science and Technology Beijing, Mine Micro Seismic Research Center (BMS), and carried out a microseismic monitoring system for the rupture height of roof plate, abnormal pressure, dynamic pressure in the mines of Fengfeng, Yanzhou, and Shuangyashan. Pan et al. [68] developed a set of the first domestic kilometer-scale destructive mining earthquake monitoring and positioning system with independent intellectual property rights in the mining area, which was applied in the prediction and forecasting of impact mine pressure in Muchengjian coal mine in Beijing. Lu et al. [69] installed seismological observation system in over a dozen mines with impact pressure phenomenon, and predicted the mine earthquake and impact pressure events for many times, which greatly reduced the possible disasters caused by the mine. Tang et al. [70] used the microseismic equipment of South Africa ISS (Integrated Seismic System) International Company to install a microseismic monitoring system for monitoring rockburst disasters in Dongguashan Copper Mine. Chen et al. [71] combined acoustic emission and microseismic monitoring techniques to study the rock damage and rockburst problems in the deeply buried tunnels of Jinping Grade II Hydropower Station. Xu et al. [72] used the Canadian ESG microseismic monitoring system to construct the first rock slope microseismic monitoring system on the left bank slope of Jinping Grade 1 Hydropower Station, initially circling the potential rock damage areas and slip surfaces that may exist on the slope. Based on the academic concept that micro rupture is the precursor characteristic of potential landslide, Ma [73] "captured" the micro rupture in the excavation process of slope rock body through microseismic monitoring technology, and researched the distribution law of microseismic activity under the action of excavation disturbance, revealing the mode of slope destabilization under the condition of excavation disturbance.

1.2.4 Study on the Effect of Reinforcement on Rock Slopes

The anti-slip reinforcement of rock slopes is a complex systematic project. When the slope safety coefficient cannot meet its stability requirements, it is necessary to synthesize all aspects of the influencing factors to formulate a set of targeted management program to ensure the safety of the project. Especially in the construction of large-scale water conservancy and hydropower projects, the study of slope stabilization method is even related to the safe operation of the whole reservoir and dam [74–76]. At present, the reinforcement methods of rock slopes are mainly two categories: active reinforcement method and passive reinforcement method. Active reinforcement method is to change the original structural form of rock body or change the size and distribution of the load of the slope body, which can reduce the downward force of the slope, in order to effectively change the strength of the rock body and improve the stability of the slope method. The passive reinforcement method exerts the potential of slip resistance only when the slope body is in close contact with the slip-resistant structure with certain deformation, i.e., its structural resistance is exerted passively [77–83].

China has increasingly applied prestressing anchorage technology for slope reinforcement after the 1980s, which is also the most widely used reinforcement measure for large-scale water conservancy and hydropower high and steep slope projects [84]. For example, the number of large-tonnage prestressing anchor cables used in Yunnan Xiaowan Hydropower Station exceeded 10,000 [85]; more than 4000 anchor cables were used in only the left bank slope reinforcement project of Jinping First Grade Hydropower Station [86]; Longtan Hydropower Station managed the 425 m high slope consisting of the intake and the dumped creeping body of the left bank by installing comprehensive treatment measures, such as prestressing anchors and over-topping anchor rods [87]; Xiaowan Hydropower Station used self-progressing anchor rods and other supportive measures to carry out comprehensive reinforcement of the left bank slope of the excavation height of up to 700 m [88]. Therefore, the mechanical mechanism of prestressed anchor (cable) reinforcement of high slope rock mass, the synergistic mechanism of anchor cable and other reinforcement measures such as shear-resistant holes and anchorage holes, as well as the long-term durability of the prestressed anchorage system have received more and more attention from scholars domestically and internationally [89–95]. Wang [96] identified the anchorage angle as the dominant factor for optimization in 1979 with an optimal anchorage angle for prestressed anchors and its application method. Zhu et al. [97] analyzed the application of active and passive anchors in the high side slopes of the Three Gorges Project locks by preloading the initial strain method. Li et al. [98] analyzed the reinforcement mechanism of prestressed group anchors using numerical simulation, pointing out that the magnitude of rock strength improvement under group anchors is much larger than the simple superposition of the improvement value for single anchors. Zhao [99] applied the classical Boussinesq solution, Cerruti solution and Mindlin solution for superposition problem in elasticity problem to obtain the distribution law of three-dimensional additional stress field in infinite medium with linear elasticity of prestressing anchors (cables). Zhang et al. [100] studied the strengths and weaknesses of active and passive anchors based on the measured data of field-tested anchor cables, concluding that the optimal anchoring method is toward the combination of the two types of anchors. Hou et al. [101] investigated the effect of roadway anchors on the improvement of mechanical parameters of the rock mass within the anchoring range through indoor modeling tests. Chen et al. [102] explored the reinforcement effect of prestressing anchors on massive rock mass by analyzing the main influencing factors affecting the anchoring effect such as anchor length, anchor spacing, pre-stressing magnitude and pad pier size, deriving the influence range and characteristics of different factors. Li et al. [103] conducted a study on the combination application of protective structures such as anchor spray, anchor pile, anchor wall and anchor beam, which analyzed the respective advantages of the related structures for more effective slope reinforcement. Cao et al. [104] considered the synergistic deformation of anchor rods (cables) for rock reinforcement as an organic structural body connecting the rock block units by utilizing the numerical manifold method. Zhang et al. [105] derived a theoretical model of nonlinear function for load transfer of anchor rods (cables) and obtained the distribution law of shear displacement along the axial direction of anchor rods and friction resistance. Ding et al. [106] conducted numerical simulations using FLAC3D to investigate the mechanical behavior of prestress anchorage cables, and pointed out that the configuration of the rock anchoring wall is influenced by both the arrangement of the anchorage cables and the magnitude of the prestress. Zhu et al. [107] studied the mechanical properties of the anchored rock body on the high side slope of the Three Gorges permanent ship lock, concluding that the rock body in a certain area after prestress anchoring mainly shows a compressive stress state. Xu et al. [108] proposed a set of ideas of shear stress exponential distribution for the calculation of anchorage segment length within a prestressed anchor cable, on the basis of which the anchorage segment length was determined. Zou et al. [109] discussed the determination of optimal azimuth and optimal anchorage length of anchored and free segments respectively according to the reinforcement mechanism of prestressed anchor cables by combining engineering examples. Wu et al. [110] analyzed the impact of the way of anchors arrangement on the strength of fractured structural rock, pointing out that the angle of arrangement and the height of arrangement have a large influence on the strength and deformation of the specimen. Liu et al. [111] pointed out the key factors affecting the design parameters of the anchor cable in the design of prestressed anchor cable reinforcement to propose the optimal design scheme and optimal design parameters of prestressed anchor cable reinforcement for this slope. You [112] based on the principle of joint deformation of geotechnical body-anchor solid and conducted an experimental study on the mechanical characteristics of the interface in anchored segments, obtaining the analytical solution of stress distribution in anchored segments of different prestressed anchor cables. Based on mathematical-mechanical principles, He et al. [113] analytically established a damage evolution model of anchored segments under tension conditions, investigating characteristics such as the lateral resistance distribution pattern and load displacement of the anchored segments. Xiong et al. [114] analyzed the value range of the optimal anchor cable direction angle according to the structural requirements, application purposes and actual construction conditions, obtaining a new method for calculating the direction angle of prestressed anchor cable considering the influence of the inclination angle between the sliding surface and the slope surface. Cao et al. [115] studied the stress characteristics of pressure-dispersed anchor ropes on the basis of summarizing previous studies combined with field test analysis, proposing a new type of anchor rope design method suitable for the reinforcement of slopes with broken rock bodies. Zhang et al. [116] applied optimization and decision-making theory to analyze the factors affecting the design parameters of rock anchors, proposing a method for determining the design parameters of rock anchors, which provides a new method for rock anchor design decision-making. Zhu et al. [117] derived the distribution function of shear stress in the anchorage section of prestressed anchor cables through a composite power function model of the axial force distribution in the anchor rods of the anchorage section. Huang et al. [118] established a three-dimensional nonlinear finite element computational model of pressure-dispersive anchor cables by taking the hazardous rock body of the slope as the object of study, revealing the interaction effect of prestressing anchor cables and the slope structure body. Li et al. [119] established the quantitative relationship between macroscopic mechanical properties and material properties of rock-anchor support system through composite material mechanical analysis. Xu [120] established an analytical expression for the calculation of the length of the anchorage section within a pressure (dispersed) type anchor cable through numerical analysis, which provided a basis for the reasonable determination of the distance between the anchorage section of the anchor cable and the group anchor cable. Relying on the actual engineering geological conditions, Liu [121] studied the performance of pressure-dispersed anchor cables and the distribution of stress–strain in the perimeter-rock reinforcement zone by comparing and analyzing the theoretical and experimental results. Ge et al. [90] explored the shear performance of anchor rod on joint face and the mechanism of pinning through indoor tests and theoretical analysis, revealing that the inclination angle of the anchor rod has a great influence on the shear strength of the anchored joint face. Yang et al. [122] calculated the constitutive relationship of anchored rock mass based on the damage evolution equation of the fracture propagation process of anchored joints. Ye [123] comprehensively studied the strength, stress–strain curve and failure behavior of anchored joints by combining triaxial compression equivalent tests and finite element numerical analysis. Yang et al. [124] derived the constitutive equation of anchored layered rock mass and made a reasonable quantitative evaluation of the reinforcement effect of anchor rod in jointed rock mass. Wang [125] equated the weak surface of layered rock mass to mechanical model of equidistant collinear multiple joints and proposed an anchoring effect formula, which effectively solves the efficiency enhancement problem of slope anchoring.

1.3 Existing Problems

Most of the studies on the sliding mode of high-steep rock slopes are limited to the geomechanical models built on the control of homogeneous bodies or tectonic surfaces, with the damage modes mostly generalized to several simple shallow damage types, such as along-structural surface damage or rotational sliding, ignoring the damage in the deeper layers of the slopes, which makes the studies highly empirical. Research on the stability of slopes under the action of excavation is even more rarely reported. In fact, slope destruction is a complex dynamic process. Damage types can also be converted to each other under certain conditions, resulting in fuzzy assessment of damage models leading to repeated failure of landslides.

The right bank slope of the Dagangshan Hydropower Station of the Dadu River, on which this chapter is based, has experienced a long period of shallow epigenetic remodeling action, which has resulted in massive gabbro veins interspersed with the rock mass within the slope. These gabbro veins are broken with poor physico-mechanical properties such as deformation resistance and shear strength. Moreover, the slope rock body has the most developed gently inclined structural surface, which destroyed the integrity of the rock body and caused the anisotropy of the rock body of the slope project, making it have the typical geological conditions of "high and steep slope body, complex geological boundary and difficult construction conditions". Especially, the slope rock body above the dam top of the right bank has obvious relaxation and pulling cracks, forming a slope structure represented by unfavorable

(a) Topography of the slope before excavatio (b) Topography of the slope after excavation

Fig. 1.1 Geologic map of the right bank slopes at the dam site [132]

structural surfaces such as unloading fissure dense zone and faults inclined to the outside of the slope, which constitutes a big threat to the slope stability, rarely seen in the construction of large hydropower projects in the past (Fig. 1.1) [126–132].

Geological survey shows that: β62, β85, β4 and other gabbro veins of the right bank arch shoulder channel slope are steeply inclined within the slope, constituting the trailing edge boundary; Fault f_{231} is off the mid-dip slope constituting a bottom slip surface with several potentially unstable blocks. The deeper areas of the slope rock body also have unloading intensive zones and other geological formations. The complexity of the structural surface in the slope and the mutual cutting effect of unfavorable combinations made the slope produce 8 relatively obvious local deformation and damage phenomena one after another during the excavation process (Fig. 1.2). Since the large-scale excavation and construction of the slopes in August 2007, the slopes have remained generally stable. On November 28, 2008, an avalanche occurred along the downstream side of the right bank slope at an elevation of 1180–1195 m, which was followed by a small avalanche along an elevation of 1150 m. In January 2009, the excavation was carried out to an elevation of 1135 m, with the excavated slopes still stable. In February 2009, the right bank below the top of the dam started blasting and excavation. The upstream slope of the right bank arch shoulder groove was excavated to the elevation of 1120 m on March 12, and to the elevation of 1110 m on March 26. The signs of deformation, such as the cracking and stripping of the concrete spray layer along part of the rock veins in the slope, the local slacking of the fault f_{231} in the PD314 flat cave, and the slight cracking and misalignment of the cement stone at the bottom of the cave, etc., appeared in the section of LP V to LP VII locally. Since entering the elevation below 1135 m, three large deformations had occurred on the upstream slope of the right bank arch shoulder channel. May 3, 2009, there were cracks on the right bank 1124–1127 m elevation (dam longitudinal 0 + 95–0 + 113 m) in the slope which has been closed by hanging net spray protection, with cracks extending intermittently, manifesting as 3 sections of cracked area. The probing window reveals that the crack of 1# probing window in the longitudinal depth of the dam 0 + 115 m is open 4 ~ 7 cm. Subsequently, the excavation was carried out on May 27 to an elevation of 1090 m, on July 2 to an elevation of 1080 m, and on July 23 to an elevation of 1070 m. On August 16, 2009, cracks appeared on the slope of the right bank from 1075 to 1092 m elevation (0 + 69 to 0 + 88 m of the dam depth). There were four cracks in total: L2-1# and L2-2# opened horizontal cracks of 0.1–0.2 cm, extending 4–6 m; L2-3# opened cracks of 1.5–2 cm, extending about 10 m; L2-4# opened cracks of 0.1–0.3 cm, extending about 6.3 m. On August 23, 2009, 9 cracks appeared in the concrete bottom slab of PD314 flat hole (EL 1107 m, dam depth 0 + 75 m), when the monitoring instrument was being laid. At about 2:10 p.m. on September 1, 2009, cracks of split-peeling appeared on the slope of the right bank at the elevation of 1135–1146 m (depth of the dam 0 + 40 to 0 + 50 m). Cracks also appeared on the side slopes at 1100–1135 m elevation. As of September 2009, when the right bank slope was excavated to an elevation of 1070 m, several cracks were generated on the slope surface, with a trend of gradual expansion to the bottom of the slope (Fig. 1.2). Obviously, with the excavation of the slip-resisting rock body on the slope surface, the medium-dipping fault f_{231} inclined out of the slope is exposed. At this time, the unloading fissure zone XL-316 is easy to be combined with the shallow slope-oriented fault f_{231}, forming a series of unstable blocks in the upstream and downstream ranges of 100–200 m, with heights of 200–300 m, and cubic volumes of 2–5 million m^3. Cracks in the sprayed concrete slope surface were caused by successive deformation damage to the slope surface. Especially in the local section of LPV-LPVII, the concrete spray layer of the slope is cracked and peeled off along part of the rock veins, the fault f_{231} in the flat hole of PD314 is locally slackened, with signs of deformation such as slight cracks and misalignments of the cement stones at the bottom of the hole. Deformation misalignment along the unloaded fissure zone XL-316 and fault f_{231} is even more frequent near the shear exit, which made the stability of the slope during the construction period a prominent issue. After geological investigation and analysis, the deformation of the slope was mainly caused by the deformation of the potential unstable blocks inside the slope. If the key blocks continue to deform, it may lead to the destruction of large block combination and jeopardize the overall stability of the slope. Therefore, after excavation to 1070 m elevation, the excavation was stopped with suspension of slope surface excavation.

As shown in Fig. 1.3, the external deformation monitoring points TP31R and TP32R produced more obvious and continuous deformation in the X-direction (the proximity direction of the excavated surface of the slope) during the excavation of the right bank slope. Both monitoring points were located in the right bank shoulder slope 1070 m elevation Madao, near the construction excavation surface, with corresponding geological structure soft zone.

At present, displacement and deformation monitoring techniques (e.g., multipoint displacement meter, inclinometer tube, osmometer, joint meter, reinforcement meter

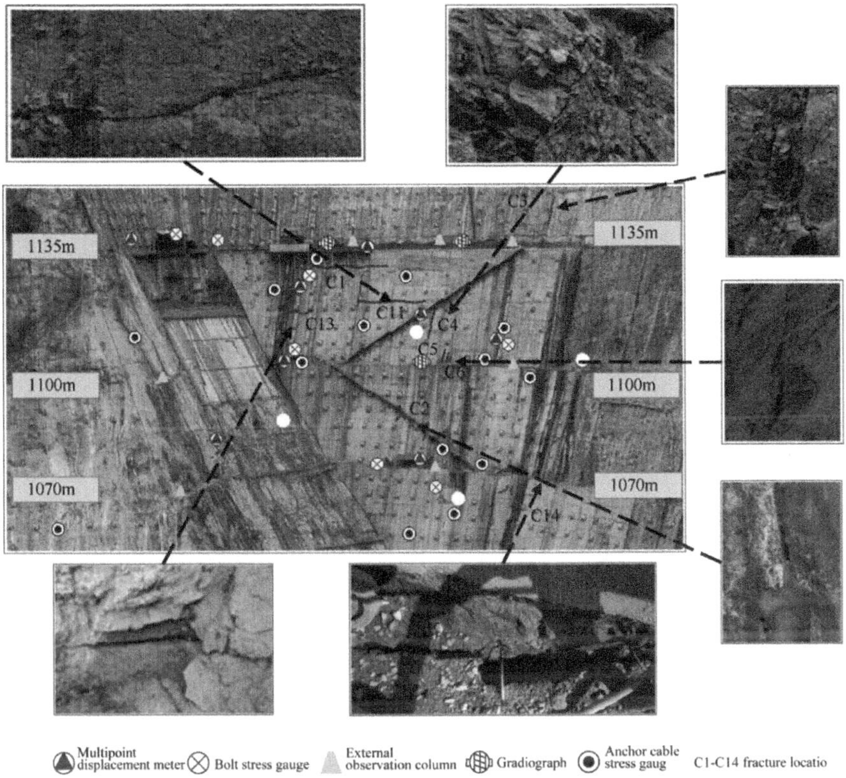

Fig. 1.2 Macroscopic cracks in the slope surface and the arrangement of regular monitoring instruments

and GPS, etc.) have been used in the stability monitoring of the right bank slope of Dagangshan Hydropower Station, which have effectively monitored the appearance and deformation of the right bank slope under the disturbances of different construction phase, with the displacement and deformation values of the slope at each point of measurement, analyzing the deformation tendency of the right bank slope under the disturbances of the excavation. As the rock slope belongs to the class of brittle materials, its deformation is small before the macro-damage, with insufficiently obvious surface displacement. The slope has some slack deformation along the direction of unloading fissure zone after excavation. The complex geological conditions such as faults, unloading fissures and the randomness of stress release and transfer cause the phenomenon that the deformation of partial rock body does not converge without obvious abnormal deformation inside the construction cavern. Conventional monitoring methods can only monitor significant "point" deformation, but difficult to effectively capture the internal microfracture of the rock mass, which brings great potential danger to the safety of engineers. Therefore, there is an urgent need for a

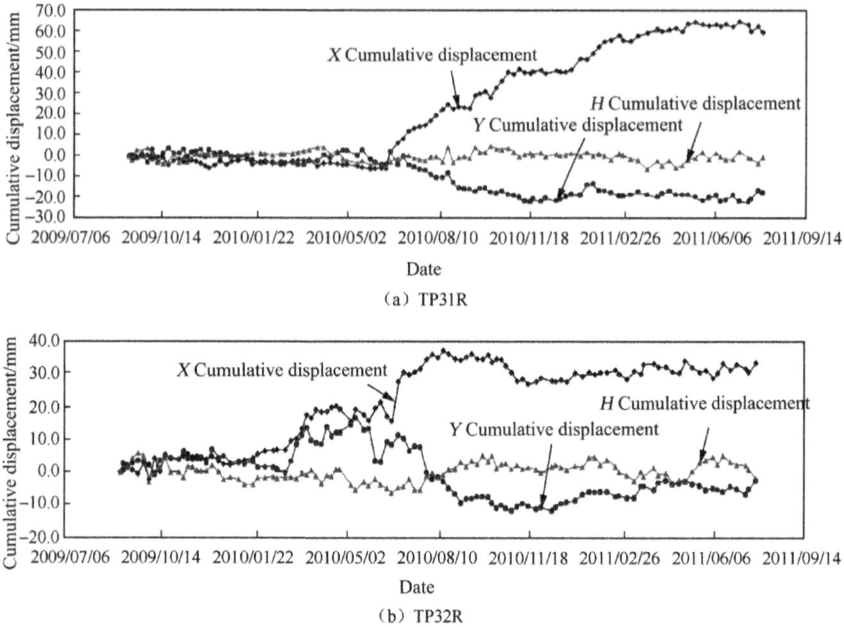

Fig. 1.3 Cumulative displacement variation curves of external deformation monitoring points TP31R and TP32R at 1070 m elevation [133]

new monitoring tool to effectively monitor the information on rock microfracture processes that mark the precursors of rock slope instability [134–138].

1.4 Research Content

The geological structure of the right bank slope of Dagangshan hydropower station is different, and the potential deformation and instability modes of the slope are complex and diverse. This section focuses on the key scientific issue of excavation deformation effect and stability evolution mechanism of high and steep slope under complex environment, and based on the academic idea that "the microseismic activity rule of slope rock mass progressive failure is the essential characteristic of slope rock mass failure and instability", took the excavation slope of right bank of Dagangshan Hydropower Station with complex geological conditions and unique slope structure as the research object. Based on a large number of geological data of field tracking and investigation, slope deformation signs during slope excavation and the spatial distribution of microseismic activity monitored in practice, the stability of slope excavation was studied by numerical simulation. Firstly, the engineering geology, which refers to the mechanical environmental conditions of the rock mass where the slope is located, was thoroughly understood. The engineering geological characteristics of the slope were studied, and the dominant structural plane that affects the stability of the slope was clearly defined. This allowed for the establishment of a numerical model that could accurately characterize the structure of the slope and provide a detailed description and modeling of the slope's rock mass. The results obtained from this study reveal the process of deformation evolution and the failure mode of the slope. By closely examining the micro-fractures that occur during the excavation of the slope's rock mass, the mechanism behind the deformation evolution of high and steep slopes under excavation was uncovered. At the same time, a three-dimensional numerical model was established to study the evolution of damage in the excavation disturbance area of a high and steep slope rock mass. This was done by employing the research method of mutual feeding between numerical calculation and field monitoring. Taking into consideration various disciplines such as engineering geology and mechanics, the geological structure of the slope, rock mass micro-fracture deformation information, and numerical simulation analysis were integrated to evaluate the slope stability during excavation. Finally, in accordance with the current slope reinforcement measures, a numerical model of a shear hole, which is a typical slope control structure, was established. This section analyzed the interaction mechanism and deformation coordination sharing mechanism between landslides and anti-shear galleries during the failure process, and verified the suitability and correctness of the anti-shear galleries landslide control structure. This study provided a theoretical reference for the selection and prevention of reinforcement schemes for high and steep slopes in large-scale water conservancy and hydropower projects with similar geological conditions. In the face of the complex rock mass structure characteristics of the slope, adopting effective new methods to address the challenges encountered in actual project construction not only provides a reference for geological investigation, excavation, unloading, and reinforcement construction of the right bank slope of the Dagangshan Hydropower Station, but also offers a new perspective for the study of dynamic stability of rock slopes under complex engineering geological conditions.

1.5 Engineering Geology and Stability Analysis of Right Bank Slope of Dagangshan Hydropower Station

According to the cybernetics of rock mass structure [139, 140], the deformation and failure of rock mass are controlled by its structure. The deformation failure of rock mass is the structural instability first and the material failure second. Rock mass structure affects and controls the failure mode of the whole rock engineering, so any research on rock mass engineering must first conduct a detailed investigation on the genetic type, occurrence and form of the structural plane in the geological environment, so as to identify the characteristics of the structural plane in the rock mass such as orientation, trace length, spacing, relief, filling, opening degree and roughness [141]. The failure of rock mass is often represented by the failure of the structural body enclosed by the structural plane. In the slope of water conservancy and hydropower engineering, the structural plane and its combination relationship not only control the stability of rock slope, but also control the shape, scale and trend of sliding body. Rock slopes are basically destabilized along the main structural plane [142–144], so determining the main structural face of slope is of great significance to analyze the type of slope failure, failure mechanism and stability evaluation [145]. The dam site of Dagangshan Hydropower Station is a typical deep "Ω" shaped valley with steep valley slopes and straight and narrow river channels, which is a good arch dam site from the perspective of landform. However, deep cracks are developed inside the slope on the right bank of the dam site, and the distribution depth of these cracks is far beyond the distribution range of the unloading zone of the rock mass in the slope in engineering experience, so this problem has actually become the key engineering geological problem restricting the smooth construction of Dagangshan hydropower station. The accurate establishment of the geomechanical numerical model of slope structure is the premise of analyzing the stability of high steep rock slope. According to the geological characteristics, the slope structure model was established, and the real strength parameters of rock mass were given to study the failure mode and stability. In this chapter, a numerical model suitable for stability analysis was generalized on the basis of engineering geological investigation data, considering the special complexity of high steep rock edge slope structure [146].

1.5.1 Engineering Geological Environmental Conditions

The dam site of Dagangshan Hydropower Station controls an area of 62,727 km^2, accounting for 81% of the total basin area of Dadu River. The average annual discharge at the dam site is 1010 m^3/s, the normal water level of the reservoir is 1130.00 m, the total storage capacity is 742 million m^3, the total installed capacity of the power station is 2600 MW, and the annual power generation is 11.43 billion kW h. The main task of the project is to generate electricity. The power station junction is mainly composed of concrete double-curved arch dam, right bank spillway tunnel and left bank water diversion and power generation building, etc. The dam top elevation is 1135 m and the maximum dam height is 210.00 m (Fig. 1.4) [132].

The dam site and reservoir head section of Dagangshan Hydropower Station is located on Huangcaoshan fault block. The block is cut by the Moxi Fault, Daduhe Fault and Jinping Fault, and is generally in the northern part of the Sichuan-Yunnan north–south tectonic belt, which is a north–south and "north-west" and "north-east" and other groups of tectonic convergence and composite area. Its tectonic units are located in the Kangdian axis of the second-order structure in the western part of the Yangtze platform, adjacent to the Yajiang Miogeosynclinal fold belt in the west, the Yangtze platform fold belt in the east, and the Longmenshan platform margin fold fault belt in the northeast. The Jinping Mountain-Xiaojin River Fault and the Moxi Fault are its western boundary, while the Jinping Fault and the Erlang Mountain Fault are to the east and northeast. The mountains on both sides of the hub area are strong, with steep valley slopes and exposed bedrock, and the Hailiu Gully on the left

Fig. 1.4 A rendering of Dagangshan Hydropower Station

bank and the Copper Trough Gully on the right bank are the larger branch gullies. The natural slope of the right bank is generally 35°–40°, the valley slope is steep, the bedrock is exposed, and the relative elevation difference is generally more than 600 m. There are only small trenches developed on the slope, and the cut is shallow, and the terrain is relatively complete. The valley is in the shape of "Ω" embedded meander. The river level in dry season is 952.6 m, the water surface width is 40–70 m, the water depth is 7–11 m, and the valley width is 380 m when the normal water level is 1130 m. The right bank is bounded by an elevation of 1220 m, above which the slope topography is relatively gentle, and most of the surface is covered with landslide deposits, with a vertical thickness of 8–20 m and a horizontal thickness of 8–22 m. Below the line is a steep cliff composed of exposed granite (Fig. 1.5).

The rock mass of the right bank slope at 1200–1300 m elevation is gray-yellow crushed stone (col + dlQ4). According to the exploration, 138 diabase dikes and 9 granitic aplite dikes are developed on the right bank. Above 1250 m elevation of the right bank slope, there are mainly dike fracture zones of β_4, β_{97} (f_{93}), β_{146}, β_{168} (f_{154}), β_{170}, β_{202} (f_{191}), β_{203} (f_{194}), etc. Most of the dike fault zones are in block fracture and cataclastic structure. γ_{L5} granitic aplite dike is developed at 1300 m elevation. The dike strikes more than SN, dip-slope (NW or SW), and dip angle is steep. Among them, γ_{L5}, β_{169} and other existing dikes are exposed on the slope. For excavation and supplementary exploration, β_{202}, β_{203}, etc. have newly revealed 26 large scale or poor properties of dikes. The rock mass is mainly composed of Chengjiang granite with hard rock quality, and core cake and sheeting phenomenon are concentrated in the unloading zone. The results of in-situ stress test show that the stress field in Dagangshan pivot area is superimposed stress field with tectonic stress as the main

Fig. 1.5 Landform of slope on right bank of dam site before excavation

stress and gravity stress as the auxiliary stress. The in-situ stress test shows that the tectonic stress roughly 50–80 m below the valley bottom bedrock surface is the main component of in-situ stress, and the direction of the stress field is NWW-SEE extrusion. The measured horizontal stress value of the rock mass is relatively low, indicating that the rock mass in the shallow part of the valley is unloaded and the in-situ stress is released to different degrees. According to the analysis of the in-situ stress test data of the valley bottom riverbed, the direction of the regional tectonic stress field is NWW-SEE extrusion, and the initial in-situ stress is as high as 26.51 MPa at the depth of 142 m, indicating that the initial in-situ stress in the Dagangshan area is relatively high. The fault zone of the slope area mostly develops along the diabase dikes, which are mainly in the steep dip angle and incline slope, belonging to the rock block and rock debris type, rock debris with mud type or mud with rock debris type. According to the exploration, 84 faults are developed on the right bank slope, and the scale of F_1, f_{191} and f_{174} is relatively large. Among them, there are 1 grade II fault, 5 grade III faults and 23 grade IV faults revealed in the previous exploration, and 3 faults of fracture bandwidth is $0.1\ m \leq b \leq 0.3\ m$. Among the 55 faults newly revealed by construction excavation and supplementary exploration, 26 faults belong to class III structural plane and 29 faults belong to class IV structural plane. Among them, 22 faults of class IV have a fracture bandwidth of 0.1 m $\leq$ b $\leq$ 0.3 m. The dikes such as diabase are hard and dense and belong to fine grain and particle structure in general. However, hidden micro-cracks are developed in the dikes, belonging to hard, brittle and broken rock mass, and faults are often developed along the dikes, accompanied by strong weathering. In addition, there are weathering and alteration phenomena of rock mass, which further reduces the quality of rock mass. Among them, biotite weathering into chlorite, vermiculite, etc., and feldspar minerals after alteration often appear kaolin, sericite. The weathering characteristics of rock mass on the right bank show obvious horizontal and vertical zonation, and its variation law gradually increases with the increase of elevation, and it has a certain degree of heterogeneity, which is the result of micro-topography, lithology, structure and other factors.

(1) Fully weathered zone: The rock mass has a cataclastic and granular structure with very poor integrity, and the rock completely loses its luster. All the rock minerals except quartz particles are weathered into secondary minerals, and the hammer has a soft feeling and pits appear. The corresponding P-wave velocity is under 2000 m/s. The right bank slope is mostly distributed above 1290 m.
(2) Strongly weathered zone: The rock mass has a fragmented structure with poor integrity, most of the rocks lose luster, most of the rock minerals except quartz particles are weathered into secondary minerals, and most of the rocks become crisp and fragile when hammered. The corresponding V_P is 2000–2500 m/s. The right bank slope is mostly developed at an elevation above 1150 m, and the depth is generally less than 20 m.
(3) Weakly weathered upper zone: The rock mass is mainly mosaic and block crack structure, and a small part is sub-block structure, with poor integrity. Most of the rock surface has discoloration phenomenon, the fracture is granular, the hammering sound is relatively dumb, and the crack is slightly tensioned. The corresponding V_P is 2500–3500 m/s, and the horizontal depth varies greatly with the elevation.
(4) Weakly weathered lower zone: The rock mass is mainly of sub-block-mosaic structure, with good integrity. The rock only has weathering phenomenon on the surface and fracture surface, and still maintains fresh and original luster on the fracture surface, and the hammering sound is crisp. The corresponding V_P is 3500–4500 m/s, and the horizontal depth varies greatly with the elevation.
(5) Slightly weathered to fresh rock mass zone: The rock mass is mainly block-sub blocky structure, partly mosaic structure, rock integrity is good, and the combination is close. The rock mass basically maintains fresh luster, and the hammering sound is crisp, corresponding V_P exceeds 4500 m/s.

The project area has deep and steep valley slope, high stress and strong unloading development. The unloading direction is from the rock mass on the bank slope to the river surface. According to the information obtained from the exploration of the flat cave, the rock mass unloading zone mainly develops along the existing structural plane, and according to the different unloading degree, the strong unloading zone and the weak unloading zone can be divided. It can be seen from Table 1.2, the

Table 1.2 Right bank slope rock mass unloading depth [132]

Elevation zoning	Hole number (elevation/m)	Strong unloading/m	Weak unloading/m
1255 m or more	PD316(1300.38)	0.0–103.5	103.5–156.7(Unexposed)
	PD321(1303.7)	0.0–138	138.0–148.0
	PD321-1	0.0–97.5	–
	PD322(1266.9)	0.0–152.0	–
	PD322-1	0.0–127.0	–
	PD10(1300.58)	22.2–79.0	79.0–103.0(Unexposed)
1070 ~ 1255 m	PD323(1220.34)	18.0–108.3	108.3–149.5(Unexposed)
	PD09(1225.11)	0.0–121.4	121.4–149.0
	PD325(1166.73)	–	0.0–43.0
	PD308(1137.55)	0.0–4.0	4.0–54.0
	PD306(1100.37)	0.0–27.8	27.8–48.1
	PD314(1107.11)	0.0–44.0	44.0–50.5
	PD303(1107.13)	0.0–7.4	7.4–52.3
	PD304(1103.53)	0.0–36.0	36.0–51.0

Note The 0–22.2 m section of PD10 flat cave and the 0–18 m section of PD323 flat cave are the cover layer. The opening of PD325 is 38 m away from the original ground

unloading depth of the rock mass on the right bank in the vertical space of the valley generally deepens significantly with the increase of the elevation, and the micro-topography, lithology and structure there are variable, resulting in a certain degree of heterogeneity in the unloading distribution.

The right bank is the crest slope above 1135 m elevation, and the spinner groove slope below 1135 m elevation. The slopes above the 1300 m elevation are covered with slope failure deposits, and upstream and downstream between 1220 and 1300 m elevation are covered with several meters thick slope failure deposits. There are more than 10 rock veins and faults exposed on the slope, and the direction of the rock veins and faults are mainly NNW and NE. The rock mass of the right bank slope is mainly composed of black cloud diorite. The surface layer of the right bank slope is covered with overburden, and the rock mass is strongly weathered and unloaded, while the upper slope is located in the fully weathered rock mass, and the degree of weathering of the rock mass still has a greater impact on the quality category of the rock mass. Micro-new unloaded fractured-mosaic granite masses and micro-new unloaded block-structured gabbro veins are class II; micro-new unloaded mosaic granite masses, weakly weathered lower unloaded sub-block-mosaic granite masses, micro-new unloaded-mosaic gabbro veins, micro-new unloaded block-sub-block-structured sodic altered granite masses are class III_1; micro-new unloaded block-cracked structural granite masses The weakly weathered lower zone-micro-new unloaded mosaic structure of gabbro veins is class III_2; micro-new unloaded block cracked-fractured structure of granite masses, weakly weathered upper zone unloaded block cracked-fractured structure of granite masses, weakly weathered-micro-new block cracked structure of gabbro veins is class IV; strongly weathered, strong unloaded fractured-bulk structure of rock masses and fault fracture zones is class V_1; full weathering of bulk-fractured structure of granite rock masses is class V_2. According to the right bank slope excavation and testing revealed that the fully weathered rock mass is bulk-fractured, easy to excavate, slope shaping is better, wave velocity value is less than 2000 m/s; Strongly weathered rock mass is fractured structure, easy to excavate, the excavation slope is locally controlled by the structural surface, the wave velocity value is 2000–2500 m/s; Weakly weathered upper zone of the rock mass is block-fractured structure, the excavation needs to be blasting, the excavation slope is controlled by the structural surface, the wave velocity value is 2500–3500 m/s. The value of wave velocity is 2500–3500 m/s. Various structural surfaces in the rock mass of the slope are divided into two categories of rigid and weak structural surfaces according to the situation of the filler. Rigid structural surface according to the gap wall contact tightness and alteration characteristics are subdivided into cemented structural surface, altered structural surface and open structural surface of three categories; weak structural surface according to its cause type, thickness of the filling material, material composition, etc. are subdivided into the rock clasts of the rock mass, rock clasts sandwiched with mud, mud sandwiched with rock clasts of the three categories. Previous experience has shown that the depth and extent of the slope unloading zone are closely related to its stress field and slope morphology [147], i.e., the higher the intensity of the stress field, the greater the depth of the slope unloading, and the higher the slope and the steeper the slope, the greater the extent of the unloading zone. A number of tensile cracks have developed in the rock mass of the right bank slope, which is because the Gongga Mountain area where the dam site is located has been rapidly uplifted recently, forming a topography with high mountain and deep valley. The rapid downcutting of the valley will inevitably lead to the strong release of high ground stress in the rock mass of the slope, driving the slope to produce unloading rebound toward the free surface. Thus, residual tensile stress environment is formed in some relatively weak facial positions (such as small faults, long cracks or crack dense zones) in the slope body, which leads to the stretching of the original structural plane, which is one of the foundations of deep tensile cracks. In particular, the crack group 5 (occurrence N0°–35°E/SE < 35°–50°) formed at the end of Middle Pleistocene produces tension during the unloading and rebound of the slope towards the free plane, and causes the tensile fracture group of the slope rock mass under the action of gravity. The right bank unloading fissure dense zone is formed by tracking the fissure outside the middle dip slope of group 5 under the combined effect of stress adjustment and gravity unloading. Its distribution and development characteristics can be summarized as two medium-dip unloading fracture dense zones XL-316 and XL-915. Lithologic conditions, structural plane and their combination with slope surface not only control the unloading range of slope, but also affect the failure mode of slope [148, 149]. Therefore, only from the geological conditions, numerical methods and other aspects of the deep fracture space development characteristics and deformation law of the slope comprehensive analysis, in order to reasonably determine such slope stability problems.

According to the geological analysis, the upper part of the slope on the right bank consists of overburden layer, fully weathered rock mass and strongly weathered rock mass, the middle part is weakly weathered rock mass, and the lower part is weakly weathered micro-new rock mass (below 1135 m elevation). In this way, the slope composed of several groups of steeply inclined structural planes may have uncertain failure modes. For example, three failure modes revealed by geological investigation are flexural-tensile deformation, slip-tensile deformation and creep-tensile deformation.

(1) Flexural-tensile deformation: the development of the fracture zone of diabase dike and the crack of group 2 in the steep slope makes the slope body have the characteristics of steep inner laminar slope, forming the bending and tension deformation of the surface of the slope with a depth of 5–15 m. Flexural-tensile deformation can be seen in the exposed β4 diabase vein upstream of PD304 (Fig. 1.6a).
(2) Slip-tensile deformation: the surface rock mass is strongly unloaded along the crack of group 5, with the fracture zone of the dike with steep dip angle as the rear edge tension fracture surface and the crack of group 5 outside the dip

(a) Bending-tensile deformation (PD304 flat hole) (b) Slip-crack deformation collapse (PD303 flat hole)

(c) Slip-tensile deformation failure (PD303 flat hole) (d) Local collapse of the overburden slope (PD10 flat hole)

Fig. 1.6 Potential failure forms of right bank slope

slope as the bottom slip surface, which may result in slip-tension deformation and failure. For example, the landslide deposits distributed at 1000–1100 m elevation along the II–III lines on the right bank may be residual deposits of slip-tension deformation and failure near PD303 flat cave with β4 as the trailing edge tension plane and the structural plane outside the intermediate slope as the slip plane in the early stage, with a thickness of 2–5 m (Fig. 1.6b and c).

(3) Creep-tensile deformation: the landslide deposits distributed above 1220 m on the right bank have a horizontal thickness of 10–20 m and a vertical thickness of 10–15 m. No signs of deformation are found in the natural state, and the slope body is stable. However, during the exploration and construction of the walkway and flat cave, due to the lack of effective support, local creep and tensile deformation occurred, and the opening of PD10 flat cave collapsed (Fig. 1.6d).

1.5.2 Two-Dimensional Numerical Simulation and Stability Analysis of Slope

Under normal circumstances, the shear strength of slope rock mass decreases gradually due to weathering, rainwater infiltration and strain softening, which leads to gradual failure of slope. The basic principle of finite element strength reduction method is introduced into rock fracture process analysis RFPA^{2D} (Fig. 1.7), and the safety factor is calculated by reducing the shear strength of the material so that the slope reaches the critical failure state. The slope strength reduction method–RFPA^{2D}-SRM is established. Using the mesoscale maximum failure primitive number as the instability criterion, the evolutionary process of progressive failure of slope structure with the deterioration of rock mass structural strength is reproduced [150]. RFPA^{2D}-SRM is used to analyze the typical section VI–VI of the right bank slope of Dagangshan Hydropower Station, and the slope failure mode and safety factor are obtained through numerical simulation, which is of great significance for understanding the formation mechanism of slope failure under natural conditions, and is conducive to guiding the reinforcement of slope from the source.

1. Determination of safety factor of RFPA^{2D}-SRM

Safety factor (F_s) is an important concept in slope stability research, which is based on the strength reserve concept. The definition of *Fs* is: the ratio of the actual shear strength to the shear strength when the critical failure is reduced, that is, the degree to which the shear strength of rock mass can be reduced. When the shear strength parameters c and φ of the material are replaced by their critical strength parameters c' and φ' respectively, the slope will be in a state of critical equilibrium [151], where,

$$\begin{cases} c' = c/F_s \\ \tan\varphi' = \tan\varphi/F_s \end{cases} \tag{1.1}$$

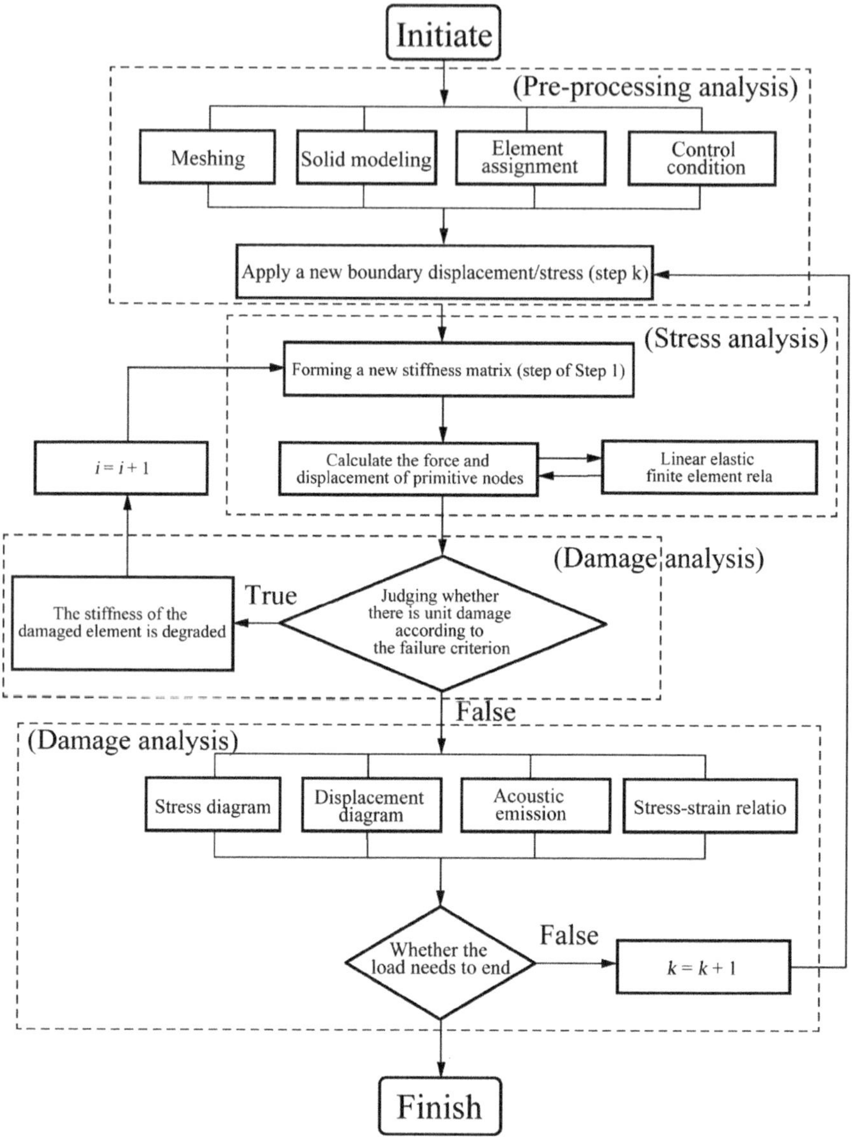

Fig. 1.7 RFPA^{2D} program flow chart [151]

In the finite element calculation, the strength parameters are reduced according to the proportion or a certain step size. Combined with the constitutive model characteristics of quasi-brittle rock materials unique to RFPA, when the calculation reaches the critical state, that is, when the slope has macroscopic failure, the corresponding safety factor of the slope is obtained.

In RFPA^{2D}-SRM, the compressive (shear) strengths and tensile strengths of the primitive elements are unified, and the initial strength f_0 is reduced according to a certain step. Every time the reduction occurs, the strength f^{trial} corresponds to a safety factor F_s^{trial}, and the reduction criterion is as follows:

$$f^{trial} = \frac{f_0}{F_S^{trial}} \tag{1.2}$$

When the failure number of basic elements reaches the maximum value, that is, when the slope is unstable, the test safety factor F_s^{trial} is the final safety factor F_s [151].

2. Stability analysis of RFPA^{2D}-SRM for typical slope sections

To study the initiation, development, expansion and penetration evolution law of micro-fracture during the failure process of the right bank slope under natural conditions, the stress–strain characteristics, failure mechanism and stability of a typical section of the slope VI–VI were calculated and analyzed (Fig. 1.8). The distribution law of stress field and acoustic emission activity in the process of slope progressive failure was studied, and the phenomena of stress accumulation, stress release and stress transfer in the process of slope instability catastrophe were reproduced. This paper attempts to study the breeding process of slope instability from the level of stress field evolution and acoustic emission distribution. A numerical model was established and relevant parameters were set for the geological structure of the slope's main faults, rock layers and unloading fracture dense zones (Fig. 1.8).

The physical and mechanical parameters of rock mass in each layer of the section are shown in Table 1.3, and the main control structural plane (mass) parameters such as unloading fracture zone XL-316, XL-915 and fault f_{231} are uniformly selected as type V_2 in Table 1.3.

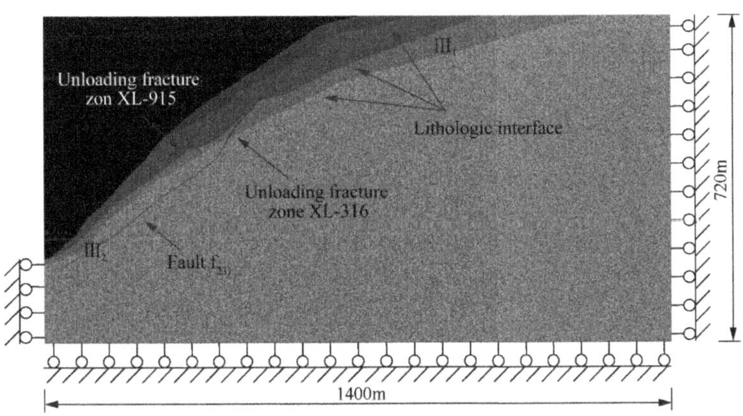

Fig. 1.8 Numerical model of section VI–VI of the right bank slope

Table 1.3 Rock physical and mechanical parameter value table [152]

Material number	Density/(g/ cm^3)	Compressive strength/MPa	Modulus of deformation GPa	Poisson's ratio	Friction factor	Cohesive force/MPa
II	2.65	70–80	18.00–25.00	0.25	1.30	2.000
III_1	2.62	40–60	9.00–11.00	0.27	1.20	1.500
III_2	2.62	40–60	6.00–9.00	0.30	1.00	1.000
IV	2.58	20–40	2.50–3.50	0.35	0.80	0.700
V_1	2.45	< 15	0.25–0.50	> 0.35	0.50	0.200
V_2	2.10	< 10	0.20	> 0.35	0.40	0.175

Considering the problem according to plane strain, RFPA^{2D}-SRM was used to calculate the elastic mode diagram, displacement vector diagram, shear stress diagram, acoustic emission diagram and failure mode of section VI–VI of the right bank slope to evaluate its stability (Fig. 1.9a–d). RFPA^{2D}-SRM recorded the number of primitive failures in the reduction calculation process. When the reduction reached the 25th step (the reduction factor is 0.01), the number of primitive failures reached the maximum, and the slope had overall instability failure, and the safety factor is 1.33. The results are consistent with the two-dimensional calculation results of Shi et al. [148] and Liu et al. [153] for the typical section of the slope on the right bank of Dagangshan Hydropower Station.

The worst stability part of the slope is determined by RFPA^{2D}-SRM simulation. As shown in Fig. 1.9, in step 25-1, obvious shear failure occurred at the foot of the slope (fault f_{231} area), and micro-fracture of rock mass began to emerge there. In step 25-17, rock mass failure of the slope continued to evolve and expand along fault f_{231} to the top of the slope. During steps 25-18, a large number of primitive failures occurred at the junction of the unloading fracture zone XL-316 and fault f_{231}, and the "rock bridge" at the junction was connected. The rupture expanded along the main control structure plane composed of the unloading fracture zone XL-316 and fault f_{231}, forming an obvious slip surface from bottom to top and connecting to the top of the slope. During steps 25-19, the "sliding body" composed of the unloading fracture zone XL-316 and fault f_{231} slides down and fails. It can be seen from the acoustic emission map that a large number of tensile failures occurred at the unloading fracture zone XL-316, and shear failure occurred at the fault f_{231} (in Fig. 1.9d, the black circle represents the tensile failure and the white circle represents the shear failure), and obvious failure also occurred at the unloading fracture zone XL-915. It shows that the main weak structures such as XL-316, XL-915 and fault f_{231} are prone to deformation and failure during the reduction process. During step 25-20, a huge sliding body with unloading fracture zones XL-316 and XL-915 as hanging wall and fault f_{231} as hanging wall was formed along the through slip zone and extrusion fracture zone, resulting in sliding failure of the whole slope. From the process of slope crack initiation to slip, it can be seen that the macroscopic instability is due to the development and evolution of the internal rock micro-fracture along the soft

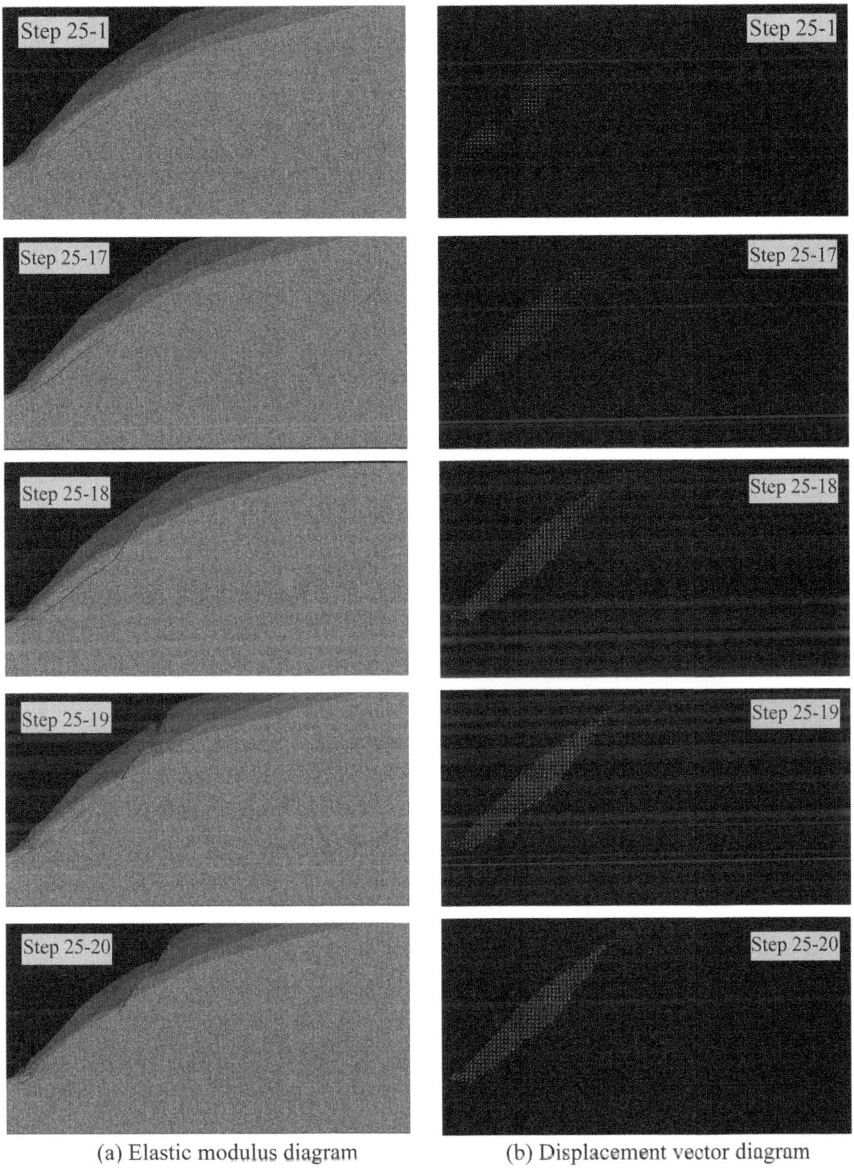

Fig. 1.9 Stability analysis results of the right bank slope by RFPA^{2D}-SRM

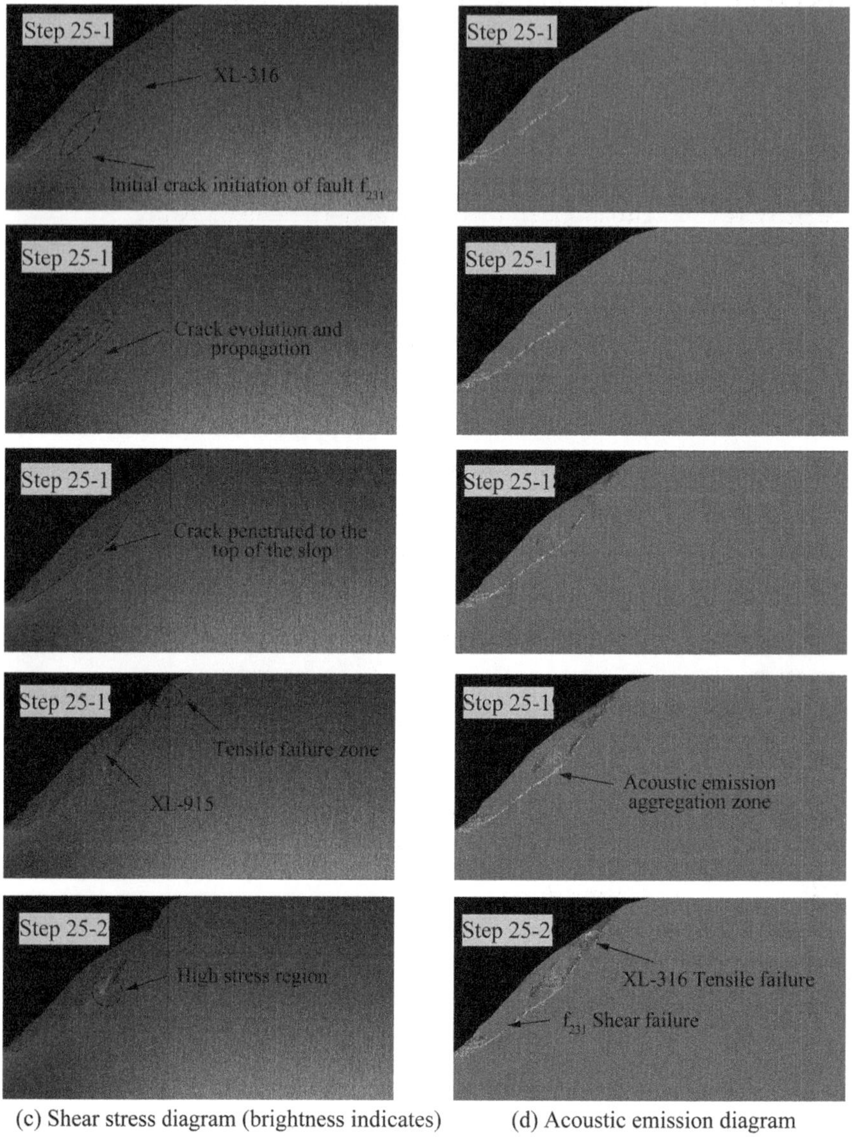

Fig. 1.9 (continued)

structural plane of the controlling slope. When the rupture zone is through, the slope body will appear obvious displacement, and when the overall slip failure occurs, the deformation failure increases sharply. It can be seen from the AE diagram that the AE distribution law in the failure process reproduces the process of continuous damage accumulation of primitives in the slope failure process. The continuous damage of primitives in the reduction process causes local damage of the slope, and the accumulated local damage is connected to form a slip strip. The AE generated in the simulation process is mainly distributed on the main control structures such as XL-316, XL-915 and fault f_{231} with dense unloading fractures. This confirms that the weak structural plane is one of the main factors controlling the failure mode of rock slope [154–158].

1.5.3 Summary

The stability evaluation of high and steep rock slope engineering must be based on qualitative geological structure. Combined with engineering geological data, RFPA^{2D}-SRM is used to analyze the stability of typical sections of the right bank slope under natural conditions. The following conclusion can be obtained.

(1) RFPA^{2D}-SRM was used to simulate the typical section, and the whole process of rock micro-fracture initiation, development, expansion, and formation of potential slip surface through penetration was reproduced. When the reduction reached the 25th step, the whole slope undergone macroscopic instability failure, and the safety factor of the slope is 1.33. In general, the slopes that are not disturbed by excavation unloading under natural conditions are still in a stable state. In the horizontal direction, the slope shows deformation in the direction of the valley, and in the vertical direction, it shows vertical downward deformation.

(2) The acoustic emission results of numerical simulation also revealed that tension failure occurred at the unloading fracture dense zone XL-316 at the top of the slope, and a large number of shear failure occurred at the fault f_{231} at the foot of the slope, forming a potential slip surface dominated by the unloading fracture dense zone XL-316 and fault f_{231}, which provided a reference for the later slope reinforcement scheme.

(3) The main dominant structural plane composed of the unloading fracture dense zone XL-316 and fault f_{231} of the right bank slope is the main factor affecting the stability of the slope, and the initial crack initiation, sliding and collapse of the slope basically develop along the main control structural plane. The unloading fracture zone XL-316 is taken as the upstream boundary, fault f_{231} is taken as the bottom slip surface, and diabase dike is taken as the trailing edge cutting surface to form the sliding failure mode. The unloading fracture zone XL-915 is also the accumulation area of damage, but with the continuous development of unloading and excavation work of the slope in the later stage, the impact on the slope stability is limited.

1.6 Microseismic Monitoring of the Right Bank Slope of Dagangshan Hydropower Station

Many scholars at home and abroad have put forward a variety of collection methods to obtain the information of rock mass structure plane. For example, photogrammetry, sampling window method, borehole core joint collection method and fine line measurement method [159–166]. Hammah et al. [167] used fuzzy clustering method to study the automatic identification of structural planes. Feng et al. [168] applied a non-reflective total station to measure discontinuous point positions, and then obtained the occurrence of exposed rock masses. Siekfo et al. [169] used a 3D laser scanner to obtain the information of rock mass structure plane. Dong et al. [170] used 3D laser measurement technology to obtain the occurrence of a single fracture in the rock mass structure of a high and steep slope. Based on the results of close-range laser scanner, Hong et al. [171] used fuzzy clustering method to make statistics and classification of rock mass structural planes. Wang et al. [172] achieved effective acquisition of geometric information of rock mass structural plane by means of digital photogrammetry. Liu et al. [173] used laser point cloud data to quickly obtain geometric information such as spacing and trace length of structural planes, which opened up a new way for the acquisition of dominant occurrence statistics and geometric information of rock mass structural planes. However, many of the above methods are either affected by terrain and environmental factors, the measurement accuracy is poor, or the heavy field workload brings great inconvenience to human operation, and it is difficult to obtain comprehensive and accurate rock mass structural plane information. Therefore, how to quickly and efficiently obtain the information of the structural plane in the slope body structure has become the bottleneck of evaluating the slope stability [174, 175]. In particular, the southwest region of our country is located in the interlacing zone of the Indian plate and the Eurasian plate. Under the action of geological structural movements such as the long-term uplift of the Qinghai-Tibet Plateau and the sharp downcutting of rivers, the deep unloading development of slope rock mass, heterogeneous engineering geological structure, and extremely poor stability of valley bank slopes and mountains [176, 177]. Under such a very complex natural environment and geological background, hydropower resources development projects of various cascades in the Dadu River, Yalong River, Jinsha River, Lancang River and other main streams in Southwest China have been carried out in different degrees, and various engineering geological problems have been more and more complex.

The deformation of rock slope is mainly manifested as slip failure along the fracture surface. Once this kind of rock quasi-brittle material is damaged, the traditional stress–strain constitutive relationship cannot characterize the deformation law. However, the displacement change of any point in the rock mass will affect the mechanical properties of the material itself, so many scholars have tried to use the stress-displacement relationship to describe the deformation and failure law in recent years. As early as 1970, Japanese scholar Saito used displacement as the prediction parameter and established a differential equation of accelerated creep to successfully predict the Gotang Mountain landslide in Japan [178]. China Railway Northwest Scientific Research Institute Co., LTD. (Formerly Northwest Branch of Scientific Research Institute of Ministry of Railways) also successfully predicted the Xujiahe landslide of Baocheng-Chengdu Railway [179]. He and Li [180] established mechanical criteria under two actual loading conditions, namely curved shear-out surface and planar sliding, for large-scale landslides in accumulative strata. Li et al. [181] established an inverse Verhulst function model based on the characteristic that the cumulative disposition-time curve of landslides was opposite to the Verhulst curve. Huang et al. [182] proposed the nonlinear theory of collaborative prediction model to study landslide prediction problems for complex problems. Romeo [183] used the Newmark stability calculation model to predict the landslide displacement caused by earthquakes, and calculated the displacement criterion for predicting the occurrence of landslides. Li [184] developed an integrated landslide prediction system based on geographic information system (GIS). Peng [185] used information fusion technology to process landslide monitoring data. Liao [186] proposed a physical power model for landslide prediction based on the principle of elastics and plasticity. Wang [187] respectively predicted the Xintang landslide, Jipickaozi landslide and Huangwaxi landslide in the Three Gorges of the Yangtze River based on various deformation and failure signs and inducing factors of landslides, and summarized a set of comprehensive landslide prediction methods and criteria. Li et al. [188] conducted a comprehensive study on the time prediction of landslide disasters by organically combining the macroscopic deformation and failure signs of slopes and precursor information. Relying on a large number of landslide deformation monitoring data, Xu et al. [189–191] comprehensively predicted and forecast landslides from the perspective of spatio-temporal evolution laws, emphasized the importance of combining spatio-temporal characteristics in landslide early warning, and obtained an improved tangent Angle early warning criterion. Based on a comprehensive review of numerous achievements in slope prediction and early warning, Jin et al. [192] developed an integrated forecasting model tailored to the slope conditions of the Jinping I Hydropower Station. Herrera et al. [193] used a variety of monitoring techniques to observe the relationship between Portalet slope displacement and rainfall, and predicted the long-term slope displacement.

Although many beneficial results have been achieved in the monitoring and forecasting of slope displacement, due to complex geological conditions, slope deformation evolution behavior is complicated, random and uncertain, so most of the methods are mainly devoted to the research of prediction methods and models or only for a certain macro-deformation validation prediction. Lack of systematic monitoring of precursor information of internal rock mass damage during slope construction disturbance. The stability of most excavated rock slopes is a process of dynamic evolution, and the instability failure is not achieved overnight, but is the result of gradual accumulation and expansion of micro-cracks caused by the accumulation of local elastic energy to a certain critical value. Only by effectively monitoring the precursor laws of microfracture initiation, development, expansion, penetration and interaction, can the stability and instability failure mode of excavated slope be accurately judged. When several macro-cracks were produced in the excavation process of the right bank slope, microseismic monitoring technology was introduced to determine the time and location of the micro-fracture of the excavated slope according to the micro-fracture signal, determine the size and concentration of the micro-fracture, and infer the trend of the macro-deformation of the rock to achieve the effect of predicting the slope instability.

1.6.1 Construction of Microseismic Monitoring System

The right bank slope of Dagangshan Hydropower Station adopts Canada ESG microseismic monitoring system, which is mainly composed of Paladin signal acquisition system, Hyperion data processing system, high-precision acceleration sensor, combined with the MMS-View 3D visualization software based on remote network transmission developed by Mechsoft Technology (Dalian) Co., LTD. (Fig. 1.10). In the caverns located at 11 elevations of the slope of the right bank, with the dam shoulder as the center, 22 sensors are arranged along the upstream and downstream 300 m in the direction of the river, the elevation 800–1400 m and the 400 m range (600 m × 600 m × 400 m) in the slope surface rock mass to monitor the excavation slope of the right bank for 24 h all-weather, and realize the precise positioning of three-dimensional events. The accuracy is within 10 m (Figs. 1.11 and 1.12). For the specific installation steps and implementation process of sensors, refer to literature [194], which will not be repeated here.

1.6.2 Spatial–Temporal Distribution of Excavation Disturbance and Microseismic Activity

The distribution law of rock mass microfracture during large-scale slope excavation is analyzed (May 2010 microseismic monitoring system entered the field until September 2011 large-scale main slope excavation construction ended). As of September 30, 2011, when the main excavation of the slope was completed, a total of 1506 microseismic events were collected. After noise identification, screening and elimination, 1337 microseismic events and 1217 blasting events were obtained in the effective area of the study. The size and color of microseismic events respectively represent the energy size and moment magnitude of microseismic events [194]. According to the occurrence frequency of microseismic events combined with the site construction situation, the distribution of microseismic events is generally stable, and there is no dangerous phenomenon of sudden increase in the number of events in a certain area in a short time, which indicates that the right bank slope is relatively stable on the whole. The microseismic events generated during the slope excavation were distributed in strip bands in the area of the three-layer shear hole on the slope (elevation of 1135–1240 m) and near the spinner trough, that is, the upper wall of the

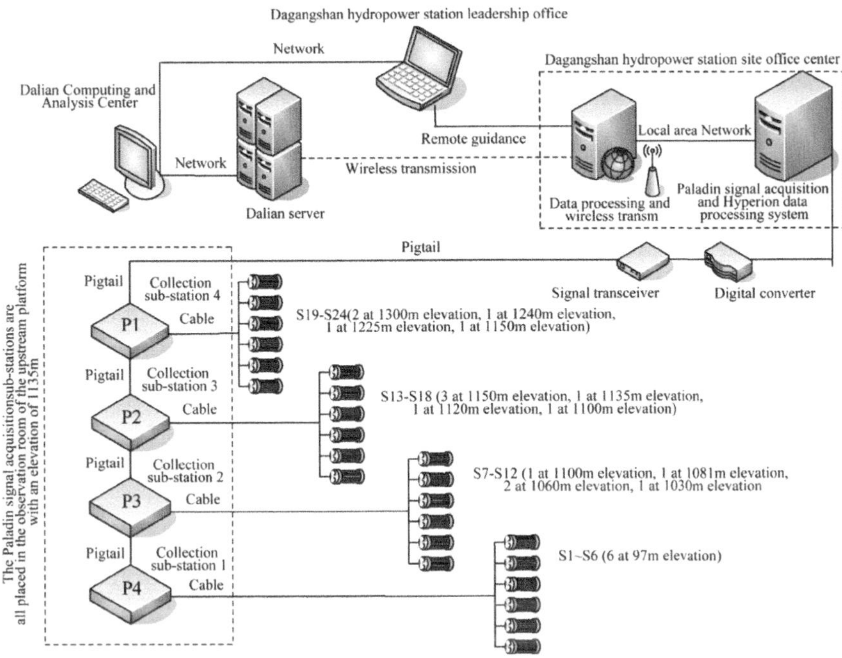

Fig. 1.10 Network topology of the microseismic monitoring system

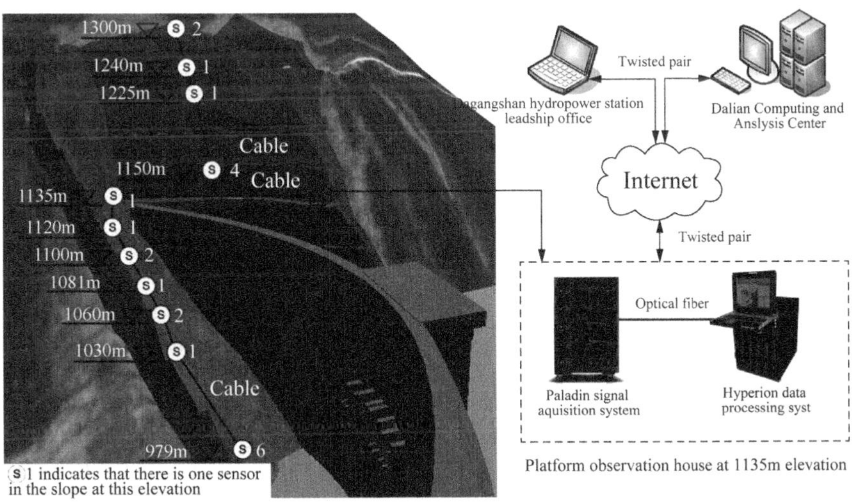

Fig. 1.11 Composition of microseismic monitoring system [195]. *Note* Ⓢ indicates that there is one sensor in the slope at this elevation

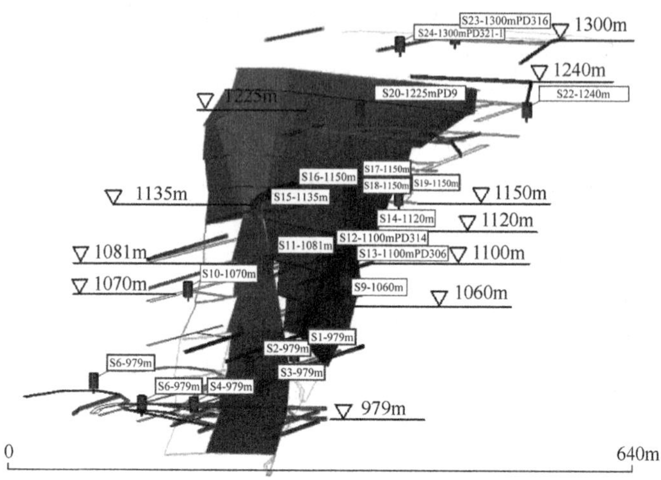

Fig. 1.12 Sensor space layout of microseismic monitoring system [73]

fault f_{231} and the lower area of the unloading fracture zone XL-316. The geological conditions of the location where the microseismic events gathered were relatively poor (Fig. 1.13).

As can be seen from the cumulative density cloud map of microseismic events in Fig. 1.14, the energy density of microseismic events continuously accumulates, releases and transfers during slope excavation, which also causes local stress adjustment. The area with large energy density of microseismic events is mainly concentrated below, 225 m elevation, along the unloading fracture zone of main control structure XL-316 and fault f_{231} through the slope. The micro-seismic energy loss caused by excavation is also constantly released, transferred and adjusted, and the energy loss area keeps increasing, and the accumulation of micro-fracture keeps extending. When the stress superimposed by the released energy in the elastic area is less than the elastic strength of the rock mass material, the damage area will no longer expand and the damage surface will no longer expand. The energy loss during the failure process was mainly concentrated in the vicinity of XL-316 and f_{231} in the unloading fracture zone of 1135–1240 m elevation and 979–1100 m elevation, and occurred at a depth of 80–150 m from the slope surface, which was consistent with the geological conditions revealed by the geological exploration data (Fig. 1.15). Combined with the comprehensive analysis of construction conditions, under the action of excavation construction disturbance, the unloading fracture zone or fault activation induces the slope unloading relaxation, which leads to the compression shear failure of the lower rock mass, so most of the micro-fracture events gather on the dominant structure. It shows that microseismic monitoring can effectively identify and delineate the deep rock mass damage area of slope induced by construction disturbance, which provides a reference for slope construction and reinforcement treatment. The moment magnitude distribution of microseismic events ranges from

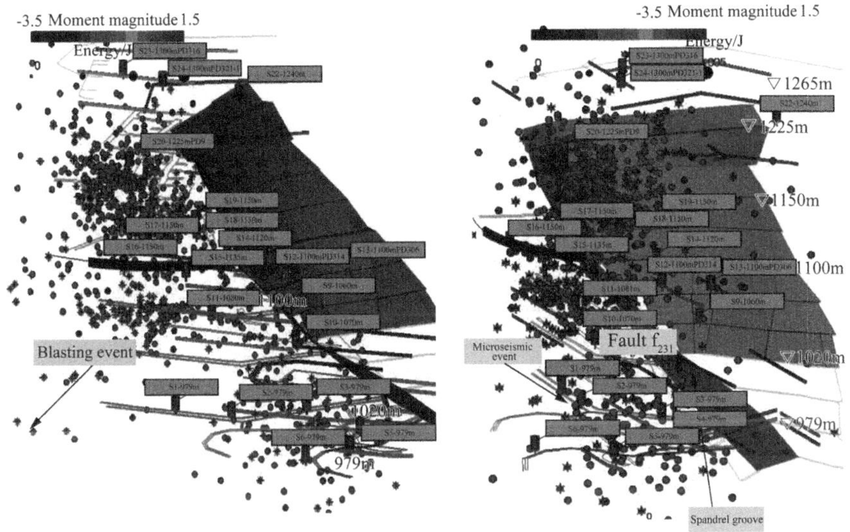

(a) Spatial distribution diagram of blasting events (b) Spatial distribution diagram of microseismic events

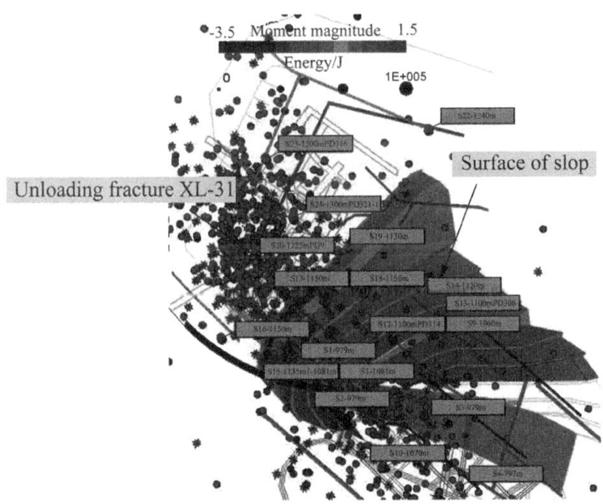

(C) Top view of the spatial distribution diagram of microseismic events

Fig. 1.13 Spatial distribution of blasting events and microseismic events and top view of spatial distribution of microseismic events

− 2.5 to 0 and fluctuates above and below − 1.0. The distribution is approximately a normal distribution with − 1.0 as the center, and the fluctuation does not change significantly, but the overall moment magnitude is still less than − 1.0 (Fig. 1.16).

Since the successful construction and operation of the system on May 4, 2010, up to the end of the excavation work of the main slope in September 2011, a total of 1337 effective microseismic events have been obtained through waveform identification,

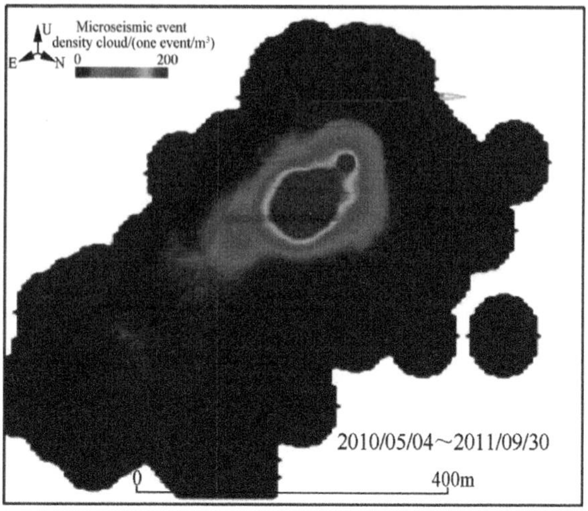

Fig. 1.14 Cumulative density cloud map of microseismic events during excavation

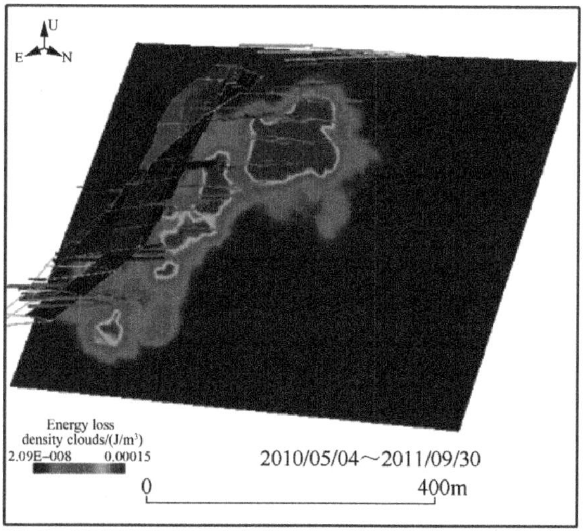

Fig. 1.15 Spatial distribution of energy loss density of microseismic events during excavation

filtering and event positioning. Figure 1.17 shows the distribution of cumulative energy release and activity rate of microseismic events in rock mass with time during the excavation of the right bank slope. The microseismic activity was strong in August 2010 and April to July 2011, and the maximum frequency of the event was 10 times/day. During this period, the cumulative energy release of microseismic rock mass in slope also increased rapidly. The slope of the change curve of cumulative energy release in July 2011 and August 2010 increased significantly, but lagged behind the high microseismic activity during this period, indicating that there were high energy (high magnitude) events in the slope, which affected the stability of the slope.

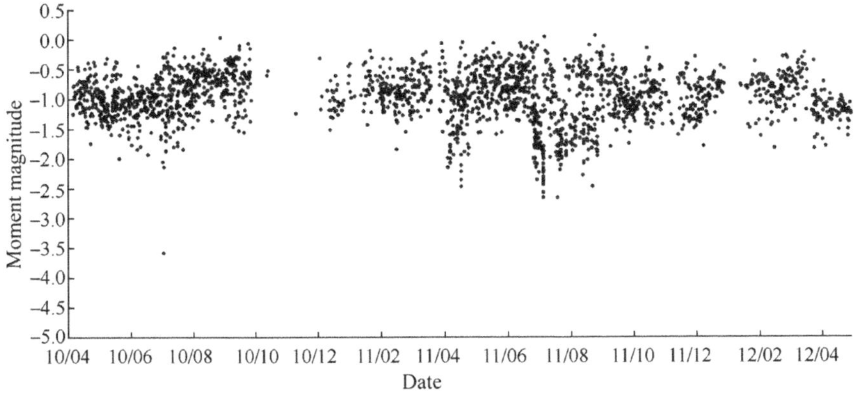

Fig. 1.16 Moment magnitude distribution of microseismic events with time

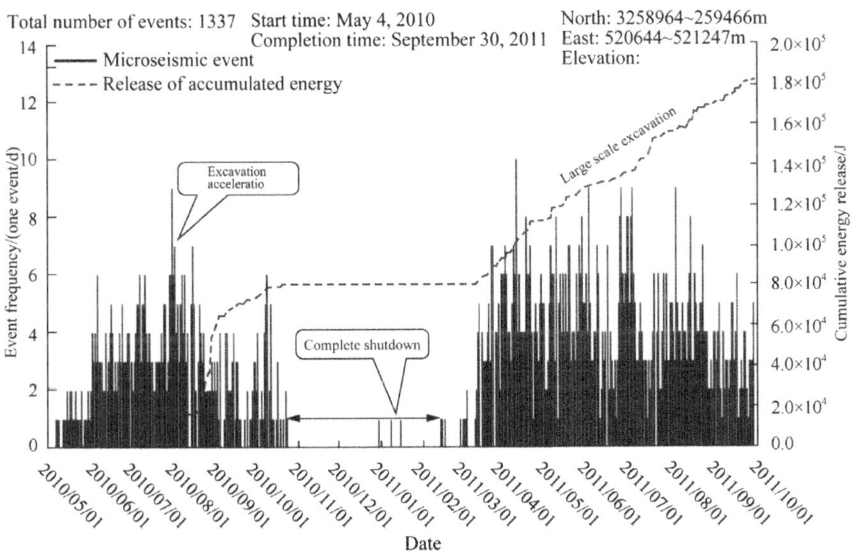

Fig. 1.17 Cumulative energy release, event frequency and time distribution of microseismic events

According to the analysis of site excavation construction, the construction progress of slope excavation was accelerated after June 2010. From the end of October 2010 to the middle of February 2011, the slope excavation construction was suspended, and the excavation construction resumed in early March and gradually accelerated. It is found that the overall microseismic activity of slope during excavation is closely related to excavation construction, and excavation unloading is the main external force that controls the evolution of slope microfracture and affects the stability of slope [73].

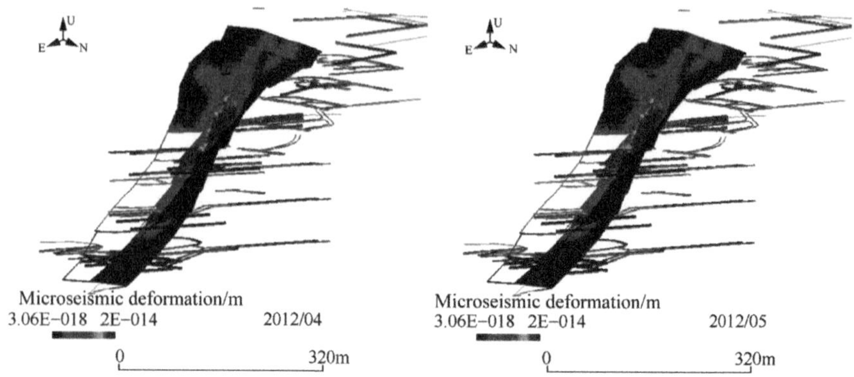

Fig. 1.18 Deformation area of slope surface rock mass induced by excavation disturbance

With the deepening of excavation, the damage range of slope rock mass expands gradually. It can be inferred that several macroscopic cracks produced by the side slope surface are the damage and failure of deep rock mass caused by excavation unloading. The variation of conventional monitoring points during excavation indicates that the displacement and stress field of slope change gradually with excavation unloading. Because the excavation unloading changes the natural shape of the slope and the boundary conditions of the structural plane, the local slope loosens slowly due to the stress adjustment, which is caused by the gradual expansion of the small dislocation between the rock particles and the deformation and failure (Fig. 1.18).

1.6.3 Identification and Delineation of Slope Main Control Structural Plane

According to the distribution law of rock microfracture during the unloading excavation of the main body of the slope, it can be seen that the micro-fracture of deep rock mass accumulates around the potential slip plane of the slope. The spatial aggregation of microseismic events is used to recognize the trend and distribution of the unknown large weak structural plane in the slope. The spatial distribution of microseismic events generated during excavation is mainly initiation and evolution at the controlled structure composed of unloading fracture zone XL-316 and fault f_{231} (Fig. 1.19a). In the process of excavation, the microseismic energy release area is mainly concentrated in the height of 1080–1240 m, and it is found that the area with large microseismic deformation is mainly concentrated in the height of 1100–1255 m, and the deep rock mass is 80–150 m away from the slope surface. The disturbance area of microseismic deformation well reflects the spatial range, distribution form and intensity of rock mass damage after slope excavation (Fig. 1.19b).

Fig. 1.19 Parameter characterization of microseismic events during slope excavation [152]

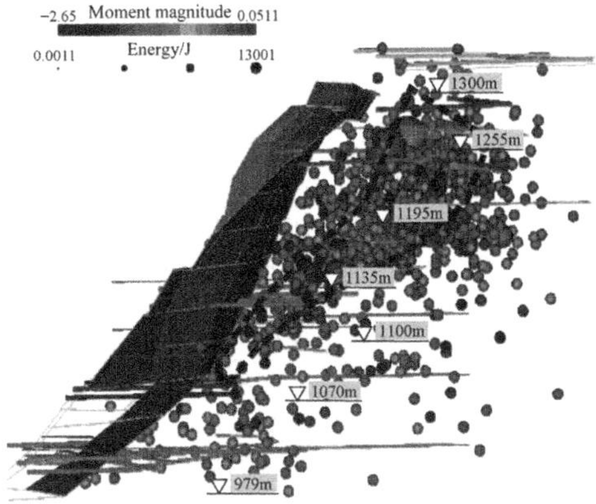

(a) Spatial distribution of microseismic events during slope excavation

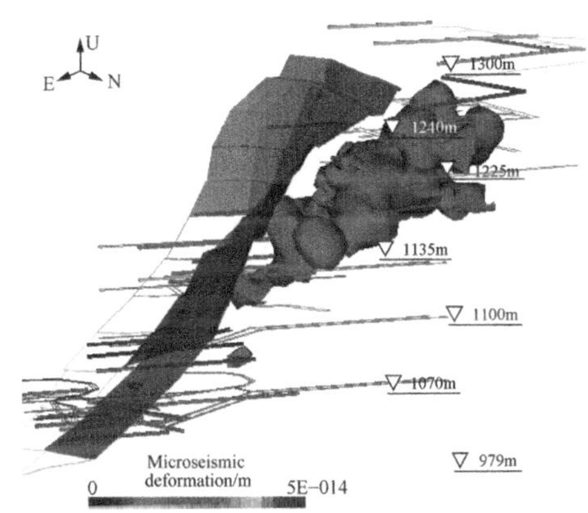

(b) Microseismic deformation distribution region

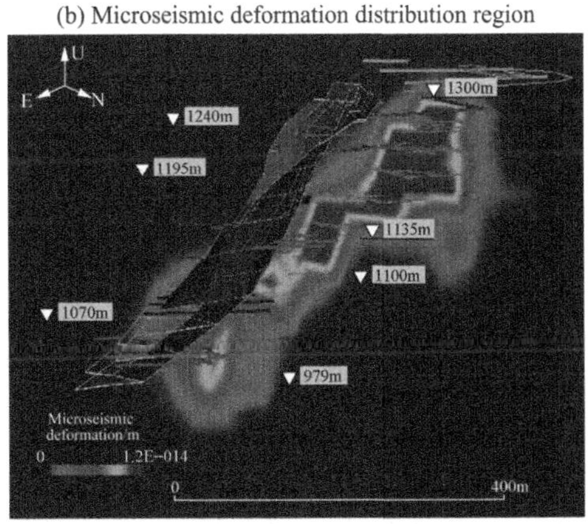

(c) Microseismic deformation cloud diagram

During slope excavation, microseismic events were mainly distributed at the height of 1070–1225 m (Fig. 1.19a), and the disturbance area of microseismic deformation and the main deformation area of microseismic deformation were concentrated at the fault f_{231} at the unloading fracture zone of 1240–1100 m and the fault f_{231} at the elevation of 979–1070 m. At a depth of about 80 m from the slope surface (Fig. 1.19b and c). Due to the small magnitude of seismic deformation, the slope is still stable, which indicates that the rock micro-fracture energy loss caused by excavation disturbance is within a reasonable range. The formation process of slip surface itself is the accumulation and evolution of micro-fracture of rock mass inside the slope. The microseismic monitoring results reveal the potential instability surface and failure area inside the slope in the global space. The energy loss density of microseismic events and the disturbance area of microseismic deformation are exactly consistent with the locations of the unloading fracture zones XL-316, XL-915 and fault f_{231} (Figs. 1.15, 1.19b and c). This indicates that microseismic monitoring information can reflect the laws of rock microfracture initiation, development, aggregation, migration and evolution expansion induced by on-site construction disturbance of right bank slope in real time from the perspective of time and space, effectively identify and delineate the activation degree of unknown fault or fracture zone in the slope, so as to accurately and quickly identify the damage area of rock mass inside the slope, and qualitatively analyze and evaluate the overall stability of the slope. Geological engineers can accurately describe the damage of deep rock mass according to microseismic monitoring information, and geotechnical engineers can also take corresponding treatment and design measures to provide reference for excavation slope construction and reinforcement.

1.6.4 Evolution Characteristics of Microseismic Energy Density

In the process of rock mass fracture and slip, elastic strain energy will be transformed into inelastic strain energy. The energy dissipated in the process of fracture and seismic wave propagation is removed, and the remaining energy is the microseismic radiation energy received by the sensor. As a measure of the intensity of microfracture events, it is widely used in the study of microseismic activity in mines [196, 197]. According to the research work of Boatwright et al. [198], the microseismic energy of the source body can be calculated by monitoring the body waves (P-wave and S-wave):

$$E = 4\pi \rho c R^2 \frac{J_c}{F_c^2} \tag{1.3}$$

where, E is the microseismic energy; ρ is the density of rock mass; c is the body wave velocity of the source body; R is the distance from the source to the sensor; J_c is the energy flux, which can be obtained by integrating the particle velocity spectrum in the frequency domain; F_c is the empirical coefficient of seismic wave radiation type, 0.52 and 0.63 for P wave and S wave respectively [199], and the sum of calculated energy of the two body waves is the microseismic energy.

According to the principle of quantitative microseismology, the volume of rock mass in the inelastic deformation area of the microseismic source can be estimated by Eq. (1.4) [200]:

$$V = \frac{M_0}{\Delta\sigma} \tag{1.4}$$

where, M_0 is the seismic moment of the microseismic event, which has the following relationship with the low-frequency amplitude Ω_0 of the far-field displacement spectrum of P wave or S wave [201], and is a reliable measurement of the seismic moment of small seismic events (mine earthquakes) [202]:

$$M_0 = \frac{4\pi\rho c^3 R\Omega_0}{F_c} \tag{1.5}$$

Since the stress drop of the source body $\Delta\sigma$ depends on the calculation model, the difference of the model will affect the corner frequency of the seismic waveform, and a small change in the corner frequency will lead to a larger change in the stress drop [203], the apparent volume V_A and apparent stress σ_A independent of the calculation model are used to replace the source volume and stress drop respectively to represent the rock mass volume and stress level of the inelastic deformation of the source [203].

$$\sigma_A = \frac{\mu E}{M_0} \tag{1.6}$$

$$V_A = \frac{M_0}{2\sigma_A} \tag{1.7}$$

where, μ is the rock mass stiffness. As is usually the case, Mendecki [204] gives a definition of Eq. (1.7). Although the seismic moment is the most appropriate description of the overall microseismic intensity, the microseismic energy can better describe the potential influence range and degree of microseismic in engineering rock mass. The microseismic energy density is defined as follows:

$$E_d = \frac{E}{V_A} \tag{1.8}$$

By substituting Eq. (1.8) into Eqs. (1.6) and (1.7), and assuming that the apparent strain ε_A of the source body is equal to the change of strain $\Delta\varepsilon$, it can be obtained that

$$E_d = 2\sigma_A \cdot \varepsilon_A \approx \Delta\sigma \cdot \Delta\varepsilon \tag{1.9}$$

Microseismic energy density, as the energy released when microfracture occurs per unit volume of rock mass, not only reflects the location and intensity of microfracture of rock mass, but also represents the storage and release process of strain energy of rock mass in the focal area, as well as the state of engineering rock mass under the influence of external disturbances. The area of high energy density distribution in slope rock mass is the area of concentrated stress and energy release of rock mass, as well as the area of slope microfracture concentration and rock mass damage. Microseismic events of slope rock mass are distributed along the deep unloading fracture zone XL-316 and fault f_{231} (Fig. 1.20a), and the magnitude and color brightness of microseismic events represent the energy and moment magnitude of microseismic events respectively [152]. At the elevation of 1225–1135 m and 80–150 m away from the slope, the deep rock mass is the area of high energy release of microearthquakes, and its distribution pattern tends to be consistent with the distribution of deep unloading fracture zone XL-316 (Fig. 1.20b). At the elevation below 1135 m, 30–50 m away from the slope surface, the microseismic energy release is high, and extends to the vicinity of the shear exit of the fault f_{231} at the elevation of

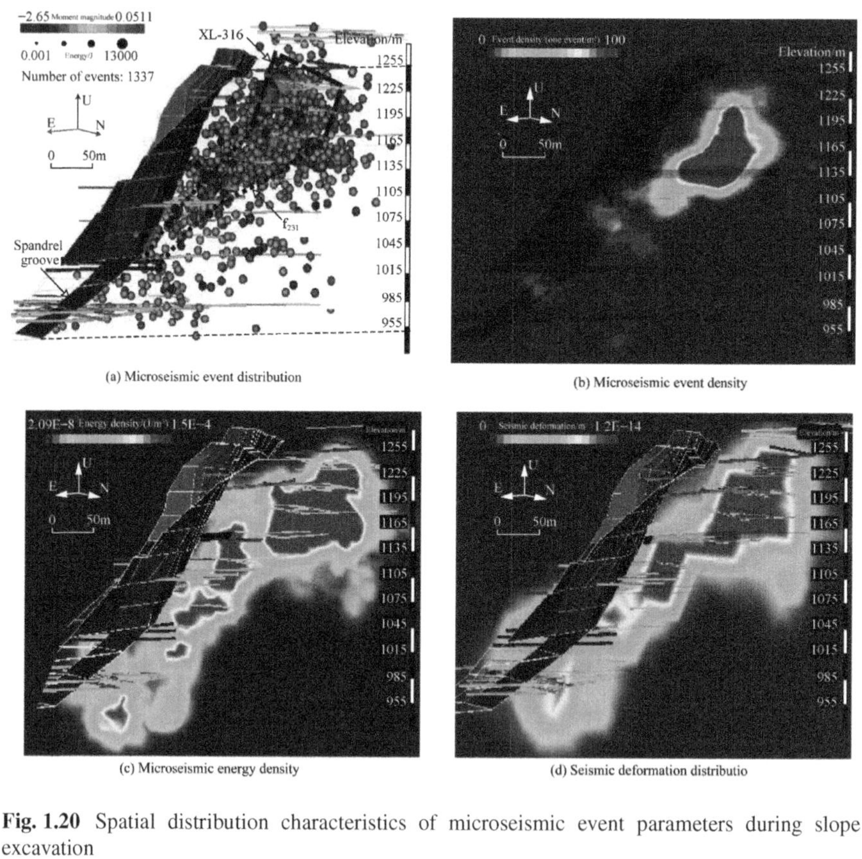

(a) Microseismic event distribution

(b) Microseismic event density

(c) Microseismic energy density

(d) Seismic deformation distributio

Fig. 1.20 Spatial distribution characteristics of microseismic event parameters during slope excavation

1070 m. By comparing Fig. 1.20b and c, it can be found that the average microseismic energy release in the area below 1135 m elevation during excavation is higher than that in the area above 1135 m elevation, indicating that the microfractures near XL-316 in the deep unloading fracture zone during excavation are mostly low-energy events with high activity rate. However, microfractures near fault f_{231} and shear outlet are mostly low-activity high-energy events, and the evolution process of microfractures is not conducive to slope stability, and there is a danger of instability along fault f_{231} at the shear outlet. The seismic deformation distribution calculated by microseismic event data also confirms the possibility of danger (Fig. 1.20d). In addition, the energy release in the 1030 and 940 m elevation area of the spandrel trough is also higher, and it is found that several diabase dikes such as β_{43}, β_{68} and β_{110} are developed in this area, and micro-fractures occur under the action of excavation blasting and unloading. This area is a possible seepage channel of the slope during the water storage period and becomes an important position for grouting reinforcement.

1.6.5 Causes of Fracture During Excavation and Unloading

In the process of excavation of the right bank slope, a number of macroscopic cracks have been produced, and the local deformation of the rock mass of the slope has caused the deterioration of the stability of the slope, forcing the construction of the slope to suspend. According to the distribution map of microseismic energy density of slope surface during excavation obtained from slope microseismic monitoring, the high energy release area is mainly concentrated in the 1 + 10 to 0 + 80 m longitudinal area of the dam at the elevation of 113–1100 m (the gray area marked by the box in Fig. 1.21a). The distribution of macroscopic cracks on the surface of the main slope during excavation is basically consistent with the distribution of micro-fracture high-energy release zone (Fig. 1.21b). This indicates that after the slope excavation, micro-fracture initiation and aggregation of rock mass occur in C1, C11 and C13 fracture areas, resulting in further damage of rock mass in fracture areas and finally the formation of macroscopic cracks. According to the geological investigation and site disclosure of the slope [132], diabase dike β_{62}, β_{85}, β_{219}, β_{j601} and fault f_{231} are developed in the range of 1 + 15 to 0 + 35 m of the right bank slope with an elevation of 1135–1100 m. As can be seen from Fig. 1.21a, the locations of the above high energy release zones are consistent with the locations of the fault f_{231} outcropping at elevations 1130–1115 m and the β_{85} diabase dikes outcropping, respectively. Due to the low strength of structural plane rock mass such as fault and dike in slope, micro-fracture damage is easy to occur. It can be inferred that the slope surface cracks are mainly caused by the stress adjustment of rock mass caused by the slope excavation disturbance, which leads to the stress concentration of the structural bodies of fault f_{231}, diabase dyke β_{j601} and β_{85}. The high energy release zone formed exceeds the strength limit of the structural rock mass and micro-fracture damage occurs, and finally several macroscopic cracks appear on the slope surface. It shows

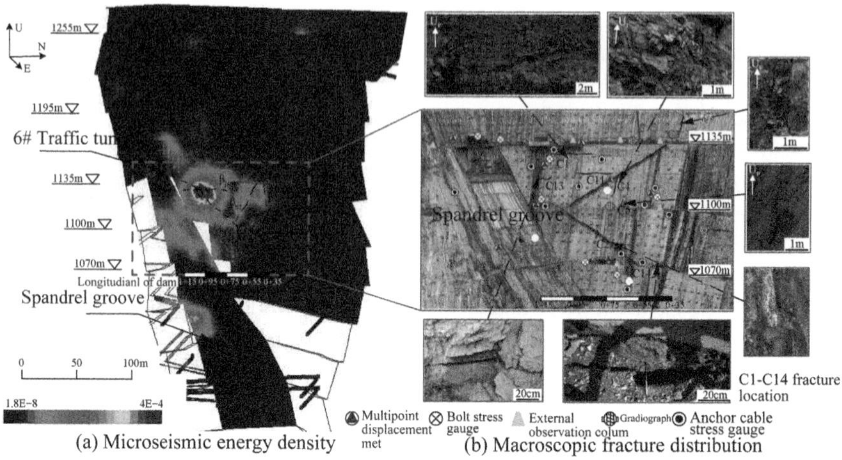

Fig. 1.21 Macroscopic crack location and microseismic energy density distribution on slope surface

that fault f_{231} is the main structural plane controlling the formation of slope surface cracks and affecting the stability of slope [154].

1.6.6 Summary

In this section, the phenomenon of slope surface crack formation is explained by microseismic monitoring results, and the temporal and spatial distribution of microfracture damage of rock mass during excavation is obtained. The analysis shows that the real reason for the occurrence of several deformation cracks at the slope surface during excavation is the result of the continuous accumulation, release and transfer of energy loss from microseismic events, and the resulting local stress adjustment makes the accumulation of micro-cracks in the rock mass in the slope continue to extend, and finally causes cracks on the side slope surface. The following conclusions are drawn.

(1) Under natural conditions, the dense zone of unloading fracture on the outside of the intermediate slope, which controls the stability of the rock mass of the right bank slope, is not exposed at the front edge of the slope body, but is not deeply buried in the slope body. After slope excavation, the structural surface of the intermediate slope is exposed on the slope surface, and the surface cracking of the slope has occurred several times, mainly due to the stress adjustment of the rock mass caused by blasting excavation. The appearance monitoring points TP31R and TP32R have a large deformation in the direction of the pointing slope facing the air, which is related to the exposure of the slope below the fault f_{231} and the failure to strengthen in time.

(2) Based on the temporal and spatial distribution of microseismic activity, the damage area of deep rock mass on the right bank slope induced by excavation disturbance is defined and recognized, that is, the damage accumulation of deep rock mass XL-316 and fault f_{231} under the action of excavation disturbance induces a large number of micro-fracture initiation and aggregation to form the slope damage area. The damage area is about 150 m away from the slope, and the damage area is the largest at 1180 and 1150 m. Below 1240 m elevation, microseismic events mainly gather, evolve and develop along the controlled soft rock structure composed of unloading fracture zone XL-316 and fault f_{231}, and gradually distribute in the direction of spinner trough. It shows that the micro-fracture of deep rock mass in slope is closely related to the location of structural plane distribution, and also proves that large weak structural plane is one of the main factors controlling slope stability.
(3) The number of micro-seismic events at or below 1210 m elevation accounts for almost 63.5% of the total number of micro-fractures in deep rock of slope in the whole study area. Obviously, with dam pouring and later water storage, micro-fractures of rock mass will inevitably increase due to slope stress migration caused by sudden rise and sudden fall of reservoir water. In particular, the rock mass in the vicinity of large unloading fracture zones XL-316, XL-915 and fault f_{231} will be weakened when exposed to water, which will aggravate the adjustment of slope stress.
(4) The slope microseismic activity is closely related to the excavation disturbance, which produces a large number of low-energy microseismic events in rock mass near XL-316, while a small number of high-energy events in rock mass near f_{231}. The high energy loss of microseismic occurs at 80–150 m away from the slope surface, and spreads along the main control structure towards the slope bottom from 1225 m elevation, reflecting the spatial range of damage of rock mass inside the slope after excavation. The microseismic deformation area is also mainly concentrated in the unloading fracture zone XL-316 and fault f_{231}.

In general, several macroscopic cracks on the slope surface during excavation are mainly due to the stress concentration of rock mass caused by excavation disturbance, which leads to a large number of micro-fracture events generated by damage and accumulation of rock mass on structural plane, indicating that the slope has a tendency of instability failure along the unloading fracture zone XL-316 and fault f_{231} as the main control structure during excavation. It can be inferred that if the disturbing force of external load is large in the later construction, the energy loss and deformation of deep rock mass of slope will increase, and the overall instability will be easily caused. Suggestions: In combination with conventional monitoring methods, close attention should be paid to the convergence and deformation at the junction of micro-seismic energy density gathering area, micro-seismic high energy damage area, unloading fracture zone XL-316 and fault f_{231}, and reasonable arrangement of construction rhythm should be made. The amount of blasting charge during excavation should be strictly controlled, and the overall instability caused by excessive slope excavation should be avoided as far as possible.

1.7 Numerical Simulation of Excavation Slope

Numerical simulation is the extraction and generalization of geological body prototype, so that it can better summarize the basic characteristics of geological body, and generalized the geological model to the standard of the numerical model, so that the numerical model has an objective basis and makes the actual monitoring have a mechanical basis [155]. A large amount of micro-fracture damage occurs in the rock mass of the main controlling structural plane of the slope due to excavation, resulting in macroscopic deformation cracks on the slope surface during construction. The special geological structure of the slope: faults f_{231}, XL-316, XL-915 and other medium-dip unloading fractures are concentrated in the development zone, which are located in the excavation line of the slope and subparallel to the slope, leading to the diversity and uncertainty of the slope failure pattern. The geological investigation and excavation monitoring show that, due to the bank slope stress adjustment caused by the Dadu River downcut, deformation and failure occurred on the shallow surface of the right bank slope, mainly manifested as flexural strain deformation, slip-strain deformation, creep slip-strain deformation, and local collapse failure [132]. The deformation stability and failure mode of the right bank slope of Dagangshan Hydropower Station during excavation and unloading have become one of the key technical problems in the project construction. On the premise of geological data, the right bank slope is generalized into 9 typical geological sections (Fig. 1.22), and numerical models are established by $RFPA^{2D}$ method and $RFPA^{3D}$ method, respectively, to calculate the slope excavation and slip surface formation process. The final failure mode of slope is determined by comparing the results of actual microseismic monitoring and numerical simulation, which lays a foundation for judging the overall stability of slope. At the same time, taking the actual field microseismic monitoring data as feedback information, $RFPA^{3D}$ method is used to put the actual monitored microseismic activity information into the background of the overall stress field of the three-dimensional calculation model to investigate (as the initial sliding surface damage), and real-time correction and adjustment of the generalized model is carried out to obtain the corresponding damage elastic modulus. The safety factor of slope considering microseismic damage effect is calculated.

1.7.1 Two-Dimensional Numerical Simulation

According to the geological investigation and exploration, the distribution and development characteristics of the unloading fracture dense zone on the right bank, the numerical model is mainly generalized as the block structure composed of two medium-dip unloading fracture dense zones, XL-316 and XL-915, and the low-dip fault f_{231}. From the geological data, it can be seen that the unloading fissure dense zone XL-316 and the slow dip fault f_{231} combine to form a huge cutting block, which has a significant influence on the stability of the slope. The typical geological section

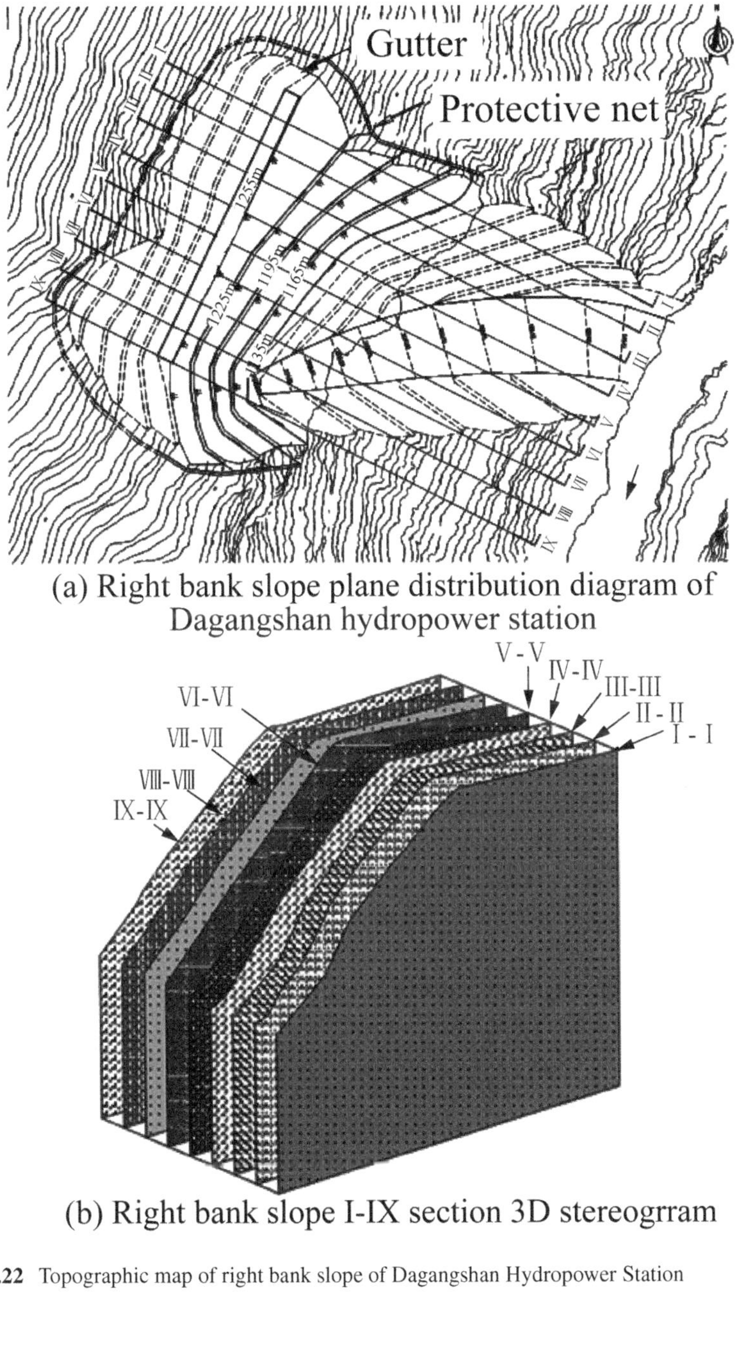

(a) Right bank slope plane distribution diagram of Dagangshan hydropower station

(b) Right bank slope I-IX section 3D stereogrram

Fig. 1.22 Topographic map of right bank slope of Dagangshan Hydropower Station

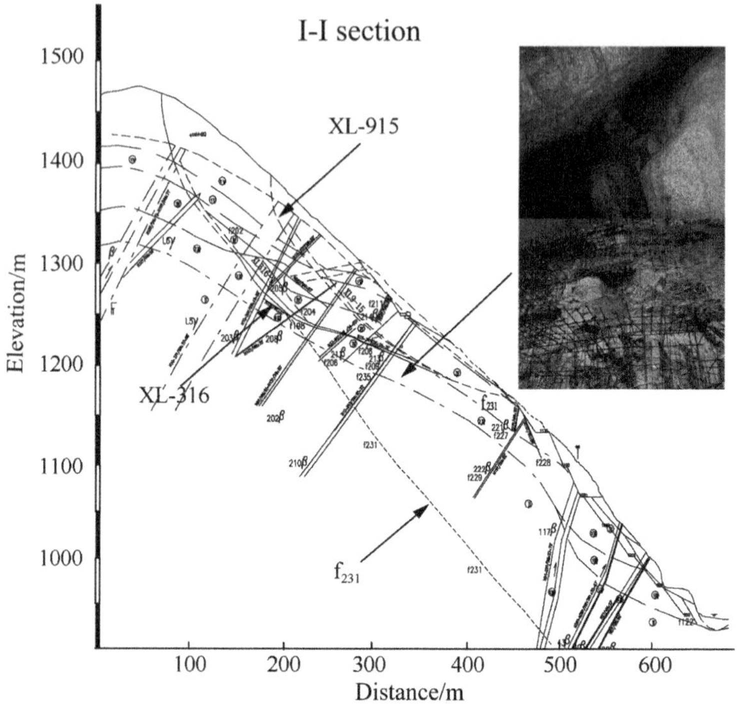

Fig. 1.23 Right bank slope I–I typical geological section

of the right bank slope I (Fig. 1.23) was selected for the numerical calculation of RFPA^{2D}-SRM.

RFPA^{2D}-SRM was used to calculate the I–I section of the right bank slope: the reduction coefficient was 0.01; step 0, the slope was not excavated; step 1, excavate according to the excavation area. In step 2, with the strength reduction, a large number of acoustic emission phenomena are gathered along the unloading fracture zone XL-316, fault f_{231} and unloading fracture zone XL-915. As can be seen from the shear figure, a through-slip plane formed along the unloading fracture zone XL-316 and fault f_{231}, and the safety factor of the slope is 1.03 (Fig. 1.24). With the increase of the number of reduction steps, the distribution form of controlled structural plane composed of slope unloading fracture zone XL-316 and fault f_{231} determines the failure mode of each section. With the increase of slope excavation and reduction, a large number of microfractures of rock mass in slope gather on the weak structural plane composed of unloading fracture zone XL-316 and fault f_{231}, and the tensile failure of unloading fracture zone XL-316 and the shear failure of fault f_{231} are the main ones (black circle represents the tensile failure, white circle represents the shear failure). The change of stress field inside the slope is continuous and uniform, and there is no sudden change, so it can be judged that the damage inside the slope is gradual accumulation. After excavation, the slope stress first accumulates at the foot

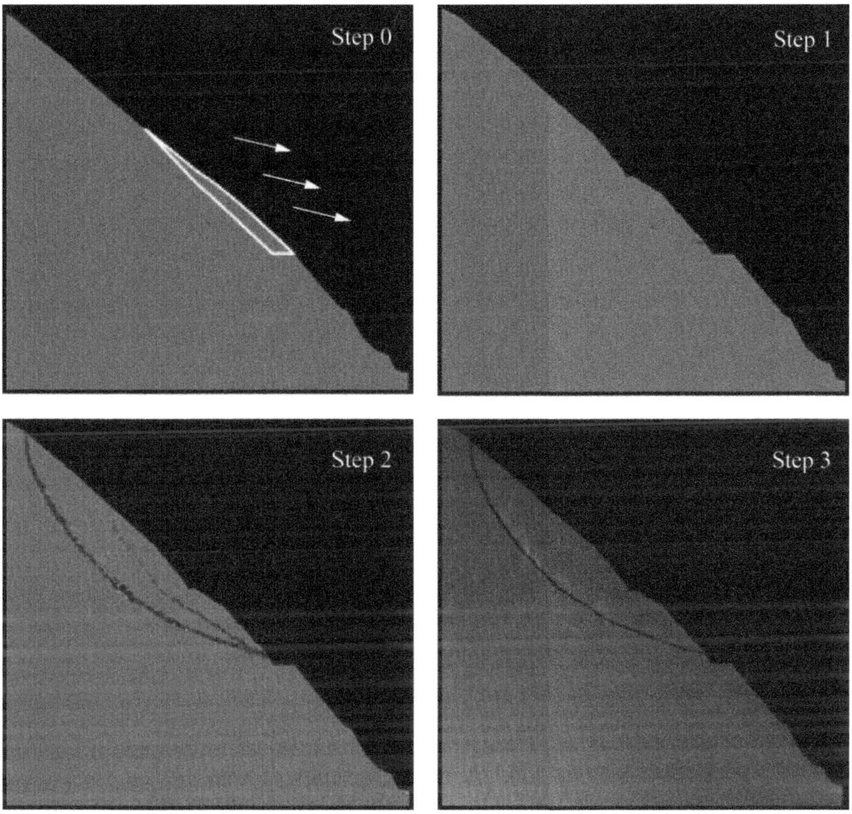

Fig. 1.24 Excavation calculation of RFPA^{2D}-SRM, a typical section of right bank slope I–I

fault f_{231}, and then develops through the slope top along the unloading fracture zone XL-316.

1.7.2 Three-Dimensional Numerical Simulation

The 3D model simulated the landform, two large unloading fracture zones XL-316, XL-915 and fault f_{231} and other main control structures. The calculation range was 690 m along the transverse river, 400 m along the river and 1060 m vertically, including the section I–IX of the cable machine platform from upstream to downstream (Fig. 1.22a and b). Different colors in the model represent different rock masses (Fig. 1.25a), and the values of macro-mechanical calculation parameters are shown in Table 1.4. The model construction follows the bottom-up modeling technology from "point → line → surface → body", and the modeling is carried out in the commercial software ANSYS. A total of 1,118,264 hexahedral units and 1,153,843

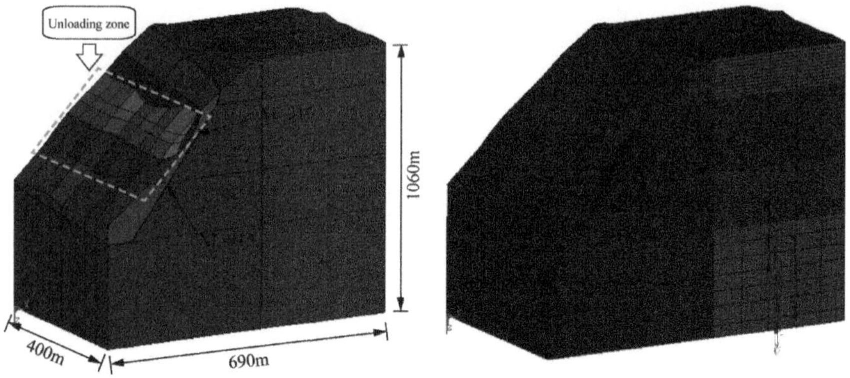

(a) 3D solid model (b) 3D finite element solid model

Fig. 1.25 Three-dimensional calculation model of right bank slope

Table 1.4 Values of physical and mechanical parameters of rock mass [152]

Material number	Density/(g/cm^3)	Compressive strength/MPa	Modulus of deformation/GPa	Poisson's ratio	Friction factor	Cohesive force/MPa
II	2.65	70–80	18.00–25.00	0.25	1.30	2.000
III_1	2.62	40–60	9.00–11.00	0.27	1.20	1.500
III_2	2.62	40–60	6.00–9.00	0.30	1.00	1.000
IV	2.58	20–40	2.50–3.50	0.35	0.80	0.700
V_1	2.45	< 15	0.25–0.50	> 0.35	0.50	0.200
V_2	2.10	< 10	0.20	> 0.35	0.40	0.175

nodes are divided, and the shallow slope part is encrypted, and the number of shallow layer units reaches about 93% of the total number. Finite element solid model and mesh generation were established (Fig. 1.26b). Normal constraints around the calculation domain, fixed hinge support at the bottom, free surface. The centrifugal loading coefficient of the calculation process is 0.005. C language is used to compile data conversion interface program A to R, and the data of nodes and units in ANSYS model is extracted and read into RFPA^{3D} program.

RFPA^{3D}-Centrifuge centrifuge loading method was used to carry out finite element calculation and analysis, and the centrifugal loading coefficient was 0.005 [205]. It is not necessary to assume the specific position of the sliding surface in advance, and the centrifugal field is used instead of the gravitational field to gradually increase the self-weight of the microscopic elementary element. Each step is automatically calculated by the finite element program to find the balance of the stress state. When the stress state of the element exceeds the strength of the element, the failure occurs, and the number of damaged elements accumulates to form the macroscopic failure slip plane. The safety factor K is defined as the ratio of the unit's self-weight when

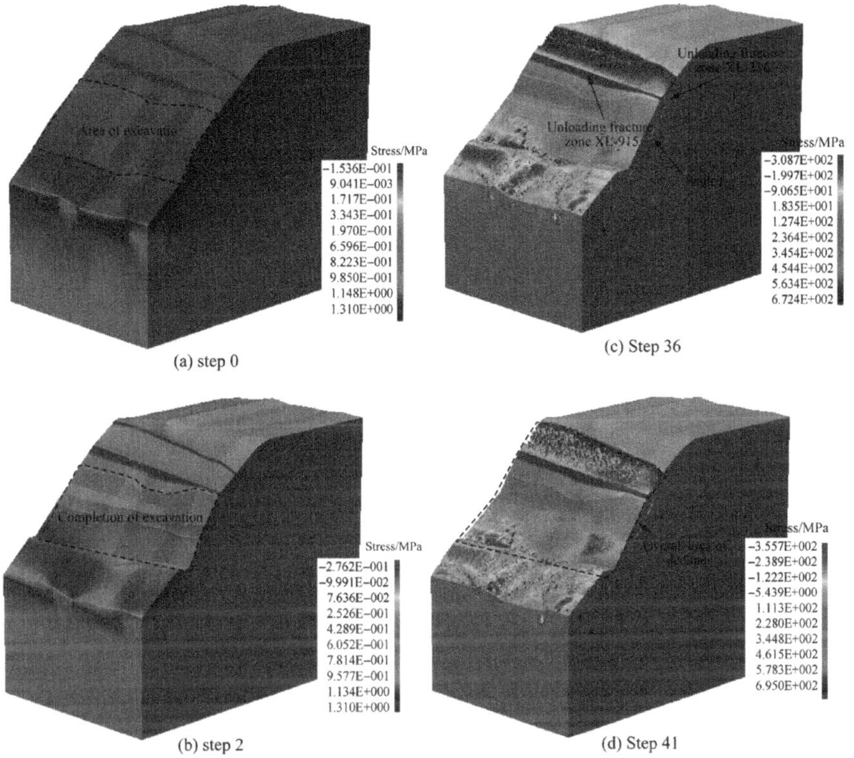

Fig. 1.26 Minimum principal stress diagram of slope failure process

the model is unstable to the initial unit's self-weight:

$$k = \frac{\gamma + \gamma\left(S_{tep} - 1\right)\Delta_g}{\gamma} = 1 + \left(S_{tep} - 1\right)\Delta_g \tag{1.10}$$

where, S_{tep} is the number of loading steps when the number of primitive failures is maximum; Δ_g is the centrifugal loading coefficient; γ is the material density.

Under the combined action of loading and excavation, the slope along the control structure is inclined to the direction of excavation. As can be seen from Fig. 1.26, with the completion of slope excavation (Fig. 1.26b), the potential slip surface of the slope first generated damage in the middle and upper areas of the slope along the unloading fracture zone XL-316. By the time of step 36, the damage of the unloading fracture zone XL-316 connected, and a large number of damage areas was also generated inside the fault f_{231}. The micro-fracture of the unloading fracture zone XL-316 and the fault f_{231} connected into a strip, and a large number of shear failure was generated at the intersection of the two, resulting in a sinking trend (Fig. 1.26c). When calculating step 41, the unloading fracture zone XL-316 and fault f_{231} were completely destroyed internally and lost their bearing capacity, forming a sliding block with the unloading fracture zone XL-316 as the upper surface and the fault f_{231} as the lower surface (Fig. 1.26d). It indicates that the controlled structure composed of the two rock masses gradually weakened downward from the top of the slope until the slope appears macroscopic instability collapse. This is due to the deformation of the weak structural plane, which provides a more favorable space for the deformation of the hard rock mass, so that the diabase structural rock mass on the weak structural plane is more likely to bend and topple to the direction of the free plane. At this time, the safety factor of the slope is 1.21, indicating that the slope is still in a safe and stable state. In the failure process, the microseismic phenomenon also showed a certain regular distribution along the potential slip surface. At the initial loading stage, a small amount of acoustic emission was generated along the unloading fracture zone XL-316 at the top of the slope (Fig. 1.27a). Subsequently, it developed and expanded along fault f_{231} from top to bottom, forming a through-slip crack surface, indicating that after slope excavation and unloading, the internal soft medium first occurred dislocation deformation, and the micro-fracture of rock mass evolved and expanded along the soft structure (Fig. 1.27b and c). At the end of the slope, the tensile failure was mainly caused by toppling deformation and tensile cracks at the top of the slope, and the shear failure was mainly shown with a small amount of tensile failure at the foot of the slope. The location of AE concentration was also the location of high stress concentration. When the stress concentration exceeded the ultimate strength of the rock mass, the deformation of the slope entered the stage of unstable development. With the gradual development of micro-fracture of rock mass, the slope suddenly lost its stable overall decline along the potential slip surface (Fig. 1.27d). Within the scope of numerical analysis, the distribution form of AE aggregation area is consistent with the spatial distribution law of microseismic events monitored by the actual microseismic monitoring system along the unloading fracture zone XL-316 and fault f_{231} (Fig. 1.20a). In general, the slope shows deformation in the direction of the horizontal direction pointing to the valley, and the deformation in the vertical direction is vertical downward.

RFPA^{3D}-Centrifuge finite element numerical calculation was used to reproduce the whole process of formation of slip surface and the initiation, development and interaction of acoustic emission in the process of slope excavation, which directly reflected the damage effect of the structure composed of unloading fracture zone XL-316 and fault f_{231} and the tensile and shear failure. The agreement between the numerical simulation and the actual microseismic monitoring results jointly verified the microfracture aggregation area of the right bank slope of the Dagangshan Hydropower Station, and revealed that the control structure composed of the unloading fracture zone XL-316 and the fault f_{231} was the key to the instability of the right bank slope of the Dagangshan Hydropower Station. Therefore, it is suggested that close attention should be paid to the convergence and deformation of the rock mass at the junction of the unloading fracture zone XL-316 and fault f_{231}.

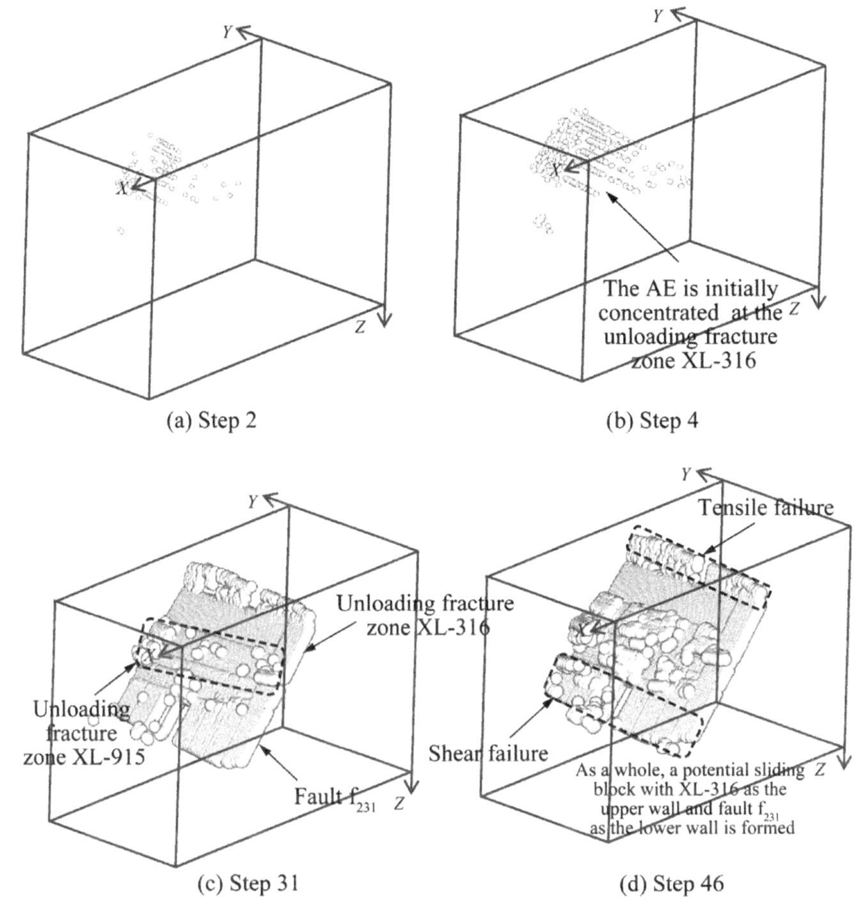

Fig. 1.27 Acoustic emission aggregation diagram of rock mass during slope failure

1.7.3 Comparative Study of Microseismic Monitoring and Numerical Calculation

A total of 1331 microseismic events were monitored in the microseismic monitoring area corresponding to the 3D calculation model during excavation. By comparing the microseismic monitoring results with the $RFPA^{3D}$ simulation results, it can be seen that the main microseismic deformation area and the microseismic energy loss density reflect the damage range of the slope rock mass, that is, the unloading fracture zone XL-316 at the elevation of 1240–1100 m and the fault f_{231} at the elevation of 979–1070 m. The distance from the slope surface is about 80 m (Fig. 1.28a and b). The three-dimensional potential slip surface of the slope and the spatial distribution law of acoustic emission calculated by $RFPA^{3D}$ centrifugal loading method are consistent with the spatial distribution characteristics of rock micro-fracture events monitored in the actual field (Fig. 1.28c). This is consistent with the failure form calculated by RFPA^{2D}-SRM numerical simulation (Fig. 1.24).

Through the coupling analysis of the actual monitored microseismic information and three-dimensional numerical simulation, the activation information and catastrophe mechanism of geological structures such as controlled weak structural planes under the disturbance of excavation construction are revealed. It is the accumulation of fault and micro-fracture of unloading fracture zone caused by excavation of slope that causes local instability and deformation of slope. It can be inferred that the cracks in the slope table during excavation are caused by the progressive failure of the slope body. The numerical simulation and microseismic monitoring results jointly reveal the largest deformation area of the slope on the right bank of Dagangshan Hydropower Station.

1.7.4 Feedback Study Considering Microseismic Damage Effects

The right bank slope of Dagangshan Hydropower Station is more than 700 m high, and the excavation slope is more than 300 m high. With high in-situ stress, complete development of faults and fractures, and large unloading depth of rock mass, the geological conditions are very complex, especially for the slope structure with soft rock interlayer, a series of engineering geological problems such as "deformation-mechanical response" and "aging deformation" will inevitably occur during excavation [126]. These structures are all constructed by faults, unloading fracture zones and dikes, which are weak in character and form potential structures with certain scale deformation in the direction of the slope in space, and play an extremely important role in controlling the stability of the slope. The spatial complexity of rock mass structure and geostress field leads to the unforeseeable spatial distribution of rock mass damage in the sliding surface region. It is a difficult problem to realize the effective mathematical representation of the rock mass mechanical parameters in the sliding surface area of slope. At present, many domestic scholars mostly focus on the field deformation monitoring and numerical analysis method for the deformation of high and steep slopes during unloading process [206–214], and the main purpose of the analysis is to identify the potential sliding surface and determine the safety factor [215–218]. In recent years, when people have increasingly realized that the occurrence of rock slope instability is inevitably related to its internal microseismic activity, microseismic technology has been indispensable in slope deformation monitoring and stability evaluation [219, 220]. The micro-fracture of rock mass in deep slope will inevitably induce damage to rock mass and weaken the mechanical parameters of rock mass. If the finite element stability analysis of slope is still calculated and simulated according to the mechanical parameters obtained in laboratory, it is bound to be different from the actual situation of rock mass in situ. On the basis of successfully obtaining the first-hand data of microseismic monitoring on the right

Fig. 1.28 Comparison between microseismic monitoring results and RFPA^{3D} numerical simulation results

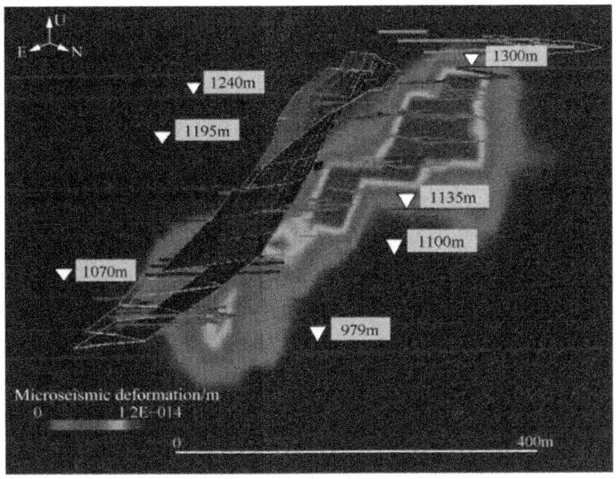

(a) Microseismic deformation cloud diagram

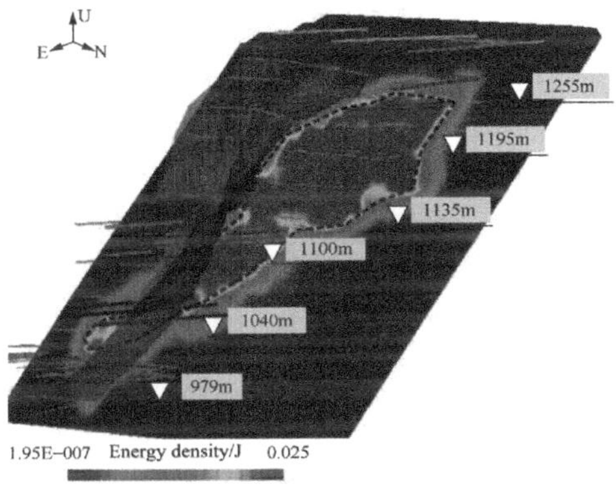

(b) Energy density cloud diagram of microseismic events

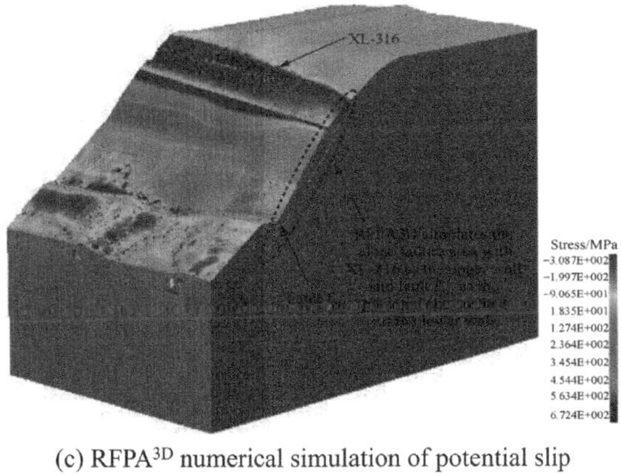

(c) RFPA^{3D} numerical simulation of potential slip surface bank slope (Fig. 1.29), in order to further quantitatively analyze the influence of rock mass damage (microseismic event accumulation area) on the stability of the slope induced by excavation and unloading, it is necessary to carry out inversion analysis on the microseismic monitoring data combined with the construction situation, and further study the influence of rock mass damage on the stability of the right bank slope. Then its stability is comprehensively evaluated. Therefore, how to "couple" the rock mass damage caused by microseismic events with the slope stability calculation is the focus of this section. The source information obtained by microseismic monitoring is extremely rich, such as magnitude, moment magnitude, energy size, energy density, static and dynamic stress drop, spatial error value, vibration frequency, vibration moment and so on, which are related to rock mass mechanical parameters. We can put the rich physical information obtained from the actual monitoring into the overall stress field of the numerical model to calibrate the mechanical parameters of the time and space coordinates of the microseismic failure point, realize the real-time correction of the microseismic monitoring results, and feedback the corrected mechanical parameters of the active surface to the slope geological model to achieve real coupling of the numerical model and monitoring analysis. The slope safety factor considering the microseismic damage effect of the actual slope rock mass is obtained.

It is assumed that the actual energy loss of heat energy and radiation energy generated by blasting is not considered in the process of excavation and unloading, and the energy of blasting event is all converted into the energy of elastic wave in the rock

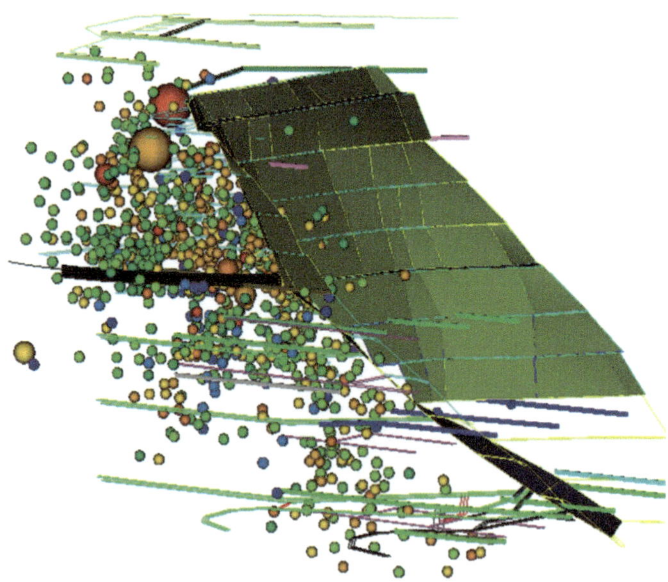

Fig. 1.29 Spatial distribution law of rock microfracture events induced by excavation disturbance [221] (May 4, 2010-September 30, 2011)

mass. Based on the principle of energy dissipation, a rock mass damage numerical model is established to study the formation mechanism and strength parameter characterization method of unloading damage zone in slope rock mass. By comparing and analyzing the energy of the blasting event collected by the microseismic system with the energy generated by the explosives at the actual blasting point, the percentage of the blasting energy measured by the microseismic monitoring system to the total energy released from the actual blasting point is obtained, which is the seismic efficiency of the right bank slope of Dagangshan Hydropower Station [222], i.e.

$$\eta = \frac{\Delta \overline{U}}{\Delta U} \tag{1.11}$$

where, η is seismic efficiency; $\Delta \overline{U}$ is the radiation energy picked up by the microseismic monitoring system after the damage of the rock mass unit within the scope of the source size; ΔU is the total energy generated by the actual damage of a single microseismic event of rock mass.

$$D = \frac{\Delta U}{U^e} \tag{1.12}$$

where, the damage variable D is the ratio of the energy obtained based on seismic radiation energy ΔU and the total energy of strain released by rock mass elements U^e [151].

$$U^e = \frac{1}{2E_0}\left[\sigma_1^2 + \sigma_2^2 + \sigma_3^2 - 2v(\sigma_1\sigma_2 + \sigma_2\sigma_3 + \sigma_1\sigma_3)\right] \tag{1.13}$$

where, σ_1, σ_2 and σ_3 are the three principal stresses of the rock mass element.

Thus, the corresponding damage elastic modulus can be expressed as:

$$E_r = (1 - D)E_0 \tag{1.14}$$

where, E_0 is the initial elastic modulus of the rock mass element, and the mechanical parameters of the rock mass are weakened in the same proportion according to the damage variable D during the numerical calculation. Based on the above energy damage criteria, the corresponding unit rock mass mechanical parameters were modified, and 1587 rock blasting events monitored by ESG microseismic monitoring system during slope blasting excavation from May 4, 2010 to September 30, 2011 were taken as the research objects. Site blasting by No. 2 rock emulsion explosive, the average amount is 72 kg, is obtained by computing the average every time the total energy of the blasting of 216,000 kJ [138], pick up ESG microseismic monitoring system to detect earthquakes can an average of 647 J as reference, the right bank slope of Dagangshan Hydropower Station unloading of excavation engineering earthquake efficiency is about 0.003%. The developed RFPA-MMS program was used which read the source information, and the actual spatial coordinates of the

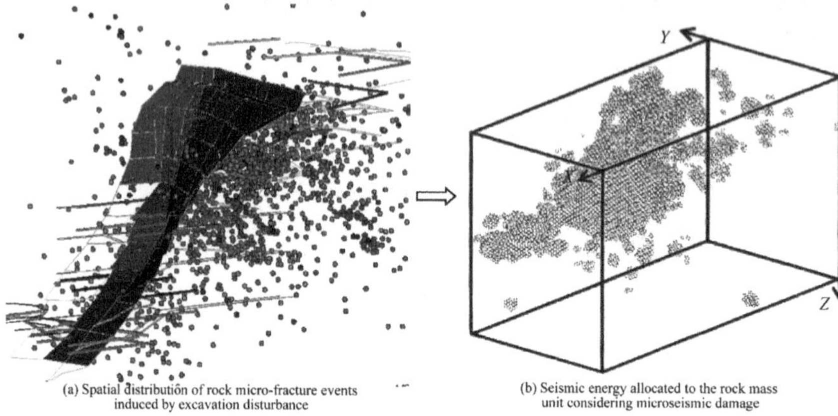

Fig. 1.30 Numerical calculation feedback analysis microseismic monitoring area

microseismic events were converted into the coordinates of the finite element model, so that the parameter information of the microseismic events collected by the micro-seismic monitoring system was imported into the above numerical model (Fig. 1.30a). According to import source influence within the scope of the mechanical parameters of rock mass damage units, namely in the source within the scope of the grid cell size evenly distributed to the earthquake. Program automatically assigned energy, all seismic energy use space of the damage of rock mass unit sphere representation (Fig. 1.30b). Combined with the modified mechanical parameters (damage elastic modulus E_r) of the damaged rock mass element of the numerical model, the re-assignment of the mechanical parameters of the three-dimensional numerical model element considering the microseismic damage effect was completed. The stability of the slope was solved by considering the damage coefficient in the process of rock mass excavation, and the relationship between the micro-seismic activity and the strength weakening of the spatial damage and deterioration of the rock mass in the process of slope instability was established.

The rock mass parameters in the model are real-time updated parameters (that is, the microseismic activity in the damage evolution process is considered, and the rock mass material heterogeneity is not considered). On the basis of this model and parameters, the slope was numerically calculated by RFPA^{3D}-Centrifuge method. The real-time safety factor considering the microseismic damage effect is 1.03, which is 15% lower than the safety factor 1.21 when the microseismic damage effect is not considered. It is obvious that the damage microseismic effect of deep rock mass of slope has great influence on the safety factor of slope.

1.7.5 Summary

This section based on the finite element method, respectively by the method of strength subtraction and centrifugal load, the main influence the stability of slope unloading fissure zone, XL-316 XL-915 and fault f_{231} calculating and analyzing the stability of two-dimensional and three-dimensional recreates the role under the influence of the stress field and acoustic emission rules of slope excavation, Moreover, the coupling analysis between the actual monitored microseismic information and the three-dimensional numerical simulation results can obtain the following conclusions.

(1) RPFA^{2D}-SRM was used to simulate and analyze the stable state and failure form of typical section of slope I–I after excavation: The micro-fracture of the structural plane is directly related to the degree of stress concentration. The stress concentration firstly gathered from the fault f_{231} at the foot of the excavation slope and generated micro-fracture. With the increase of the reduction step, a large number of micro-fractures gradually emerged along the unloading fracture zone XL-316 to the slope top and evolved into a through slip surface. Finally, a potential sliding block with the unloading fracture zone XL-316 as the hanging wall and the fault f_{231} as the footy wall was formed.
(2) The RFPA^{3D}-Centrifuge numerical calculation reveals the activation of structural plane caused by excavation disturbance of rock slope. The deformation of the slope in the horizontal direction is in the direction of the valley, and the deformation in the vertical direction is in the vertical direction. The failure of the slope after excavation is not achieved in one move, but a gradual process from the micro-fracture of the rock mass on the structural plane to the gradual accumulation of local failure and the development of the overall slip. With the progress of excavation, the micro-fracture initially occurred at fault f_{231}, and then the micro-fracture evolved and expanded along the unloading fracture zone XL-316 to the top of the slope to form a through-slip surface. The rock mass with fault f_{231} as the bottom slip boundary produced a large amount of shear failure accompanied by a small amount of tensile failure, and the rock mass with unloading fracture zone XL-316 as the trailing edge boundary produced tensile failure, and the slope safety factor was 1.21. The numerical simulation results reveal that the control structure composed of the unloading fracture zone XL-316 and the fault f_{231} is the key to the instability of the right bank slope of the Dagangshan Hydropower Station. It is suggested to pay close attention to the convergence and deformation of the rock mass at the junction of the unloading fracture zone XL-316 and the fault f_{231} in the subsequent excavation process of the slope.
(3) Based on the physical background of "consistency between acoustic emission rate and damage variables", the corresponding relationship between microseismic information and damage variables was established along the research idea of "numerical analysis-field monitoring-inversion analysis-forward analysis", so as to provide real-time corrected mechanical parameters for the calculation and analysis of three-dimensional geomechanical numerical model. The

three-dimensional spatial distribution of damage variables was obtained to calibrate the spatial–temporal reference of the breeding mode and development trend of the potential sliding surface of the slope. The stress field of the progressive failure process of the rock slope obtained by numerical simulation was coupled with the abnormal activation information of geological structures such as construction disturbance, fault and fracture zone obtained by microseismic monitoring. After considering the microfracture damage of the rock deep in the slope, the safety factor of the slope was 1.03, which was 18% lower than that of the three-dimensional slope without considering the microseismic effect.

(4) The whole process of formation of slip surface during slope excavation is reproduced by numerical simulation, and the damage effect and tensile and shear failure of the structure composed of unloading fracture zone XL-316 and fault f_{231} are reflected. The actual microseismic deformation area and microseismic energy loss density monitored in the field are also concentrated in the unloading fracture zone XL-316 at the elevation of 1240–1100 m and the fault f_{231} at the elevation of 979–1070 m, about 80 m away from the slope surface. It can be inferred that it is the slope excavation that aggravates the damage of the internal structural plane of the slope body, induces the continuous deformation of the key block of the slope, and leads to the macroscopic cracks in the slope surface.

1.8 Analysis of Reinforcement Effect of Anti-shear Galleries

For super-scale-controlled structures with defined boundaries, micro-expanded concrete is used to replace the weak structures for anti-shear galleries to increase the stiffness and strength of the structural plane, so as to prevent progressive failure in the process of slope excavation [223]. Along the direction of the unloading fracture zone XL-316 and fault f_{231}, anti-shear galleries with the same section size but different lengths were arranged at six elevation levels of slope 1240 m, 1210 m, 1180 m, 1150 m, 1120 m and 1060 m respectively, and the weak rock mass on the structural plane was replaced (Fig. 1.31). In this section, the stability of the slope before and after the reinforcement of the anti-shear galleries was studied from the perspectives of microseismic monitoring and numerical calculation, and the changes in the safety of the slope were quantitatively analyzed. At the same time, the rationality of anti-shear galleries measures was analyzed, trying to provide basis and explanation for the large deformation reinforcement problem in the process of rock slope excavation. The anti-shear galleries should be arranged as far as possible in the press-shear zone in the middle and lower part of the main sliding surface of the slope to better play the shear effect. In the range of (cable machine) 0 + 053.00 m to 0 + 189.50 m and (cable machine) 0 + 074.30 m to 0 + 220.60 m, anti-shear galleries were arranged along the unloading fracture zone XL-316, and anti-shear galleries were used to replace the unloading fracture zone XL-316. The main displacement objects are the unloading fracture zone XL-316 and diabase dikes at 1240 and 1210 m elevations.

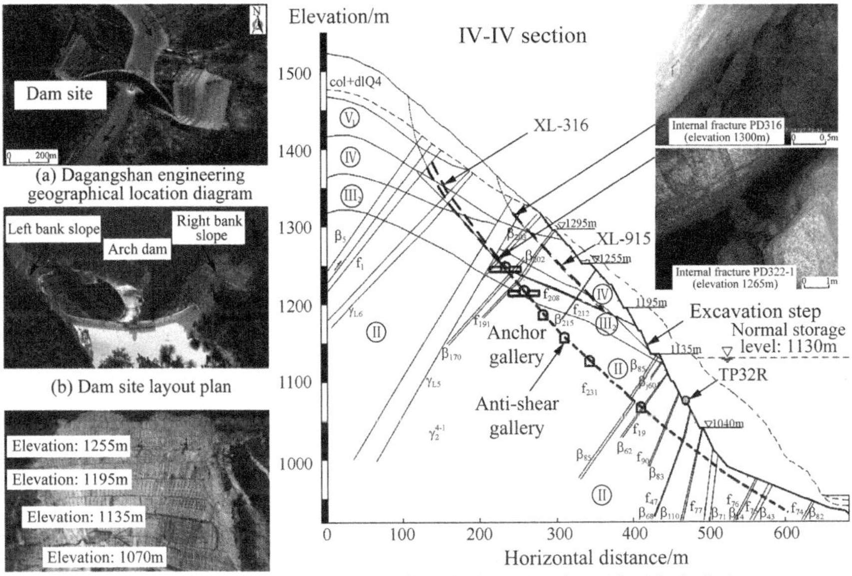

(c) Excavation diagram of right bank slope (d) Typical IV-IV section of the right bank slope

Fig. 1.31 Overall geomorphologic map of the dam site

The unloading fracture zone XL-316 and diabase dike at the high elevation are the tensile fracture surface of the deformation tensile fracture body of the right bank slope, not the bottom slip surface. Therefore, in the (cable machine) 0 + 074.60 m to 0 + 219.30 m range of 1180 m elevation, (cable machine) 0 + 070.40 m to 0 + 171.80 m range of 1150 m elevation, (cable machine) 0 + 081.10 m to 0 + 148.40 m range of 1120 m elevation, (Cable machine) 0 + 055.20 m to 0 + 109.50 m range of 1060 m elevation, for the weak structure with the bottom slip surface of fault f_{231} at the low elevation, anti-shear galleries were arranged respectively for reinforcement (Fig. 1.32a). The anti-shear galleries use micro-expanded concrete to implement consolidation grouting backfilling at a certain depth to reduce the contact surface gap caused by concrete shrinkage [10], and the concrete is appropriately equipped with rebar and section steel to increase the shear stiffness (Fig. 1.32b). Along the strike of the unloading fracture zone XL-316 and fault f_{231}, a complete set of anti-shear galleries reinforcement systems for super-large soft rock structures were established at 1240 m, 1210 m, 1180 m, 1150 m, 1120 m and 1060 m elevation of the slope respectively [221] (Fig. 1.32). The cross-section of the anti-shear gallery adopts straight wall arch, the size is 8 m (width) × 9 m (height), and the cross-section of the anti-shear gallery is large. In order to quickly play the role of the anti-shear galleries and achieve the effect of strengthening the slope, the backfill of the anti-shear galleries was directly carried out after the excavation and support was completed. The grouting corridor was set in the first stage concrete, and the size of the corridor is 3 m (width) × 3.5 m (height). After the backfilling consolidation grouting was completed, the grouting corridor was backfilled and the joint grouting was completed. The backfill concrete is all C25 concrete with low heat and micro-expansion (Fig. 1.33).

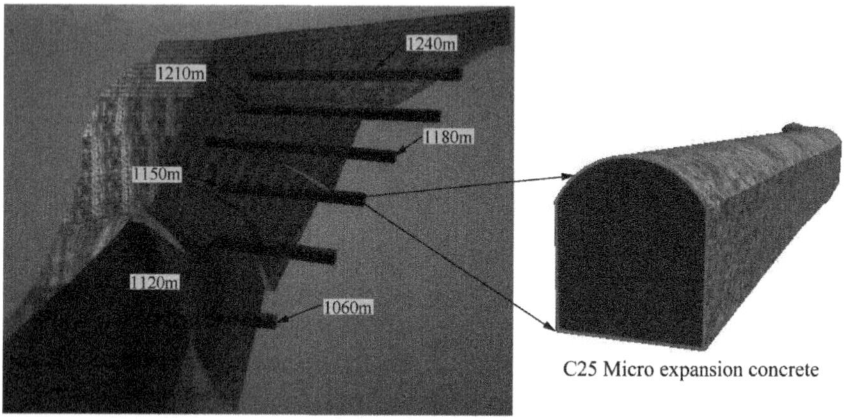

(a) Overall layout of anti-shear galleries (b) Enlarge view of anti-shear gallery

Fig. 1.32 Reinforcement scheme for anti-shear galleries of slope

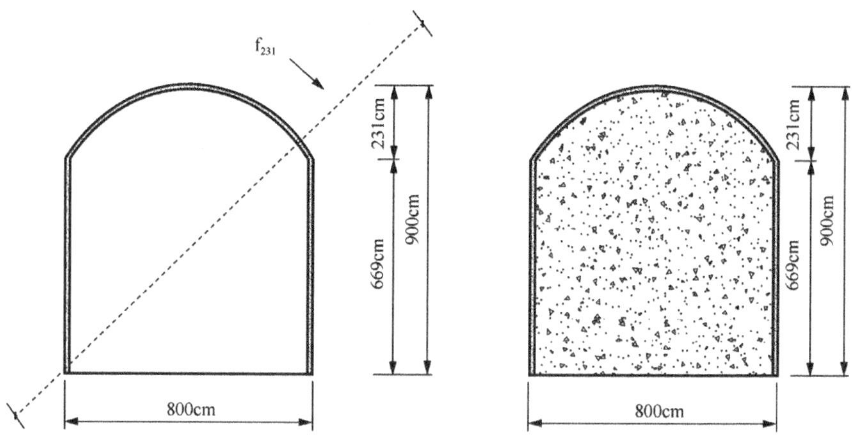

(a) Before coagulation backfill replacement (b) After concrete fill replacement

Fig. 1.33 Construction section of anti-shear galleries

1.8.1 Spatial–Temporal Distribution Characteristics of Microseismic Energy in Anti-shear Gallery

The microseismic monitoring system of the right bank slope of Dagangshan Hydropower Station has been in operation since May 4, 2010. As of May 31, 2012, the cumulative energy and event frequency of the rock mass in the anti-shear galleries reinforcement area within 20 m from the shear hole distribution location over time are shown in Fig. 1.34. In August 2010 and April to July 2011, the microseismic activity was strong, and the cumulative energy release of microseismic slope rock mass increased rapidly. The slope of the change curve of the cumulative energy release of microseismic in August 2010 and July 2011 was significantly larger, but lagged behind the high microseismic activity in this time, indicating that large-scale excavation of slope induced high energy events in this area. After September 30, 2011, the activity rate of microseismic events in the anti-shear galleries reinforcement area decreased significantly, and the cumulative energy of microseismic events increased slowly. The microseismic activity in this area was dominated by low-energy microseismic events, indicating that the development of slope microfracture was inhibited after the anti-shear galleries reinforcement was completed.

Figure 1.35 shows the spatial distribution law of microseismic energy density of slope rock mass before and after anti-shear gallery reinforcement up to May 31, 2012. The overall microseismic energy density distribution of the slope does not change

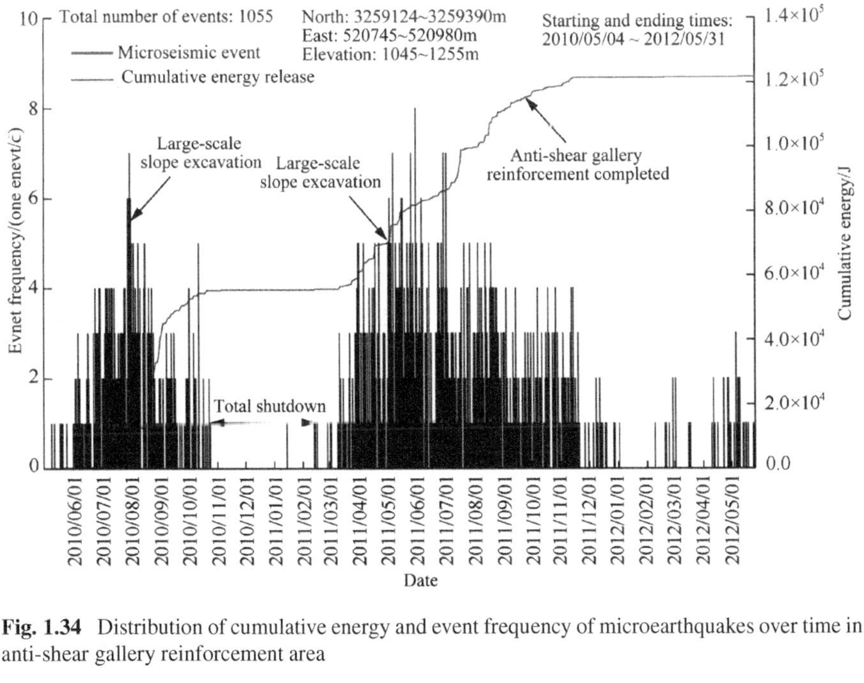

Fig. 1.34 Distribution of cumulative energy and event frequency of microearthquakes over time in anti-shear gallery reinforcement area

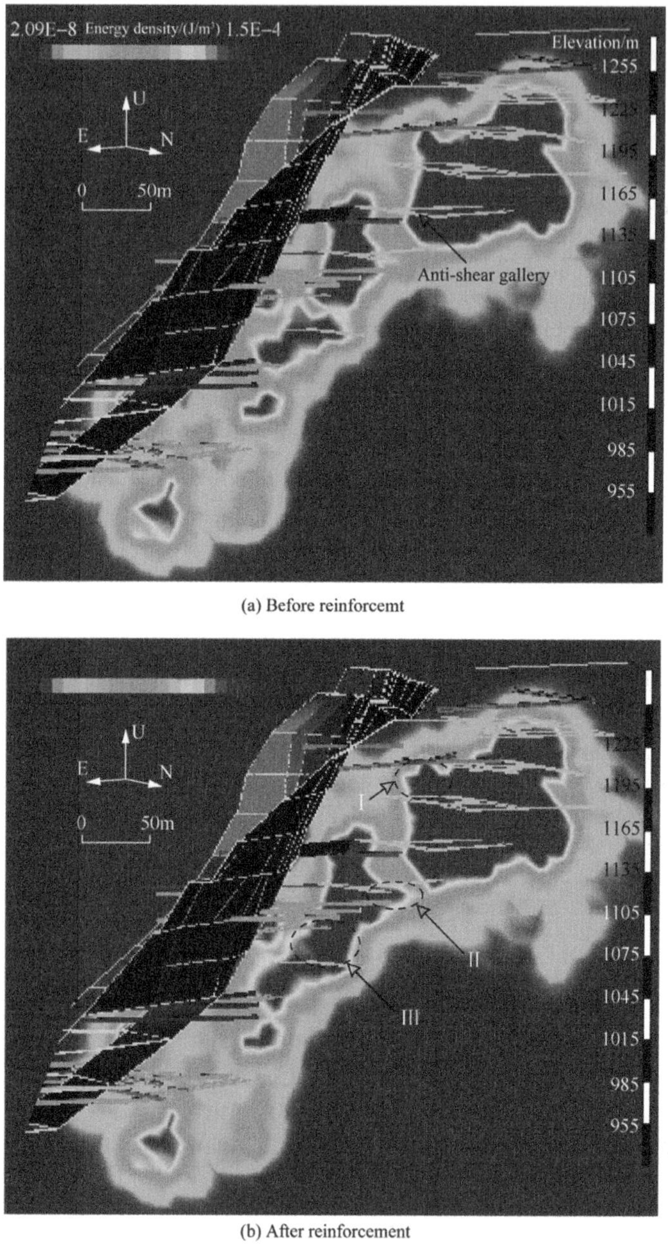

Fig. 1.35 Energy density distribution of microseismic events before and after anti-shear gallery reinforcement

significantly before and after the anti-shear gallery reinforcement. The distribution pattern of the high energy release area is basically consistent with the distribution of the deep unloading fracture dense zone XL-316 and fault f_{231}, and extends to the vicinity of the shear exit of fault f_{231} at 1070 m elevation. After the reinforcement of the anti-shear gallery, the south area between the anti-shear galleries at 1210 and 1180 m elevation, the north area at 1120 m elevation (II in Fig. 1.35b), and the upper area at 1060 m elevation (III in Fig. 1.35b) expand the high-energy release area, all of which are about 10 m. These areas are distributed at the edge of the anti-shear galleries, and are very small compared with the distribution range of the high energy release area before reinforcement. It can be found that after reinforcement of the anti-shear gallery, the slope stress is transferred to the edge area of the anti-shear galleries, and a small amount of low energy events are induced in this area. The anti-shear gallery plays the role of reinforcement and force transmission, and effectively inhibits the evolution process of slope micro-fracture.

1.8.2 Evolution Characteristics of Microseismic Energy Density in Elevation Region of Anti-shear Gallery

According to the research of Boatwright et al. [198], the microseismic energy of the source body can be calculated by monitoring the acquired body waves (P wave and S wave). The microseismic energy can better describe the potential impact range and degree of microseismic in engineering rock mass. As the energy released when a unit volume of rock mass occurs micro-fracture, the microseismic energy density can comprehensively reflect the position and strength of rock mass micro-fracture. The area of high energy density distribution in slope rock mass is the area of high energy release after the stress and energy concentration of rock mass reach the strength of rock mass, and it is also the area of concentration of slope micro-fracture and rock mass damage. The microseismic energy density distribution of rock mass at elevation 1240, 1210, 1180, 1150, 1120 and 1060 m of the slope during excavation is shown in Fig. 1.36. In the Fig. 1.36, the positions of the six anti-shear galleries, the main structural planes and the geodetic coordinates are projected onto the energy density map of the corresponding elevation one by one. It can be seen from Fig. 1.36a that at 1240 m elevation of the slope during excavation, the microseismic energy density of rock mass in the area of 3,259,210–3,259,260 m (north) and 520,788–520,823 m (east) in geodetic coordinates (1980 National geodetic coordinate system) is relatively high, which is located between No. 1 and No. 2 anchor holes on the south side of the 1240 m elevation anti-shear gallery. The anti-shear gallery basically covers the high energy release area (the dark area at the anti-shear gallery location in Fig. 1.36a). The elevation is 3,259,260 m (north), 35 m away from the slope is also a high energy release area, due to the small range, and the anchor cable is set here, its influence on slope stability is relatively weak. The microseismic energy density of the rock mass in the 3,259,190–3,259,245 m (north) and 520,785–520,835 m (east)

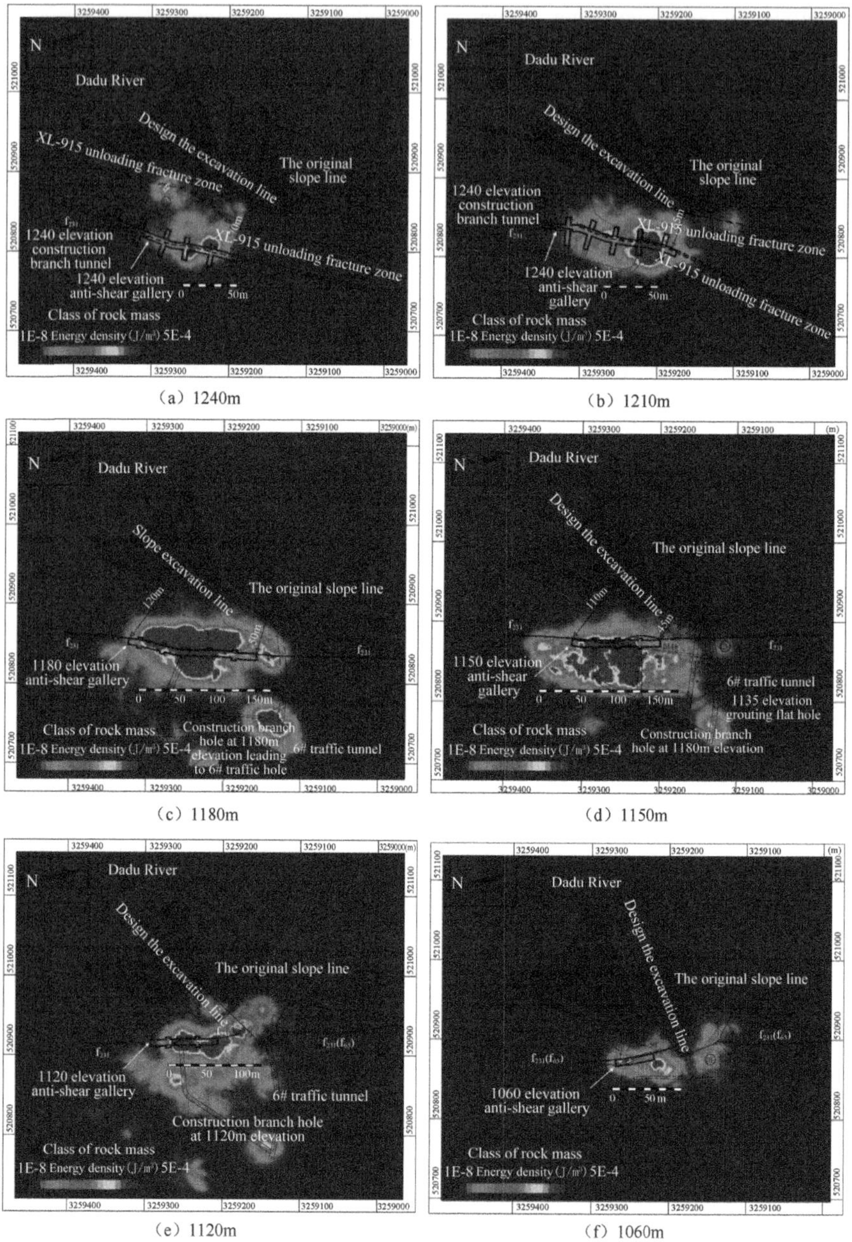

Fig. 1.36 Distribution of microseismic energy density at the elevation of each anti-shear gallery

at 1210 m elevation is relatively high, and shows a trend of extending to 3,259,270 m. The high energy release area is about 50 m away from the slope surface and is replaced by No. 1 and No. 2 anchor holes on the south side of the anti-shear gallery at this elevation (Fig. 1.36b). There are two high energy release areas at 1180 m elevation (Fig. 1.36c), one at 3,259,180–3,259,310 m (north) and the other at 520,805–520,870 m (east), with a distribution area of nearly 5000 m^2 and a maximum distance of 30 m from the slope. The other diabase dikes β_{170} and granitic dikes γ_{L6} cut off in the steep slope, located 130 m from the slope directly above the No. 6 traffic tunnel. The area enclosed by the 1150 m elevation anti-shear gallery and the construction branch hole is the concentrated area of microseismic energy release, which is about 60 m away from the slope surface (Fig. 1.36d). However, the high energy release zone at the elevation of 1120 m extends southward from 3,259,270 m to the slope surface (Fig. 1.36e), which is consistent with the high microseismic energy density of the slope surface in Fig. 1.21a, indicating again that the appearance of the slope surface cracks is the result of the accumulation of micro-fracture damage in the internal rock mass of the slope. The microseismic energy density of the north and southwest end of the 1060 m elevation anti-shear gallery is relatively high, with an extension range of about 40 m and generally distributed along the direction of the anti-shear gallery (Fig. 1.36f). Comparing Fig. 1.36a–f, except for two small areas with high microseismic energy density at 1150 m and 1120 m elevation, the distribution of other high energy release areas is closely related to the development location of XL-316 and XL-915 dense unloading fracture zones in the deep slope and fault f_{231} outside the mid-dip slope. Specifically, the unloading fracture dense zone XL-316 is disturbed under slope excavation blasting and unloading, resulting in micro-fracture initiation and aggregation, resulting in high microseismic energy density of rock masses in the above areas at 1240 m and 1210 m elevations. The XL-915 is less disturbed by excavation, and only generates a local high-energy release zone at an elevation of 1240 m and 35 m from the slope. However, the fault f_{231} outside the mid-dip slope is disturbed by excavation in a large area, resulting in the development of micro-fractures along the fault distribution direction at 1180, 1150, 1120 and 1060 m elevation, forming a high-energy release zone. The reason for the uneven distribution of the above high energy release area is that the fault f_{231} outside the mid-dip slope is constantly exposed at an elevation below 1135 m, and multiple stable control blocks with it as the bottom slip surface are damaged and deformed under the action of excavation, resulting in serious micro-fracture damage. At the same time, the granite fine-grained dikes γ_{L6}, as the stable control block, developed micro-fracture damage by the construction disturbance of No. 6 traffic tunnel and construction branch tunnel, and produced two high energy release areas at the elevation of 1150 and 1120 m. The impact range of micro-fracture damage induced by excavation on the right bank slope reaches nearly 150 m inside the slope, and the damage area is the largest at the elevation of 1180 and 1150 m. Based on the above analysis of the microseismic energy density at each elevation of the right bank slope during excavation, the anti-shear galleries at 1240, 1210, 1120 and 1,060 m elevations basically cover the high energy release area. The 1180 m elevation anti-shear gallery passes through the micro-fracture damage zone, and the 1150 m elevation anti-shear gallery is arranged on the slope side of the high-energy release zone, which cannot completely replace the rock mass in the micro-fracture damage zone, but can improve its mechanical performance. Therefore, the six elevation anti-shear galleries can play a good role in reinforcement and force transmission, and the reinforcement is more targeted to provide a guarantee for slope stability.

1.8.3 Stability Analysis of Slope Excavation and Reinforcement Based on Magnitude Frequency

The Gutenberg-Richter relation (G-R relation for short) was proposed by Gutenberg et al. [224] when they studied the characteristics of seismic activity in California, USA:

$$\lg N = a - bM \tag{1.15}$$

where, b is a statistical analysis parameter with certain physical significance [225], which usually corresponds to a small b (generally less than 0.8) with seismic activity. The size of b is not only related to the focal mechanism, but also changes with the evolution of engineering rock mass state. The increase of high magnitude events in rock mass and the decrease of low magnitude events in rock mass led to a sharp drop of b [226]. Therefore, b can be used as an evaluation index of rock mass engineering stability and disaster risk [227]. In this book, two rapid excavation periods of slope from May 4, 2010 to October 20, 2010 and from March 1, 2011 to September 30, 2011 were selected respectively, as well as microseismic event data from October 1, 2011 to May 31, 2012 after slope shear hole reinforcement. b [228] was calculated by least square fitting. During the two periods of rapid excavation, the b values of the slope are all around 1.16, and it can be judged that during the period of rapid excavation, the microseismic events are mainly stress transfer type, which is consistent with the excavation blasting and unloading action leading to stress disturbance and micro-rupture of the main control structure plane of the unloading fracture dense zone XL-316 and fault f_{231} (Fig. 1.37a and b). However, from October 1, 2011 to May 31, 2012, the slope microseismic event b was 0.77, and the abscissae intercept (a/b) of G-R relationship curve was − 0.6 (Fig. 1.37c), which means that the maximum magnitude of microseismic event in the slope was − 0.6 [229]. It mainly induces low magnitude fracture-slip type microrupture, which has little effect on slope stability. In this book, the cumulative data of microseismic events from May 4, 2010 to September 30, 2011 during slope excavation and from May 4, 2010 to May 31, 2012 after anti-shear gallery reinforcement were selected, and the relationship curves of magnitude and frequency were plotted respectively (Fig. 1.38). To further analyze the evolution process of slope stability in the whole life cycle after excavation and reinforcement. During the main excavation of the slope (from May 4, 2010 to September 30, 2011), a total of 1337 microseismic events were generated, with a *b* of 1.07 and a maximum possible magnitude of 0.1 (Fig. 1.38a). Compared with the *b* (1.16) in Fig. 1.37a and the maximum possible magnitude of 0, *b* decreased and the maximum possible magnitude increased. It shows that during the excavation of the slope, the low magnitude microseismic events decrease, the high magnitude microseismic events increase, the overall stability of the slope decreases, and the risk of instability disasters increases. During the whole life cycle of slope excavation and reinforcement (from May 4, 2010 to May 31, 2012), a total of 1625 microseismic events were generated, with a *b* of 1.48 and a maximum magnitude of − 0.1. It shows that after anti-shear gallery reinforcement, *b* increases significantly due to the increase of low-magnitude events, which improves the overall stability and reduces the disaster risk (Fig. 1.38b).

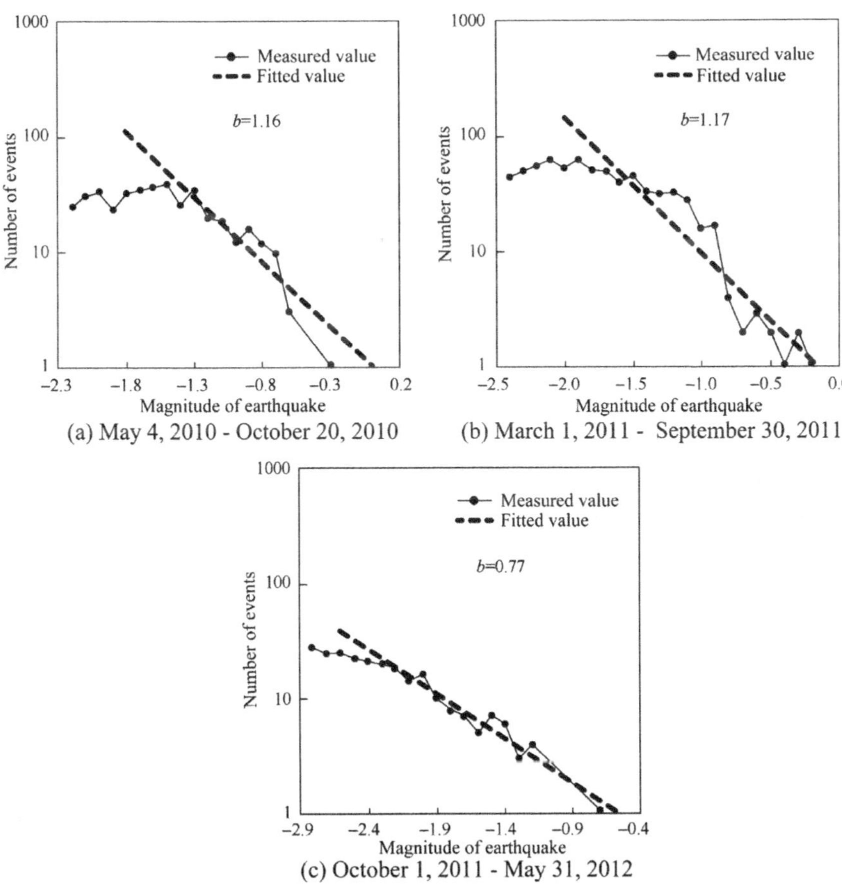

Fig. 1.37 Relation curves between magnitude and frequency of microseismic events on the right bank slope in different periods

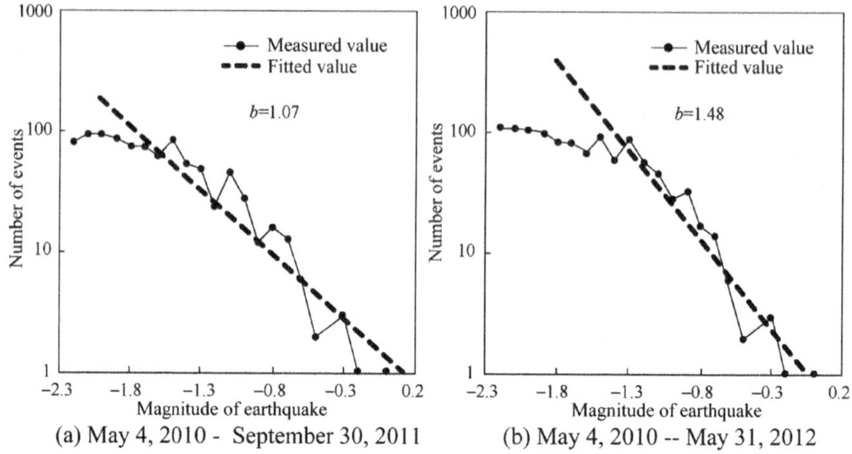

Fig. 1.38 Relation curve between magnitude and frequency of cumulative microseismic events on the right bank slope

1.8.4 Numerical Model Analysis of RFPA^{2D}-SRM for Anti-shear Gallery Reinforcement

Typical section VI–VI of the slope within the layout range of the microseismic monitoring system was selected as the research object (Fig. 1.39). RFPA^{2D}-SRM was used to conduct detailed modeling of large-scale controlled dominant structural plane such as slope weathering boundary, rock mass quality classification boundary, unloading fracture zone XL-316, XL-915 and fault f_{231}. The model size is 1400 m × 720 m, and the calculation range is 1400 m in the transverse river (1400 m is taken from the riverbed center line to the right bank slope) and 720 m in the vertical direction. 375 × 180 = 67,500 units was divided (Fig. 1.40). In the numerical calculation of RFPA, the improvement of structural plane parameters by consolidation grouting was considered, and the corresponding physical and mechanical parameters of C25 concrete were applied to the anti-shear galleries after replacement as the new value of the material. The values of other related rock mass parameters are shown in Table 1.5.

Figure 1.41a and b respectively show the shear stress and acoustic emission diagrams of the anti-shear gallery reinforcement in section VI–VI of the right bank slope. According to the numerical simulation analysis of RFPA^{2D}, it can be seen that, in general, after the reinforcement treatment of section VI–VI of the slope, the deformation in the horizontal direction is still oriented to the river valley, and the deformation in the vertical direction is vertical downward.

It can be seen from the shear stress diagram (Fig. 1.41a) that the non-uniform distribution of stress field is also influenced by the main control structural plane. Before the rock slope instability failure occurred, a large stress concentration phenomenon occurred at the foot fault f_{231} at step 68-11. At step 68-13, the stress concentration caused the rock mass damage at f_{231} fault at the bottom of the slope at first, resulting

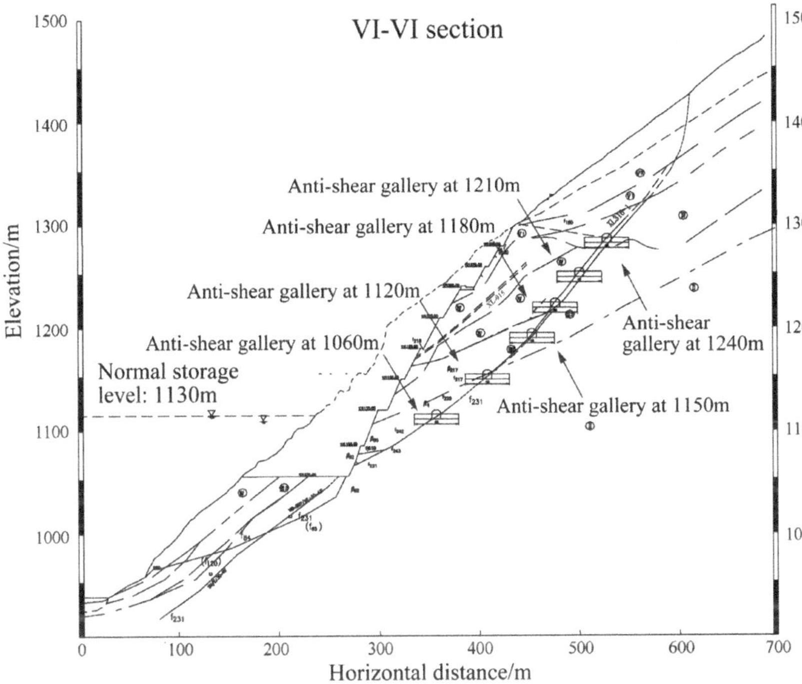

Fig. 1.39 Geological section view of anti-shear galleries reinforcement in section VI–VI of right bank slope

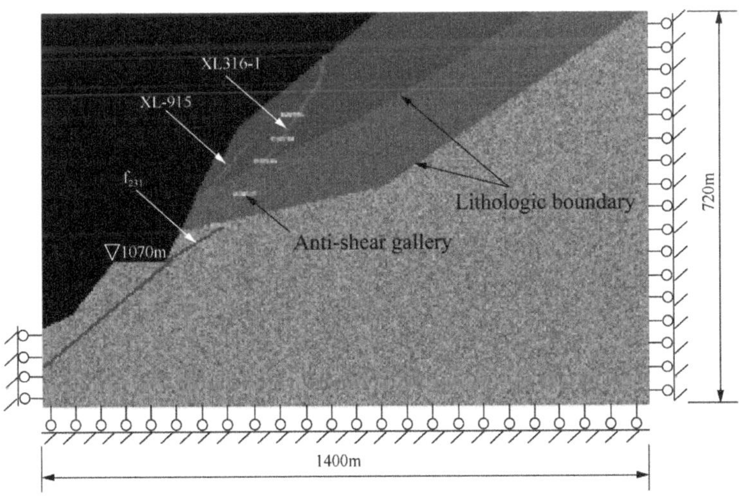

Fig. 1.40 Numerical calculation model of right bank slope VI–VI of Dagangshan Hydropower Station

Table 1.5 Values of macroscopic mechanical parameters of rock mass physics [152]

Type of rock	Density/(g/cm^3)	Compressive strength/MPa	Modulus of deformation/GPa	Poisson's ratio	Friction factor	Cohesive force/MPa
II	2.65	70–80	18–25	0.25	1.3	2.0
III_1	2.62	40–60	9–11	0.27	1.2	1.5
III_2	2.62	40–60	6–9	0.30	1.0	1.0
IV	2.58	20–40	2.5–3.5	0.35	0.8	0.7
V_1	2.45	< 15	0.25–0.5	> 0.35	0.5	0.2
V_2	2.1	< 10	0.2	> 0.35	0.4	0.175
C25 concrete	2.40	25	30.00	0.25	1.25	1.250

in a large number of micro-fractures, and initiation evolution along the unloading fracture zone XL-915. At step 68-15, the micro-fracture accumulated and gradually expanded along the slope bottom fault f_{231} to the slope top unloading fracture zone XL-316, and finally expanded into a single through slip plane, which caused great harm to the stability of the right bank slope. At steps 68-18, the potential slip plane was further penetrated, and the slope had a downward trend. At step 68-20, the slope deformation occurred along the dominant structural plane. This shows that the large weak structural plane in the rock slope controls the deformation and failure mode of the slope. Through calculation, the safety factor of the slope at this time was 1.515, which is about 47% higher than the safety factor of 1.03 before the slope reinforcement. It shows that the anti-shear gallery replacement of weak structural plane rock mass has a significant reinforcement effect, indicating that the current anti-shear gallery reinforcement treatment plan is feasible, can effectively improve the local and overall stability of the slope, and enhance the safety reserve of the slope (Fig. 1.41a). It can be seen from the RFPA^{2D} AE diagram (Fig. 1.41b) that the variation mode of AE of the whole slope was completely controlled by the geometric characteristics of the main structural plane. The tensile failure was mainly concentrated in the unloading fracture zone XL-316 above 1135 m elevation at the top of the slope, and the shear failure was mainly concentrated in the fault f_{231} at the foot of the side slope, and the main control structural plane was the dominant zone of energy accumulation. It was the accumulation and release of energy that caused the stress concentration at fault f_{231} and produced a large number of micro-fractures, which led to the deformation of rock mass. Subsequently, the energy was transferred to the unloading fracture zone XL-316, resulting in stress concentration, resulting in micro-fracture, and then causing damage and fracture of the slope top rock. It can be judged that 1060 and 1120 m elevation anti-shear galleries are effective for reinforcing fault f_{231}, and 1150, 1180, 1210 and 1240 m elevation anti-shear galleries are effective for reinforcing unloading fracture zone XL-316 (Fig. 1.41b).

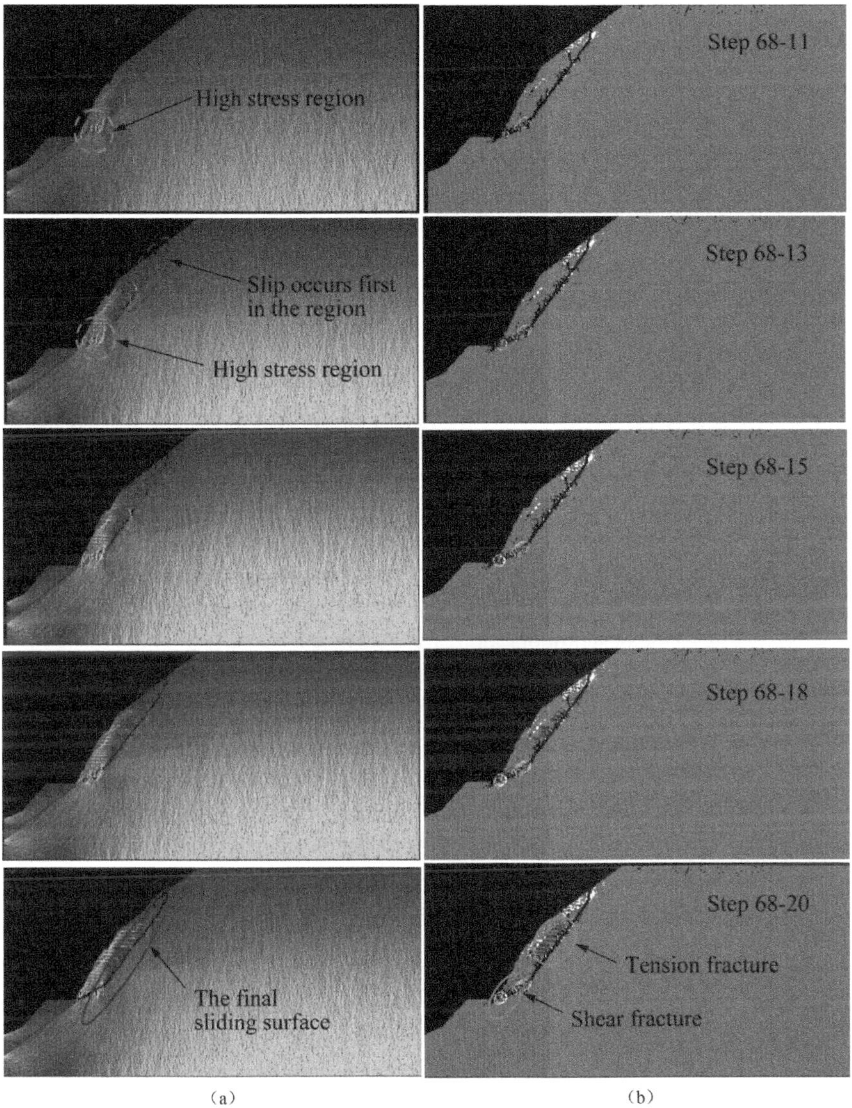

Fig. 1.41 Shear stress and acoustic emission diagram of two-dimensional progressive failure process after reinforcement of anti-shear galleries. *Note* The brighter the color in **a**, the larger the shear stress value; in **b**, the black circle represents the tensile failure and the white circle represents the shear failure

1.8.5 Numerical Simulation Analysis of Reinforcement Effect of Anti-shear Gallery

In this section, according to the distribution and development characteristics of the unloading fracture dense zone on the right bank, the numerical model was generalized to a mountain structure composed of two medium-dip unloading fracture dense zones, XL-316 and XL-915, and the low-dip fault f_{231}. The calculation range is 690 m along the transverse river, 400 m along the longitudinal river and 1060 m vertically, including the section I–IX of the cable machine platform from upstream to downstream (Fig. 1.42a) and the spatial arrangement of the anti-shear gallery of the slope (Fig. 1.42b). Different colors in the model represent different lithologies, and the macroscopic mechanical calculation parameters are selected according to Table 1.5. The model construction followed the bottom-up modeling technology from "point → line → surface → body" and was carried out in the commercial software ANSYS. A total of 1,118,264 hexahedral elements were divided, and the number of nodes was 1,153,843. The shallow slope part was processed with encryption, and the number of shallow elements was about 93% of the total. The surface was free, and the data such as the unit was extracted and read into the RFPA^{3D} program. The modeling steps and calculation process can be found in reference [73] and will not be repeated here.

According to the basic topographic and geological conditions of the slope on the right bank, the engineering measures mainly designed to resist anti-shear galleries are determined as effective control means to ensure the overall stability of the slope (Fig. 1.43a). At 0 + 053.00 to 0 + 189.50 m and 0 + 074.30 to + 220.60 m (cable

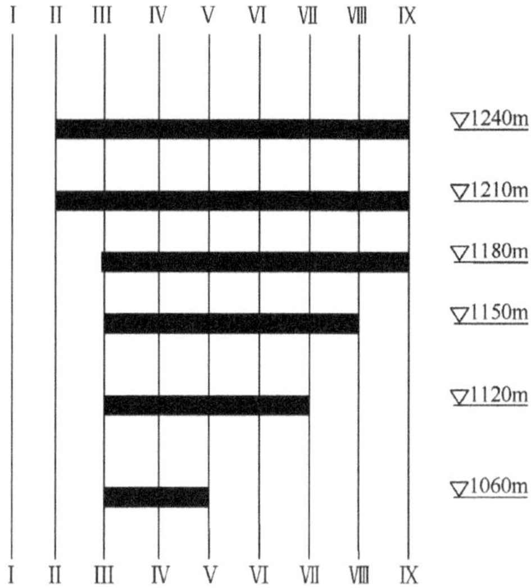

(a) Floor plan of anti-shear gallery reinforcement scheme

(b) Spatial distribution diagram of anti-shear galleryreinforcement structure

Fig. 1.42 Reinforcement plan of anti-shear galleries of right bank slope machine), anti-shear galleries are arranged along the unloading fracture zone XL-316, and anti-shear galleries are used to replace the unloading fracture zone XL-316. The main displacement objects are the unloading fracture zone XL-316 and diabase dikes at 1240 and 1210 m elevations. The size of the anti-shear gallery is 8 m (width) × 9 m (height). The high elevation unloading fracture zone XL-316 and diabase dive are the tensile fracture surface of the deformation tensile fracture body of the right bank slope, not the bottom slip surface. In this way, 1180 m elevation between (cable machine) 0 + 074.60 to 0 + 219.30 m, 1150 m elevation between (cable machine) 0 + 070.40 to 0 + 171.80 m, 1120 m elevation between (cable machine) 0 + 081.10 to 0 + 148.40 m, 1060 m elevation between (cable machine) 0 + 055.20 m to 0 + 109.5, anti-shear galleries were set respectively for the weak structures with the bottom slip surface of fault f_{231} at low elevation for reinforcement (Fig. 1.43b). The anti-shear gallery used micro-expanded concrete to carry out consolidation grouting backfilling at a certain depth to reduce the contact surface gap caused by concrete shrinkage [229]. The concrete was properly equipped with steel bars and section steel to increase the slope sliding resistance and improve the safety margin. The cross-section structure of anti-shear gallery is shown in Fig. 1.43c. In this way, a complete set of reinforcement system for super large soft rock structure was established along the unloading fracture zone XL-316 and fault f_{231} strike.

It can be seen from the maximum main stress diagram (Fig. 1.44) that after the excavation was completed, the slope instability still first expanded along the unloading fracture zone XL-316, and then turned to crack along the boundary of the

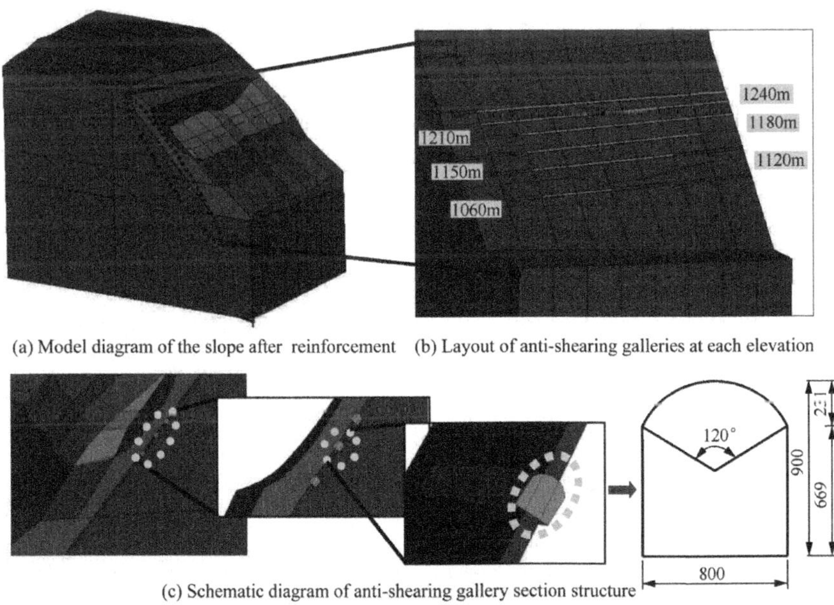

(a) Model diagram of the slope after reinforcement (b) Layout of anti-shearing galleries at each elevation

(c) Schematic diagram of anti-shearing gallery section structure

Fig. 1.43 Slope anti-shear gallery reinforcement model diagram (unit: mm)

unloading fracture rock layer, with obvious stress concentration on both sides of the anti-shear gallery. Obviously, the stress gradually "migrated" to the anti-shear gallery position during the loading process. Due to the huge tensile stress, the stress of the rock mass around the anti-shear gallery was generally distributed in the form of a split "knife edge" (Fig. 1.44a). When loading to step 167, the potential slip fracture surface of the slope was connected under, and a potential bulk body with the block formed by the unloading fracture zone XL-316 as the hanging wall and the block formed by the slope fault f_{231} as the footrace was formed on the whole. With the strengthening of the overall downward trend, many small micro-cracks appeared in the side walls on both sides of the anti-shear gallery and gradually expanded around (Fig. 1.44b). The above analysis shows that the boundary of soft rock strata, fault and unloading fracture zone are still the main factors leading to slope instability. According to the potential slip surface, it can be seen that the potential sliding block first slides down in the unpenetrated part of the anti-shear galleries at the elevation of 1240, 1210, 1180, 1150, 1120 and 1060 m (Fig. 1.44c and d), while the anti-shear gallery part of the backfill concrete in the penetration area does not cause damage, and the reinforcement effect is very significant. The AE variation curves obtained by numerical simulation before and after anti-shear gallery reinforcement are shown in Fig. 1.45. As can be seen from Fig. 1.45, the overall safety factor of the slope increases with the increase of the number of anti-shear galleries. The safety factor of the existing reinforcement scheme (6-layer anti-shear galleries) is 1.83, which is about 51.2% higher than that of 1.21 without anti-shear galleries. The change of microseismic events before and after anti-shear gallery reinforcement is shown in Fig. 1.46.

As can be seen from Fig. 1.46, by the end of May 2012, only 42 effective microseismic events were detected in the same month, as the grouting and plugging of the lower three anti-shear galleries (1150, 1120 and 1060 m elevations) and the consolidation and grouting construction of the upper three anti-shear galleries (1240, 1210 and 1180 m elevations) had been completed. Compared with the 125 microseismic events (including blasting events) before the formation of the anti-shear gallery in May 2010, the number decreased by about 66.4% (Fig. 1.46). It shows that the implementation of the anti-shear gallery makes the damage trend of the deep rock mass of the slope decrease, the microseismic activity gradually weakens, and the damage induced by the internal unloading of the slope gradually tends to stop. Combined with numerical calculation and actual monitoring analysis, it can be seen that the anti-shear gallery reinforcement treatment scheme is feasible, which can improve the stability of the slope and enhance the safety reserve of the slope. This is basically consistent with the failure mechanism and reinforcement results obtained in the slope stability analysis and reinforcement measures study of the right bank hub area of Dagangshan Hydropower Station conducted by Liu et al. [153].

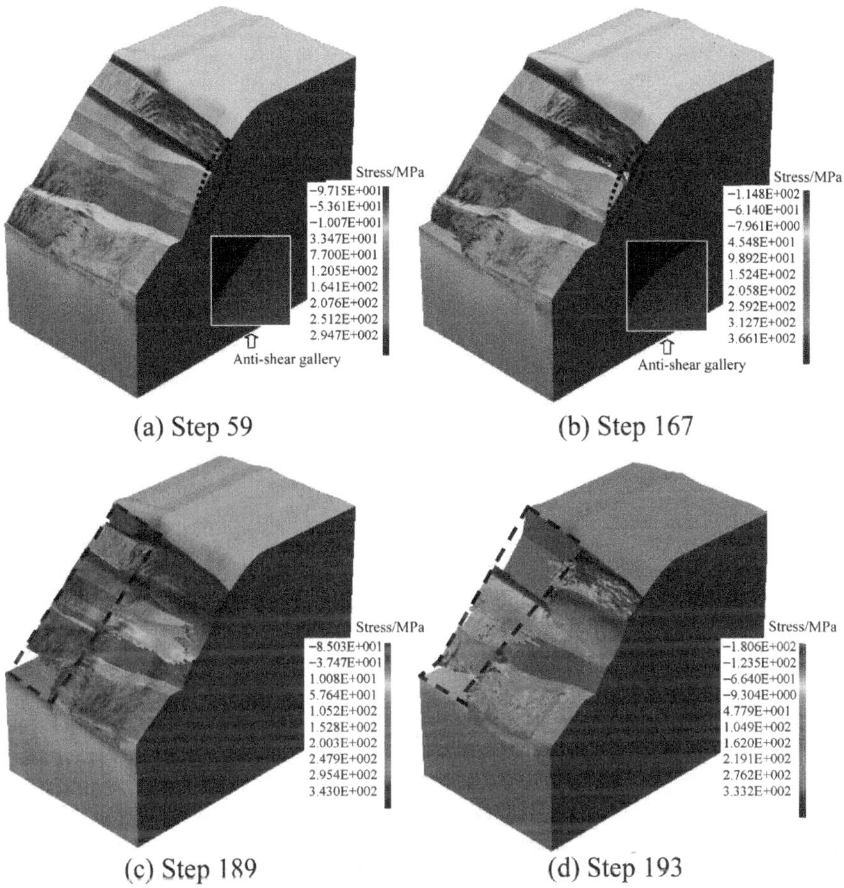

Fig. 1.44 Maximum principal stress during slope failure

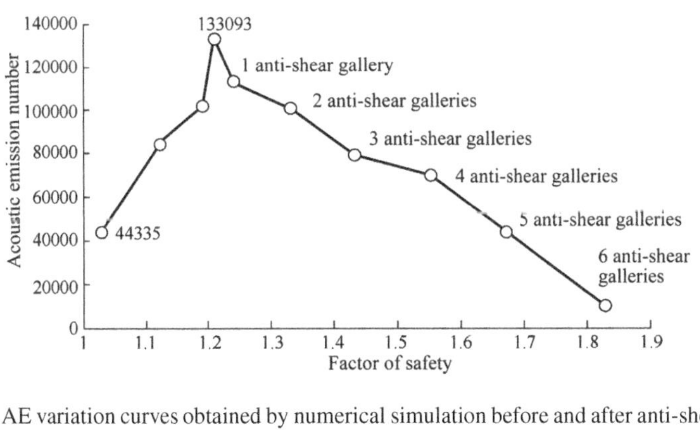

Fig. 1.45 AE variation curves obtained by numerical simulation before and after anti-shear gallery reinforcement

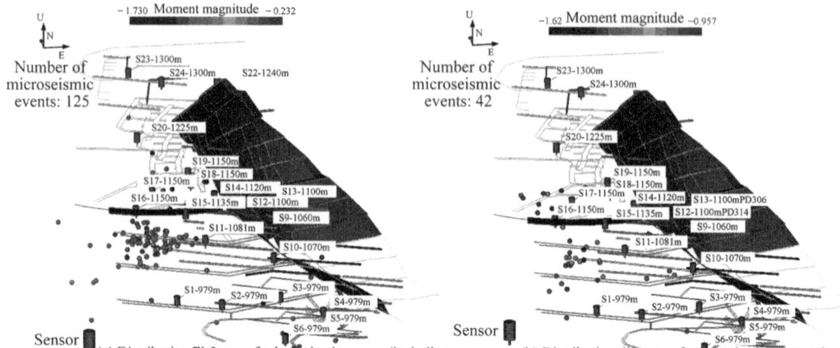

(a) Distribution diagram of microseismic events (including blasting events) before anti-shear gallery reinforcement (May 2010)

(b) Distribution diagram of microseismic events after anti-shear gallery reinforcement (May 2012)

Fig. 1.46 Changes of microseismic events before and after anti-shear gallery reinforcement [152]

1.8.6 Summary

Based on microseismic monitoring and RFPA^{3D} numerical simulation, this section analyzes the slope excavation and shear hole support design on the right bank of Dagangshan Hydropower Station, studies the deformation mode and instability mechanism of the slope after anti-shear gallery reinforcement, explains the formation mechanism of the crack on the slope surface, and comprehensively determines the stability changes of the slope after excavation and anti-shear gallery reinforcement through *b* and concludes as follows.

(1) The failure modes of slope before and after anti-shear gallery reinforcement are different. Without adding anti-shear galleries, the slope slides along the control structure composed of unloading fracture zone XL-316 and fault f_{231} in a circular arc. When the anti-shear gallery is reinforced, the "zoning" failure form of the time effect is produced, that is, the unconnected part of the anti-shear gallery in the section I–I and II–II at the elevation of 1240 and 1210 m was damaged firstly. However, there was no damage to the penetration part of the anti-shear gallery in the 1240 and 1210 m elevation sections III–IX. Obviously, the "rigid" passive reinforcement method of anti-shear gallery resistance can restrain the deformation of the main control structure plane composed of unloading fracture zone XL-316 and fault f_{231}.

(2) The anti-shear gallery improves the shear strength and shear stiffness of the rock mass on the main control structure plane, and the shear resistance and slip force are greatly improved. The safety factor of the reinforced slope is 1.83, which is about 51.2% higher than that of the non-reinforced slope, which is 1.21. It shows that the reinforcement measures of replacing large weak structures with anti-shear galleries filled with concrete greatly increase the safety of the slope.

(3) In May 2012, only 42 effective microseismic events were monitored in the month after the shear hole reinforcement, which was about 66.4% less than the 125 microseismic events (including blasting events) before the anti-shear gallery

was formed in May 2010. After reinforcement, the activity rate of microseismic events in the reinforcement area is significantly reduced, the cumulative energy of microseismic events increases slowly, the slope stress is transferred to the edge of the anti-shear gallery, and a small number of low energy events are induced in this area. The implementation of the anti-shear gallery basically covers and penetrates the rock mass in the damage area, better plays the role of reinforcement and force transmission, improves the mechanical performance of the rock mass in the structural plane, and the reinforcement is more targeted.

1.9 Summary

This chapter focuses on the key scientific problem of excavation deformation effect and stability evolution mechanism of high and steep slopes in complex environment. Based on the academic thought that "the microseismic activity law of progressive failure of slope rock mass is the essential characteristic of the precursor of slope rock mass failure and instability", Taking the right bank slope of Dagangshan Hydropower Station, an important power point project of "power transmission from west to east" with complex geological conditions and unique rock structure as the research object, the relationship between the slope body structure and the deformation and failure mode is analyzed through numerical simulation, and the control effect of slope body structure on the overall deformation and stability of the slope is determined.

The different geological structures of the slope on the right bank of Dagangshan Hydropower Station make the potential deformation and instability modes of the slope complex and diverse, and large-scale excavation on this basis has become a key technical problem affecting the success or failure of the project construction. In this chapter, through the study of the engineering geological characteristics of the soft rock belt of the slope structure, the slope body structure controlling the stability of the slope was determined. In this chapter, a deep fracture geomechanical model that can characterize the slope structure was established to reveal the slope deformation evolution process and failure mode. Firstly, combined with the deformation signs of the slope surface in the actual excavation process, the microseismic monitoring technology was used to "catch" the small cracks in the slope rock mass excavation process, and the deformation evolution law of the high and steep slope under excavation was revealed from the actual monitoring perspective. Secondly, the three-dimensional geological numerical model of damage evolution in the excavation disturbance zone of high and steep slope was established by using the research method of combining numerical calculation and field monitoring mutual feed, and the micro-fracture damage of rock mass and slope stability in the unloading evolution process were analyzed. From the perspective of geoscience, engineering science and numerical simulation, the mechanism model of slope deformation and failure and the evaluation of stability were studied. It is verified that the deformation of slope excavation disturbance area is essentially the evolution process of rock mass cumulative damage, and the causes of local deformation and failure in the process of excavation failure and the law of disaster. Finally, according to the typical slope control structure, the numerical model of slope anti-shear gallery resistance was established, which revealed the interaction mechanism between landslide and anti-shear gallery resistance and the deformation coordination and sharing mechanism in the process of slope failure, and expounded the suitability and correctness of the landslide control structure of anti-shear gallery resistance, which provided theoretical support for the selection and prevention of high and steep slope reinforcement schemes of large water conservancy and hydropower projects with similar geological conditions. The main research results of this chapter are as follows.

(1) Based on the academic idea that micro-rupture is the precursor of potential landslide, according to the deformation and failure signs reflected in the process of slope excavation, it was proposed to evaluate the stability of the excavated slope through micro-seismic monitoring, and a micro-seismic monitoring system that can continuously "catch" the micro-rupture of the excavated slope for 24 h was constructed, which realized the monitoring of the three-dimensional "damaged body". It overcame the limitation that traditional deformation monitoring methods are difficult to reflect the evolution process of rock slope instability, and the damage area induced by excavation disturbance in the deep rock mass of the right bank slope was delineated and recognized by the spatial and temporal distribution law of microseismic activity. The monitoring results show that the microseismic events gather, evolved and developed mainly along the controlled soft rock structure composed of the unloading fracture zone XL-316 and the fault f_{231} below the elevation of 1240 m, and gradually distributed in a strip in the direction of the spandrel groove. This shows that the small fracture of the deep rock mass of the slope is closely related to the location of the distribution of the structural plane, and also confirms that the large weak structural plane is one of the main factors controlling the stability of the slope. The research results provide a new idea for the monitoring and early warning of the excavated rock slope.
(2) According to the engineering geological development characteristics of the slope and the microseismic monitoring results, the main control dominant structure plane that affects the stability of the slope was introduced into the numerical model to realize the fine description and modeling of the complex rock mass structure of the slope, so as to reveal the disaster pregnancy process under the effect of slope excavation. The results show that the slope structure is the main factor affecting the deformation of slope rock mass, and the large-scale excavation unloading is the decisive factor controlling the stability of slope. Combined with the law of microseismic activity and the numerical analysis results, it is concluded that the reason for the occurrence of several deformation cracks on the slope surface during excavation is that the energy loss of microseismic events continues to accumulate, release and transfer, which leads to the accumulation of a large number of small failures on the unloading fracture zone XL-316 and

fault f_{231}, which are the main control dominant structural plane affecting the slope stability, leading to the deformation of local small rock masses.

(3) Through the study of magnitude and frequency, it is found that the stress transfer type was the main microseismic event generated during the excavation of the slope, and the low magnitude fracture-slip type microfracture was the main event after the anti-shear gallery reinforcement. During excavation, low magnitude events of the slope decreased, while high magnitude events increased, *b* both decreased, slope stability decreased, and the risk of instability disaster increased. After anti-shear gallery reinforcement, *b* increased significantly due to the increased of low magnitude events, which improved slope stability and reduced the risk of instability disasters. This change indicates that the micro-fracture deformation of the rock mass in the anti-shear gallery reinforcement area increases during the slope excavation, the slope stability decreases, and the risk of instability disaster increases. After the anti-shear gallery reinforcement, the micro-fracture deformation is suppressed, the stability is improved, and the disaster risk is reduced. The research results reveal the interaction mechanism between the landslide body and the anti-shear gallery structure and the coordination and sharing mechanism of deformation. At the same time, it is verified that the anti-shear gallery landslide control structure is an effective reinforcement measure to control the slope failure.

(4) The stability of excavated rock high slope is a dynamic evolution process, which is accompanied by complex aging deformation and rupture phenomena. In essence, the macroscopic nonlinear deformation characteristics are characterized by the accumulation of small fractures of rock mass. The potential sliding surface formed in the failure process is also continuously gestate, develop, evolve and expand, and progressive macroscopic failure finally occurs. Its stability evaluation is not only a strength problem, but also a deformation problem. In the process of large-scale excavation, attention should be paid to the control of small deformation of slope rock mass. Once the small deformation is controlled, the evolution of potential sliding surface will end in the "gestation" or "germination" stage, and it will not develop into a progressive macroscopic instability that is "out of control". It can be predicted that with the pouring of the dam and the beginning of water storage in the later period, the micro-fracture of rock mass caused by slope stress migration caused by the sudden rise and fall of reservoir water will inevitably increase, especially the phenomenon of large-scale unloading fracture zones XL-316 and XL-915, and the weakening of rock mass near weak structures such as fault f_{231} when water is encountered will aggravate the adjustment of slope stress.

(5) Combined with conventional monitoring methods, it is suggested to pay close attention to the deformation convergence of the microseismic energy density accumulation area, the microseismic high-energy damage area, and the unloading fracture zone XL-316 and fault f_{231} junction. In the process of excavation, the amount of blasting charge should be strictly controlled to minimize the disturbance and damage to the slope, and the consolidation grouting should be carried out in time according to the design requirements after excavation. It

is hoped to provide a reference for the construction and reinforcement design of water conservancy and hydropower slope engineering under similar geological conditions.

References

1. Zheng SR (2006) Discussion on hydropower resources development and environmental ecological protection in China. Strategic Study of CAE 8(6):1–6
2. Huang RQ (2005) Main characteristics of high rock slopes in southwestern China and their dynamic evolution. Adv Earth Sci 20(3):292–297
3. Xu WY (2000) Study of rock mechanics and engineering in slope and landslide environment. China Environmental Science Press, Beijing
4. Li JL, Wang LH, Liu J et al (2006) Rocky slope engineering. China Water Conservancy and Hydropower Press, Beijing
5. Gu CS, Li XH (1999) Trace analysis on behavior of accident position for Meishan multiple arch dam in 1962. Water Resour Power 17(3):5–8
6. Zhang CH (2002) Frontiers of water resources and hydropower engineering science. Tsinghua University Press, Beijing
7. Zhang YT (2003) Analysis on several catastrophic failures of hydraulic projects in view of rock hydraulics. J Hydraul Eng 5:1–10
8. Editorial Review Board of China Hydroelectric Engineering (2000) Hydropower engineering in China (engineering geology volume). China Electric Power Press, Beijing
9. Huang RQ, Xu Q, Tao LJ et al (2002) Research on simulation and engineering control of geological disaster processes. Science Press, Beijing
10. Chengdu Survey and Design Institute of China Hydropower Consultant Group (2009) Stability analysis report of the right bank slope of Dadu river Dagangshan hydropower station. Chengdu Survey and Design Institute of China Hydropower Consultant Group. Chengdu
11. Sun ZF (2006) Research on technology of horizontal cables with high accuracy and its application in the three gorges project. Jilin University, Changchun
12. Lu SQ (2007) Systematical study on engineering geology of the excavation slopes stability of Nuozadu hydro-power station in Lancang River. Chengdu University of Technology, Chengdu
13. Wu AQ, Zhu JB, Fu J et al (2010) Research on deformation measurement of whole process for the middle rock wall of the permanent shiplock at TGP. Chin J Rock Mech Eng 2010(Supplement 1):1649–1653
14. Zhu JB, Wu AQ, Zhou M (2005) Deformation and geostress change characterization of deep rockmass of permanent shiplock slope for the three gorges project. Chin J Rock Mech Eng 24(20):3749–3753
15. Huang ZQ, Liao DH, Zhang CC et al (2002) Analysis and forecast on rock deformation of middle-isolated pier of permanent shiplock in the three gorges project on Yangtze River. Chin J Rock Mech Eng 21(8):1162–1167
16. Xia XL, Zhou HM, Sheng Q et al (1999) Excavation loosened zone and mechanical behaviors of rock mass of lock's high slope of TGP. J Yangtze River Sci Res Inst 816(4):l–5
17. Zhao MH (1996) Analysis and management of high slope dumping damage at the left bank lock of Wuqiangxi hydropower station. Cental South China Water Power 3:1–5
18. Xia QF, Lu JY (1986) Analysis of the causes of landslides at the Tianshengqiao secondary hydropower station site. Bull Soil Water Conserv 4:35–40
19. Wang SK (1994) Controlled blasting and monitoring of undercut footing after landslide treatment on the left bank of Manwan power station. Water Resour Hydropower Eng 10:35–37
20. Meng H, Hu HT (1996) Disasters of landslides, rockfalls, and mudflows induced by human engineering in China. J Eng Geol 4(4):69–74

21. Chengkun Railway Technical Summary Committee (1980) Chengkun Railway (volume 2): lines, engineering geology and roadbeds. People's Railway Press, Beijing, pp 99–115, 249–250
22. He K, Jiang CS (2004) Yunnan Yuanmao expressway road graben high slope and landslide remediation work. Subgrade Engineering 2004(l):49–51
23. Che XJ, Zou ZS, Guo CM (1986) A static-gcomechanical model study on the stability of the rock excavation slopes for the navigation lock of the three gorge project. J Changjiang River Sci Res 1:57–64
24. Deng JH, Wang H, Jiang QH et al (2002) Monitor of disturbed zone for rock slope by sliding deformeter. Chin J Rock Mech Eng 21(2):180–184
25. Sheng Q (2002) Study on mechanical properties of deep excavated rocky slopes in the excavation disturbance zone and engineered rock mass. Chinese Academy of Sciences, Beijing
26. Xiao SG, Zhou DP (2003) Determination and numerical analysis method of relaxation region for cutting slope. J Southwest Jiaotong Univ 38(3):318–322
27. Wang LS, Li TB, Zhao QH (1994) Asashio chrono tectonics and human engineering. Geological Publishing Press, Beijing
28. Zheng YR, Zhao SY, Deng WD (2003) Numerical simulation on failure mechanism of rock slope by strength reduction FEM. Chin J Rock Mech Eng 22(12):1943–1952
29. Zhou GY, Li TC (2010) Slope aseismic stability analysis method based on static and dynamic finite elements. Rock Soil Mech 31(7):2303–2308
30. Zhou CY, Liu ZQ, Dong LG et al (2003) Large deformation fem analysis of slopes failure. Rock Soil Mech 24(4):644–652
31. Cao P, Zhang K, Wang YX et al (2011) Numerical simulation of failure mechanism of multilayer slope. Rock Soil Mech 32(3):872–878
32. Kou XD, Zhou WY, Yang RQ (2001) Stability analysis on the high slopes of three-gorges shiplock using FlAC-3D. Chin J Rock Mech Eng 20(1):6–10
33. Yin YP (2010) Mechanism of apparent dip slide of inclined bedding rockslide—a case study of Jiweishan rockslide in Wulong, Chongqing. Chin J Rock Mech Eng 29(2):217–226
34. Hart R, Cundall PA, Lemos J (1988) Formulation of a three-dimensional distinct element model—Part II. Mechanical calculations for motion and interaction of a system composed of many polyhedral blocks. Int J Rock Mech Min Sci Geomech Abstr 25(3):117–125
35. Bhasin R, Kaynia AM (2004) Static and dynamic simulation of a 700 m high rock slope in western Norway. Eng Geol 71(3):213–226
36. Cheng QG, Hu HT (2000) Discrete element simulation of full-course kine matics of rocky high-speed landslide. J Southw Jiaotong Univ 35(1):18–22
37. Mao YL, Hu GT, Mao XH et al (2001) Mechanism of set-out violent-slide of slope mass during earthquake and its simulation by using discrete element method. J Eng Geol 9(1):74–80
38. Cao YB, Dai FC, Xu C et al (2011) Discrete element simulation of deformation and movement mechanism for Tangjiashan landslide. Chin J Rock Mech Eng 30(Supplement 1):2878–2887
39. Sitar N, MacLaughlin MM (1997) Kinematics and discontinuous deformation analysis of landslide movement. Invited Keynote Lecture II Panamerican Symposiam on Landslides, Rio de Janeiro
40. Chen G, Ohnishi Y (1999) Slope stability analysis using discontinuous deformation analysis method. Rock Mecgan Indus 1:535–541
41. Hatzor YH, Arzi AA, Zaslavsky Y et al (2004) Dynamic stability analysis of jointed rock slopes using the DDA method: King Herod's Palace, Masada, Israel. Int J Rock Mech Min Sci 41(5):813–832
42. Wu AQ, Ding X, Lu B et al (2008) Validation for rock block stability and its application to rock slope stability evaluation using DDA method. Chin J Rock Mech Eng 27(4):664–672
43. Sun DY, Peng YJ, Wang YJ (2002) Application of DDA method in stability analysis of topple rock slope. Chin J Rock Mech Eng 21(1):39–42
44. Rockfield Software Ltd (2004) ELFEN 2D/3D numerical modelling package. Rockfield Software Ltd., Swansea

45. Zhang GX, Zhao Y, Shi GH et al (2007) Toppling failure simulation of rock slopes by numerical manifold method. Chin J Geotech Eng. 29(6):800–805
46. Liu HD, Mao CL, Guan YP (2010) Analysis method and numerical simulation for slope stability. J Water Resour Arch Eng 8(3):101–103
47. Fruneau B, Achache J, Delacourt C (1996) Observation and modeling of the saint-etienne-de-tine landslide using SAR interferometry. Tectonophysics 265(3–4):181–190
48. Tan HH, Fu HL (2010) Testing study of application of time domain reflectometry to highway slope monitoring. Rock Soil Mech 31(4):1331–1336
49. Tu XB, Dai FC, Lu XJ et al (2007) Toppling and stabilization of the intake slope for the Fengtan hydropower station enlargement project, Mid-South China. Eng Geol 91(2):152–167
50. Toshitaka K (1998) Monitoring the process of ground failure in repeated landslides and associated stability assessments. Eng Geol 50(1–2):71–84
51. Dai HC (2004) Key technologies for safety monitoring of high-rock slope of permanent shiplock of three gorges project. Chin J Rock Mech Eng 23(17):2907–2912
52. Zhang JL, Xu WY, Jin HY et al (2009) Safety monitoring and stability analysis of large-scale and complicated high rock slope. Chin J Rock Mech Eng 28(9):1819–1827
53. Meng YD, Xu WY, Liu ZB et al (2010) Analysis of 3D visualization of safety monitoring for complicated high rock slope engineering. Chin J Rock Mech Eng 29(12):2500–2509
54. Wu K, Sheng Q, Zhang YH et al (2010) Development of real-time remote monitoring and forecasting system for geological disasters at subgrade slopes of mountainous highways and its application. Rock Soil Mech 31(11):3683–3687
55. Zhao MH, Liu XP, Feng HB et al (2006) Monitoring and analysis of a high-slope stability in Xiaowan hydropower station. Chin J Rock Mech Eng 25(Supplement 1):2746–2750
56. Mccreary R, Mcgaughey J, Potvin Y et al (1992) Results from microseismic monitoring, conventional instrumentation, and tomography surveys in the creation and thinning of a burst-prone sill pillar. Pure Appl Geophys 139(3–4):349–373
57. Wang HL, Ge MC (2008) Acoustic emission/microseismic source location analysis for a limestone mine exhibiting high horizontal stresses. Int J Rock Mech Min Sci 45(5):720–728
58. Ge MC (2005) Efficient mine microseismic monitoring. Int J Coal Geol 64(1–2):44–56
59. Trifu CI, Shumila V (2010) Microseismic monitoring of a controlled collapse in field II at Ocnele Mari, Romania. Pure Appl Geophys 16(1):27–42
60. Hirata A, Kameoka Y, Hirano T (2007) Safety management based on detection of possible rock bursts by AE monitoring during tunnel excavation. Rock Mech Rock Eng 40(6):563–576
61. Bariar R, Michelets S, Baumgartner J et al (2005) Creation and mapping of 5000 m deep HDR/HFR reservoir to produce electricity. In: Proceedings of the world geothermal congress 2005, Antalya, Turkey
62. Tezuka K, Niitsuma H (2000) Stress estimated using microseismic clusters and its relationship to the fracture system of the Hijiori hot dry rock reservoir. Eng Geol 56(1–2):47–62
63. Cai M, Morioka H, Kaiser PK (2007) Back-analysis of rock mass strength parameters using AE monitoring data. Int J Rock Mech Min Sci 44:538–549
64. Young RP, Collins DS, Reyes MJM et al (2004) Quantification and interpretation of seismicity. Int J Rock Mech Min Sci 41(8):1317–1327
65. Kaiser PK, Vasak P, Suorineni FT (2005) New dimensional in seismic data interpretation with 3-D virtual reality visualisation for burst-prone mines. In: Controlling seismic risk-proceedings of sixth international symposium on rockburst and seismicity in mines. Australian Center for Geomechanics, Nedlands, pp 33–45
66. Li SL, Yin XG, Zheng WD et al (2005) Research of multi-channel microseismic monitoring system and its application to fankou lead-zinc mine. Chin J Rock Mech Eng 24(12):2048–2053
67. Jiang FX, Ye GX, Wang CW et al (2008) Application of high-precision microseismic monitoring technique to water inrush monitoring in coal mine. Chin J Rock Mech Eng 27(9):1932–1938
68. Pan YS, Zhao YF, Guan FH et al (2007) Study on rockburst monitoring and orientation system and its application. Chin J Rock Mech Eng 26(5):1002–1011

69. Lu CP, Dou LM, Wu XR et al (2008) Experimental and empirical research on frequency-spectrum evolvement rule of rockburst precursory microseismic signals of coal-rock. Chin J Rock Mech Eng 27(3):519–525
70. Tang LZ, Yang CX, Pan CL (2006) Optimization of microseismic monitoring network for large-scale deep well mining. Chin J Rock Mech Eng 25(10):2036–2042
71. Chen BR, Feng XT, Zeng XH et al (2011) Real-time microseismic monitoring and its characteristic analysis during TBM tunneling in deep-buried tunnel. Chin J Rock Mech Eng 30(2):275–283
72. Xu NW, Tang CA, Sha C et al (2010) Microseismic monitoring system establishment and its engineering applications to left bank slope of Jinping I hydropower station. Chin J Rock Mech Eng 29(5):915–925
73. Ma K (2014) Study on the potential failure mechanism monitoring and controlling methods of rock slope under excavation. Dalian University of Technology, Dalian
74. Chen ZY, Yang J (2004) Development of prestress anchor technology in rock—soil engineering. Guizhou Water Power 18(5):5–10
75. Chen ZY, Wang XG, Yang J et al (2005) Stability analysis of rocky slopes: principles, methods, procedures. China Water Conservancy and Hydropower Press, Beijing
76. Zheng YR, Chen ZY, Wang GX et al (2007) Slope and landslide engineering management. People's Communications Press, Beijing
77. Hoek E, Bray WJ (1983) Rocky slope works. Translated by, Lu SZ, Li ZC, Lei HN et al. Metallurgical Industry Press, Beijing
78. Geotechnical Engineering Manual Writing Committee (1994) Geotechnical engineering handbook. China Construction Industry Press, Beijing
79. Landslide Analysis and Control Editorial Board (1996) Landslide analysis and control. Sichuan Science and Technology Press, Chengdu
80. Zhang YT, Zhou WH (1999) Deformation and stability of rocky high slopes. China Water Conservancy and Hydropower Press, Beijing
81. Yang CZ, Tan DY, Han GJ (2000) A theoretical calculation and analysis of geotechnical slopes reinforced by spray anchors. New advances in geotechnical anchoring technology. People's Communications Press, Beijing, pp 61–66
82. Professional Committee on New Technology of Rock Anchoring and Grouting, Chinese Society of Rock Mechanics and Engineering (2002) New anchoring and grouting technology. China Electric Power Press, Beijing
83. Zhao MJ, He CG (2003) Slope engineering treatment technology. People's Communications Press, Beijing
84. Yan MM, Xu PX, Su ZY (2004) Handbook of geotechnical anchoring techniques. People's Communications Press, Beijing
85. Lin F, Huang RQ, Yan M (2009) Stability checking and anchoring force research of water-intake slope at Xiaowan hydropower station. J Eng Geol 17(1):70–75
86. Wang S (2010) Analysis on geological defects and reinforcement technology on left bank resisting rock of Jinping I hydropower. Chengdu University of Technology, Chengdu
87. Nie Q, Li PC (2008) Design and construction of high slope treatment for Longtan hydropower station left bank creeping body. Yangtze River 39(6):68–70
88. Ma HQ (2009) Technical difficulties in the construction of Xiaowan hydropower station. Water Power 35(9):17–21
89. Li SC (1996) Modelling and theoretical study of fracture damage in anchored intermittent jointed rock bodies. Wuhan Institute of Geotechnics, Chinese Academy of Sciences, Wuhan
90. Ge XR, Liu JW (1998) Study on the shear resistance behaviour of bolted rock joints. Chin J Geotech Eng 10(1):8–18
91. Chen ZY (1998) On Pan's principles of soil and rock stability analysis. J Tsinghua Univ 38(1):1–4
92. Chen AM, Gu JC, Shen J et al (2002) Model testing research on the variation of tension force of anchor cable with time in reinforcement of soft rocks. Chin J Rock Mech Eng 22(2):251–256

93. Li N, Zhang P, Yu C et al (2007) Research on stability of left abutment slope of Jinping hydropower station and reinforcement measurements of prestressed anchorage cable. Chin J Rock Mech Eng 26(1):36–43
94. Li YY (2008) Study on long term stability of geotechnical prestressed anchor system. Beijing Jiaotong University, Beijing
95. Chen X, Zhao WQ (2009) Field tests on the anchor effect of pressure and tensile anchor cable structures. Water Power 35(3):47–50
96. Wang JS (1979) Optimum anchorage angle for prestressing anchor cables and its application. Underground Eng 11:4–8
97. Zhu WS, Zhang YJ (1996) Effect of anchor reinforcement of enclosing rock and its application to the high slope of three gorges locks. In: Proceedings of the international symposium on anchorage and grouting engineering technology. China Construction Industry Press, Beijing
98. Li N, Zhao YH, Han X et al (1997) An analysis of strengthening mechanism of the pre-stressed multi-bolts and its application to engineering. J Xi'an Univ Technol 13(3):33–36
99. Zhao CY (1999) The mechanisms of prestressed rockbolts reinforcement. J Beijing Univ Civil Eng Arch 15(2):84–88
100. Zhang FM, Shao WX (1999) Prestressed anchor cables used in high rock slopes. J Hohai Univ 27(6):75–78
101. Hou CJ, Gou PF (2000) Mechanism study on strength enhancement for the rocks surrounding roadway supported by bolt. Chin J Rock Mech Eng 19(3):342–345
102. Chen AM, Gu JC, Shen J et al (2000) Model experimental study on the reinforcing effect of prestressing anchors on massive rock masses. Tunnel Construc 4:1–6
103. Li DF, Zhang YL, Chen CQ (2000) Design of prestressed-cable sommer for slope-reinforcing engineering. Rock Soil Mech 21(2):170–172
104. Cao WG, Su BY (2001) Simulation of numerical manifold method for supports by reinforced bolt and its application. Chin J Geotech Eng 23(5):581–583
105. Zhang JR, Tang BF (2002) Hyperbolic function model to analyze load transfer mechanism on bolts. Chin J Geotech Eng 24(2):188–192
106. Ding XL, Sheng Q, Han J et al (2002) Numerical simulation testing study on reinforcement mechanism of prestressed anchorage cable. Chin J Rock Mech Eng 21(7):980–988
107. Zhu JB, Han J, Cheng LK et al (2002) Research on rockmass properties near anchor with prestressing for tgp's permanent shiplock. Chin J Rock Mech Eng 21(6):853–857
108. Xu NF, Mou CX, Wang L (2002) Learning strategies of artificial neural networks for preventing over-training and their application. J Changjiang River Sci Res Inst 19(3):45–47
109. Zou JF, Li L, Ruan B (2003) Optimized design of prestressed cable anchor used for slope solidification. Min Eng 1(6):18–20
110. Wu DH, Zeng YX, Deng AF et al (2003) Model experimental study of single-anchor anchor reinforcement of fractured structural rock body. Chin J Undergr Space Eng 23(2):158–161
111. Liu ZQ, Zhou CY, Shang W et al (2004) Optimum design of prestressed anchor cables for $BIII_2$ landslide in water supply reconstruction project from Dongjiang to Shenzhen. Chin J Rock Mech Eng 23(6):1020–1024
112. You CA (2004) Theory and application study on stress-transfer mechanism of anchoring system. Shandong University of Science and Technology, Qingdao
113. He SM, Zhang XG, Wang CH (2004) Non-linear analysis on prestressed cable. Chin J Rock Mech Eng 23(9):1535–1541
114. Xiong WL, He ZG, Chen SH (2005) Optimum design of direction angle of prestressed anchor cable in slope reinforcement. Chin J Rock Mech Eng 24(13):2260–2265
115. Cao XS, Zhou PD (2005) Design method of fixed anchor unit for compression dispersion-type anchor. Chin J Geotech Eng 27(9):1033–1039
116. Zhang FM, Chen ZY, Liu N et al (2005) Optimization for determination of parameters for prestressed cable design. Prestress Technol 6:19–22
117. Zhu Y, Wei J, Liao CH (2005) The combined-power model method on determining the bounding length of prestressed anchoring rope. J Wuhan Univ Technol 27(8):60–63

118. Huang JM, Xu YF, He JD (2006) Analysis of reinforcement for slope unstable rock by distributed stress type pre-stressed anchor cable. Technol Highway Transp 4:17–22
119. Li XP, Song GH, Chen XF (2006) Research on the composite material mechanical properties of anchoring rock mass. J Wuhan Univ Technol 28(4):79–86
120. Xu YF (2005) Research on prestressed (distribution) Tendon's mechanism study and using in engineering. Sichuan University, Chengdu
121. Liu SH (2006) Study on structure and loaded status of Compression dispersion-type prestressed anchor. Jilin University, Changchun
122. Yang YY, Wang SY (1995) Study on toughening of jointed rock masses reinforced with bolts. Chin J Geotech Eng 17(1):9–17
123. Ye JH (1995) Mechanism and anchor characteristics of jointed rock masses. J Hydraul Eng 9:68–74
124. Yang SL, Xu WY, Zhu HC (2002) Reinforcement of bolt in joints. Rock Soil Mech 23(5):604–607
125. Wang C (2005) Fracture mechanical principles of anchorage for layered rock mass slope. Chin J Rock Mech Eng 24(11):1900–1904
126. Huang RQ (2008) Geodynamic process and stability control of high rock slope development. Chin J Rock Mech Eng 27(8):1525–1544
127. Li SJ, Yu H, Liu YX (2008) Results from in-situ monitoring of displacement, bolt load and disturbed zone of a powerhouse cavern during excavation process. Int J Rock Mech Min Sci 45(8):1519–1525
128. Weng MC, Tsai LS, Liao CY (2010) Numerical modeling of tunnel excavation in weak sandstone using a time-dependent anisotropic degradation model. Tunnel Undergr Space Technol 25(4):397–406
129. Tsiambaos G, Saroglou H (2010) Excavatability assessment of rock masses using the geological strength index (GSI). Bull Eng Geol Env 69:13–27
130. Hui HC, Huai Z, Bo JZ et al (2012) Finite element analysis of steep excavation slope failure by CFS theory. Earthq Sci 25(2):177–185
131. Lu Q, Chin LC, Bak KL (2012) Probabilistic evaluation of ground-support interaction for deep rock excavation using artificial neural network and uniform design. Tunnel Undergr Space Technol 32:1–18
132. Huang YK, Shao JD, Zhu KJ et al (2009) Research on slope stability of right bank of Arch Dam of Dadu River Dagangshan hydropower station engineering geological report. Chengdu Survey, Design and Research Institute of China Hydropower Consulting Group, Chengdu
133. Liu ZY, Ding XZ, Zhang L et al (2009) The external deformation monitoring project of right bank slope of Dagangshan Hydropower Station, Dadu River, Sichuan Province. In: 2009 annual monitoring report. Kunming Survey, Design and Research Institute of China Hydropower Consulting Group, Kunming, p 12
134. Dixon N, Spriggs MP (2007) Quantification of slope displacement rates using acoustic emission monitoring. Can Geotech J 44(8):966–976
135. Thomas S, Hansruedi M, Alan GG et al (2007) Microseismic investigation of an unstable mountain slope in the swiss alps. J Geophys Res 112(B7):1–25
136. Arosio D, Longoni L, Papini M et al (2009) Towards rockfall forecasting through observing deformations and listening to microseismic emissions. Nat Hazard 9(4):1119–1131
137. Cheon DS, Jung YB, Park ES et al (2011) Evaluation of damage level for rock slopes using acoustic emission technique with waveguides. Eng Geol 121(1–2):75–88
138. Ma K, Xu NW, Liang ZZ (2018) Stability assessment of the excavated rock slope at the Dagangshan hydropower station in China based on microseismic monitoring. Adv Civil Eng 1:1–16
139. Gu DZ (1979) Foundation of engineering geomechanics of rock mass. Science Press, Beijing
140. Sun GZ (1988) Structural Mechanics of rock mass. Science Press, Beijing
141. Zhang ZY, Wang ST, Wang LS (1997) Principles of engineering geological analysis, 2nd edn. Geology Press, Beijing

142. Geological Bureau of Guangdong Province. Geological survey report of Xijiang Bridge, Zhaoqing City, Guangdong Province
143. Shanley RJ, Mathtab MA (1976) Delineation and analysis of clusters in orientation data. Malhematieal Geol 8(l):9–23
144. Goodman RE, Kieffer DS (2000) Behavior of rock in slopes. J Geotech Geoenviron Eng 8:675–684
145. Cao YJ (2006) Systematic engineering geological study on the stability of high slope with soft rock—taking the Zipingpu water conservancy project in Minjiang River as a Case. Chengdu University of Technology, Chengdu
146. Zhou DP, Zhong W, Yang T (2008) Stability analysis of rocky slope based on slope structure. Chin J Rock Mech Eng 27(4):687–695
147. Qi SW, Wu FQ, Lan HX (2002) Study on the mechanism of the deep fractures of the left abutment slope at the Jinping first stage hydropower station. Chin J Geotech Eng 24(5):596–599
148. Shi YC, Ji F (2009) Special report on stability analysis and reinforcement measures of right bank slope of Dagangshan hydropower station in Dadu River, Sichuan Province. Chengdu University of Technology, Chengdu
149. Liu Y (2012) Study on deformation mechanism and static-dynamic response for the Dam Area slope of Dagangshan hydropower station. Chengdu University of Technology, Chengdu
150. Tang CA, Li LC, Li CW et al (2006) RFPA strength reduction method for stability analysis of geotechnical stability analysis engineering. Chin J Rock Mech Eng 25(8):1522–1530
151. Xu NW (2011) Study on microseismic monitoring and stability analysis of high SteepRock slope. Dalian University of Technology, Dalian
152. Ma K, Tang CA, Li LC et al (2013) Reinforcement effects of anti-shear gallery of Dagangshan right bank slope based on microseismic monitoring and numerical simulations. Chin J Rock Mech Eng 32(6):1239–1247
153. Liu YR, Lu QC, Pan YW et al (2009) Special report on stability analysis and reinforcement measures of right bank slope of Dagangshan hydropower station in Dadu River, Sichuan Province. Tsinghua University, Beijing
154. Xu NW, Tang CA, Zhou Z et al (2013) Stability analysis of hydraulic rock slope based on three-dimensional numerical simulation and microseismic monitoring. Chin J Rock Mech Eng 32(7):1373–1381
155. Yang TH, Zhang FC, Yu QL et al (2011) Research situation of open-pit mining high and steep slope stability and its developing trend. Rock Soil Mech 32(5):1437–1472
156. Chris HO, Livio LT, Nina LL (2011) Constraints on mechanisms for the growth of gully alcoves in Gasa crater Mars, from two-dimensional stability assessments of rock slopes. Icarus 211(1):207–221
157. Cheng YM (2003) Locations of critical failure surface and some further studies on slope stability analysis. Comput Geotech 30:255–267
158. Cheng YM, Lansivaara T, Wei WB (2007) Two-dimensional slope stability analysis by limit equilibrium and strength reduction methods. Comput Geotech 34:137–150
159. Terzaghi RD (1965) Source of error in joint survey. Geotechnique 15(4):287–304
160. Zhao W, Tang CA (1988) The measure theory of discontinuities spacing and trace length. China Mining Mag 1:36–38
161. Massanobu O (1982) Fabric tensor for discontinuous geological materials. J Jap Soc Soils Found 22(4):96–108
162. Kulatilake PHSW, Wu TH (1984) Estimation of mean trace length of discontinuities. Rock Mech Rock Eng 17(3):215–232
163. Rouleau A, Gale JE (1985) Statistical characterization of the fracture system in the Stripa granite, Sweden. Int J Rock Mech Mining Sci Geomech Abstr 22(6):353–367
164. Rahn PH (1996) Engineering geology: an environmental approach, 2nd edn. Prentice Hall, New Jersey
165. Zhang L, Einstein HH (1998) Estimating the mean trace length of discontinuities. Rock Mech Rock Eng 31(4):217–235

166. Jimenez AR, Ceres R, Puns JL (2000) Avision systerm based on a laser range-finder. Mach Vis Appl 11:321–329
167. Hammah RE, Curran JH (1999) On distance measures for the fuzzy K-means algorithm for joint data. Rock Mech Rock Eng 32(1):1–27
168. Feng Q, Sjgren P, Stephansson O et al (2001) Measuring fracture orientation at exposed rock faces by using a non-reflector total station. Eng Geol 59(1–2):133–146
169. Slob S, van Knapen B, Hack R et al (2005) Method for automated discontinuity analysis of rock slopes with three-dimensional laser scanning. Transp Res Rec 1913(1):187–194
170. Dong XJ, Huang RQ (2006) Application of 3D laser scanning technology to geologic survey high and steep slope. Chin J Rock Mech Eng 25(2):3629–3635
171. Hong ZE, Feng ZY, Wu ZJ (2007) Measurement of potential energy of rock slope outcrop by 3D laser scanning. J Soil Water Conserv 39(3):247–267
172. Wang FY, Chen JP, Fu XH et al (2008) Study on geometrical information of obtaining rock mass discontinuities based on VirtuoZo. Chin J Rock Mech Eng 27(1):169–175
173. Liu CJ, Ding LQ, Sun DY (2011) Automatic fuzzy clustering analysis and geometric information acquisition of rock mass discontinuities based on laser point cloud data. Chin J Rock Mech Eng 30(2):358–364
174. Feng XT, Yang CX (1999) Intelligent rock mechanics (2)—intelligent recognition of input parameters and constitutive models. Chin J Rock Mech Eng 18(3):350–353
175. Liang NH, Qu WB, Cao XS (2008) A back analysis method for fault surface parameters in slope using immunization-genetic algorithm. J China Coal Soc 33(9):977–982
176. Editorial Board of Lithosphere Dynamics Atlas of China (1991) Introduction to lithosphere dynamics in China. Earthquake Press, Beijing
177. The Institute of Geology, Chinese Academy of Geological Sciences (1991) Engineering geomechanics of rock mass (V). Science Press, Beijing
178. Li XZ (2004) Study on time prediction and evaluation of landslide hazard. Chengdu University of Technology, Chengdu
179. Wang NQ, Wang YF, Luo DH et al (2008) Review of landslide prediction and forecast research in China. Geol Rev 54(3):355–360
180. He KQ, Li XZ (1996) Mechanical conditions and comprehensive displacement mechanical criteria for shear outlet formation of large accumulation layer landslide. Geotech Investig Surv 5:13–16
181. Li TB, Chen MD (1996) Time prediction of landslide using verhulst inverse-function model. J Geol Haz Environ Preserv 7(3):13–18
182. Huang RQ, Xu Q (1997) Synergetic prediction model of slope instability. J Mount Res 15(1):7–12
183. Romeo R (2000) Seismically induced landslide displacements: a predictive model. Eng Geol 58(3–4):337–351
184. Li YR (2003) Research on development and application of landslide prediction system based on GIS. Chengdu University of Technology, Chengdu
185. Peng JB (2005) Study on the technique of information fusion applied to the prediction of landslide. Chengdu University of Technology, Chengdu
186. Liao XP (1994) Probe into the new theory of time prediction of landslides. J Geol Haz Environ Preserv 5(3):25–29
187. Wang SQ (1998) Monitoring and forecasting of landslides in the three Gorges of Yangtze River. Geology Press, Beijing
188. Li XZ, Xu Q, Huang RQ et al (2003) Research on prediction criterion for temporary prediction of landslide. Chin J Geol Haz Control 14(4):5–11
189. Xu Q, Tang MG, Xu KX et al (2008) Research on space-time evolution laws and early warning-prediction of landslides. Chin J Rock Mech Eng 27(6):1104–1112
190. Xu Q, Zeng YP (2009) Research on acceleration variation characteristics of creep landslide and early-warning prediction indicator of critical sliding. Chin J Rock Mech Eng 28(6):1099–1106
191. Xu Q, Zeng YP, Qian JP et al (2009) Study on a improved tangential angle and the corresponding landslide pre-warning criteria. Geol Bull China 28(4):501–505

192. Jin HY, Xu WY, Meng YD et al (2008) Research on comprehensive prediction for slope stability at left bank of Jinping I hydropower station. Chin J Rock Mech Eng 27(10):2058–2063
193. Herrera G, Fernandez-Merodo JA, Mulas J et al (2009) A landslide forecasting model using ground-based SAR data: the Portalet case study. Eng Geol 105(3–4):220–230
194. Xu NW, Tang CA, Wu SH et al (2010) Application of microseismic monitoring technology in right bank slope of Dagangshan hydropower Station. J Disast Prev Mitig Eng 30(Sl):216–221
195. Ma K, Tang CA, Liang ZZ et al (2017) Stability analysis for the right bank slope of Dagangshan hydropower station during excavation and after reinforcement based on microseismic monitoring. Eng Geol 218:22–38
196. Snelling PE, Laurent G, Mckinnon SD (2013) The role of geologic structure and stress in triggering remote seismicity in Creighton Mine, Sudbury, Canada. Int J Rock Mech Min Sci 58:166–179
197. Hudyma MR (2008) Analysis and interpretation of clusters of seismic events in mines. University of Western Australia, Perth, Australia, pp 50–51, 148–150, 205–206
198. Boatwright J, Fletcher JB (1984) The partition of radiated energy between P and S waves. Bull Seismol Soc Am 74(2):361–376
199. Akik K, Richards PG (1980) Quantitative seismology: theory and methods,vol I and II. W. H. Freeman and Company, San Francisco
200. Zhuang DY (2019) Study on seepage channel identification and stability of the large-scale water-sealed underground oil storage caverns subjected to excavation. Dalian University of Technology, Dalian
201. Aki K (1968) Seismic displacements near a fault. J Geophys Res 73:5359–5376
202. Gibowicz SJ, Kilko A (1994) An introduction to mining seismology. Academic Press Inc., San Diego, California, pp 2–8, 15–22.
203. Senatorski P (2007) Apparent stress scaling and statistical trends. Phys Earth Planet Inter 160(3):230–244
204. Mendecki AJ (1997) Seismic monitoring in mines. Chapman & Hall, London
205. Yang Y, Xu NW, Li T et al (2018) Stability analysis of left bank rock slope at Baihetan hydropower station based on RFPA^{3D} software and microseismic monitoring. Rock Soil Mech 39(6):2193–2202
206. Sheng Q (2003) Excavation disturbed zone of deep cutting rock slopes and mechanics behaviour of engineering rock mass. Chin J Rock Mech Eng 22(10):1761
207. Xue XG, Chen JP, Huang RG (2006) Numerical simulation and supporting measure of high slope excavation on left abutment of a large hydropower station. Rock Soil Mech 27(S1):222–226
208. Chen YF, Zhou CB, Yu ZX et al (2007) 3D elastoplastic finite element simulation on excavation and reinforcement of the diversion tunnel outlet slope at the left bank of Jinping-I Hydropower Project. Rock Soil Mech 28(8):1565–1570
209. Cai GJ, Huang RQ, Yan M et al (2008) Physical simulation study on deformation and failure response of an anti-inclined slope during excavation. Chin J Rock Mech Eng 27(4):811–817
210. Wang RH, Li JL, Jiang YZ et al (2008) Quality evaluation of unloaded slope rock mass. Rock Soil Mech 29(10):2741–2746
211. Liu J, Qiao LP, Li PJ et al (2008) Study of excavation disturbance and anchoring effects of high-geostress dam foundation slope of laxiwa hydropower engineering. Chin J Rock Mech Eng 27(6):1094–1103
212. Huang RQ, Huang D (2010) Experimental research on affection laws of unloading rates on mechanical properties of Jinping marble under high geostress. Chin J Rock Mech Eng 29(1):21–33
213. Zhu JL, Huang RQ, Zhang SY et al (2010) Deformation response of high-order slope excavation of a large hydroelectric station in China. Chin J Geotech Eng 32(5):784–791
214. Tian B, Lu XC, Huang YY et al (2010) Excavation disturbance and its influential factors on material-yard slope of Guandi hydropower station in Yalong river. Chin J Rock Mech Eng 29(S1):3199–3207

215. Xu WY, Luo QX, Xie SY (1999) Three-dimensional numerical simulation of underground rock excavation in maya rock slope for Shuibuya hydropower project. J Eng Geol 7(1):89–94
216. Deng JH, Li ZF, Ge XR (2001) Disturbed zones and displacement back analysis for rock slopes. Chin J Rock Mech Eng 20(2):171–174
217. Zhou HM, Sheng Q, Li WS (2004) Excavation-disturbed zone and weaken degree of mechanical properties for rockmass of tgp shiplock slope. Chin J Rock Mech Eng 23(7):1078–1081
218. Li N, Qian QH (2010) Four criteria of stability analysis and assessment of high rock slope. Chin J Rock Mech Eng 29(9):1754–1759
219. Guri VG, Guro G (2008) Geological model of the aknes rockslide, Western Norway. Eng Geol 102:1–18
220. Xu NW, Tang CA, Li LC et al (2011) Microseismic monitoring and stability analysis of the left bank slope in Jinping first stage hydropower station. Int J Rock Mech Min Sci 48:950–963
221. Ma K, Tang CA, Xu NW et al (2012) Conclusion report of microseismic monitoring project on right bank slope of Dagangshan Hydropower Station in Dadu River, Sichuan Province. Dalian University of Technology, Dalian
222. Gutenberg B, Richter CF (1956) Magnitude and energy of earthquakes. Ann Geophys 9.1–15
223. Shao J (2007) Study on progressive failure of the excavated Slope and the pre-reinforced measurement with pile and anchor. Southwest Jiaotong University, Chengdu
224. Gutenberg B, Richter CF (1944) Frequency of earthquakes in California. Bull Seismol Soc Am 34(4):185–188
225. Li QL, Yu T, Hao BL et al (1979) Earthquake frequency: spatiotemporal scanning of earthquake magnitude relationship. Seismological Press, Beijing
226. Wang CL, Wu AX, Liu XH et al (2009) Study on fractal characteristics of b value with microseismic activity in deep mining. Proc Earth Planet Sci 1(1):592–597
227. Xu NW, Dai F, Zhou Z et al (2014) Study of characteristics of b value for microseismic events in high rock slope. Chin J Rock Mech Eng 33(S1):3868–3874
228. Lu P, Li ZX, Tao BZ et al (2010) Precursory significance of the deviation of the magnitude-frequency from the G-R relation. Earthquake 26(4):1–8
229. People's Republic of China National Industry Standard Writing Group. Design specification for slope of hydropower and water conservancy project: DL/T 5353—2006. China Electric Power Press, Beijing

Chapter 2
Microseismic Monitoring and Stability Analysis of Large Underground Water-Sealed Petroleum Caverns

Abstract This chapter explores the microseismic monitoring and stability analysis of large underground water-sealed petroleum caverns. It highlights the background and significance of the study, reviews the research progress both domestically and internationally, and analyzes numerical methods, safety monitoring techniques, and microseismic monitoring approaches for these structures. Key challenges in current research are identified. Using Jinzhou underground water-sealed oil caverns as a case study, the chapter examines regional geological and hydrological conditions, the establishment of a microseismic monitoring system, and the accuracy of microseismic location and signals. The spatial and temporal distribution of microseismic activity and the local stability of surrounding rock are assessed. Additionally, numerical simulation based on the Complex Distinct Element Method (CDEM) is conducted, incorporating the project's profile, the method's principles, and simulation results. Comparative analysis between numerical findings and microseismic monitoring data further reveals insights into the stability and safety of underground caverns. This comprehensive study offers valuable theoretical and practical support for the design, monitoring, and maintenance of underground water-sealed petroleum storage systems.

2.1 Background and Significance of Study

As global industrial process continues to advance, nations are increasingly focusing on their oil reserves. Employing stable groundwater to seal and store oil (or natural gas) within subterranean rock caverns–underground water-sealed oil storage caverns, offers several benefits, including a minimal space, enhanced security, reduced maintenance expenses, and suitability for strategic military preparations. It is the most foremost storage solution for oil (or natural gas) in the world, and is recognized globally as a "reserve with highly strategic security". Numerous successful construction projects employing this method have been documented in Sweden, Japan, and South Korea [1–5].

K. Ma and C. Tang, *Microseismic Monitoring and Stability Analysis of Large Rock Mass Engineering*, https://doi.org/10.1007/978-981-96-6483-2_2

Due to the sustained and rapid economic and social development, oil consumption in China has increased proportionally. Year after year, China oil imports have shown a substantial upward trend, resulting in a significant level of external dependency. Consequently, the imperative task facing China is the construction of large underground water-sealed oil storage caverns. During the second phase of the oil reserve base construction project, the National Development and Reform Commission has granted approval for the implementation of the underground water-sealed cavern storage model. By the end of 2020, China has established 12 national oil reserve facilities in Tianjin, Shanshan, Zhoushan, Huangdao (comprising both ground and caverns storage), Dushanzi, Zhenhai, Huizhou, Dalian, Lanzhou, Jinzhou (caverns storage), Jintan, and Zhanjiang. These bases collectively boast a total storage capacity of 70 million m^3, equivalent to 60 million tons, which can be further translated into approximately 440 million barrels (1 ton = 7.33 barrels). Based on available data and considering other reserves, China total oil reserve capacity amounts to 85 million tons, equivalent to almost 600 million barrels. Given 2021 national consumption of 720 million tons and imports totaling 500 million tons, these reserves only can sustain 30 days of domestic consumption and provide a 40 days buffer for import protection [6]. Anticipations suggest that additional underground water-sealed oil storage cavernswill be developed in the foreseeable future [7]. However, China has not yet achieved full mastery of the core technology required for underground water-sealed oil storage caverns. The fundamental reasons are as follows: ① The distinctive structure of these caverns, characterized by their unlined construction, high sidewalls, big volume, large span, and continuous exposure to dynamic groundwater environment, presents substantial challenges to the research of construction technology. ② The current domestic construction technologies in this field are completely imported from foreign sources. For instance, the construction of the four caverns in Shantou, Ningbo, Zhuhai, and Huangdao relied on foreign companies that closely guarded their core construction techniques. Therefore, China construction technology for underground water-sealed oil storage caverns remains in a groping and experimental phase. Notably, there is insufficient understanding of the characteristics of surrounding rock deformation and fractures in water-sealed caverns during excavation and operation. The theoretical framework and methods for controlling surrounding rock deformation damage are also in the early stages of development. In engineering practice, the viewpoint of "bringism" is frequently embraced, and many engineering challenges are quite common. A case in point is the instance of a small-scale underground water-sealed oil storage project in Huangdao Island (150,000 m^3) that was ultimately abandoned due to issues of crude oil leakage and exorbitant operational costs, rendering the cavern's advantages ineffective. Given the highly stringent selectivity and the reliance on geological conditions, it is impractical to replicate existing engineering experiences and techniques in the construction of underground water-sealed oil storage within rock caverns. Consequently, during the period of vigorous development of energy strategic reserve bases and large petrochemical enterprises underground raw material storage facilities in China, undertaking targeted and innovative research on the stability of underground water-sealed oil storage caverns, aligning with engineering requirements, carries substantial engineering significance.

According to the storage principles, underground water-sealing oil storage caverns are generally constructed in areas with good geological conditions (usually at depths ranging from 70 to 160 m below the surface). This kind of engineering project possesses unique characteristics, such as the continuous injection of high-pressure water, the continuous infiltration of high-pressure water from the cavern walls into the cavity. This ensures that the permeating pressure of water in the existing rock fractures is greater than the dynamic internal pressure generated by frequent extraction of the storage medium. Due to the long-term effects of the water curtain system, the exchange of water between rock blocks and fractures, chemical interactions between rock blocks and fracture water, and the mechanical coupling between seepage stress and rock deformation all influence the overall (local) stability of underground water-sealing oil storage caverns. Rock damage is the fundamental reason for the instability of caverns and constitutes the core of studying two key fundamental scientific issues related to underground water-sealing oil storage caverns: "water sealing effectiveness" and "surrounding rock stability." Scholars both domestically and internationally have conducted extensive research on these two issues, leading to numerous valuable achievements [8–13]. However, underground water-sealing oil storage caverns, characterized by high conductivity and large fractures and fault structures, are subjected to long-term coupled interactions in a multiphase and multi-field environment of water–rock–oil (gas) under pressure. The probability of damage (secondary microfractures) occurring is significantly increased. Moreover, the cavern cross-section is large, the spaces within the caverns are interconnected, there are numerous excavation faces, and the excavation and unloading actions are intense. The stability analysis and assessment of its construction and operational period have always been highly complex due to the strong dependence on engineering geological conditions. As is well-known, both the airtightness of the caverns and the stability of the surrounding rock ultimately come down to the issue of rock damage. Throughout the entire process from loading to failure, rock undergoes either the absorption of external energy (mechanical or thermal energy) transforming into its internal energy or the release of internal strain energy in some way. Its damage and failure represent a state of instability driven by energy [14]. In recent years, as a three-dimensional spatial monitoring technique for rock microfractures, microseismic monitoring technology can accurately and effectively monitor rock microfractures, thereby revealing the patterns of energy release in rocks under external disturbances. Numerous scholars both domestically and internationally have utilized microseismic monitoring technology to analyze various engineering projects such as mines [15–18], hydropower steep slopes [19], tunnels [20], hot dry rock power generation method [21], and underground powerhouse [22]. They have researched the initiation and evolution patterns of microfractures, providing valuable data for the monitoring and early warning of engineering rock stability. Hong et al. [23] successfully applied microseismic technology in the monitoring of underground water-sealing oil storage caverns. They conducted a preliminary analysis of the mechanism behind microfracture events in the blasting-damaged zone and provided an empirical evaluation criterion for judging the stability of the cavern surrounding rock based on microseismic monitoring results. This demonstrates the feasibility of using microseismic monitoring technology in underground water-sealing oil storage caverns. Utilizing microseismic monitoring to "capture" internal minor rock damages and assess the degree of rock damage will become a necessary requirement for the construction of large-scale underground water-sealing oil storage caverns [24, 25].

This chapter is based on the fundamental understanding that microfractures occurring during the geological deterioration caused by excavation disturbances represent the common precursor characteristics of instability in large underground water-sealing oil storage caverns. It breaks away from the traditional approach of predicting instability disasters in underground water-sealing oil storage caverns based on apparent monitoring information such as displacement and stress. Instead, it delves into the essence of cavern instability caused by excavation-induced deterioration in the cavern rock mass. Focusing on the large-scale underground water-sealing oil storage cavern in Jinzhou, Liaoning Province, the study employs microseismic monitoring and numerical simulation as research methods. The research primarily investigates the microseismic damage effects in underground water-sealing oil storage caverns induced by excavation disturbances and their instability mechanisms. The study attempts to identify the evolution patterns of microfractures in the surrounding rock of underground water-sealing oil storage caverns under excavation unloading, elucidate the intrinsic causes of instability, describe their energy release characteristics and formation mechanisms, and identify and delineate potentially hazardous areas within the cavern. The research not only addresses practical engineering problems related to the studied object but also makes progress in new technologies and methods for the deformation and stability study of underground water-sealing oil storage caverns. This chapter proposes a set of stability assessment mechanisms and disaster early warning design methods applicable to large underground water-sealing oil storage caverns under construction disturbances. The aim is to comprehensively reveal the essence of the surrounding rock state and instability during the excavation process of underground water-sealing oil storage caverns. This provides valuable references for the design and construction of such structures.

2.2 Research Status and Analysis at Domestic and International

2.2.1 Numerical Analysis Methods for Underground Water-Sealed Oil Storage Caverns

Gnirk et al. [26] investigated the stability of large underground caverns for compressed gas energy storage in hard rock and developed a numerical simulation model for assessing cavern stability. Yang et al. [27] examined the seepage field of an underground water-sealed oil storage cavern using finite element methods and formulated a numerical simulation model for groundwater seepage. Liu et al.

[28] employed Visual Modflow to construct a three-dimensional numerical simulation model for underground water-sealed oil storage caverns, simulating the seepage field around the cavern. Zhang et al. [29] utilized FLOW-3D seepage software to calculate the average hydraulic gradient of an underground cavern for liquefied petroleum gas (propane) storage in Shantou. They also assessed the function of a water curtain system. Wu et al. [30] applied Visual Modflow software to analyze the two-dimensional seepage field in an underground water-sealed oil storage in rock cavern in Jinzhou. They determined the extent and velocity of groundwater levels under horizontal and vertical water curtain conditions. Lu [31] conducted simulations of surrounding rock deformation under high pressure (52 MPa) for underground gas storage in rock caverns, based on the Mohr–Coulomb criterion. Lee et al. [32] explored the stability of underground water-sealed oil storage cavernsemploying the finite element analysis, block theory, inverse analysis, and fracture network simulation techniques. Wang et al. [33] employed a viscoelastic rheological model in FLAC3D to analyze the long-term stability of underground water-sealed oil storage caverns. Chen [34] utilized FLAC3D to simulate the cavern excavation process in the Huangdao underground water-sealed oil storage caverns. The analysis encompassed cavern surrounding rock deformation, stability, and the interaction between caverns during group excavation. Shi [35] conducted numerical simulations of Huangdao's water-sealed liquefied petroleum gas (propane) storage using the unsaturated groundwater flow and heat flow transport program TOUGH2. The results showcased the effectiveness of the artificial water curtain system in averting underground oil cavern leakage. Wang et al. [36] scrutinized the natural water sealing stress-seepage coupling and rock stability within China inaugural large-scale underground water-sealed oil storage caverns. The study employed ABAQUS and generated several valuable findings, including the influence of hydraulic head on water sealing conditions due to the hydraulic gradient and rock permeability. Xu et al. [37] used the finite difference method (FDM) to examine water inflow rates in underground water-sealed oil storage cavernsand discerned significant patterns within the groundwater seepage field. Kim et al. [38] carried out an investigation into the factors affecting the hydraulic gradient in underground water-sealed oil storage caverns. This was achieved through the use of a stochastic simulation method, leading to the identification of the critical hydraulic gradient for unlined caverns. Tezuka et al. [39] examined the stability of surrounding rock in Japan, a region prone to earthquakes and volcanoes. Their analysis centered on underground water-sealed oil storage caverns. Lian [40] employed ANSYS for numerical analysis to assess the surrounding rock stability in a sizable underground water-sealed oil storage cavern located in Jinzhou, Liaoning Province. This analysis facilitated the visualization of the cavern's surrounding rock stability. Yu et al. [41] based on the actual characteristics of Liaoning Dalian underground water-sealed oil storage caverns, employed the 3DEC three-dimensional model to count the number of joints and production information, thus inversely obtained the preliminary seepage field. The research results mentioned above have established a robust foundation for the study of stability in underground water-sealed oil storage caverns.

2.2.2 *Safety Monitoring Methods for Large Underground Water-Sealed Oil Caverns*

The stability monitoring of large underground cavern systems is a research process characterized by continuous feedback loops between on-site monitoring and theoretical calculations. It is a matter of widespread concern in both the scientific and engineering communities. The stability analysis and evaluation of large underground cavern systems are determined by complex geological conditions and engineering factors. Advanced monitoring techniques and appropriate mechanical models are necessary for accurate assessment. Only by integrating on-site monitoring with theoretical analysis can made the prediction of cavern deformation and instability more objective [42]. From the current analysis of cavern deformation monitoring, the monitoring methods and instruments commonly adopted are as follows: ① Surface monitoring (level, theodolite, rangefinders, total stations, etc.); ② Displacement meter; ③ GPS; ④ Infrared remote sensing; ⑤ Optic fiber sensoring; ⑥ Interferometric Synthetic Aperture Radar; ⑦ Micro-displacement of laser monitoring technology; ⑧ Bore-hole tilter meter; ⑨ Anchor and anchor cable dynamometer; ⑩ acoustic emission monitoring technology. In the field of large cavern monitoring, scholars both domestically and internationally have conducted extensive research and exploration, leading to a series of significant achievements [43–45]. Especially the comprehensive studies on surrounding rock deformation monitoring in several large hydroelectric engineering underground cavern complexes, such as the Jinping I hydropower project, Monkey Rock, Guandi, etc. [46]. Huang et al. [47] used on-site stress and displacement monitoring data to analyze the deformation, cracking characteristics, and mechanical mechanisms of the surrounding rock during the construction period of the underground power plant at the Jinping I Hydropower Station. Shi et al. [48] used a multi-point displacement monitor to study the deformation characteristics of the underground power plant surrounding rock masses in the high-stress zone of the Laxiwa Hydropower Station. Xia et al. [49] proposed a monitoring method for large unstable underground caverns with rock fragmentation. They utilized an improved convergence meter as the primary monitoring tool and successfully forecasted roof collapses and collapses during the cavern construction period. Tan [50] introduced the use of intelligent total stations for measuring large underground caverns, focusing on monitoring surrounding rock convergence and cross-sectional inspection during cavern excavation. Zhao et al. [51] applied fiber Bragg grating (FBG) sensors to monitor the deformation of tunnel surrounding rock. Zhang [52] conducted on-site deformation monitoring of surrounding rock and support systems during the extra-large cross-section high-speed railway tunnel's excavate process. This enabled reasonable predictions and early warnings regarding the tunnel's safety.

Most of the methods mentioned above are based on conventional monitoring techniques such as stress and displacement measurements. They can only provide monitoring results for significant deformations or macroscopic instability that has already occurred at specific localities in the rock mass. However, these methods are unable to detect the precursor information of microscopic rock fractures that occur before the instability of the rock cavern, which are imperceptible to the human eye. They fail to serve as the "eyes" of geotechnical engineering in monitoring these subtle changes. Additionally, their monitoring effectiveness is limited in providing a comprehensive macroscopic evaluation of the stability of large-scale rock masses. Currently, traditional surface monitoring methods focused on destruction and deformation are shifting towards modern monitoring technologies aimed at detecting internal damage and fractures within the rock mass.

2.2.3 Micro-seismic Monitoring and Stability Analysis of Large Underground Water-Sealed Oil Storage Caverns Group

In underground caverns, rock heterogeneity serves as the origin of surrounding rock instability and the mechanical basis for the predicting instability [53]. Utilizing optimized monitoring network technology to capture elastic waves and stress waves generated by micro-ruptures of the rock mass within the study area, invert the calculation of micro-rupture of the three elements of "spatiotemporal strength". This approach facilitates the identification and delineation of potential hazard zones within the oil cavern surrounding rock, enables the assessment of rock stability during construction. The three-dimensional spatial monitoring technique tracking of elastic waves resulting from micro-fractures in the rock mass is referred to as micro-seismic monitoring (MS) [54]. In contrast to conventional monitoring methods, MS techniques have the capability to detect rock mass damage at an earlier stage, thereby facilitating early warnings of rock mass instability and the implementation of appropriate preventive measures. This technology has emerged as one of the most efficient tools for tracking rock mass movements in the mining [55–57]. In recent years, micro-seismic monitoring technology, a spatiotemporal dynamics three-dimensional monitoring method, has been capable of timely capturing micro-fractures within rock bodies and analyzing real-time crack initiation, development, expansion, evolution, penetration, and the formation of macroscopic slip surfaces. Consequently, it has become a pivotal monitoring tool for evaluating rock mass stability [58]. Micro-seismic monitoring technology was applied in the mining field as early as the 1970s by countries such as the Soviet Union, Canada, and the United States, resulting in substantial economic and societal advantages. Currently, micro-seismic monitoring has proven to be an effective approach for ground pressure monitoring in several countries, including Germany, South Africa, the United States, Poland, the United Kingdom, Japan, Canada, and Australia [59]. Research on micro-seismic monitoring techniques has yielded significant results in various fields, including tunneling, hot dry rock power generation, and underground nuclear waste laboratories [60]. Numerous scholars have conducted fruitful research by applying micro-seismic monitoring technology to various areas, including the rockburst, coal and gas outburst, and water inrush in the coal field; the hydraulic fracturing problem in the petroleum field; underground caverns, tunnel rockbursts and high slope instability problems in water conservancy and hydropower fields. Tang et al. [61] established the rockburst real-time disaster early warning system by employing the South African ISS micro-seismic equipment in Dongguashan copper mine; Zhuang et al. [62] applied a self-developed mine seismic monitoring and positioning system to the rockburst prediction in the Beijing Muchengjian coal mine; Xu et al. [63] constructed the first hydropower slope micro-seismic monitoring system by using the Canadian ESG micro-seismic monitoring system, which was successfully applied to the left bank slope of Jinping I hydropower station, China; Lu et al. [64] used the Polish SOS micro-seismic monitoring system to successfully predict several mine earthquakes and rockbursts in many mines; Chen et al. [65] applied the principle of micro-seismic monitoring to study the rockburst problem of high stress and high burial depth tunnels in Jinping II hydropower station. The micro-seismic monitoring mentioned above has yielded promising results in applied research within the field of engineering, as well as in the prediction and forecasting of large-scale geological disasters.

Nevertheless, research pertaining to micro-seismic monitoring in similar underground caverns has primarily centered on the mining and hydraulic engineering. Presently, there is no prior application of rock micro-seismic technology in monitoring the underground water-sealed oil storage caverns in China. With the gradual increase in the scale of the underground water-sealed oil storage caverns projects, the stability of the cavern surrounding rock has emerged as a critical factor affecting the project safety. Owing to its distinctive engineering attributes, such as towering sidewalls, extensive spans, and elusive structural features, studying the state of cavern rock during excavation, construction, and operation has presented significant challenges. To date, there is no unified cognition of the surrounding rock stability occurrence mechanism in complex geological and operational conditions of the underground water-sealed oil storage caverns.

As for the stability analysis of large cavern groups, some scholars began to study this aspect as early as the late 1970s. Mo et al. [66] believe that the failure of surrounding rock is mainly characterized by tensile failure, and the radial tensile strain of rock around the cave can be used as the stability evaluation index of surrounding rock. According to local failure phenomenon and deformation monitoring results, Zhang [67] analyzed the stability of surrounding rock of Ertan underground powerhouse system in high ground stress area. Shi [68] analyzed the stability of surrounding rock mass of underground power house cave group of Ertan Hydropower Station based on the structure and quality of rock mass, stress concentration degree and deformation around the cave. Ding et al. [69] proposed that radial tensile strain around the cave, calculation of rock yield zone around the cave and stress state of the supporting structure were simultaneously used as stability analysis criteria. Chen et al. [70] simulated the construction excavation process of the underground power house area of Qingjiang Shuibuya Water Project by using three-dimensional fast Lagrange method, and analyzed the stability of surrounding rock group by using the distribution of surrounding rock deformation, stress and plastic zone. Zhang et al. [71] applied the critical block theory to the stability analysis of the rock mass of the underground powerhouse of Baise Water project, and analyzed the stability of the block around the tunnel and the required anchorage force by taking the safety coefficient of the block anti-sliding stability as the evaluation index. Yang et al. [72] comprehensively analyzed the stability of surrounding rock of Longtan underground cavern group by taking the distribution characteristics of deformation field, stress field, plastic zone field and anchor shaft force as evaluation indexes. Wang et al. [73] studied the stability of surrounding rock of the underground power house cave group of Nuozhadu Hydropower Station through the stress field, displacement field and distribution of plastic failure zone obtained by finite element calculation. Yu et al. [74] studied the influence of phased excavation on the surrounding rock stress of hard rock cavern, and pointed out that full-section excavation is best for not too large cavern. Even for large cavern, the number of excavations should be reduced as far as possible if the construction conditions permit. Zhu et al. [75] used the area of the damaged area around the cave as an evaluation index to study the optimal construction scheme of the cavern group, and gave the calculation result of the plane problem. Xiao [76] proposed a three-dimensional finite element numerical analysis method for dynamic simulation of large-scale underground cavern construction process according to excavation procedures and anchoring construction methods. Based on rock mass failure characteristics and the principle of elastic–plastic mechanical dissipative energy, the excavation optimization method based on unloading, plastic zone, cracking, total failure volume and plastic dissipative energy evaluation indexes was proposed. The influence of excavation blasting on the stability of surrounding rock is also considered. Wang et al. [77] adopted three-dimensional elastic–plastic finite element analysis and took circumcave displacement as the optimization index to carry out an overall optimization of the construction of the underground workshop of Tianhuangping Pumped Storage Power Station. Zhu et al. [78] proposed a constitutive model of energy dissipation considering unit damage, plasticity and tensile failure, took plastic area and maximum horizontal displacement as optimization indexes, and combined with the stability problem of Xiluotu hydropower project tunnel group, made a comparative analysis of construction sequence optimization of four schemes. Chen et al. [79] applied the three-dimensional fracture damage finite element method and the basic principles of construction process mechanics to study the stability and optimal construction scheme of large underground chamber groups excavated in steep rock strata of Longtan Hydropower Station according to the size of the damaged zone of surrounding rock and the displacement of key points. An [80] proposed a weighted synthesis of the maximum displacement of the key point of the cav and the ratio of the difference between the volume of the damaged area and the design standard as an evaluation index to optimize and analyze the soft rock replacement scheme of the underground workshop in Shuibuya. Jiang [81] optimized the anchoring parameters of the underground workshop of Shuibuya by using the weighted comprehensive optimization indexes of comprehensive displacement around the tunnel, volume of the tensile zone, volume of the plastic zone, number of joints with large deformation, and support cost. Therefore, it is of great practical significance to carry out microseismic monitoring and stability analysis of large underground water-sealed oil caverns, and then analyze the mechanical characteristics of rock mass to comprehensively evaluate the stability of surrounding rock of underground water-sealed oil caverns. The complex geological conditions and special engineering structure of underground water-sealed oil cavern together decide that it needs to combine microseismic monitoring technology and reasonable mechanical model to make the prediction and prediction of deformation and instability more objective [42, 82].

This chapter summarizes a large number of research results made by predecessors, and concludes the feasibility of microseismic monitoring technology in the stability analysis of oil underground water-sealed oil caverns. In July 2014, the first set of 24 h all-weather real-time microseismic monitoring and analysis system for the stability evaluation of oil underground water-sealed oil caverns in China was successfully constructed in Jinzhou, Liaoning Province. Combined with the comparative analysis of numerical calculation methods, the potential instability area of rock mass and the degree of deformation damage of surrounding rock mass are determined from different angles, which undoubtedly has important reference significance for the stability evaluation and instability prediction of large underground water-sealed oil caverns.

2.3 Existing Problems

(1) Traditional monitoring methods for underground cavern groups typically focus on parameters such as deformation magnitude, deformation velocity, and deformation acceleration. However, this approach often overlooks critical "spatiotemporal precursory features" within the rock structure. It is challenging to elucidate the mechanism causing macro-instability resulting due to the gradual rupture of large underground water-sealed oil storage caverns through apparent deformation monitoring methods reliant on temporal judgements. In fact, the destabilization of large caverns is inevitably associated with internal microfracture. By the time surface deformation of the caverns becomes detectable, a considerable number of microfractures have already developed within the rock mass. Undoubtedly, the destabilization of large caverns is intricately tied to the formation of internal microfractures. Importantly, when surface deformation of the caverns becomes detectable, a significant number of microfractures have already formed within the rock mass. Consequently, the deformation-based displacement monitoring method is inadequately equipped to provide early warnings of cave instability. Microfractures within the rock structure serve as precursors to instability in large underground water-sealed oil storage caverns. Traditional displacement monitoring methods, relying on deformation indicators, struggle to detect the characteristics of rock microfracture activity preceding observable damage to underground water-sealed oil storage caverns. Moreover, it does not elucidate the fundamental mechanisms contributing to its destabilization. Traditional displacement monitoring (using deformation indicators) struggles to capture the characteristics of rock microfracture activity preceding macroscopic damage to underground water-sealed oil storage caverns. Furthermore, it fails to reveal the fundamental mechanisms behind its destabilization.

(2) The majority of contemporary numerical computational methods applied to large caverns are based on static analyses, making it challenging to capture the entire progression of internal damage leading to macroscopic instability in cavern rock formations. In particular, the numerical simulation of the initiation, development, expansion, and interaction of microfractures within rock bodies. The damage disturbance areas produced due to the significant unloading effects during the excavation of large underground water-sealed oil storage caverns play a pivotal role in governing cavern deformation and instability. Most contemporary numerical computational methods used for large caverns rely on static analyses, posing challenges in capturing the entire progression of internal damage that leads to macroscopic instability in cavern rock formations. Specifically, numerical simulations struggle to capture the initiation, development, expansion, and interaction of microfractures within rock formations. Specifically, numerical simulations struggle to capture the initiation, development, expansion, and interaction of microfractures within rock bodies. Damage disturbance areas resulting from significant unloading effects during the excavation of large underground water-sealed oil storage caverns significantly influence cavern deformation and instability. Nevertheless, there is an existing research gap concerning the excavation loosening zone within underground water-sealed oil storage caverns, especially in theoretical studies requiring the integration of innovative theories and methodologies. The numerical simulation method using micro-seismic (microfracture) activity can illustrate material property degradation resulting from excavation and illuminate the entire process of sudden rock damage. This aspect holds significant importance for characterizing mechanical strength parameters related to rock damage within the excavation loosening zone. The numerical simulation method based on micro-seismic (microfracture) activity can capture the degradation of material properties due to excavation and provide a comprehensive sequence of sudden rock damage. This aspect is crucial in characterizing mechanical strength parameters associated with rock damage within the excavation loosening zone.

2.4 Research Contents

(1) An optimization strategy is essential for deploying a spatial array of sensors within a micro-seismic monitoring system during both the excavation and operational phases of large underground water-sealed oil storage caverns. The intensive excavation unloading and operational activities induce stress variations within these caverns, resulting in the significant generation and accumulation of microfracture events. Consequently, this process considerably heightens the risk of instability within local or overall cavern regions, potentially leading to cavern failure. To address these challenges, an optimized micro-seismic monitoring system entails augmenting the number and type of sensors, specifically three-component velocimeters, within the water curtain tunnel. This

strategic enhancement aligns with the network arrangement plan and aims to cover the entire cavern and the water curtain system. Leveraging this approach during the excavation unloading phase involves considering the unique structural attributes of the cavern, thereby strengthening the monitoring system's effectiveness. Expanding on the previous analysis, our research delves into distinct studies concerning the equivalent wave velocity throughout the phases of intense unloading excavation and storage operations. The rock properties inside underground water-sealed oil storage caverns undergo continual changes due to excavation disturbances, prompting adjustments in the equivalent wave velocity model utilized by the micro-seismic monitoring system. Consequently, manual tapping serves as an essential process for iteratively refining the existing model, ultimately bolstering the precision and accuracy of the system's positioning.

(2) Investigation of the "precursor" mechanisms leading to instability in substantial underground water-sealed oil storage caverns utilizing micro-seismic monitoring technology. Micro-seismic activity serves as a direct indicator of damage within the underground water-sealed oil storage caverns. Consequently, by analyzing the micro-seismic activity characteristics and observed trends within the caverns during the excavation and operation phases, researchers explored the patterns of micro-seismic activity and potential indicators of instability throughout this period in the underground water-sealed oil storage caverns. This chapter proposes integrating micro-seismic monitoring with traditional conventional monitoring data and geological information within the broader framework of the underground water-sealed oil storage caverns structure. This integration aims to establish a correlation between areas susceptible to instability disasters and micro-seismic activity. Additionally, the study investigates the impact of microfracture development and evolution on significant structural damage within these caverns, specifically examining their response to substantial excavation unloading. It includes the interpretation of precursor phenomena of instability, such as changes in micro-seismic activity characterized by calming, shifting, and acceleration, observed within the underground water-sealed oil storage caverns.

(3) Numerical simulation method for large underground water-sealed oil storage caverns. Building upon the aforementioned research points and considering the correlation between acoustic emission and damage within the realm of physical mechanics, the study proposes an investigation into the time–frequency domain of microseismic signals monitored on-site. This investigation aims to determine corresponding source sizes and locations, establishing the initial damage threshold and critical damage threshold of the monitored rock mass in the area. To achieve this, a three-dimensional geomechanical model of the microseismic monitoring area within underground water-sealed oil storage caverns will be developed for a detailed characterization of the active microseismic region. By incorporating microseismic data obtained during the monitoring process (such as source energy, magnitude, moment magnitude, stress drop, seismic moment, etc.), a numerical model utilizing the Continuous-Discontinuous Element Method (CDEM) specific to underground water-sealed

oil storage caverns will be constructed. This numerical model aims to perform computational analyses on the evolutionary process of potential hazard areas. Consequently, it seeks to propose a dynamic stability evaluation method tailored for large underground water-sealed oil storage caverns.

This chapter is founded on the academic premise that the gradual development of microseismic activity within rock masses manifests essential characteristics indicating macroscopic instability and pre-failure signs. It commences by examining the deterioration of rock masses induced by excavation, closely analyzing their corresponding microseismic activity. The chapter integrates contemporary microseismic monitoring technology with numerical computational analysis to unveil the intrinsic dynamics and precursor patterns of rock mass damage within expansive underground water-sealed oil storage caverns. By utilizing alterations in the mechanical parameters of damaged rock masses monitored through microseismic monitoring, this study aims to assess the potential destabilization trends within large underground water-sealed oil storage caverns. Interpretation of physical information obtained from actual microseismic monitoring serves as a foundation to gauge the potential for destabilization. The primary focus lies in exploring the mechanisms governing microfracture evolution during intense unloading excavation disturbances and throughout subsequent operational phases in these caverns. This chapter endeavors to establish fundamental links between stress fields, microseismic activity, and construction activities, delineating the macroscopic failure processes within large underground water-sealed oil storage caverns. Consequently, it proposes a comprehensive set of mechanisms for microseismic monitoring stability evaluation and early warning prediction methods applicable throughout the entire life cycle of these caverns. As China continues to develop large-scale projects like underground water-sealed oil storage caverns, stability issues during excavation disturbances become increasingly pronounced. Therefore, further comprehensive research into the stability challenges encountered during both construction and operational phases of these caverns within complex stress environments holds substantial scientific and practical engineering value. The research aims to unveil the fundamental causes of instability within underground water-sealed oil storage caverns, elucidate conditions leading to instability-induced damage, and provide crucial scientific insights essential for dynamic stability assessment, early warning systems, and predictive methodologies.

2.5 Microseismic Monitoring of Jinzhou Underground Water-Sealed Oil Caverns

Jinzhou National Petroleum Reserve is located in the denudation hill on the southeastern edge of the Liaoxi Mountains, with a surface elevation of 15.3–43.0 m and a groundwater level of 10–25 m. It adopts water seal technology to store oil. The total designed capacity of Jinzhou underground water-sealed oil cavern is 3 million m^3. The oil storage facilities of the underground oil cavern mainly include 4 groups of 8 oil storage cavern arranged in parallel from east to west, water curtain system (including water curtain roadway and water curtain hole), water curtain roadway, construction roadway, oil inlet and outlet shaft and sealing plug, etc. (Fig. 2.1). The size of the oil storage chamber is 934 m × 19 m × 24 m (length × width × height), the elevation of the bottom of the cave is − 80 m, and the elevation of the roof is − 56 m. The dimensions are shown in Fig. 2.2a.

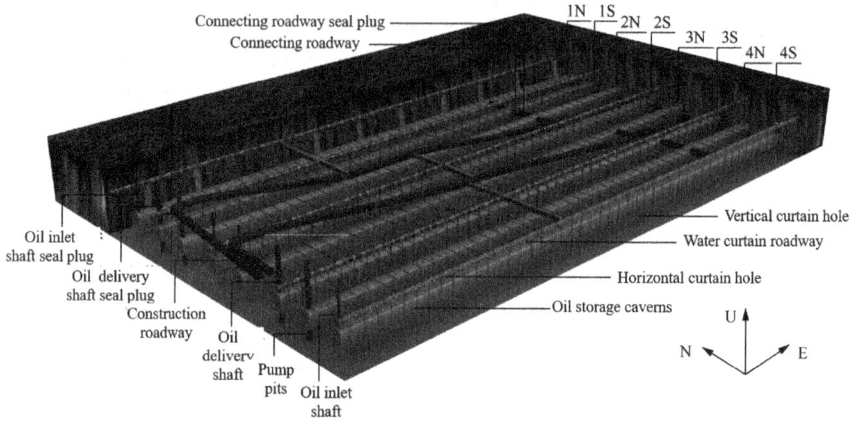

Fig. 2.1 Schematic diagram of an underground sealed oil vault in Jinzhou

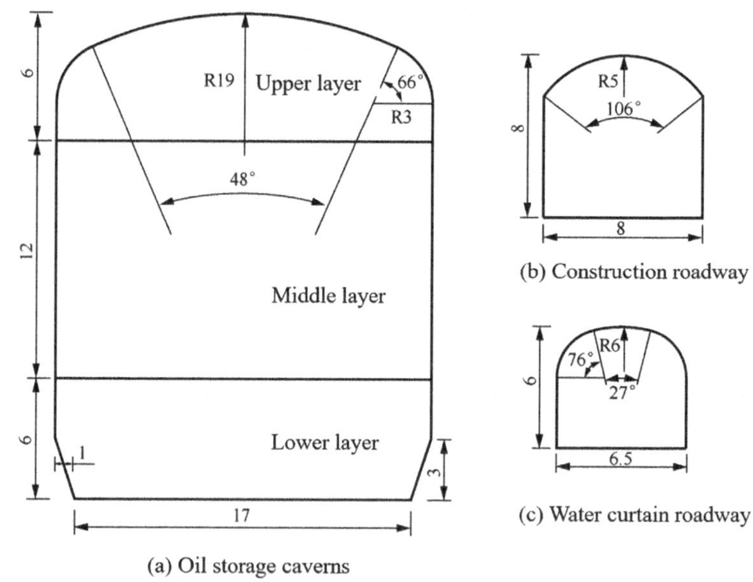

Fig. 2.2 Dimensions of oil storage chamber, construction roadway and water curtain roadway

The water curtain system is set 24 m above the top of the oil storage chamber to ensure the water sealing effect of the oil storage chamber. The water curtain system consists of two parts: one is the water curtain roadway arranged in the north–south direction and the east–west direction respectively, and the bottom elevation is − 32 m; The other part is the horizontal and vertical water curtain holes connected with the water curtain roadway. The five water curtain tunnels arranged in an east–west direction are 20 m beyond the scope of the oil storage chamber, and the water curtain tunnels 1 to 5 are arranged from north to south. The water curtain tunnels 6 arranged in a north–south direction are divided into three sections, and the north and south ends are 14.5 m beyond the scope of the oil storage chamber. The horizontal water curtain holes are arranged in an east–west direction, with a diameter of 100 mm and a spacing of 20 m between the two holes. Vertical water curtain hole diameter is 100 mm, hole spacing is 20 m, hole depth is 58 m, and bottom elevation is − 90 m [83]. The two construction roadways are 8 m × 8 m straight wall arched caverns (Fig. 2.2b), which are connected with the water curtain roadway 6 and the water curtain roadway (Fig. 2.2c), and the exit is located on the west side of the site.

2.5.1 Monitoring Regional Geological and Hydrological Conditions and Construction Plans

1. Monitor regional geological conditions

Jinzhou is located in the north temperate sub-humid monsoon climate zone. Winters are dry and cold, and summers are mild and rainy. Located in the southeastern edge of the western Liaoning Mountains, the geomorphic unit belongs to the low mountain denudation hills. The site is located in the Tianqiao granitic body of Yanshanian tectonic magmatic activity. The granite body is a crust-derived remelting type medium coarse-grained alkali-feldspar granite with mineral composition of potassium feldspar, plagioclase and quartz. The dark minerals are mainly hornblende in the north and biotite in the south. The plane of the rock mass is NM-SE elliptic; the area is about 7.6 km^2, and the intrusive contact surface is slow wave. There is no significant large fracture within 10 km of the site. The largest earthquake in the history of Jinzhou was magnitude 5. Since its own seismic detection, the largest earthquake that occurred near the site was a magnitude 2.4 earthquake near Huludao on August 20, 1988 (about 12 km away from the site in a straight line). The basic seismic fortification intensity of the site is 6 degrees. Interspersed with hornblende porphyrite, granite porphyry, diabase and other dikes. The thickness of rock mass above the strong regolith is 10–30 m.

According to the results of in-situ drilling, in-situ testing, in-hole testing and laboratory tests, the surface layer of the site is residual soil layer, and the underlying bedrock is medium coarse-grained granite. There is no obvious large-scale fault within 10 km of the field area, but there are more small-scale dikes, mainly diabase dikes, aplite and hornblende porphyrite. The strike, dip and dip Angle of the dikes are basically consistent with the granite joints, the dominant strike is NNW and NNE, and the dip Angle is mainly concentrated in 70°–80° and 40°–50°. The physical and mechanical property indexes of the granite rock mass measured are shown in Table 2.1 [84]. At the site of dike intrusion, the granite near the dike contact zone is generally broken due to hydrothermal alteration and joint development. Groundwater in the field area is dominated by bedrock fissure water, which is mainly divided into network fissure water in total regolith, strong regolith and vein fissure water under it according to its occurrence conditions [83]. Field flow statistics show that the average water inflow of a single oil storage cavern during the construction period is 295 m^3/d, and the average water inflow of the whole underground water-sealed oil cavern is nearly 4800 m^3/d. As of September 2014, during the construction period, the average groundwater level of the site decreased by 11.05 m, and the water level of the west and northeast end of the site decreased by 49.77 m, which was nearly 30 m lower than the design allowable water level. The decrease of groundwater level will weaken the strength of rock mass in dike and fracture zone, and cause the deterioration of stress state and stability of surrounding rock of underground water-sealed oil caverns.

Two boreholes were selected in the reservoir area to measure the ground stress by hydraulic fracturing method at different depths. The direction of the measured horizontal principal stress is near the east–west direction, which is relatively consistent with the overall direction of the maximum horizontal principal stress in the peripheral area of the measuring point [3]. The maximum and minimum horizontal principal stress is much larger than the vertical stress at the corresponding depth. The relationship of the three-way principal stress is as follows: the maximum horizontal principal stress SH_{max} > the minimum horizontal principal stress SH_{min} > the vertical principal stress SV, and the horizontal structural stress is the main stress. The test results are shown in Table 2.2 [84].

The rock types exposed in the reservoir area are mainly Yanshanian granite. Yanshanian granite is of medium coarse-grained structure and massive structure, which is widely distributed in the study area. The surface weathering is strong,

Table 2.1 Physical and mechanical property indexes of granite rock mass

Index	Medium regolith	Breezy regolith	Unweathered regolith
Saturated density/(g/cm^3)	2.59	2.61	2.61
Saturated uniaxial compressive strength/MPa	64.78	86.97	89.00
Compressive strength/MPa	6.63	7.50	8.63
Shear strength *C*/MPa	12.49	17.71	15.03
Internal friction angle/(°)	54.23	50.73	58.85
Softening coefficient	0.83	0.91	0.82
Elastic modulus/GPa	43.40	55.53	56.89
Poisson's ratio	0.21	0.24	0.21
Shear modulus/GPa	21.32	27.36	28.85

Table 2.2 Measurement results of hydraulic fracturing stress

Drill	Depth/m	Stress test result			Stress direction
		SH_{min}/MPa	SH_{max}/MPa	SV/MPa	
ZK1	78.5	5.95	11.71	2.06	N78°E
	79.5	6.35	10.51	2.09	
	89.5	4.87	8.25	2.35	
	102.00	6.58	11.55	2.68	
	115.50	7.11	12.47	3.04	N33°E
	132.50	7.68	12.64	3.48	N72°E
ZK3	28.00	6.23	10.20	0.74	N73°E
	41.00	5.57	9.94	1.08	
	71.00	4.68	6.67	1.87	N79°E
	85.00	4.62	6.80	2.24	
	101.00	6.37	10.54	2.66	N86°E
	115.00	7.10	11.46	3.02	

mostly in the shape of loose sand, and the thickness of strong weathering is generally 20–30 m. There are many dikes distributed in the reservoir area, but the scale is small. The general trend of dikes is NEE and NW. The lithology of the dikes is mostly granite-aplite and diabase. There are many joints in the granite layer, and the development of joints is closely related to dikes. Most of the faults in the study area are compressive and compression-torsion, and the fault scale is generally small, all of them are stable faults, and the water conductivity is poor.

Formation Lithology

According to on-site drilling, in-situ testing and on-site palm surface sketch, the study area is mainly exposed to medium coarse-grained granite, interspersed with diabase, hornblende porphyrite, aplite and other dikes. According to the core characteristics, RQD data statistics, wave velocity test results and ultrasonic imaging data, the strata in the site are comprehensively divided into residual soil layer, full regolith, strong regolith, medium regolith, breezed layer and unweathered layer according to the weathering grade and integrity of the rocks. Interspersed with hornblende porphyrite, granite porphyry, diabase and other dikes. The thickness of rock mass above the strong regolith is 10–30 m.

(1) Residual soil layer. Residual granite structure can be seen at the bottom of this layer. The original rock structure has been destroyed, has weathered into soil, dry drilling is easy to drill.
(2) Total regolith. Medium coarse-grained granite: yellow–brown, the main mineral composition is quartz, potassium feldspar, plagioclase, amphibole, and a very small amount of biotite, the original rock structure is basically destroyed, but still recognizable, with residual structural strength. Hand pinch is broken, can be dug spade pick, dry drilling can be drilled. Diabase: grayish white, clay-like,

structural composition is basically broken, but still recognizable, with residual structural strength. Hand pinch is broken, can be dug spade pick, dry drilling can be drilled.

(3) Strong regolith. Medium coarse-grained granite: light flesh red, medium coarse-grained granite structure, massive structure, the main mineral composition is quartz, potassium feldspar, plagioclase, hornblende and a very small amount of biotite, the core is broken, fragmentary, local more complete. Most of the original rock structure is destroyed, weathered cracks are very developed, can be excavated with spade picks, dry drilling is not easy to drill. Diabase: gray-green, diabase structure, block structure, the main mineral composition is pyroxene, plagioclase, hornblende, etc., the core is broken to more broken, the core is generally long and short columns or blocks, and the local sheet. Granite aplite: gray white, fine crystalline structure, massive structure, the main mineral composition of quartz, potassium feldspar, plagioclase, hornblende, and a very small amount of biotite. The core is generally long and short columnar, and partially fragmented.

(4) Medium regolith. Medium coarse-grained granite: light flesh red, medium coarse-grained structure, massive structure, the main mineral composition of quartz, feldspar and a small amount of biotite. The core is mainly in the shape of long columns, and the local is in the shape of short columns, flat columns and fragments, and the RQD value is 25–75%. Diabase: gray-green, diabase structure, massive structure, the main mineral composition is pyroxene, plagioclase, hornblende, etc., the core is broken to more broken, generally long and short columns or blocks. Granite aplite: gray white, fine crystalline structure, massive structure, the main mineral composition of quartz, potassium feldspar, plagioclase, hornblende, and a very small amount of biotite. Joint fissure is very developed ~ relatively developed, the core is generally long and short columnar, local fragmentation.

(5) Slightly regolith. Medium coarse-grained granite: variegated ~ light flesh red, medium coarse-grained structure, massive structure, the main mineral composition is quartz, feldspar and a small amount of biotite, amphibole and so on. The core is mainly long columnar, but partially short columnar and fragmentary, and the RQD value is 75% - 90%. Diabase: gray-green, diabase structure, block structure, the main mineral composition is pyroxene, plagioclase, hornblende, etc., the core is broken to be more complete, generally long and short columns or blocks. Granite aplite: gray white, fine crystalline structure, massive structure, the main mineral composition of quartz, potassium feldspar, plagioclase, hornblende, and a very small amount of biotite. The core is generally in the shape of long and short columns, and local fragments.

(6) Unweathered regolith. Medium coarse-grained granite: variegated ~ light flesh red, medium coarse-grained structure, massive structure. The main mineral composition is quartz, potassium feldspar, plagioclase, hornblende and a small amount of mica. The geological sketch surface is fresh, the joint cracks are not developed, and the fracture surface is mostly tightly closed, without filling. In some sections, diabase, granite aplite, dike thickness and distribution have

no obvious regularity. The core of this layer is mainly long columnar, and the RQD value is generally 75–100%. Comprehensive evaluation of rock integrity. Diabase: gray-green, diabase structure, massive structure, the main mineral composition is pyroxene, plagioclase, hornblende, etc., relatively broken, the core is short columnar. Granite aplite: gray white, fine crystalline structure, massive structure, the main mineral composition of quartz, potassium feldspar, plagioclase, hornblende, and a very small amount of biotite. Joint fissure development - not development, the core is generally long and short columnar, local fragmentation.

The excavation area of oil storage caverns is mainly composed of deep breezy layer and unweathered granite. There are two types of soft zones in the granite body. One is the local hydrothermal alteration zone formed by underground hydrothermal rising along the fissure, and most of the deep soft zones belong to this type. The second is the deep weathering zone caused by water due to joint development. The two kinds of weak zones have no obvious regularity with depth. Due to their irregular scattered distribution, the integrity of the aeolian layer and non-regolith granite is reduced to a great extent, and more attention should be paid to the evolution law of microfracture of rock mass in microseismic monitoring.

Fault

After field verification, it was found that faults were exposed to the surface in the near field area, but no signs of the quaternary system were found on the faults, and no obvious faults were found on the surface. It shows that these faults have not been active since the Quaternary period and are not active faults. According to geotechnical investigation and excavation geological sketch in the feasibility study stage, small faults and joint fracture zones are developed in the study area of microseismic monitoring. In the study area, a small fault with a strike of NE56°, compression-torsion, and a width of about 5 m was found.

Joints and Dikes

According to the measured data, the joint strike rose diagram, joint pole diagram and joint isodense diagram were drawn respectively (Fig. 2.3).

According to statistical data, the joint trend of – 80 to – 50 m elevation section is mainly NNW and NNE, and the main development is 270°∠30°, 150°∠75°, 85°∠60°, 330°∠30° and near-horizontal joints. According to core drilling and surface mapping data, the coarse-grained granite body in the study area is interspersed with diabase, aplite and other dikes. At the site of dike intrusion, the granite near dike and contact zone is generally broken due to hydrothermal alteration and joint development. The statistical results of the surveying and mapping of dikes show that the strike, dip and dip Angle of the dikes are basically consistent with the joint, and the strike is mainly NNW and NNE.

2. Monitoring regional hydrogeology

The site is located in the southern part of the river system on the west coast of the Bohai Sea near the estuary. In the west coast of Bohai Sea, groundwater flows from north to

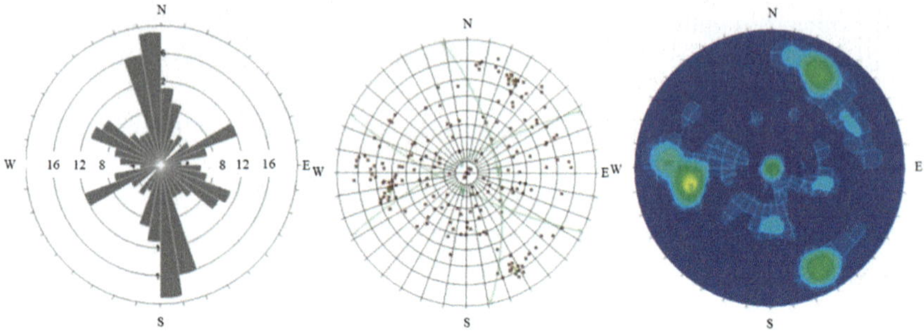

Fig. 2.3 Joint strike rose diagram, joint pole diagram and joint isodense diagram in the study area

south and from northeast to south into Liaodong Bay. Groundwater recharge is mainly composed of vertical recharge of atmospheric precipitation, surface runoff recharge and infiltration recharge of farmland irrigation. Drainage is mainly by evaporation and westward runoff. The quaternary overlying layer in the field area is thin, and the bedrock in some areas is exposed. The occurrence conditions of groundwater are mainly fracture water in bedrock such as weathering fracture, structural fracture and vein primary joint fracture. According to the different conditions of rock and soil body water and water-bearing medium, it can be divided into pore water of quaternary loose rock and bedrock fissure water. Water chemistry type heavy calcium carbonate sodium type, fresh water, good water quality. According to the permeability test data of the local rock mass, it is estimated that the water inflow of 400,000 m^3 caverns is about 120 m^3/d, which is not large. Because the structural fractures are mostly enlarged and connected by weathering, the distribution of the network fracture water is relatively uniform, with a unified groundwater level, slightly fluctuating with the terrain, and the hydraulic connection is relatively close. The vein fissure water is distributed along the joint fissure, and the upper part is connected with the network fissure water, which is directly supplemented by the upper network fissure water. The open fissure in deep rock mass can be supplemented by network fissure water, while the unopened fissure has no adverse effect on the underground water-sealed oil caverns, and the water sealing condition of the site is good. The type of groundwater in the microseismic monitoring area is mainly bedrock fracture water, which mainly exists in the form of network fracture water and vein fracture water. The water richness of groundwater is uneven, and the water yield of a single well is less than 100 m^3/d. In the depth range of cavern excavation, the main type of groundwater is bedrock fissure water. During the excavation, the groundwater level of the existing exploration holes was monitored. The groundwater level in the study area was near 20 m height, and the water level was relatively stable. Groundwater recharge is mainly through atmospheric precipitation and water curtain hole injection. The groundwater runoff mainly flows to the excavated water curtain tunnel 1, water curtain tunnel 2, water curtain tunnel 6 and the excavated oil storage tunnel 1N and 1S. The average flow rate of water curtain hole is 200 m^3 per day per chamber.

3. Monitoring regional rock mass quality

The main structural plane of the rock mass in the field is the joint plane, and the joint fissure in the NNW direction is mainly developed in the breezed layer and unweathered granite, and the spacing of the structural plane is mostly greater than 1 m. The RQD value of the breezed layer and unweathered granite is mostly greater than 75%, and the quality of the rock mass is good to very good. The overall strength of the rock mass is high, and the compressive strength is 79.1–148.5 MPa, which belongs to the hard rock. The structure type of rock mass is monolithic block structure, and the structural planes contain each other, and the integrity of rock mass is good. According to the "Engineering Rock Mass Classification Standard", most of the BQ values of the breezed layer and non-regolith granite are in the range of 450–650, which belongs to the I–II level of surrounding rock. According to the Code for Geological Investigation of Water Conservancy and Hydropower Engineering, the total score T of the aeolized layer and unweathered granite is 65–90, which belongs to the I–II level surrounding rock. According to the calculation of the Q system method proposed by Barton, the Q value of the aeolian granite is generally greater than 10, which belongs to Class II surrounding rock, and the Q value of the non-regolith granite is mostly greater than 40, which belongs to class I surrounding rock. The rock in this area belongs to the hard rock, the rock mass is relatively complete to complete, and the quality of the rock mass basically belongs to the class I to II. There are densely-distributed dikes and joint zones locally, and the quality of the rock mass is III–IV. In general, the evaluation results of different evaluation methods are quite consistent.

Combined with the construction experience of underground water-sealed oil caverns at home and abroad, Q system method is adopted in the classification of rock mass quality in most projects. The quality evaluation index of rock mass in Jinzhou underground water-sealed petroleum caverns is mainly Q value. According to the geological exploration and geological sketch data analysis, it can be concluded that the quality of rock mass in the microseismic monitoring research area is generally good, but the quality of rock mass in some areas is poor. The Q value of the surrounding rock of the water curtain roadway and the oil storage cavern is not uniform in the elevation direction, but it is generally greater than 10, and most of it is greater than 40, and the rock mass quality is good. The Q value is less than 10 or even less than 1 in local areas, especially in the joint development area and the location of dikes. For example, the Q value of rock mass generally decreases in the multi-group joint cutting area in the study area. In the exposed area of granite aplite dike, the average Q value is 1–4, or even lower. In the area of diabase dyke development, the Q value is generally less than 1, and the Q value of rock mass in local area is less than 0.1, and the quality of rock mass is very poor. In the microseismic monitoring area, diabase dikes and granitic aplite dikes are developed locally, which have a great influence on the quality of rock mass. In the following microseismic monitoring analysis, we will focus on it.

4. Construction overview

A microseismic monitoring and early warning system for underground water-sealed oil caverns is constructed based on the successful experience of microseismic monitoring system for engineering rock mass. Since the successful construction of the system on July 28, 2014, it has realized the continuous monitoring and analysis of the microfracture of rock mass within the range of 300 m × 200 m × 100 m in the underground water-sealed oil caverns. The microseismic monitoring system mainly covers the micro-fracture initiation and evolution of rock mass within 1 + 50 m–4 + 00 m distance of oil storage caverns 1N and 1S during unloading excavation, and delineates the potentially dangerous areas of surrounding rock of underground water-sealed oil caverns. The following introduces the construction of underground water-sealed oil caverns in the process of system monitoring. Microseismic monitoring is constructed in water curtain tunnel 1, water curtain tunnel 2 and water curtain tunnel 6, and the construction of water curtain tunnel has been completed. The coverage area of the monitoring system is mainly the mileage range of 1 + 50 m–4 + 00 m of the oil storage cavity 1N and 1S. The large section excavation of the oil storage cavity shall be carried out step by step in the upper, middle and lower layers, and smooth blasting shall be carried out around the oil storage cavity. When the fault fracture zone is encountered, the excavation section is reduced, and the excavation is carried out according to short footage and weak blasting. The upper, middle and lower parts of the south and north oil storage caverns are constructed at the same time. In order to prevent the group reservoir effect, the adjacent driving faces are staggered at a certain distance during construction to ensure construction safety. In the monitoring area, oil storage cavity 1N should be excavated in the middle layer, and oil storage cavity 1S should be excavated in the lower layer. The middle layer of the 1N chamber and the lower layer of the 1S chamber are drilled by a truck, and the optical explosion layer is reserved on both sides to carry out the middle groove excavation. The optical explosion layer is drilled by a three-arm truck, and the bottom of the lower layer is reserved about 0.5 m by artificial drilling and blasting to pick up the bottom to ensure that the bottom is smooth, so as to facilitate the loading of slag by forklift truck and the cleaning of the cave. Excavators are used to remove the loose rocks around the chamber after blasting and the remaining stone residue on the palm surface after slagging. Two front-loading side-unloading loaders with 3m3 bucket capacity are used to load slag at the same time to ensure the loading speed.

During the monitoring period (July 29, 2014 to August 24, 2014), the excavation footage schedule of oil storage caverns 1N and 1S is shown in Table 2.3. Combined with the data in the table, the evolution law of rock micro-fracture is analyzed next, and the relationship between the construction process and the micro-fracture evolution of the surrounding rock of the underground water-sealed oil tunnel is explained.

The excavation of each roadway and oil storage chamber of the cavern is carried out by truck drilling and drilling and blasting. Blasting cycle footage is determined for each roadway and cavern according to the level of surrounding rock. Full section excavation is carried out in the water curtain roadway. Due to the large excavation

Table 2.3 Schedule of excavation footage for 1N and 1S of oil storage caverns

Date	Oil storage cavern 1N Middle layer	Oil storage cavern 1N Lower layer	
July 29th	2 + 66 m	4 + 18 m	3 + 95 m
August 1st	2 + 54 m	4 + 13 m	3 + 95 m
August 5th	2 + 41 m	3 + 53 m	3 + 95 m
August 7th	2 + 30 m	3 + 39 m	3 + 95 m
August 9th	2 + 17 m	3 + 39 m	3 + 48 m
August 11th	2 + 06 m	3 + 25 m	
August 13th	1 + 95 m	3 + 08 m	
August 15th	1 + 84 m	2 + 91 m	
August 17th	1 + 73 m	2 + 73 m	
August 19th	1 + 63 m	2 + 55 m	
August 22th	1 + 52 m	2 + 20 m	
August 24th	1 + 42 m	1 + 96 m	

section, step-by-step excavation method is adopted for the construction of the oil storage cavity, and smooth surface-controlled blasting is implemented around the tunnel. The upper, middle and lower layers of the oil storage caverns are excavated at staggered distances at the same time, and the excavation heights are 6 m, 12 m and 6 m respectively (Fig. 2.4). For each layer, according to the level of surrounding rock, the excavation methods of full section, left and right half section, step method and other excavation methods are adopted. The excavation of the upper, middle and lower layers of the oil storage cavity shall be staggered at the same time, and the adjacent driving face shall be staggered at a certain distance to ensure the construction safety. As of July 28, 2014, all the water curtain roadway, construction roadway and connecting roadway have been excavated, the upper layers of 8 oil storage caverns have been excavated, the middle and lower layers are being excavated, and the entire underground water sealed oil caverns have up to 17 excavation faces, as shown in Fig. 2.4. The middle layer is blasted once every 3–4 days, with a footage of about 10 m, and each excavation volume is up to 4000 m^3. The lower layer has a daily footage of 5–10 m, forming construction characteristics of fast footage, multi-section, large excavation and strong unloading.

2.5.2 *Construction of Microseismic Monitoring System for Underground Water-Sealed Oil Caverns*

Considering a series of features of the water-sealed oil caverns, such as high side wall, large span, interlaced caverns, and multiple excavation sections [85], considering the excavation progress of the oil storage caverns 1N and 1S, combined with the requirements of the Canadian ESG microseismic monitoring system, and based

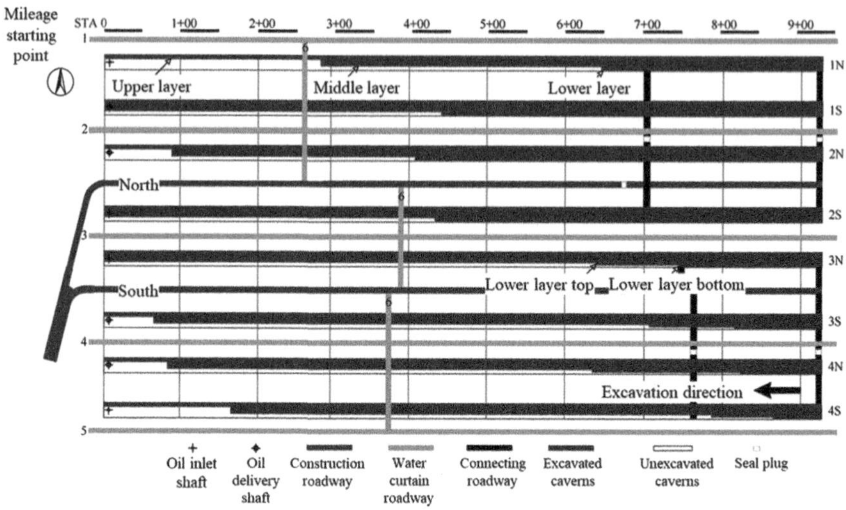

Fig. 2.4 Construction overview of an underground sealed oil tunnel in Jinzhou

on the spatial–temporal distribution law of micro-fracture of surrounding rock of the water-sealed oil caverns under the action of excavation and unloading, The microseismic monitoring system for underground water-sealed oil caverns is arranged, as shown in Fig. 2.5. The system mainly includes single-axis acceleration sensor, Paladin signal acquisition system, Hyperion data processing system, data analysis center and remote transmission system. The diameter of the acceleration sensor is 32 mm, the acquisition frequency range is 50–5000 Hz, the sensitivity is 30 V/(m s), the sampling frequency of the Paladin signal acquisition system is 20,000 Hz, 24-bit analog-to-digital conversion, and the threshold trigger is adopted. Considering the characteristics of high side wall, large span, numerous working faces, strong excavation and unloading of underground water-sealed oil caverns, and the requirements of microseismic monitoring system, as of July 28, 2014, the excavation of water curtain roadway has been completed, the middle layer of oil storage caverns 1N has been excavated to 2 + 80 m mileage, and the lower layer 1N and 1S have been excavated to 6 + 47 m mileage and 4 + 38 m mileage respectively. Therefore, the 6-channel microseismic monitoring system in the water curtain roadway can not only monitor the micro-fracture of rock mass under the action of heavy excavation and unloading of oil storage caverns in real time, but also ensure the safe and stable operation of the system. The monitoring range of the six acceleration sensors distributed in array type covers the area consisting of oil storage tunnel 1N and 1S, water curtain tunnel 1, water curtain tunnel 2 and water curtain tunnel 6, which is nearly 300 m in the north–south direction, 200 m in the east–west direction, and nearly 100 m in elevation (the enlarged area in Fig. 2.5). The microseismic monitoring system continuously monitored the surrounding rock microfracture events under excavation and unloading of underground water-sealed oil tunnel for 24 h, eliminated the noise events by

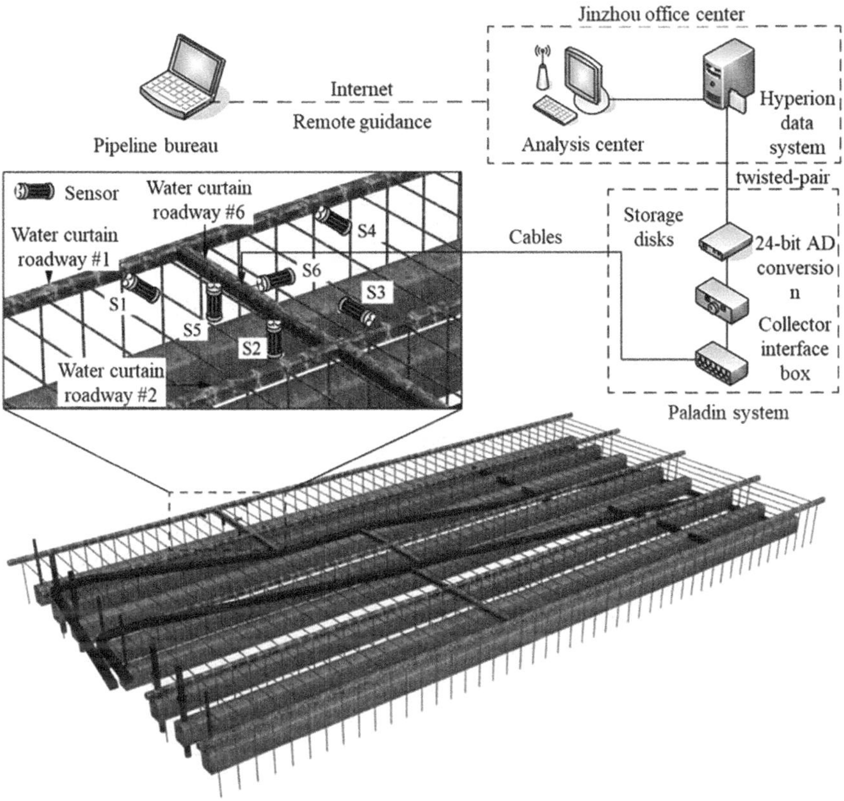

Fig. 2.5 Network topology of microseismic monitoring system

time–frequency analysis and filtering processing, identified and stored the effective microrupture event waveform and spectral map, and retrieved the spatio-temporal distribution, magnitude, energy and other source parameters of microseismic events.

2.5.3 Identification of Microseismic Location Error and Signal

The long-distance location damage error of microseismic is mainly determined by the spatial distribution of microseismic network, arrival accuracy, P-wave and S-wave velocity models, source location methods and other factors [86]. Many scholars have tried to reduce the long-distance location damage error of microseismic by increasing the number of sensors in the network and optimizing the location algorithm [87, 88]. However, due to the highly nonlinear and non-uniform nature of rock engineering, the precision of microseismic location error is different. For example, the long-distance positioning error of microearthquakes in mines can be controlled within 12 m, and by combining with optimization algorithms, the positioning error can reach 3.78 m [89]. The location accuracy of long-distance microseismic damage in downhole fracturing of oil fields is controlled within 50 m [90]. The microseismic location error in key areas of deep buried rockburst tunnels is less than 10 m [91]. After the optimization of the sensor array of the underground powerhouse and hydropower slope, the error of the positioning algorithm can be controlled within 10 m [22]. In large-scale rock engineering, when the source is located in the center area of the monitoring network, the time error determines the positioning error. When the source is far from the center of the network, the wave speed becomes the dominant factor affecting the positioning error [92].

Due to the special structural characteristics of underground water-sealed oil caverns, the propagation of elastic waves caused by micro-fracture of surrounding rock induced by excavation and unloading of underground water-sealed oil caverns is complicated. It is of great importance to accurately determine the propagation velocity of elastic waves in surrounding rock for the location of microseismic events and the calculation of focal parameters [86]. Based on the simplified overall wave velocity model, it is assumed that the propagation velocity of microseismic signal in the medium is equivalent to a global wave velocity model, and the wave velocity model of rock mass in the study area is determined by artificial fixed-point tapping test. According to the geological survey data of the cave [93], the elastic wave propagation velocity of the rock mass in the site is 4400–6100 m/s. Through the analysis of percussion waveform, 18 different wave speeds with P-wave speed of 4400–6100 m/s of the system were set respectively, and the average positioning error of percussion test at different time and wave speed on July 29, 2014 was calculated (Table 2.4). Figure 2.6 shows the error curve between the striking point and microseismic monitoring results at different P-wave velocities. It can be seen that when the P-wave velocity is 5200 m/s, the positioning error of the microseismic monitoring system reaches the minimum, 7.5 m, and the positioning error is all < 10 m, meeting the requirements for microseismic monitoring errors of engineering rock mass.

The excavation of caverns is carried out by drilling and blasting. Construction disturbances such as excavation blasting and air drilling produce a lot of noise signals, which results in the mixing of effective microseismic signals and noise signals. The accurate identification of microseismic signals is the prerequisite for the correct analysis of microseismic activity. Depending on the source of the event, the corresponding signal waveform will be quite different. By using the time–frequency signal analysis technology to convert the time series recorded by the instrument into frequency fast Fourier transformation (FFT), the amplitude spectrum and phase spectrum of the corresponding signal waveform can be obtained by transforming the digital signal sequence in the time domain to the digital signal sequence in the frequency domain. According to the amplitude-frequency characteristics of the signal, the actual signal types and characteristics of the Jinzhou underground water-sealed oil tunnel can be quickly analyzed and distinguished.

Table 2.4 Striking point positioning error at different P-wave speeds (July 29, 2014)

Number	Time	Wave velocity 4400 m/s		Wave velocity 4500 m/s		Wave velocity 4600 m/s		Wave velocity 4700 m/s		Wave velocity 4800 m/s		Wave velocity 4900 m/s		Wave velocity 5000 m/s		Wave velocity 5100 m/s		Wave velocity 5200 m/s	
		Error/m	Mean value/m	Error/m	Mean value/m	Error/m	Mean value/m	Error/m	Mean value/m	Error/m	Mean value/m	Error/m	Mean value/m	Error/m	Mean value/m	Error/m	Mean value/m	Error/m	Mean value/m
1	15:51:36	14.9	13.4	14.3	12.5	13.6	13.6	13.0	10.8	12.4	10.0	11.8	9.3	11.3	9.6	10.9	7.9	10.5	7.5
2	15:52:09	10.3		9.5		20.5		8.3		7.7		7.3		7.0		6.9		7.6	
3	16:00:30	13.8		13.4		13.0		12.7		12.4		12.2		18.1		12.0		11.9	
4	16:06:24	9.4		8.6		8.0		7.3		6.6		6.0		5.4		4.9		4.4	
5	16:08:04	10.8		10.1		9.3		8.7		8.0		7.4		6.8		6.2		5.6	
6	16:50:58	20.9		19.0		17.1		15.0		12.9		11.0		9.0		6.8		5.1	
Number	Time	Wave velocity 5300 m/s		Wave velocity 5400 m/s		Wave velocity 5500 m/s		Wave velocity 5600 m/s		Wave velocity 5700 m/s		Wave velocity 5800 m/s		Wave velocity 5900 m/s		Wave velocity 6000 m/s		Wave velocity 6100 m/s	
		Error/m	Mean value/m	Error/m	Mean value/m	Error/m	Mean value/m	Error/m	Mean value/m	Error/m	Mean value/m	Error/m	Mean value/m	Error/m	Mean value/m	Error/m	Mean value/m	Error/m	Mean value/m
1	15:51:36	10.2	8.2	10.1	7.9	15.7	8.7	14.1	8.3	12.9	8.1	12.2	7.9	12.1	8.4	12.0	8.2	11.9	8.1
2	15:52:09	9.1		9.0		8.9		8.9		8.9		8.9		8.9		8.8		8.8	
3	16:00:30	16.1		15.6		15.1		14.8		14.5		14.4		14.4		14.6		14.6	
4	16:06:24	4.2		4.2		4.6		4.9		5.1		5.1		5.2		5.0		5.1	
5	16:08:04	5.0		4.4		3.9		3.5		3.2		3.2		6.3		5.4		5.2	
6	16:50:58	4.5		4.3		4.1		3.9		3.8		3.6		3.5		3.3		3.2	

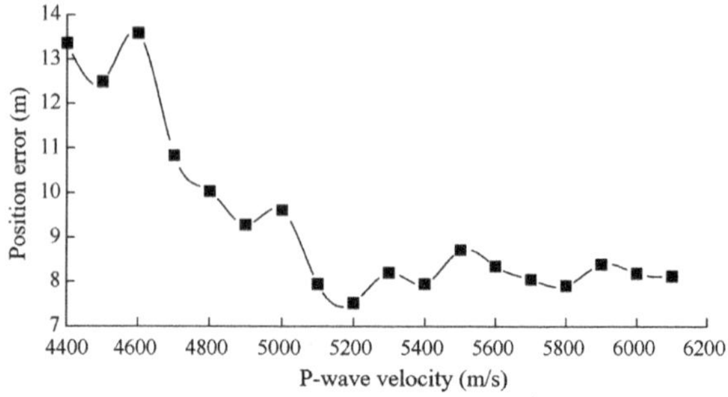

Fig. 2.6 Error curves of striking points and microseismic monitoring results at different P-wave speeds

$$X(k) = \sum_{n=0}^{N-1} x(n) \exp\left\{-j\frac{\pi}{2N}k\right\} \tag{2.1}$$

where, $x(n)$ represents the time series of the signal; $X(k)$ represents the frequency sequence of the signal; N is the number of sampling points; $k = 1, 2\ldots, N-1$.

The drilling and blasting method are used in Jinzhou underground water-sealed oil tunnel. The noise information such as excavation and blasting, mechanical slagging, air drilling and pumping machine generates vibration signals, which results in a lot of noise mixed with effective microseismic signals. Rapid identification of surrounding rock microfracture signals is the prerequisite for correct analysis of microseismic activity. The events detected by microseismic monitoring system during the construction of oil storage caverns in Jinzhou underground water-sealed oil caverns can be divided into three categories: excavation blasting event, surrounding rock micro-rupture event and other noise event by using time–frequency analysis technology.

1. Excavation blasting events

Figure 2.7 shows the typical excavation blasting waveform and amplitude-frequency curve of the oil storage chamber of the underground water-sealed oil caverns. The horizontal coordinate of the curve represents the time and frequency respectively, and the vertical coordinate represents the amplitude corresponding to the time and frequency respectively. The waveform in the time domain is relatively neat, and there are generally multiple peaks along the time axis, which is caused by the use of differential blasting technology in construction. The amplitude of the waveform measured by the sensor is about 4000 mV (generally greater than 1000 mV), and the moment magnitude is 1.6. The intensity of the blasting signal is large, the moment magnitude is generally greater than 0.5, the energy is generally greater than 1000 J, the vibration duration is short, about 300 ms, the attenuation is fast, the tail wave is not developed, and the vibration frequency distribution range is wide, but mainly the middle and low frequency vibration below 1000 Hz.

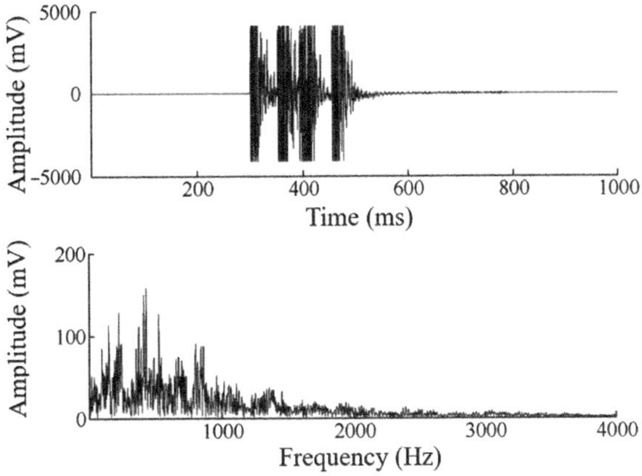

Fig. 2.7 Waveform and amplitude-frequency curve of excavation blasting

2. Microfracture event of surrounding rock

The microfracture waveform and amplitude-frequency curve of typical underground water-sealed oil caverns are shown in Fig. 2.8. No blasting operation was carried out when the signal was generated, the signal lasted about 350 ms, the attenuation was slow, the tail wave was relatively developed, and the amplitude was relatively small. As can be seen from the amplitudefrequency curve, the signal wave component is relatively simple, and the frequency band distribution is narrow, mainly distributed within 400 Hz. The moment magnitude of rock microfracture event in underground water-sealed oil caverns is generally less than 0.5, and the order of energy released is generally 100 J.

3. Other noise events

In the process of excavation and construction of underground water-sealed oil tunnel, various kinds of noise such as manual knocking, air drilling, slag extraction truck,

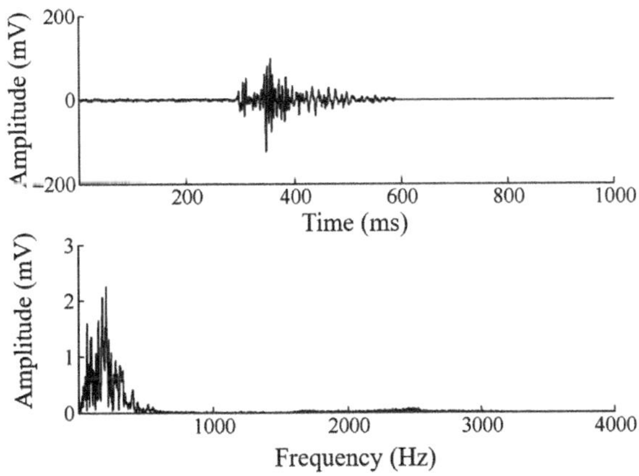

Fig. 2.8 Microfracture waveform and amplitude-frequency curve of surrounding rock

pumping machine, current oscillation and unknown signal are mixed in the wave of surrounding rock micro-fracture event. Figure 2.9a shows the waveform of manual knocking noise and its amplitude-frequency curve. The analysis shows that the waveform components are mainly high-frequency vibration above 1000 Hz, and there is an obvious peak in the time domain and rapid attenuation. Figure 2.9b shows the noise waveform and amplitude-frequency curve of air drilling in the excavation process. There are multiple peaks in the waveform, and the components of the waveform are similar to manual knocking noise events, which are also dominated by relatively high frequency vibration above 1000 Hz. Considering that the noise waveform of manual tapping and air drilling has obvious characteristics in the time domain, artificial filtering combined with Butterworth filter can be used to filter. Because the project adopts trackless slagging, the noise of the slagging car always exists in the whole monitoring process, and its waveform and amplitude-frequency curve are shown in Fig. 2.9c. Through a certain frequency range (1500–3000 Hz) band-stop filter can filter out most of the slag car noise. In addition, the waveform and amplitude-frequency curve of pump noise and current oscillation noise are shown in Fig. 2.9d and e respectively, and their characteristics are relatively obvious in the time domain. Suitable band-pass or band-stop filters can be selected for noise filtering. It should be pointed out that for the unknown signal waveform shown in Fig. 2.9f, the amplitude and frequency curve analysis show that such signals have a wide frequency band distribution and are mainly low-frequency vibration below 1000 Hz, which is similar to the amplitude and frequency curve of excavation blasting waveform. Considering the characteristics of elastic wave propagation in rock mass, high frequency attenuation is fast and low frequency attenuation is slow, it is inferred that the elastic wave is unknown excavation blasting waveform far away from the sensor.

2.5.4 Spatial Distribution of Microseismic Activity in Underground Water-Sealed Oil Storage Caverns

Since the establishment of the microseismic monitoring system on 28th July 2014, by 26th August 2014, the system had recorded a total of 492 valid events, consisting of 361 microseismic events, 123 blasting events, and 8 percussion events. During intense excavation and unloading, micro-fractures in the cavern surrounding rock primarily concentrate within the space formed by water curtain roadways 1 and 2, along with oil storage caverns 1N and 1S, delineating two concentrated strips of distribution. One strip is positioned approximately 30 m east of water curtain roadway 6 (at the 2 + 63 m mark), exhibiting predominantly horizontal distribution, while the other strip spans from 2 + 40 m to 2 + 60 m between oil storage caverns 1N and 1S, characterized by a near-vertical distribution (refer to Fig. 2.10a). These two zones primarily accumulate microseismic events, with the highest event density observed between 2 + 40 m and 2 + 60 m mark within the oil storage cavern 1N (refer to Fig. 2.10b). This region signifies a potentially hazardous area prone to rock destabilization during the

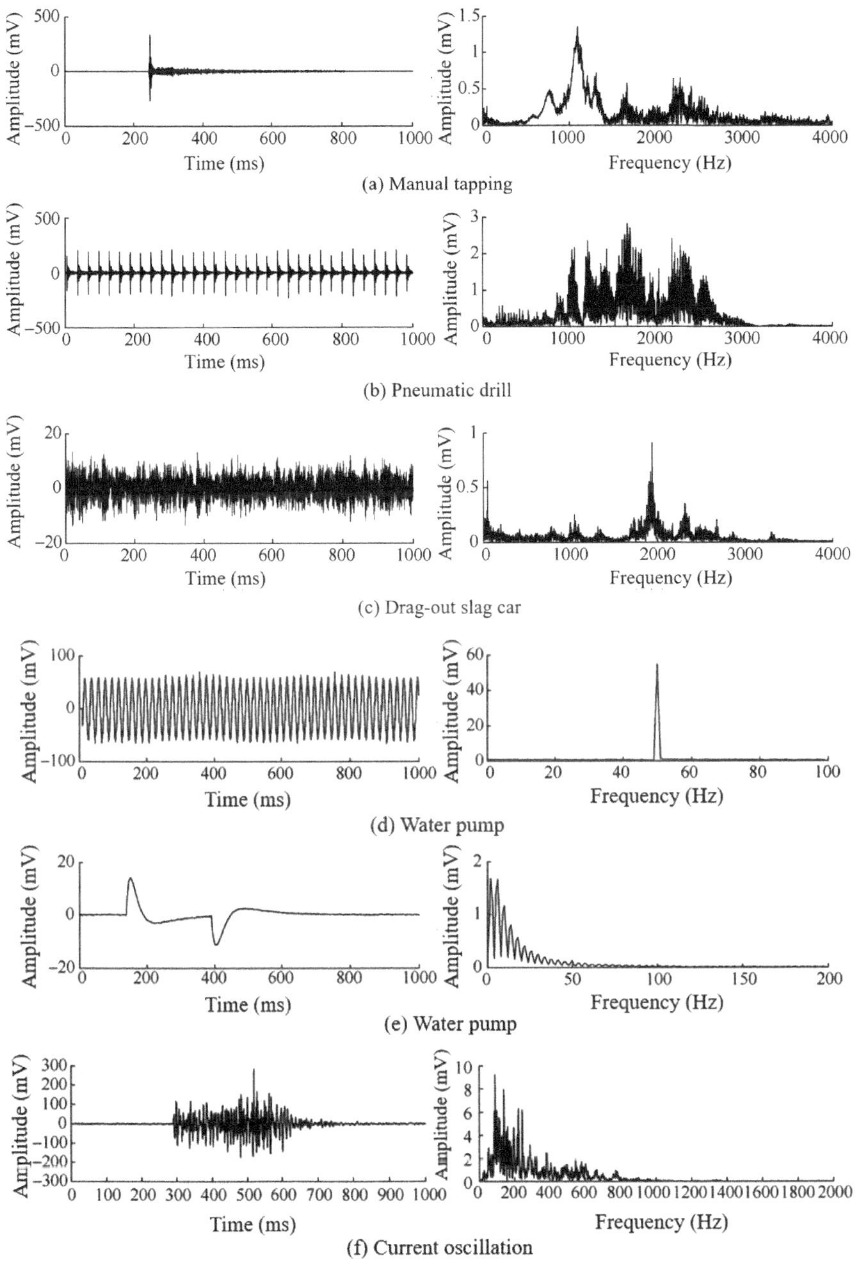

Fig. 2.9 Waveform and amplitude-frequency curve of other noise events

construction process. Upon analyzing the data collected from the site investigation, it was found that the granite in this area contains interspersed diabase dikes, granite aplite dikes, and other types of dikes (refer to Fig. 2.11). Furthermore, the primary concentration area of microfractures in oil storage caverns 1N and 1S aligns more closely with the distribution pattern of diabase dikes within the surrounding rock (240°∠85°). It indicates that the damage of dikes in the surrounding rock is more intense under the strong excavation and unloading of the oil storage cavern.

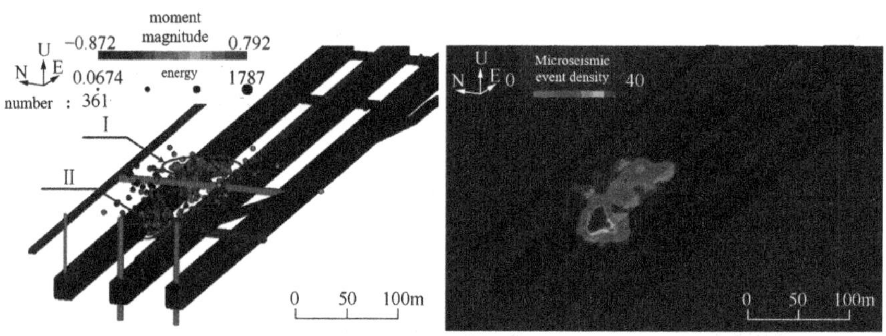

(a) Microseismic events distribution (b) Density of microseismic events

Fig. 2.10 Spatial distribution of microseismic events in underground water-sealed oil storage caverns (July 28, 2014–August 26, 2014)

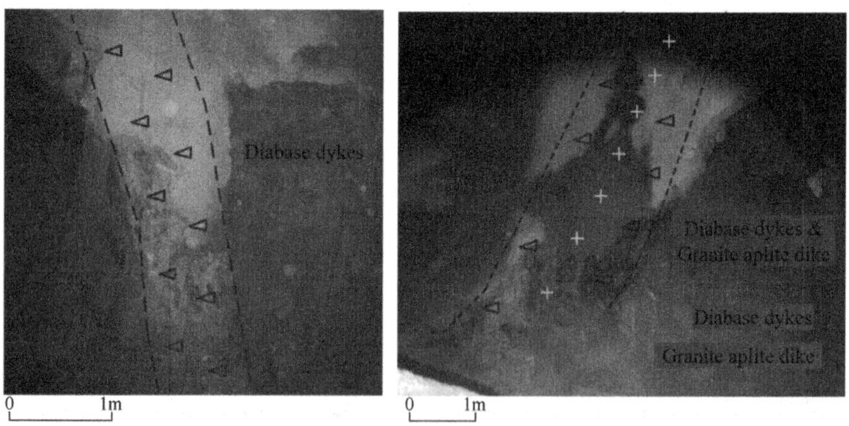

(a) 1N 2+45m South side wall (b) 1S 2+80m North side wall

Fig. 2.11 Exposure of diabase dikes in parts 1N and 1S of oil storage cavern

2.5.5 Time Distribution of Microseismic Activity in Surrounding Rock of Underground Water-Sealed Oil Storage Caverns

Following the successful installation and operation of the microseismic monitoring system for the Jinzhou underground water-sealed oil storage cavern on 28th July 2014, the system has functioned effectively, experiencing only sporadic power outages at the site on 2nd, 6th, and 20th August. It has effectively accomplished continuous microfracture monitoring across the study area encompassing oil storage caverns 1N, 1S, and water curtain roadways 1, 2, and 6 (an area detailed in Fig. 2.5). Figure 2.12 illustrates the microseismic activity within the cavern surrounding rock and the pattern of cumulative energy release changes over time throughout the monitoring period. The equivalent excavation footage was determined through specific steps. Initially, the extent of blasting damage from the cavern excavation is computed [94]. Subsequently, the extent of blasting footage impacting the monitoring area is quantified and matched to the lower excavation footage, considering the relationship between the area of each excavation section and the lower excavation surface area of the oil storage cavern. For instance, if the middle excavation surface is 12 m high, while the lower excavation surface is 6 m high, then 1 m of the middle excavation footage corresponds to 2 m of the lower excavation footage. By 15th August, the overall count of monitored microseismic events remained relatively low, averaging approximately 10 events per day. However, after 16th August, there was a substantial surge in micro-fractures within the surrounding rock, resulting in an average of nearly 20 microseismic events per day. Post 15th August, there was a notable surge in the equivalent excavation footage. This indicates a clear correlation between microfractures in the surrounding rock and excavation unloading, with a direct relationship where increased excavation footage corresponds to heightened microfracture activity in the surrounding rock. Analysis of the cumulative energy release curve for microseismic events reveals a consistent correlation between the energy release from microfractures in the surrounding rock and the impact of excavation unloading on microseismic activity. Furthermore, upon examining the excavation footage alongside microseismic activity and its cumulative energy release distribution, it is evident that a substantial excavation footage for the oil storage cavern corresponds to a pronounced excavation unloading effect. Interestingly, immediate high microseismic activity in the surrounding rock is not observed in these cases. This suggests that during excavation unloading, the microfracture response of the surrounding rock in the oil storage cavern undergoes a cycle involving energy accumulation, release, and transfer, showcasing a temporal effect [95].

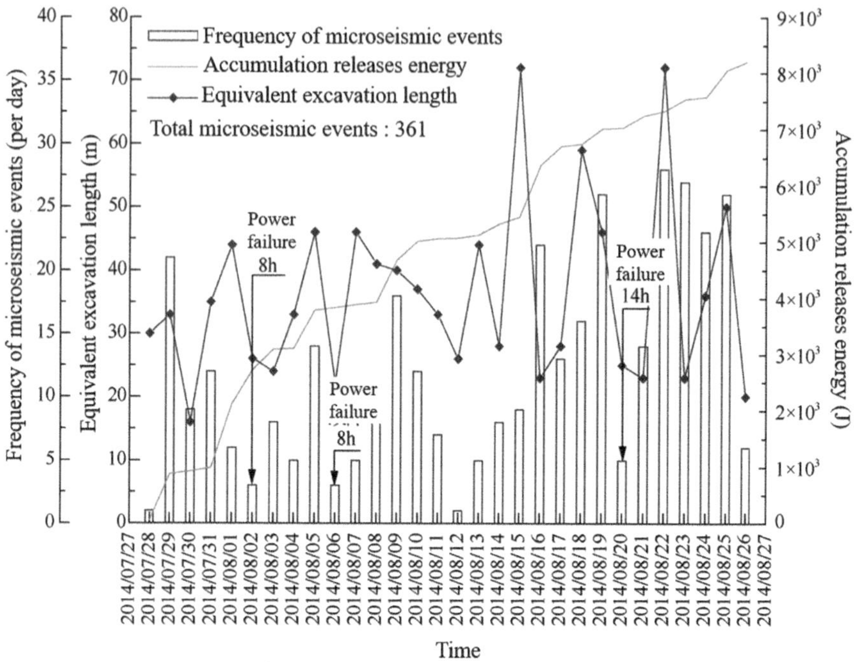

Fig. 2.12 Microseismic activity of surrounding rock and its cumulative release energy change with time during the monitoring period

2.5.6 *Local Stability Analysis of Underground Water-Sealed Oil Storage Caverns*

Figure 2.13 illustrates the distribution of microseismic events and a density cloud map depicting the surrounding rock under excavation unloading. The density cloud map of microseismic events indicates the degree of aggregation of rock microfractures: higher density (darker color) signifies greater rock microfracturing and relatively higher microseismic energy and the pre-fracture stress value of the rock body [95]. Initially, during the monitoring period, excavation at the middle layer of oil storage cavern 1N reached a depth of 2 + 66 m, leading to the emergence of microfractures near the working face and an inclination to congregate at the north side wall (Fig. 2.13a and b). By 9th August, the working face at the middle layer of oil storage cavern 1N progressed to the 2 + 17 m mark, while the lower layer of 1S underwent excavation to a depth of 3 + 39 m. During this phase, micro-fracture events congregated at the south side wall of 1N, spanning from 2 + 45 m to 2 + 55 m. Simultaneously, micro-fractures gradually developed in the surrounding rock, extending from 2 + 70 m to 3 + 10 m (Fig. 2.13c). Intense excavation unloading induces a redistribution of stress in the surrounding rock, leading to stress transfer towards the contact zone between diabase dikes and granite aplite dikes, consequently fostering increased aggregation of microfractures (Fig. 2.13d). Upon excavating the middle layer of oil storage cavern 1N to a depth of 1 + 52 m, microfracture aggregation within the oil storage chamber displays a banded distribution pattern (Fig. 2.13a and e). Concurrently, multiple instances of rockfall phenomena were observed on the south side wall of the middle layer of oil storage cavern 1N, spanning from 2 + 45 m to 2 + 55 m.

Figure 2.14a–c depict the microseismic event density cloud maps illustrating the surrounding rock of the underground water-sealed oil storage cavern reservoir during three distinct monitoring periods. These maps represent the degree of micro-fractures aggregation within the surrounding rock. Typically, greater event density (represented by darker shades) correlates with increased microfractures and subsequent energy release [53]. At the onset of the monitoring period, the middle layer of the 1N oil storage cavern was excavated to 2 + 66 m mark. Micro-fractures emerged near the

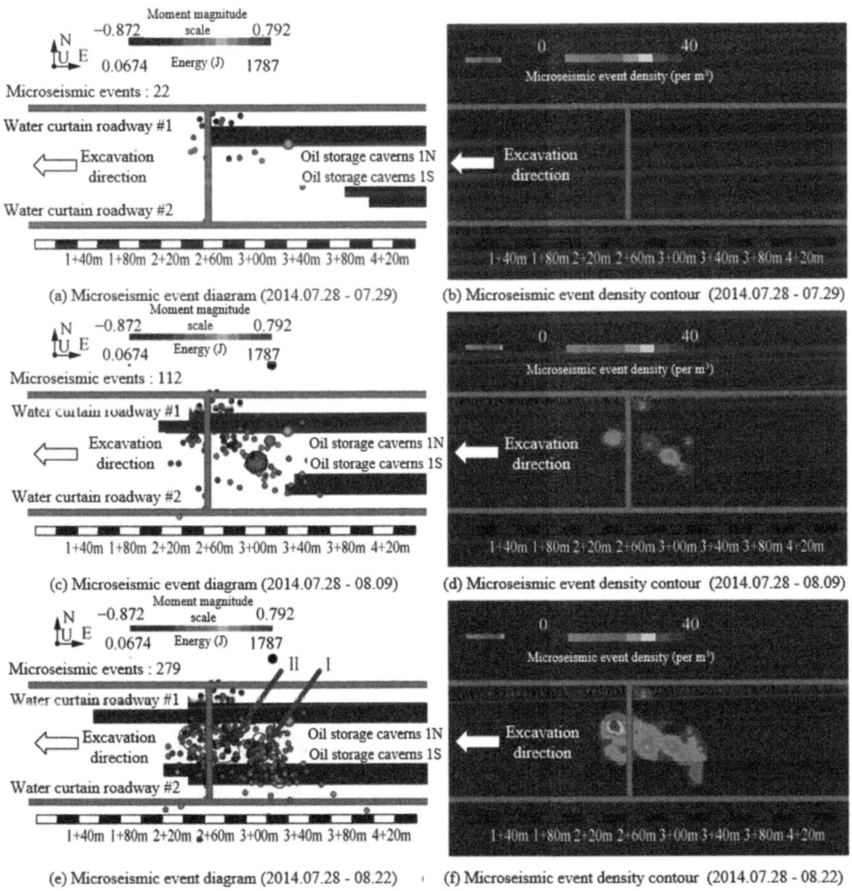

Fig. 2.13 Distribution and density cloud map of microseismic events in surrounding rock of underground water-sealed oil storage caverns under unloading excavation

working face and exhibited a tendency to gather along the north side wall (Fig. 2.14a). By August 9th, the working face of the middle layer in the 1N oil storage cavern had progressed to 2 + 17 m. Concurrently, the 1S lower layer had been excavated to a depth of 3 + 39 m. Notably, micro-fracture events on the south side wall of the 1N cavern, spanning from 2 + 45 m to 2 + 55 m, coalesced, forming a zone characterized by a significant energy release. Simultaneously, during this period, micro-fractures gradually developed in the peripheral rock, extending from 2 + 70 m to 3 + 10 m (Fig. 2.14b). As illustrated in Fig. 2.14c, the continuous advancement of the working face leads to a significant increase in micro-fractures within the surrounding rock near oil storage cavern 1S. Subsequent field investigations revealed the emergence of a "collapse cavity" measuring about $5m^3$ on August 9th along the south side wall in the middle layer of oil storage cavern 1N, specifically within the 2 + 45 m to 2 + 55 m range (Fig. 2.14d). This occurrence stems from the unloading during the excavation of the oil storage cavern, resulting in heightened stress in the surrounding rock that surpassed its strength limit, subsequently releasing elastic strain energy. Simultaneously, the excavation unloading prompted the transfer of the high-energy zone within the surrounding rock to oil storage cavern 1S.

Subsequently, micro-fractures spanning from 2 + 45 m to 2 + 55 m intensified, culminating in the occurrence of localized instability characterized by multiple "rockfalls" (Fig. 2.14e). Analysis correlating with Fig. 2.12 indicates a relatively tranquil phase in microseismic activity and energy release within the surrounding rock before a substantial surge in cumulative energy on August 9th, aligning with findings on acoustic emissions in existing literature [96]. This observation suggests that continuous excavation of the oil storage cavern led to escalated damage to the diabase dikes within the surrounding rock. Following this relatively calm period, the unstable occurrence of "rockfalls" transpired multiple times (Fig. 2.15).

Microseismic monitoring operations commenced on July 28th. Micro-fractures appeared in the area near the 2 + 45 m to 2 + 55 m segment of the south side wall of oil storage cavern 1N. As excavation progressed to 241 m mark in 1N, the number of microseismic events in this area increased to 12. The corresponding cloud map exhibited a substantial rise in event density, indicating over-excavation in this zone (Fig. 2.16). Focusing on the rupture process of the rock body from 2 + 45 m to 2 + 55 m on the south side wall of the middle layer of oil storage cavern 1N, we replicated this process by considering the evolution of microfracture damage within the rock body. This replication provided an initial illustration of the construction-related perturbation influence on the damage and stability of the surrounding rock during the excavation of the oil storage cavern.

The construction of an underground water-sealed oil storage cavern constitutes a comprehensive project. During the extensive excavation process, the stress within the cavern surrounding rock continuously undergoes dynamic adjustments. By consistently monitoring the evolution of micro-fractures within the disturbed area of the rock body, we gain insights into the nature of damage incurred by the cavern structure. Consequently, this process aids in determining the stress development trends within the rock body during excavation and enables speculation about potential water-conducting pathways. These findings offer valuable references for subsequent

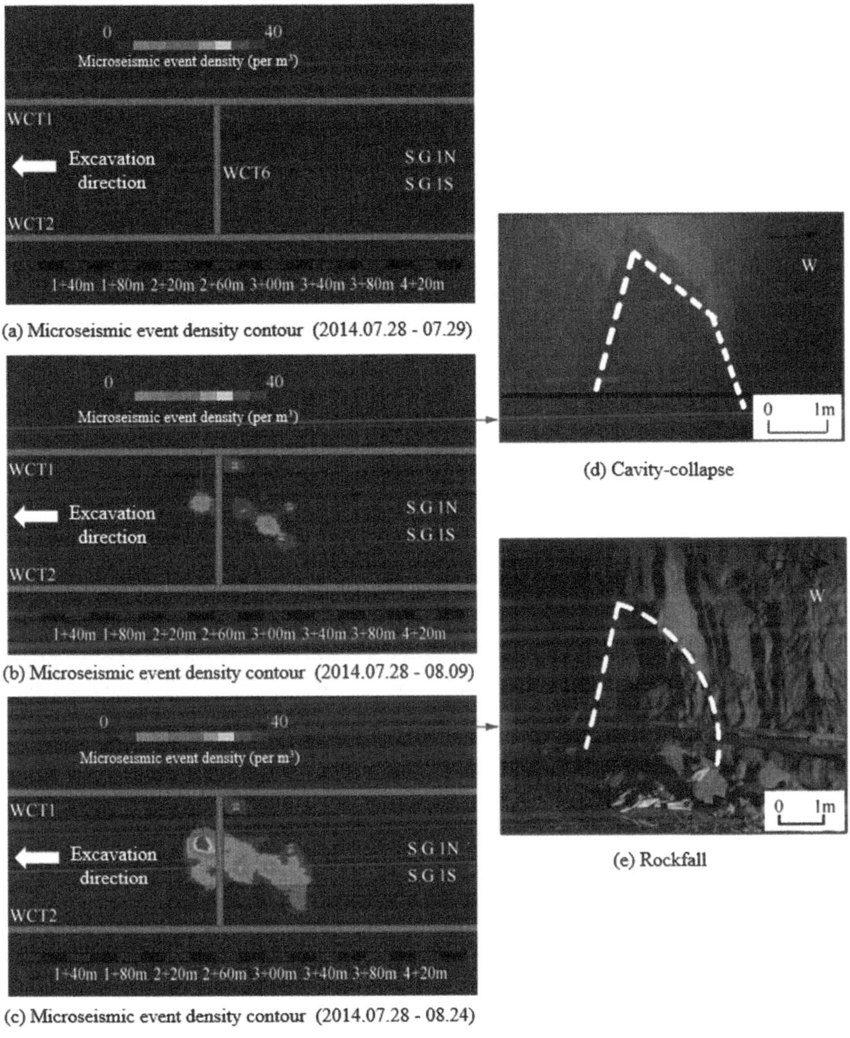

Fig. 2.14 Microseismic event density cloud map of surrounding rock in the study area under excavation unloading

stages involving blasting excavation of the main cavern, as well as for implementing grouting support and construction in later phases.

Figures 2.17a–l and 2.18a–l display the top and right views of microseismic event density maps within the designated monitoring area along the southern wall at the mid-level of reservoir cavern 1N, spanning from 2 + 45 m to 2 + 55 m. The color gradient from grey to black signifies varying densities of micro-fracture events within the rock mass. In stress analysis, as the cloud map approaches black, it indicates increased stress in the rock mass compared to its original strength, hence suggesting

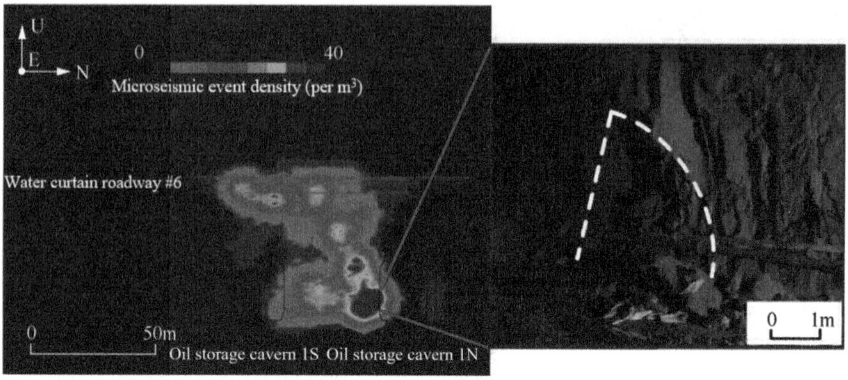

Fig. 2.15 Collapse of the south side wall of the oil storage cavern with a distance of 1N 2 + 45 m

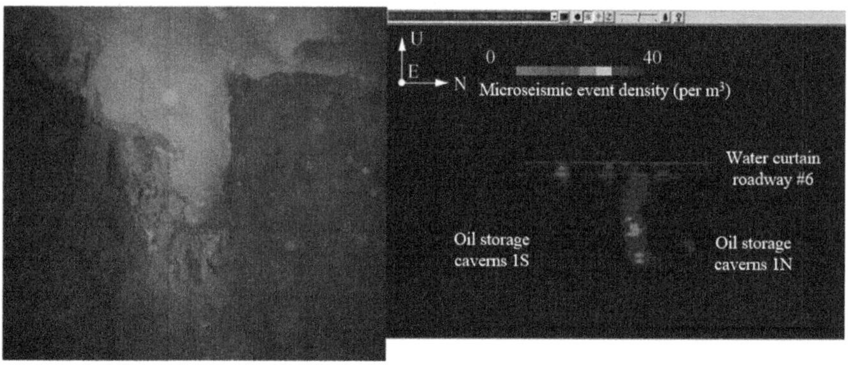

Fig. 2.16 South side wall of 1N Middle level 2 + 45 m mileage (taken on August 6, 2014)

a higher likelihood of fracture. Analysis of Figs. 2.17 and 2.18 reveals the initial occurrence of rock micro-fractures in the upper region at the 2 + 80 m mark along the northern wall of oil storage cavern 1N, potentially instigated by fissures resulting from blasting activities at the 2 + 66 m mark. Excavation continually alters stress within this monitored zone. Notably, rock micro-fractures between 2 + 45 m to 2 + 55 m along the mid-level of the southern wall in oil storage cavern 1N are consistently accumulating. Some rock energy dissipates through fractures, while the remainder transfers deeper into the dike system. The analysis results indicate a correlation between the micro-fracture damage area within the microseismic monitoring zone and the distribution pattern of diabase dikes in the area. Furthermore, the micro-fracture damage zone within the rock body, situated between caverns 1N and 1S, notably expands close to the 2 + 40 m mark and exhibits a penetrating trajectory. Examination of geological exploration data reveals the presence of developed granite aplite dikes near the 2 + 40 m mark within oil storage cavern 1S. The contact interface between the dikes and granite exhibits considerable fragmentation, aligning with heightened microseismic activity, potentially serving as a conduit for groundwater flow. This discovery holds significant implications for guiding grouting design and construction.

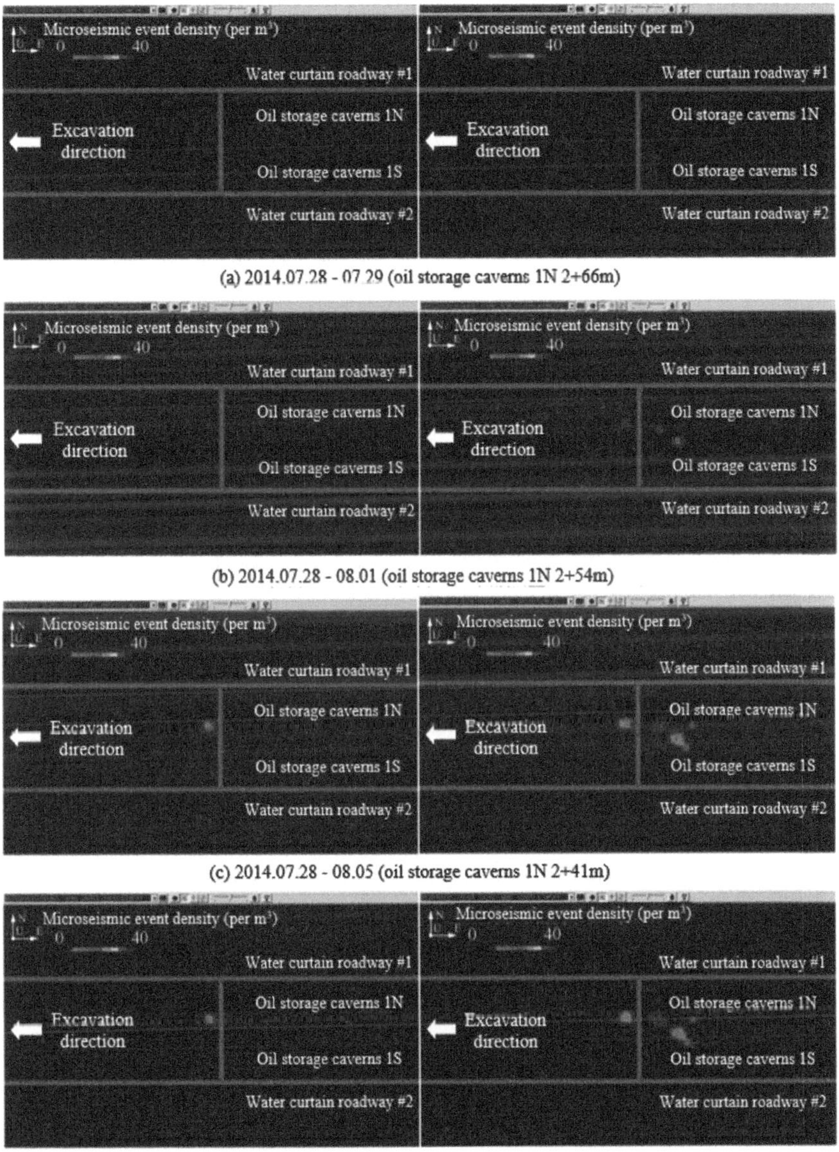

Fig. 2.17 Density map of microseismic events in the case area and microseismic monitoring research area (top view)

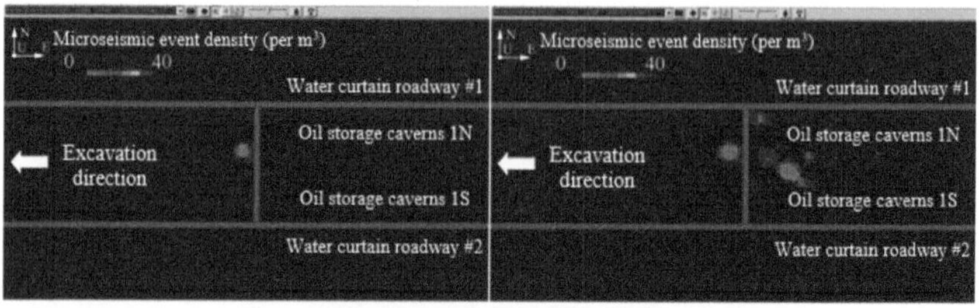

(e) 2014.07.28 - 08.09 (oil storage caverns 1N 2+17m)

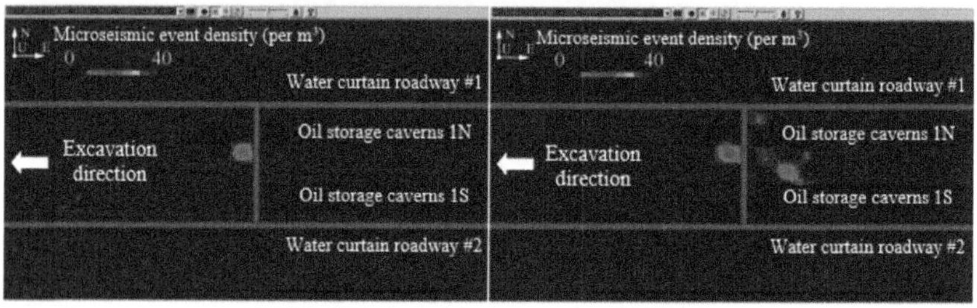

(f) 2014.07.28 - 08.11 (oil storage caverns 1N 2+06m)

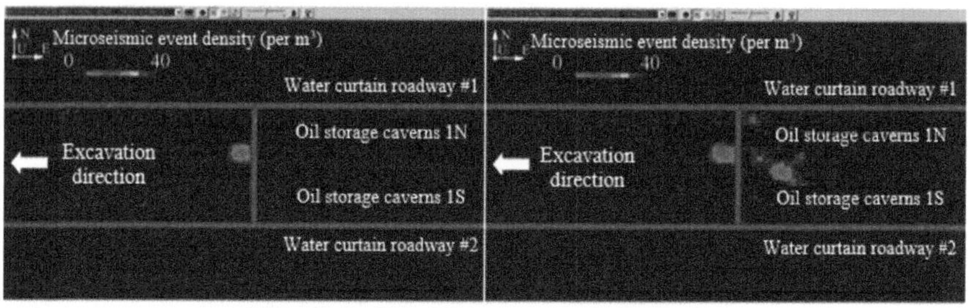

(g) 2014.07.28 - 08.13 (oil storage caverns 1N 1+95m)

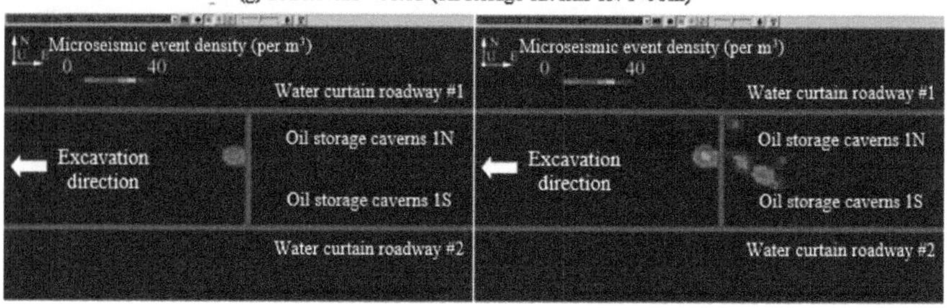

(h) 2014.07.28 - 08.15 (oil storage caverns 1N 1+83m)

Fig. 2.17 (continued)

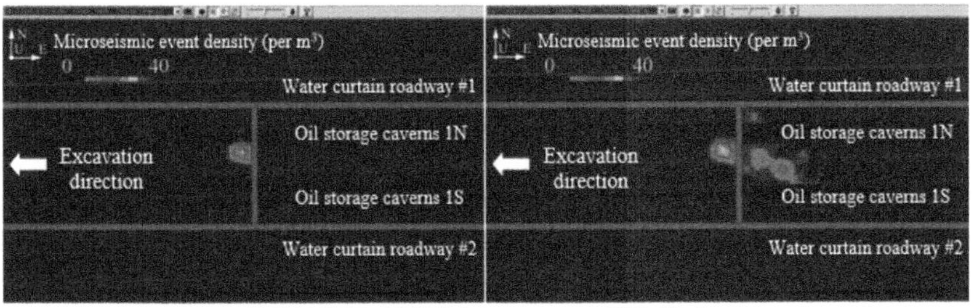

(i) 2014.07.28 - 08.17 (oil storage caverns 1N 1+73m)

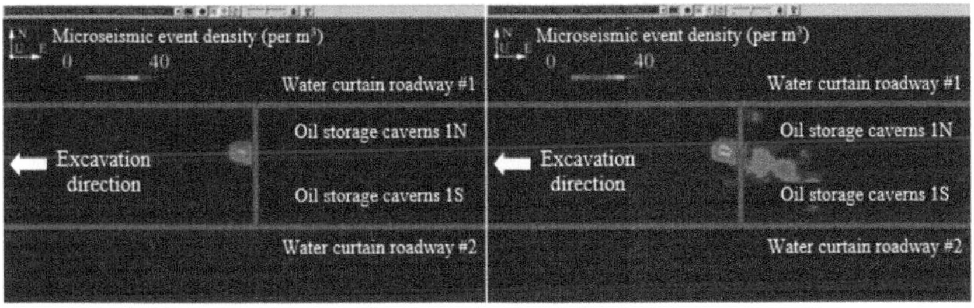

(j) 2014.07.28 - 08.19 (oil storage caverns 1N 1+63m)

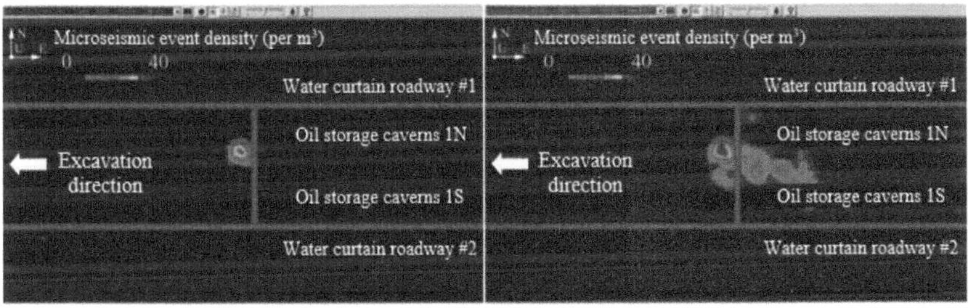

(k) 2014.07.28 - 08.22 (oil storage caverns 1N 1+53m)

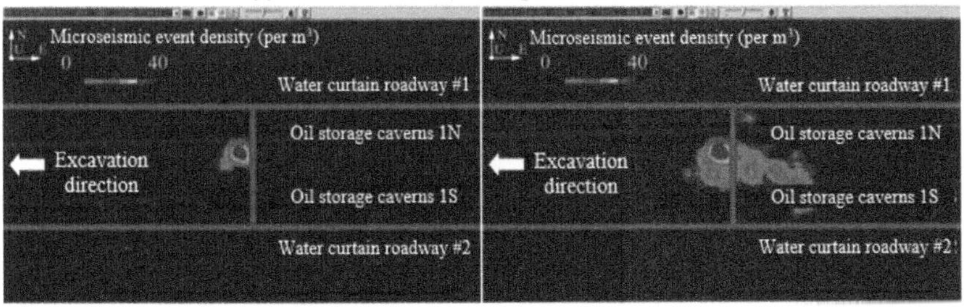

(l) 2014.07.28 - 08.24 (oil storage caverns 1N 1+43m)

Fig. 2.17 (continued)

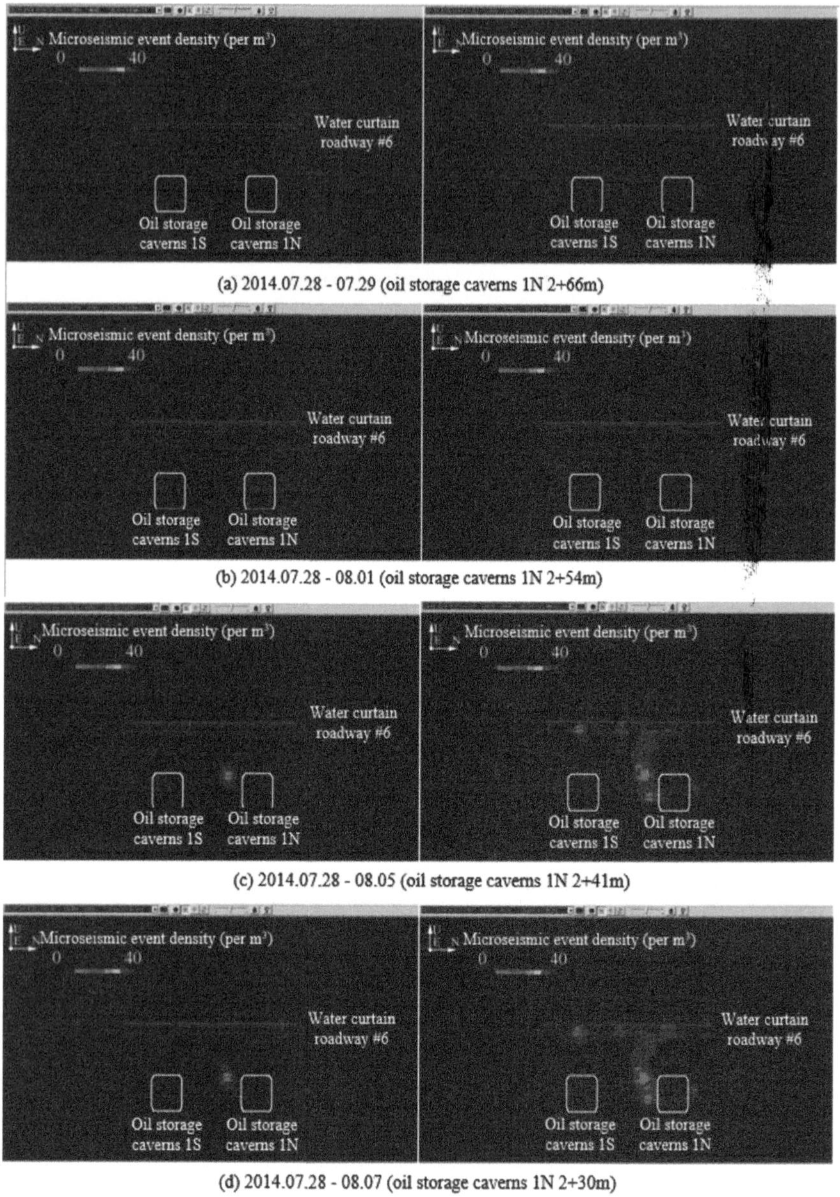

Fig. 2.18 Density map of microseismic events in the case area and microseismic monitoring research area (right view)

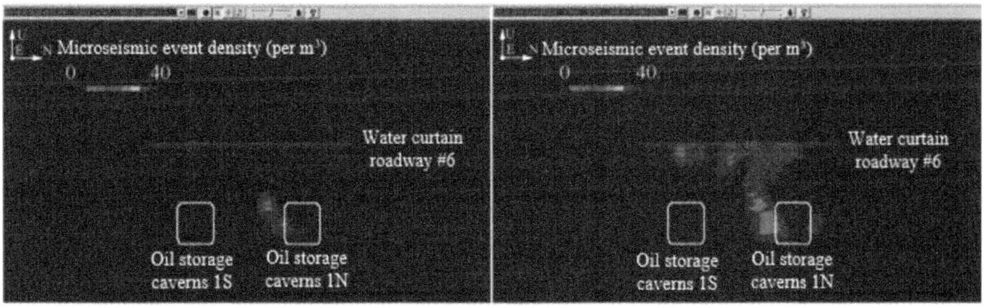

(e) 2014.07.28 - 08.09 (oil storage caverns 1N 2+17m)

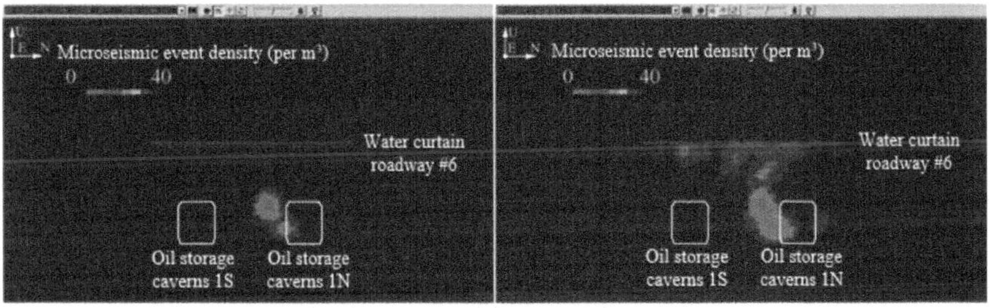

(f) 2014.07.28 - 08.11 (oil storage caverns 1N 2+06m)

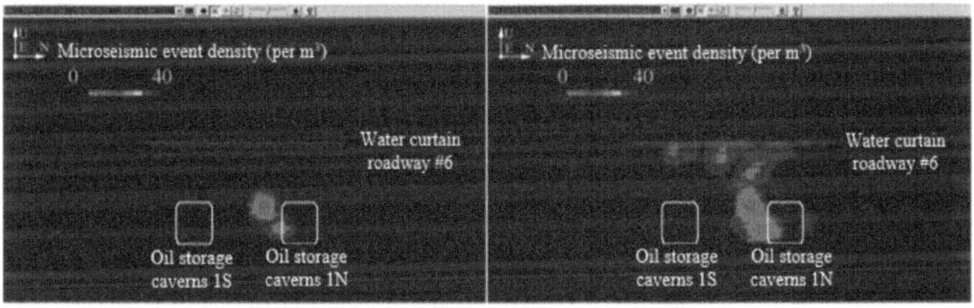

(g) 2014.07.28 - 08.13 (oil storage caverns 1N 1+95m)

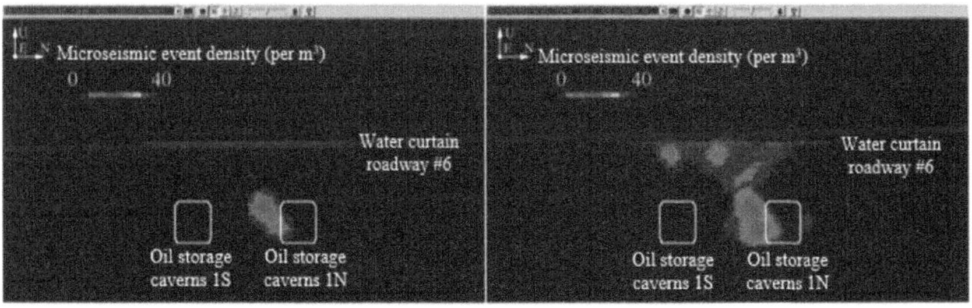

(h) 2014.07.28 - 08.15 (oil storage caverns 1N 1+84m)

Fig. 2.18 (continued)

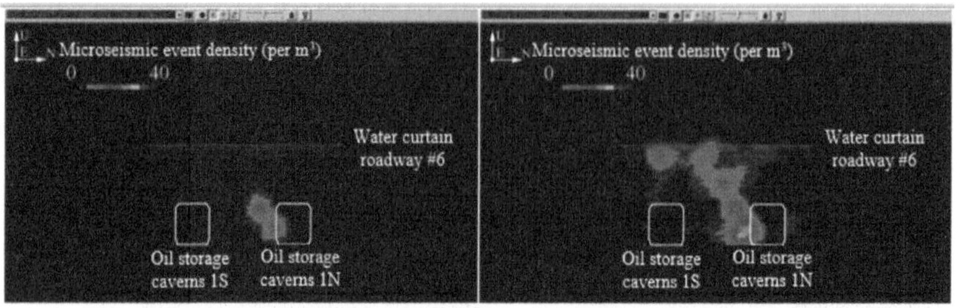

(i) 2014.07.28 - 08.17 (oil storage caverns 1N 1+73m)

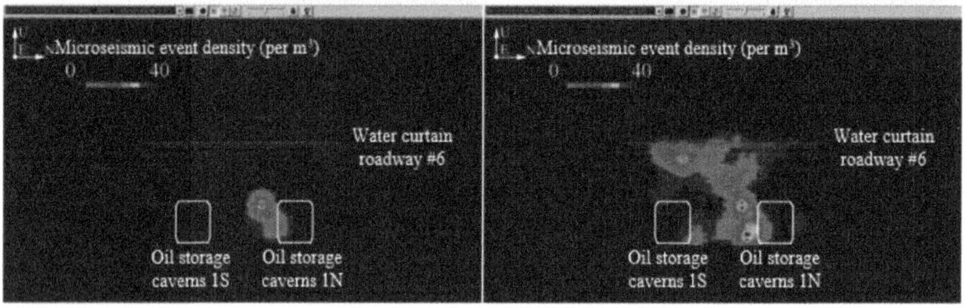

(j) 2014.07.28 - 08.19 (oil storage caverns 1N 1+63m)

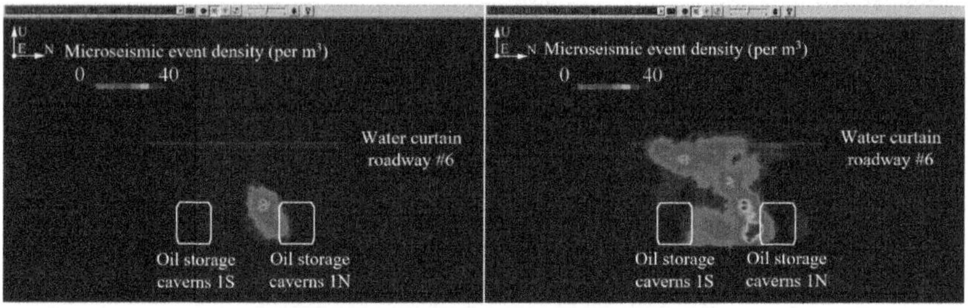

(k) 2014.07.28 - 08.22 (oil storage caverns 1N 1+52m)

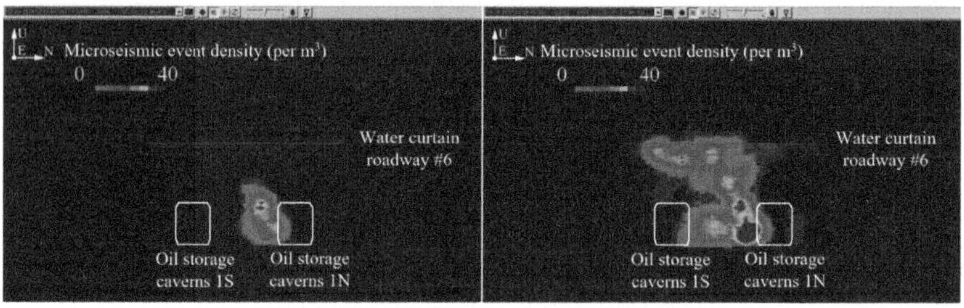

(l) 2014.07.28 - 08.24 (oil storage caverns 1N 1+42m)

Fig. 2.18 (continued)

During the excavation of the mid-level of oil storage cavern 1N to the 1 + 52 m mark, the aggregation of micro-fractures within the surrounding rock exhibited two distinct bands (Fig. 2.19a): one situated near water curtain tunnel 2N at the 3 + 10 m mark, displaying near-horizontal distribution (I in Fig. 2.19a); the other observed between oil storage caverns 1N and 1S spanning from 2 + 40 m to 2 + 60 m (II in Fig. 2.19a). These two zones also present heightened density of microseismic events, with the highest concentration of micro-fractures in the south side wall of oil storage cavern 1N between 2 + 40 m and 2 + 60 m (Fig. 2.19b). The local occurrences of "cavity collapse" and "rockfall" in this region stem from the high-energy accumulation, release, and transfer within the rock mass due to significant unloading during excavation. This process results in micro-fractures within the surrounding rock, leading to localized instability. The distribution of microseismic energy release density within the monitoring period in the study area of Jinzhou underground water-sealed oil storage cavern is depicted in Fig. 2.19c. This figure includes the placement of diabase dikes and granite aplite dikes alongside the energy release density envelope area. The area of energy release concentration aligns more closely with the spatial distribution of microseismic events, revealing two concentrated zones. The microseismic energy release emanating from the rocks between the 2 + 40 m and 2 + 80 m marks within oil storage caverns 1N and 1S exhibits heightened concentration, extending vertically upward from the cavern base for approximately 40 m. This concentrated energy release zone locally expands towards the granite aplite dikes near the 2 + 55 m mark in cavern 1S. However, it correlates more closely with the distribution pattern of diabase dikes and other dikes within the surrounding rocks (240°∠85°). This alignment suggests that excavation and unloading activities have caused the concentration of energy release along the diabase dikes within the surrounding rocks. Consequently, this unloading-induced concentration of energy release along the diabase dikes contributes to the localized instability phenomenon within the 2 + 45 m to 2 + 55 m range in the mid-level of oil storage cavern 1N. The northern section of water curtain roadway 2, near the 3 + 10 m mark, constitutes another area of concentrated energy release. Analyzing its correlation with the development of diabase dikes and granite aplite dikes reveals that the western portion (adjacent to water curtain roadway 6) aligns more closely with the proliferation of diabase dikes. Conversely, the opposite side exhibits a parallel distribution with water curtain roadway 6, primarily due to the presence of the water curtain wall on the north side of water curtain roadway 2, housing the water curtain holes at the 3 + 07 m and 3 + 27 m marks. The injection of water induces micro-fractures in the surrounding rock of the water curtain roadway, resulting in a more concentrated energy release. Additionally, the two areas of energy release in the surrounding rock show a trend of penetration. Given the potential hazards during cavern excavation and unloading processes, it is crucial to closely monitor their energy release patterns, understand their developmental trends, and utilize this understanding for engineering design and construction purposes.

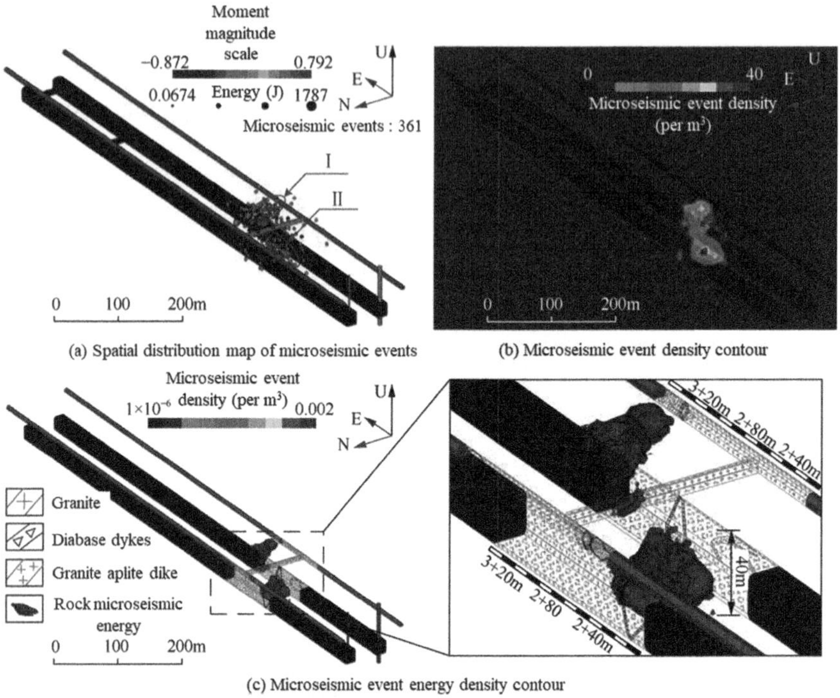

Fig. 2.19 Spatial distribution of microseismic events in the study area (July 28, 2014–August 26, 2014)

Upon comprehensive analysis, it was observed that the blasting during the construction of the oil storage cavern led to the accumulation of stress within the microseismic monitoring area, causing fractures at weak structural surfaces, particularly diabase dikes. Consequently, some energy was discharged through microfractures, while another portion propagated along these surfaces, penetrating the surrounding rock interior, thereby instigating new fractures within the rock weak areas until the dissipation of energy occurred. The microseismic monitoring system replicates the evolutionary process involving the sprouting, development, and eventual macrofracturing of the surrounding rock during intense excavation unloading in the underground water-sealed oil storage cavern. It identifies and delineates potential hazardous zones during the cavern's construction phase, suggesting the need for reduced loading, controlled feeding, and timely cavern support during the excavation process of the oil storage cavern.

2.5.7 Summary

This section represents the inaugural attempt to establish a three-dimensional 'body' monitoring method, termed microseismic, within the water curtain tunnel of an underground water-sealed oil storage cavern. The purpose was to monitor real-time micro-fractures within the surrounding rock of the cavern during excavation of the chamber located 24 m below it. The subsequent conclusions drawn are as follows:

(1) The overall equivalent P-wave velocity within the study area rock body is 5200 m/s, with the monitoring system demonstrating positioning errors of less than 11 m horizontally, less than 4 m vertically, and an average error below 8 m, ensuring precision aligning with engineering requirements.
(2) Examining the spatial distribution of microseismic activity within the surrounding rock of the underground water-sealed oil storage cavern reveals two distinct zones of micro-fractures. One zone is positioned on the eastern side of water curtain roadway 6, situated between water curtain roadways 1 and 2, exhibiting a gentle inclination concerning the horizontal plane. The other zone spans the 2 + 40 m to 2 + 60 m area between oil storage caverns 1N and 1S, approximately corresponding to the spread of diabase dikes in that vicinity. The excavation process facilitates the gradual transfer of energy release towards the diabase dikes and granite aplite dikes, resulting in heightened energy release leading to 'excessive' damage in these dikes, inducing numerous micro-fractures. These findings affirm the utility of the microseismic monitoring system in identifying previously unknown dikes and weak structural surfaces within the peripheral rocks of the reservoir.
(3) The investigation into the changes in microseismic event density during the excavation process of the underground water-sealed oil storage cavern enabled the identification of hazardous areas prone to surrounding rock instability. This study facilitated round-the-clock monitoring and safety alerts for the surrounding rock throughout the excavation and unloading stages of the underground water-sealed oil storage cavern. Furthermore, it serves as a foundational reference for developing and deploying a comprehensive large-scale microseismic monitoring system tailored for underground water-sealed oil storage caverns.

2.6 Numerical Simulation of Underground Water-Sealed Oil Storage Caverns Based on CDEM

Compared to surface storage, underground water-sealed oil storage caverns offer numerous advantages, including increased storage capacity, prolonged usability, and decreased resource consumption. Additionally, underground storage caverns are immune to threats such as fire and explosions, thereby enhancing their safety. Regarding underground water-sealed oil storage caverns, the groundwater sealing diminishes the potential for crude oil leakage into surrounding rock formations. As a result, underground water-sealed oil storage caverns are extensively utilized in multiple countries for the storing strategic energy reserves, including crude oil, liquefied petroleum gas, or liquefied natural gas. Two fundamental principles must be observed irrespective of the energy type stored in underground water-sealed oil storage caverns. Firstly, the underground water pressure surrounding the caverns must surpass the pressure of the stored energy to prevent energy leakage. Secondly, the rock formations surrounding the caverns must exhibit stability. Nevertheless, the stability of these caverns is significantly threatened by high sidewalls, extensive span characteristics, and discontinuities in the surrounding rock. Several scholars have conducted research focusing on stability aspects during the construction of underground water-sealed oil storage caverns. Gnirk and Fossum [26] developed a numerical model employing a probabilistic design program to analyze the stability of caverns storing compressed air. Lindblom [97] proposed a set of standards for rock stability to assess the integrity of underground water-sealed oil storage caverns during full-load operation. Ren et al. [98] described the spatial distribution characteristics of joints using the discrete element method, analyzed the depth and position of unstable blocks in caverns, and validated the discrete element method suitability for assessing the stability of surrounding rock in water-sealed oil storage caverns. Li et al. [99], based on the theory of discrete medium fluid–structure interaction, successfully obtained rock mechanics and hydraulic parameters through triaxial tests on rocks and shear-seepage tests on structural surfaces. They calculated displacements and stresses in various caverns to precisely evaluate the stability of the surrounding rock. Yang et al. [100] investigated the influence of seepage-stress coupling effects on the anisotropy of surrounding rock formations in underground water-sealed oil storage caverns. However, owing to the uncertainty surrounding geological structures and rock properties, there remains no unified understanding of the stability mechanism of underground water-sealed oil storage caverns. With the gradual expansion of underground water-sealed oil storage caverns, the stability of surrounding rock formations during excavation has emerged as a critical determinant of the success of such projects. The unloading effects during excavation significantly affect the stability of these caverns. Therefore, it is imperative to utilize various methods collectively to analyze both the stability and failure mechanisms of underground water-sealed oil storage caverns during the construction process. Continuous numerical simulation methods like the finite difference method (FDM), the finite element method (FEM), and the boundary element method (BEM) provide substantial advantages in analyzing issues associated with continuum mechanics. Discontinuous methods, such as the discrete element method (DEM) and the DDA method, are essential for analyzing discontinuities within the surrounding rock of caverns. However, there is a scarcity of research focusing on the transition from continuous to discontinuous states concerning the damage and fracture evolution of surrounding rock in underground water-sealed oil storage caverns. Munjiza et al. [101], Li et al. [102], and Wang et al. [103] integrated the benefits of continuous and discontinuous methods, proposing coupled methods like the finite discrete element method (FDEM) and the continuous-discontinuous element method (CDEM). In research conducted by Cai et al. [104], the continuous method FLAC3D was combined with the discontinuous method PFC to investigate the acoustic emission phenomenon during large-scale underground excavations. Lisjak et al. [105] employed FDEM to simulate the development of fractures resulting from the excavation of circular tunnels in argillaceous rock. The CDEM not only addresses continuous and discontinuous problems but also replicates the progressive damage and material evolution. This study section will employ CDEM to simulate the excavation process of underground water-sealed oil storage caverns. Microseismic monitoring technology has proven successful in various engineering projects for detecting micro-fractures within surrounding rock. This section intends to employ an advanced microseismic monitoring system to validate the computational results of CDEM. The system includes microseismic sensors installed within the rock, a Paladin signal acquisition system, and a data analysis center.

This section explores the mechanisms causing damage to the surrounding rock during the construction of two contiguous underground water-sealed oil storage caverns. The primary objectives comprise simulating the progression of surrounding rock damage, validating CDEM results using microseismic monitoring data, and assessing the stability of the underground water-sealed oil storage caverns.

2.6.1 Project Profile

This section performs a stability analysis of an underground water-sealed oil storage cavern employing a combined approach involving numerical simulation and microseismic monitoring. The project is located in Jinzhou City, Liaoning Province, China, with a total crude oil storage capacity of $300 \times 104\ m^3$. It consists of eight oil storage caverns labeled 1N to 4N and 1S to 4S (Fig. 2.20), positioned at a depth of 100 m underground. The caverns' top sits at an elevation of − 53 m, while the bottom is at an elevation of − 76 m. Each cavern has a cross-sectional width of 19 m and a height of 24 m. Along the east–west axis, the caverns span 946 m, with a gap of 48 m between two adjacent caverns. This section concentrates on the construction process of caverns 1N and 1S, analyzing the mechanisms causing damage to the surrounding rock formations to evaluate the stability of the caverns.

Based on field surveys and laboratory experiments, the surface is covered by residual soil, and the bedrock comprises coarse-medium-grained granite. Within a 10 km radius of the site, no extensive fractures have been detected. However, there are occurrences of diabase, aplite, and diorite dikes (Fig. 2.21) trending in the NNW and NNE directions, inclined mostly at angles between 70° to 80° or 40° to 50°. As a result of hydrothermal alteration, joints have formed near the rock dikes, and groundwater primarily exists within the fracture network and these joints.

Because of the large cross-section of the oil storage caverns, a three-stage controlled blasting method is utilized for excavation. The excavation sequence for these storage caverns initiates from the upper part, progresses to the middle section, and finally advances to the lower part (Fig. 2.22). Each working face stands at a height

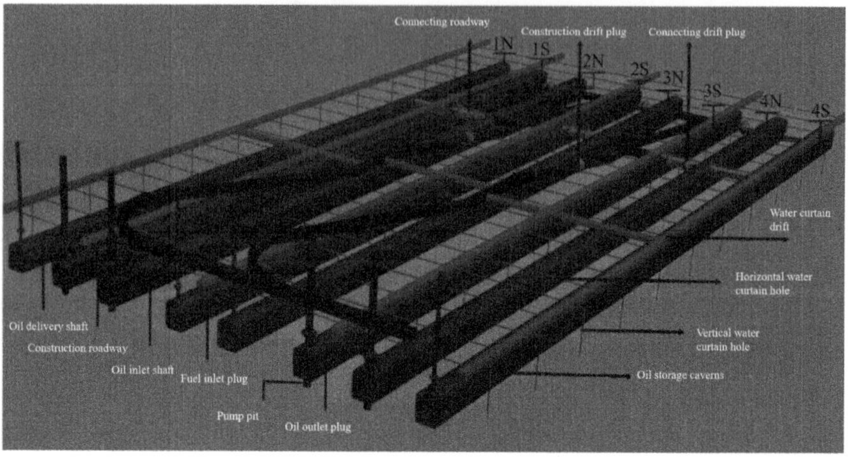

Fig. 2.20 Overview of Jinzhou Underground water-sealed oil storage caverns project

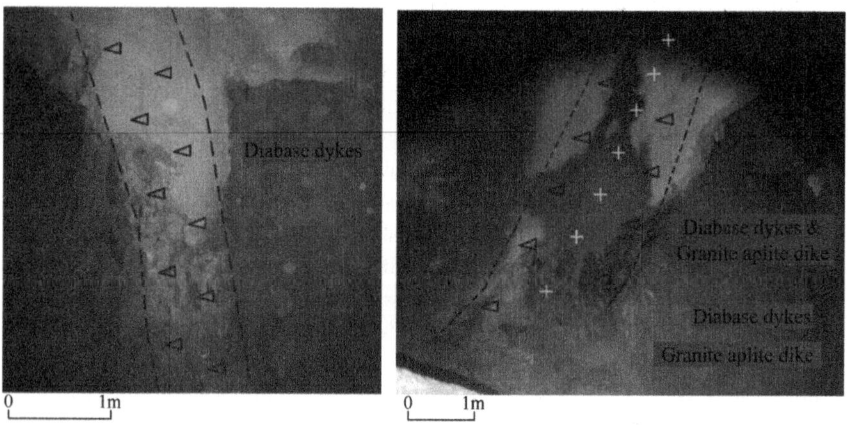

(a) 1N 2+45m South side wall

(b) 1S 2+80m North side wall

Fig. 2.21 Diabase dikes in parts 1N and 1S of oil storage cavern

of 8 m, and specific excavation methods are employed based on rock classification. Simultaneous excavation of the upper, middle, and lower working faces is conducted while maintaining a specific distance to ensure safety. Blasting for the middle-level face, at a depth of 10 m and a volume of 4000 m^3, occurs every 3–4 days. The excavation speed for the lower working face varies from 5 to 10 m per day. Excavating multi-level caverns is characterized by a rapid excavation rate, extensive excavation range, and concurrent robust unloading processes.

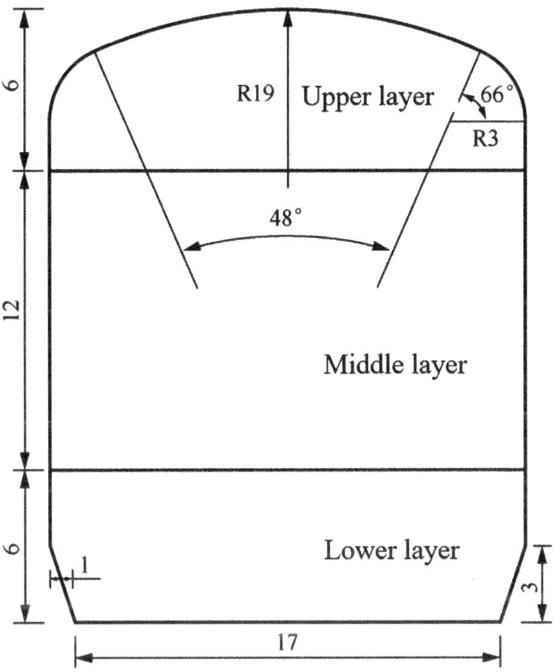

Fig. 2.22 Excavation section of oil storage cavern (unit: m)

2.6.2 *The Basic Principle of Continuous-Discontinuous Element Method*

CDEM governing equations are referenced [106], including dynamic equilibrium equation, linear elastic constitutive equation and strain–displacement relation.

1. Calculation model

The computation area is discretized into finite units, which can use continuous, discontinuous, or partially continuous models. These three models correspond to FEM, DEM and CDEM regions respectively, as shown in Fig. 2.23, and are used to solve different types of problems. In general, the FEM region is used to solve completely continuous problems, the DEM region is used to solve completely discontinuous problems, and the CDEM region is used to simulate the problem from continuous state to discontinuous state. According to the tensile strength criterion or Mohr–Coulomb strength criterion [107], the contact interface fracture is judged, and the fracture evolution process of the rock mass is simulated. As shown in Fig. 2.24, the CDEM comprises two types of elements: block elements and contact elements. Essentially, CDEM can be regarded as a hybrid method that integrates the FEM and the DEM. The FEM is employed to compute the stress distribution within block elements, while the DEM is used to evaluate the contact forces at the interfaces between adjacent blocks [108]. Block elements can be represented by four-node tetrahedrons, six-node wedges, or eight-node hexahedrons. The three-dimensional

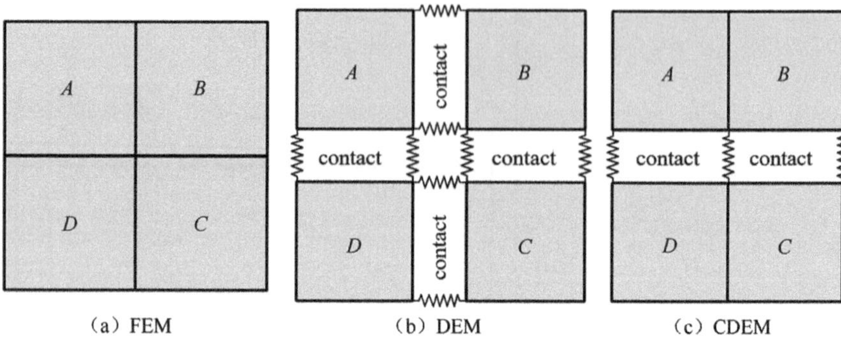

Fig. 2.23 Comparison of calculated regions

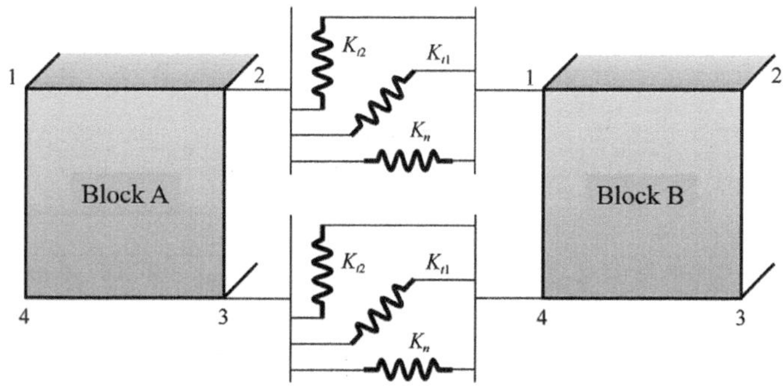

Fig. 2.24 Three-dimensional contact in CDEM: Contact A2-B1 and A3-B4

contact elements, which connect two adjacent nodes of neighboring blocks, consist of three contact springs [109]: one normal spring (K_n) and two tangential springs (K_{t1}, K_{t2}), as illustrated in Fig. 2.24. These springs are mutually orthogonal and exhibit mechanical behavior analogous to that used in DEM. Due to the incorporation of a sufficient number of breakable springs at the contact interfaces, CDEM is capable of accurately capturing the micro-fracturing behavior of rock masses.

2. Fracture criterion

The maximum tensile stress rule and Mohr–Coulomb strength criterion are used to describe the microfracture process of rock mass. Before using the fracture rule to judge the rock mass fracture, firstly calculate the spring force in the contact surface [109]:

$$\begin{cases} F_n^j = -K_n^j \Delta u_n^j \\ F_t^j = -K_t^j \Delta u_t^j \end{cases} \tag{2.2}$$

where, superscript j represents the jth spring; F_n and F_t represent normal spring force and tangential spring force respectively. K_n and K_t are normal stiffness and tangential stiffness respectively. Δu_n and Δu_t denote normal relative displacement and tangential relative displacement, respectively.

The initiation and expansion of microcracks occur along the contact surface of two adjacent elements. For tensile cracks or frictional cracks, normal or tangential spring forces need to satisfy the following inequalities, respectively:

$$F_n \geq TA \tag{2.3}$$

$$F_t \geq cA + F_n \tan \phi \tag{2.4}$$

where, A represents node area; T stands for maximum tension; c and stand for cohesion and internal friction angles respectively.

With the development of microfissure, the rock mass changes from continuous to discontinuous. Therefore, the normal or tangential spring force should be corrected to follow:

$$\begin{cases} F_n = 0 \\ F_t = F_n \tan \phi \end{cases} \tag{2.5}$$

3. Explicit CDEM program

The solid deformation equation is obtained according to the equilibrium equation of the node acting on the force, so a set of block equilibrium equations is obtained:

$$[M]^e\{\ddot{u}\}^e + [C]^e\{\dot{u}\}^e + [K]^e\{u\}^e = \{F\}^e_{ext} \tag{2.6}$$

where, superscript e represents the unit; $[M]$ represents the concentrated mass matrix; $[C]$ stands for damping matrix; $[K]$ represents the stiffness matrix; $\{u\}$ represents the displacement vector; $\{F\}_{\mathrm{ext}}$ represents external force vectors, including physical force $\{F\}_{\mathrm{b}}$, spring contact force $\{F\}_{\mathrm{s}}$, and boundary force $\{F\}_{\mathrm{t}}$:

$$\{F\}^e_{ext} = \{F\}^e_b + \{F\}^e_s + \{F\}^e_t \tag{2.7}$$

CDEM uses a hybrid explicit method for time integration. Acceleration is solved by central difference scheme and velocity by backward difference scheme. The solution formats are as follows:

$$\{a\}^n = \{\ddot{u}\}^n = \frac{\{u\}^{n+1} - 2\{u\}^n + \{u\}^{n-1}}{(\Delta t)^2} \tag{2.8}$$

$$\{v\}^{n+1} = \{\dot{u}\}^{n+1} = \frac{\{u\}^{n+1} - \{u\}^n}{\Delta t} \tag{2.9}$$

where, {a}, {v} and {u} represent acceleration, velocity and displacement respectively; n represents the NTH time step; Delta t is the time step. The explicit iteration format can be further converted from Eqs. (2.8) and (2.9) to the following form:

$$\{\dot{u}\}^{n+1} = \{\dot{u}\}^{n} + \{\ddot{u}\}^{n}\Delta t \tag{2.10}$$

$$\{u\}^{n+1} = \{u\}^{n} + \{\dot{u}\}^{n+1}\Delta t \tag{2.11}$$

The explicit iterative flow in CDEM is shown in Fig. 2.25 and can be summarized as the following steps.

(1) Calculate the nodal external forces of cell e, including physical forces, possible spring forces from adjacent cells, and boundary forces.
(2) Judging whether micro-cracks occur based on fracture criteria. If micro-cracks occur, the spring force is corrected based on Eq. (2.5).
(3) Calculate the internal force of the node

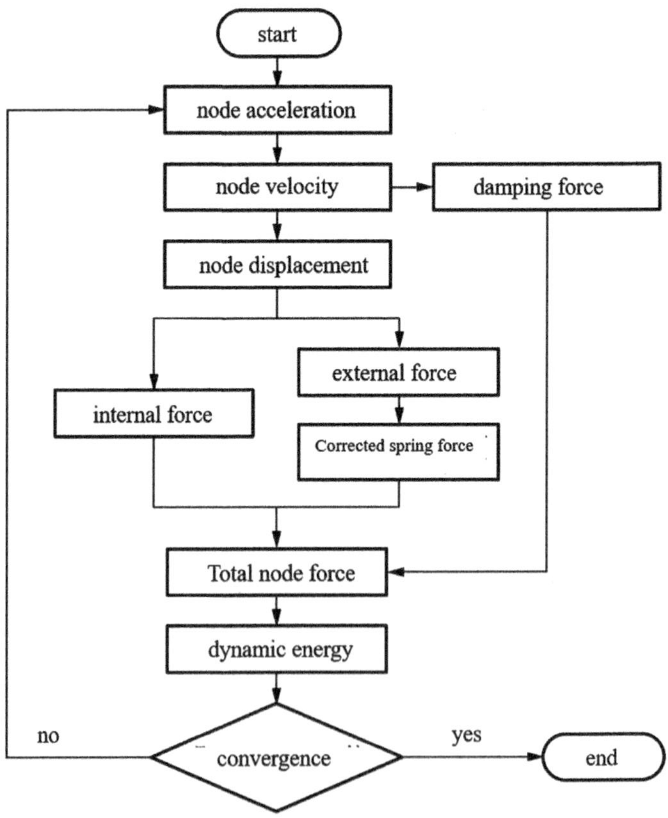

Fig. 2.25 CDEM solution process

$$\{F\}_{int}^{e} = [K]^{e}\{u\}^{e} + [C]^{e}\{\dot{u}\}^{e} \tag{2.12}$$

(4) Calculate the total force of the node

$$\{F\}_{tot}^{e} = \{F\}_{ext}^{e} - \{F\}_{int}^{e} \tag{2.13}$$

(5) Node acceleration is calculated according to Newton's second law

$$\{a\}^{e} = \{\ddot{u}\}^{e} = \left([M]^{e}\right)^{-1}\{F\}_{tot}^{e} \tag{2.14}$$

(6) Node velocity and displacement are calculated by Eqs. (2.10) and (2.11) respectively.
(7) Calculate the total kinetic energy of the system

$$E_k = \sum_{e} \frac{1}{2}\{v\}^{eT}\{v\}^{e} \tag{2.15}$$

(8) Repeat the above steps until the kinetic energy reaches the convergence condition.

According to the Courant-Friedrichs-Lewy (CFL) stability condition, the critical time step is:

$$\Delta t_{cr} = \frac{L_{min}}{C_{\rho}}, \quad C_{\rho} = \sqrt{\frac{E}{\rho}} \tag{2.16}$$

where, $L_{\min}$ represents the minimum unit size; C_{ρ} represents p-wave velocity; ρ is the density; E is the elastic modulus.

4. CDEM verification

Li et al. [110] initially verified CDEM by comparing it with the uniaxial compression test results of rock samples. In the study, the numerical results obtained by CDEM simulation agree well with the experimental data. Zhang et al. [111] verified the simulation of fractured rock by CDEM through slope sliding experiment. The slope slip pattern predicted by CDEM is similar to the experimental data. Li et al. [112] used CDEM to predict the tipping failure mode of an open pit mine, and the numerical results of microfracture development were consistent with the monitoring data of GB-InSAR. Li et al. [112], Ma et al. [113] and Wang et al. [103] reviewed many applications of CDEM in engineering, introduced the parallelization of CDEM based on graphics processing unit (GPU), and expanded its application. In this section, validated CDEM is used to study the development of microfractures around underground water-sealed oil caverns.

2.6.3 Numerical Simulation of CDEM

Based on the on-site investigation, a numerical model was established, illustrated in Fig. 2.26. The model comprises various soil and rock layers. The model encompasses three oil storage caverns, labeled 1N, 1S, and 2N, respectively. The profile under study measures approximately 338 m in length and 187 m in height. The predominant composition of the rock mass is granite, with occasional instances of diabase and granite aplite dikes. The rock mass can be classified into five layers from top to bottom: completely weathered layer, highly weathered layer, moderately weathered layer, slightly weathered layer, and unweathered layer (Fig. 2.26a). With the exception of the dikes, each rock layer comprises moderate to coarse-grained granite exhibiting varying degrees of weathering. Extensive fractures of a large scale have not been found in this area. The physical and mechanical properties of diverse granite layers were acquired via field investigations and laboratory experiments [114], displayed in Table 2.5. The properties listed in the table refer to disturbed rock mass properties, while the rest denote in-situ rock properties, all supplied by the China Engineering and Consulting Association. Additionally, it's essential to consider the in-situ rock stress field. As depicted in Table 2.6, the original rock stress field exhibits a certain degree of variation in the horizontal direction. During numerical simulations, the horizontal stress is derived from the average in-situ rock stress field, whereas the vertical stress is computed using the following formula:

$$\sigma_v(h) = \int_0^h \rho g \mathrm{d}y \tag{2.17}$$

where, h is the depth from the ground; y is the numerical direction coordinate.

A pseudo-three-dimensional numerical model, comprised of wedge-shaped pentahedron elements, accurately depicts the study area specifics, encompassing the five layers of granite and the occurrences of diabase and granite aplite dikes. The f granite aplite dikes precisely intersect three oil storage caverns, whereas the diabase dikes traverse the water curtain cavern (Fig. 2.26b). The positioning, dimensions, and orientations of these dikes ascertain the distribution of major fractures. Significantly, the stability issues in this project mainly arise due to the proximity of these dikes to the caverns. Considering the scarcity of similar cases in the literature, there is an increased necessity to concentrate on the stability of water-sealed oil storage caverns in this region. Therefore, the establishment of a precise numerical model for conducting comprehensive studies is crucial.

Using this numerical model, simulate the progression of rock micro-fractures during the excavation of underground water-sealing oil storage caverns. The excavation sequence starts from 1N and advances to 2S, excavating each oil storage cavern successively from the upper layer to the lower layer. Consequently, a total of 9 excavation stages are simulated, potentially leading to a significant unloading effect. The

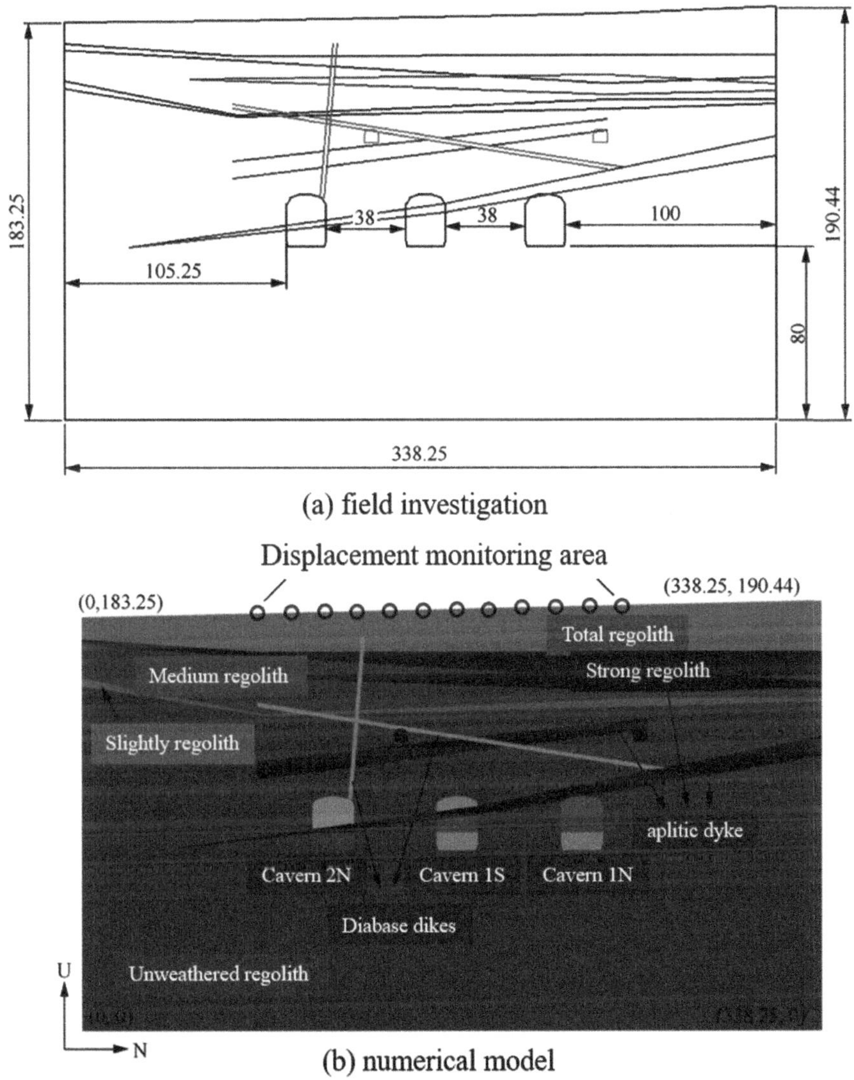

Fig. 2.26 Section model in this study (unit: m)

CDEM (Complex Distinct Element Method) can precisely simulate intricate excavation processes, unveiling unloading mechanisms, and evaluating the stability of the oil storage caverns. It is important to note that the simulation does not consider the excavation of the water curtain tunnel, subsequent to the excavation of the oil storage caverns. Moreover, the unloading effect along the excavation direction is significantly smaller compared to the vertical excavation direction, allowing an assumption that

Table 2.5 Physical and mechanical properties of rock mass

Rock mass	Elastic modulus	Poisson's ratio	Density	Cohesion	Friction	Tensile strength
	E/GPa	v	ρ/(kg/m^3)	c/MPa	ϕ/(°)	T/MPa
Completely weathered granite	20	0.32	2550	0.20	20	0.17
Heavily weathered granite	25	0.28	2582	0.40	25	0.34
Moderately weathered granite	30	0.24	2614	0.60	30	0.51
Slightly weathered granite	35	0.20	2646	0.80	35	0.68
Unweathered granite	40	0.16	2678	1.00	40	0.85
Fine grained dikes	30	0.20	2618	0.50	25	0.43
Diabase dikes	30	0.20	2788	0.40	25	0.26

Table 2.6 Distribution of stress field in the study area

Depth/m	Horizontal stress range/MPa	Mean horizontal stress/MPa	Vertical stress/MPa
≤ 65	3.14–5.52	3.82	Calculate according to Eq. (2.17)
> 65	3.63–9.02	6.17	

the deformation of the surrounding rock during excavation aligns with the plane strain assumption.

A Canadian ESG microseismic monitoring system has been installed in proximity to the caverns. It comprises multiple acceleration sensors, a Paladin signal acquisition system, and a Hyperion data processing system. A single-component sensor has a sensitivity of 30 V/(m s) within a frequency range of 50–5000 Hz. The Paladin signal acquisition system employs analog-to-digital conversion and trigger thresholds for sampling, operating at a frequency of 20 kHz. Given the intricate nature of this project, a six-channel micro-seismic monitoring system has been deployed. This system facilitates real-time monitoring of micro-cracks while ensuring a stable operational state. The micro-seismic monitoring system encompasses the square area illustrated in Fig. 2.5, covering a monitoring area of 300 m × 200 m × 100 m. Six acceleration sensors are sequentially positioned to continuously monitor microseismic events. All these sensors are single-component devices, and their monitoring direction is determined by their placement inclination. Sensors are set at

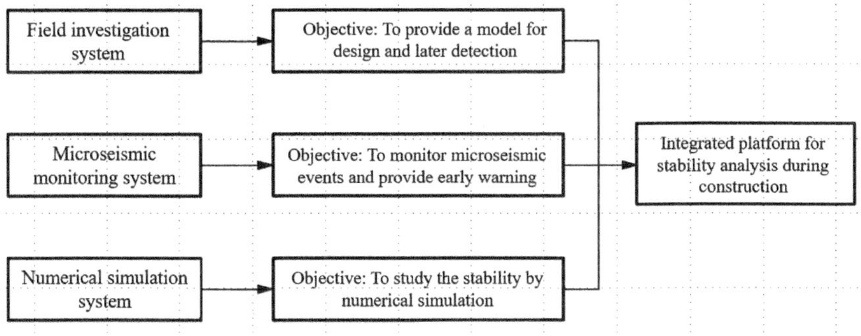

Fig. 2.27 Integrated system

various inclinations to capture micro-seismic signals from multiple directions. Techniques involving threshold setting, filtering, and bandwidth detection are employed to identify waveforms and archive spectral analysis charts [114].

The integration of on-site investigations, micro-seismic monitoring, and numerical simulations occurs on a unified platform to analyze the stability throughout the construction of subterranean water-sealing oil storage caverns, illustrated in Fig. 2.27. The primary objective of the on-site investigation is to design the subterranean water-sealing oil storage caverns, establishing sites for micro-seismic monitoring and the numerical simulation model. The micro-seismic monitoring system provides real-time data on micro-seismic events and alerts regarding potential threats. In case of potential threats, the numerical simulation can predict the development of micro-cracks and evaluate potential risks to the caverns.

The distribution of post-excavation fractures, acquired via CDEM simulation, is depicted in Fig. 2.28. In general, fractures predominantly occur in four areas (Fig. 2.28a), encompassing the zones surrounding the two caverns and the space between them. On-site investigations reveal the presence of diabase and fine-grained rock veins traversing through these areas (Fig. 2.28b). The generation of these fractures is primarily attributed to the post-unloading rock veins, a deduction corroborated by analogous findings from micro-seismic monitoring results (Fig. 2.28c).

Figure 2.29 presents a detailed observation of cavern 1N, showcasing the distribution of two distinct banded regions of fractures within the enclosed area. The upper fracture band above the cavern displays an almost horizontal distribution, while the fractures south of cavern 1N exhibit an almost vertical orientation. The enclosed area above the cavern primarily contains micro-cracks, whereas the area to the left of the cavern exhibits larger fractures. Besides the rock collapse from the cavern's top, the majority of the damage sustained by the rocks in the enclosed area above the cavern did not lead to visible fractures. Nevertheless, significant fractures appear in the rock on the cavern's left-hand side within the enclosed area, attributed to the unloading effect. The majority of the fractures depicted in Fig. 2.29 are concentrated within this enclosed area. Numerical simulations also indicate rock collapse on the

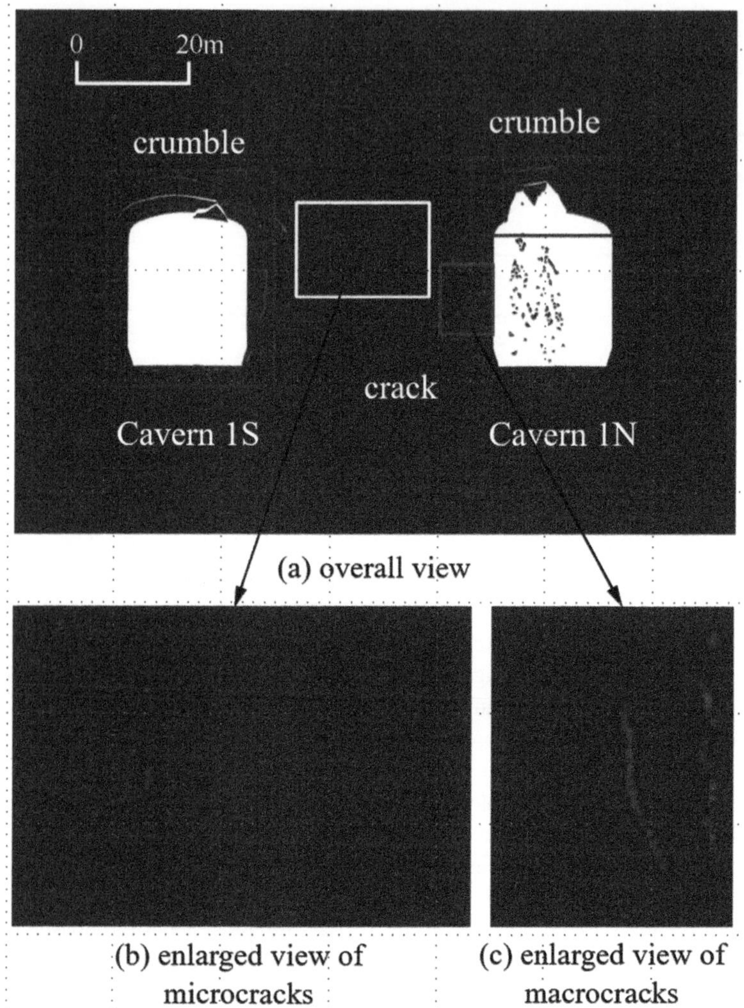

Fig. 2.28 Fracture distribution simulated by CDEM

south side of cavern 1N, illustrated in Fig. 2.30a. This observation is corroborated by on-site investigations, depicted in Fig. 2.30b. This is attributed to the unloading effect resulting from excavation, causing stress concentration in this specific area. When stress surpasses the strength of the surrounding rock mass, fractures occur, releasing the stored elastic energy within the rock mass.

Surface subsidence obtained through numerical simulations is also represented in Fig. 2.31. The surface subsidence observation points are depicted in Fig. 2.26b, with X-coordinates ranging from 80 to 260 m. These monitoring points, positioned above the three underground water-sealing oil storage caverns, track the deformation of the

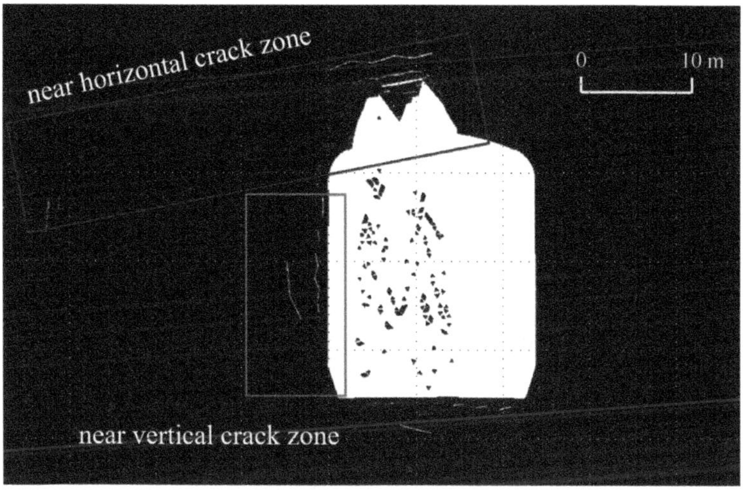

Fig. 2.29 Crack distribution of vault 1N attachment

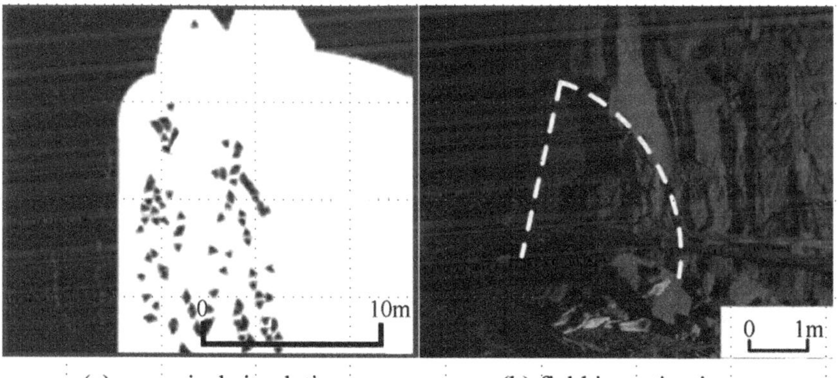

Fig. 2.30 Rock collapse in the south of cave 1N

rock mass subsequent to cavern excavation. Figure 2.31 reveals that the subsidence ranges from 6 to 14 mm, with an average of 10 mm. The greatest subsidence occurs between cavern 1N and cavern 1S, signifying the highest risk of instability in the vicinity of these caverns. Micro-seismic monitoring data yields comparable results. Hence, surface subsidence can be an indicative factor for evaluating cavern stability.

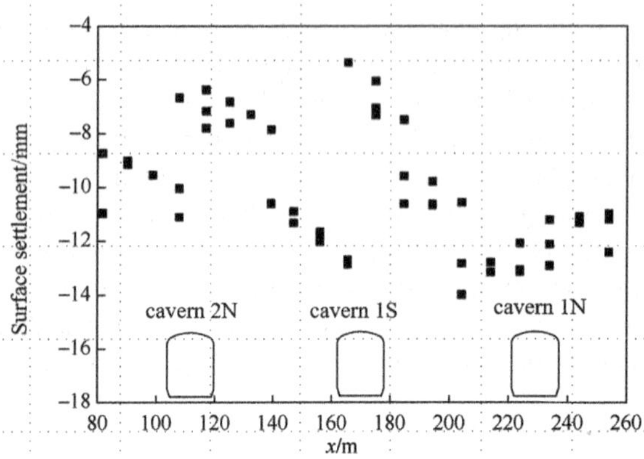

Fig. 2.31 Surface settlement after excavation simulated by CDEM

2.6.4 Microseismic Monitoring Results

Since the installation of the microseismic monitoring system on June 28, 2014, 492 active microseismic events have been detected as of August 26, 2014, including 361 microfracture events, 123 blasting events, and 8 strike events. The micro-cracks caused by excavation and unloading are mainly distributed in and around the caverns 1N and 1S, as shown in Fig. 2.32a. A band is distributed nearly horizontally, located at distance 2 + 63 m; The other band is distributed almost vertically, between the miles 2 + 40 m and 2 + 60 m. The microseismic events are mainly concentrated in these two belt areas, and the maximum density occurs in the distance of 2 + 40 m to 2 + 60 m, as shown in Fig. 2.32b. The area of maximum density is the most dangerous area. Due to unloading, the rock mass is completely broken. As shown in Fig. 2.33, this is similar to the numerical simulation results in Fig. 2.29. At the same time, the unloading effect causes the stress field around the rock mass to be redistributed, and the stress is transferred to the diabase and aplite veins (Fig. 2.34). Stress transfer caused more micro-cracks, forming fracture bands (Figs. 2.28, 2.29 and 2.30). Continuous excavation is the main cause of dike damage and instability of surrounding rock mass.

The progressive failure and fracture development of underground water-sealed oil storage caverns have not been well studied, and their failure mechanism and fracture development mechanism have not been clearly studied. CDEM provides a method to understand progressive failure and fracture development, as well as the failure mechanism and fracture development mechanism of large-scale underground excavation unloading. In fact, CDEM can also simulate purely continuous or purely discontinuous problems, such as elastoplastic problems and rock joint contact problems, just like FEM and DEM. What makes CDEM even more advantageous is the introduction of the fracture mechanism, which can reproduce the progressive failure process of the material from continuous to non-continuous. Therefore, CDEM can simulate failure, fracture growth and joint problems, which is of great significance in

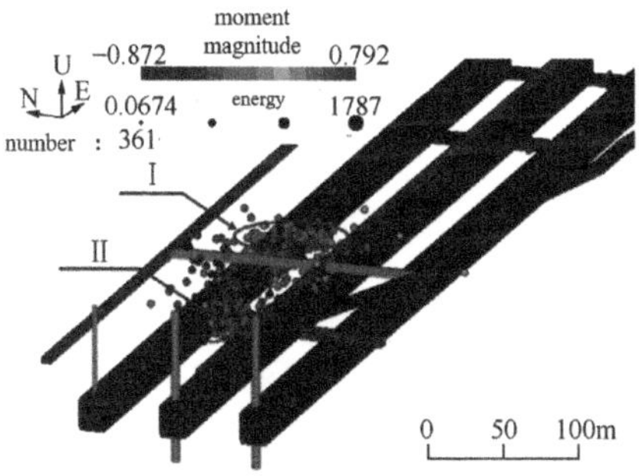

(a) Microseismic events distribution

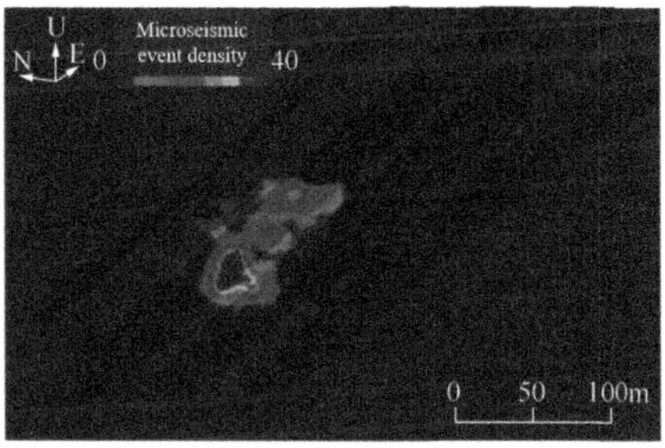

(b) Density of microseismic events

Fig. 2.32 Spatial distribution of microseismic events in underground water-sealed oil storage cavern (July 28, 2014–August 26, 2014)

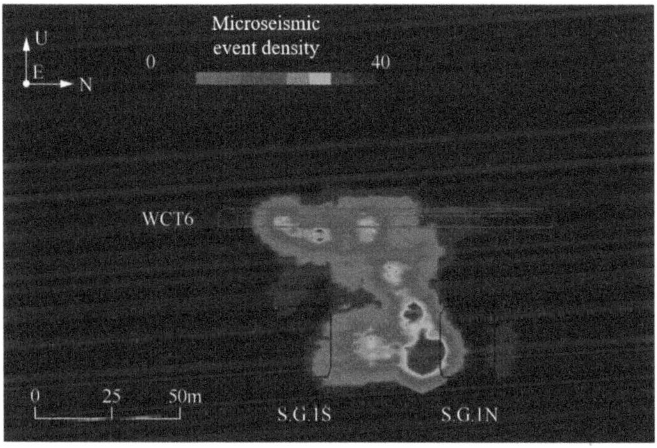

Fig. 2.33 Microseismic events near 1N and 1S of the cave

engineering applications. More specifically, CDEM is able to simulate the fracture state of the material well and takes into account joint distribution and size. Moreover, GPU-based parallel computing enables CDEM to handle large computational problems at the engineering scale. However, CDEM has some limitations in simulating underground water-sealed oil storage caverns, for example, it cannot estimate the time of rock mass rupture well. In addition, the liquid flow effect in rock mass and fracture has not been introduced in this study. These are the issues that need to be further studied by CDEM in the future.

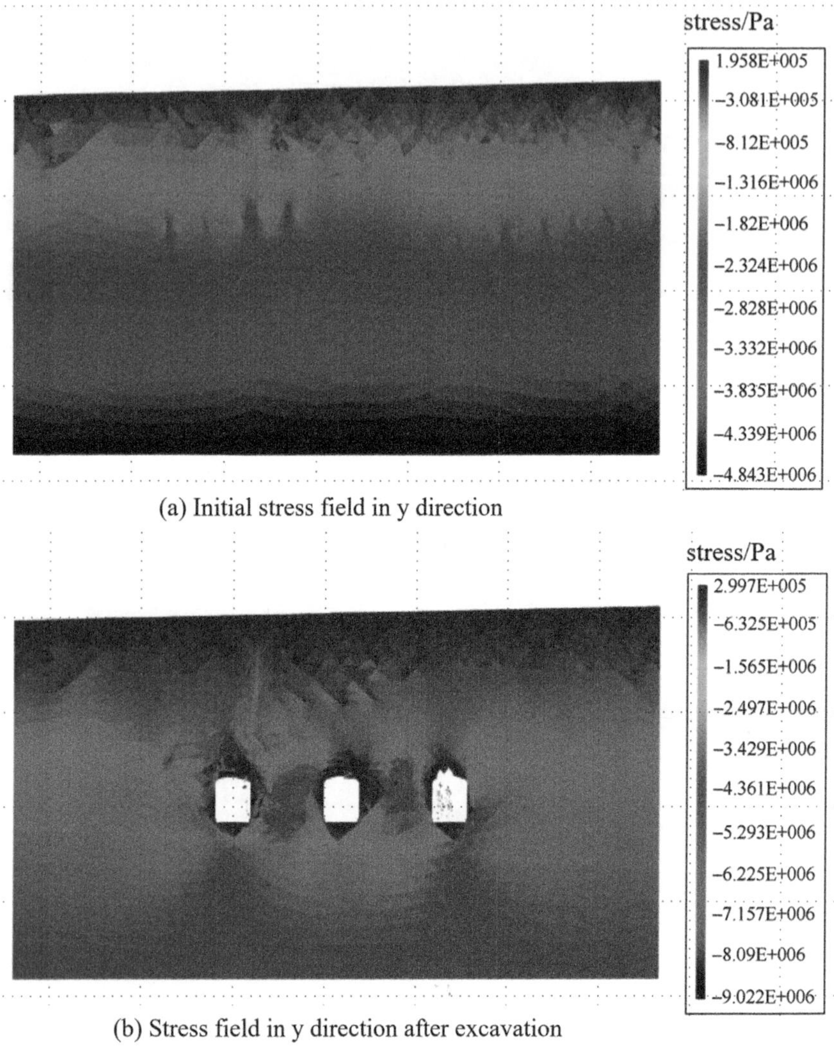

Fig. 2.34 Stress transfer caused by excavation

2.6.5 Summary

The main research content of this section is to analyze the stability of underground water-sealed oil storage caverns. A new numerical simulation method, continuous-discontinuous element method, and a six-channel microseismic monitoring system were integrated into a unified platform to study the microseismic events and the development of microfractures. Specifically, on the one hand, CDEM is used to model microcracks caused by excavation and unloading; on the other hand, microseismic monitoring system is used to monitor real-time microseismic events during the construction of underground water-sealed oil storage caverns. Meanwhile, the simulation results are confirmed by microseismic monitoring data, and the following conclusions are obtained.

(1) The integrated platform of CDEM numerical simulation and microseismic monitoring system is very effective in capturing micro-cracks in underground water-sealed oil storage caverns. The fracture region obtained by CDEM is consistent with the microseismic monitoring data. The microcracks are mainly located in two zones, one nearly horizontal distribution and the other nearly vertical distribution. The microseismic events are mainly distributed in two zones, and similar phenomena can also be obtained by CDEM simulation.
(2) The numerical simulation shows that rock collapse occurs in the south of cave 1N, which is consistent with the field observation and is mainly caused by excavation unloading and stress redistribution. High stress leads to rock mass rupture and release of elastic energy, continuous excavation leads to the destruction of dikes and local instability of surrounding rock mass. In addition, the stress field around the vault is redistributed, transferring the stress to the diabase and aplite dikes, thus forming micro-cracks in the dike area.
(3) Numerical simulation results confirm that excavation causes ground settlement. The subsidence reveals that the rock mass around the region between vault 1N and vault 1S has the greatest instability risk, which can be used as a warning for stability analysis and risk assessment.

2.7 Conclusion

In this chapter, microseismic monitoring technology is applied to the excavation and unloading process of underground water-sealed oil storage caverns. Microseismic monitoring of the research area of underground water-sealed oil storage caverns is used to study the spatio-temporal evolution law of surrounding rock microfracture, and to reveal the energy release law of underground water-sealed oil storage caverns under the action of excavation and unloading and its relationship with the stability of surrounding rock. Identifying and delineating the potential dangerous areas of surrounding rock of underground water-sealed oil storage caverns can provide reference for scientific design and construction of underground water-sealed oil storage caverns. The main conclusions are as follows.

(1) Based on the research and analysis of the characteristics of underground water-sealed oil storage caverns, combined with the requirements of microseismic monitoring system, focusing on the energy release law of surrounding rock microfracture of underground water-sealed oil storage caverns under the action of excavation and unloading of oil storage caverns, a microseismic monitoring system suitable for underground water-sealed oil storage caverns is successfully

constructed. The identification of micro-fracture aggregation area of underground water-sealed oil storage caverns and the evaluation of surrounding rock stability are realized.

(2) The time–frequency combined analysis method was adopted to analyze and identify the noise information of excavation blasting, mechanical slags, and air drilling in the process of microseismic monitoring, successfully filter the noise, accurately and quickly obtain the micro-fracture signal of surrounding rock of underground water-sealed oil storage caverns during excavation and unloading, and lay a foundation for further analysis of the energy release law and stability of surrounding rock of caverns. At the same time, the unknown signal characteristics mentioned in previous studies are analyzed, and it is inferred that it is the unknown excavation blasting waveform far away from the sensor.

(3) Microseismic, a three-dimensional "volume" monitoring method, is constructed in the water curtain roadway of the underground water-sealed oil storage caverns, and real-time monitoring of the micro-fracture of surrounding rock of the underground water-sealed oil storage caverns during the excavation of the oil storage caverns 24 m below it can be obtained: The whole equivalent P-wave velocity of rock mass in the study area is 5200 m/s and the average error is less than 8 m.

(4) The distribution of microseismic activity in the surrounding rock of Jinzhou water-sealed oil caverns is closely related to the excavation activities of the water-sealed oil caverns. On the whole, the frequency of rock microfracture events increases with the increase of excavation scale, and the release of microfracture energy also increases. From the perspective of time, under the action of excavation and unloading, the micro-fracture response of the surrounding rock of the oil storage tunnel may experience a cycle of energy accumulation, energy release and energy transfer, and there is a certain time effect. After August 15, 2014, the microseismic activity of surrounding rock was significantly enhanced. It should be noted that while controlling the amount of blasting charge in the excavation process, the monitoring of cavity deformation should also be strengthened.

(5) From the spatial distribution of microseismic activity in the surrounding rock of Jinzhou underground water-sealed oil storage caverns, the microfracture of surrounding rock in the monitoring area is clustered in two bands. One is located in the east side of water curtain roadway 6, the range of water curtain roadway 1 and water curtain roadway 2, and has a gentle inclination to the horizontal surface; The other one is located in the 2 + 40 m–2 + 60 m mileage area within the 1N and 1S range of the oil storage chamber, which is roughly consistent with the distribution of diabase veins in this area. Local instability phenomena such as "caving" and "rockfall" occur in the gathering area of microseismic events, which is due to the gradual transfer of energy release to diabase and granite aocrystalline dikes due to excavation, and the release of high energy leads to "excessive" damage of these dikes and induces a large number of microfractures. The results show that the microseismic monitoring system can help to delineate the distribution of unknown dikes and weak structural planes in the surrounding rock of caverns.

(6) During the monitoring period, the concentration area of microseismic energy release and the spatial distribution of microseismic events in the surrounding rock of the cave are similar in the study area, and are closely related to the development of diabase dike and local granite aplite dike in the surrounding rock. As a whole, the distribution of diabase dikes in region I and region II (240°∠85°) is relatively consistent. Excavation and unloading lead to concentrated energy release of surrounding rock along the diabase dikes, resulting in local instability near 2 + 45 m distance in the 1N side of the oil storage chamber. The other two areas show the trend of passing through, which are potentially dangerous areas in the excavation process of oil storage caverns. It is necessary to pay close attention to their development trend to ensure the construction safety and stability of surrounding rock during the excavation process.

(7) A new numerical simulation method, continuous-discontinuous element method, and a six-channel microseismic monitoring system are compared to study the microseismic events and the development of microfractures in underground water-sealed oil storage caverns. It can be seen that the fracture region obtained by CDEM is consistent with the microseismic monitoring data, and the numerical simulation shows that rock collapse occurs in the south of 1N of the caverns. This is consistent with the field observation, which is mainly caused by excavation unloading and stress redistribution. The numerical simulation results confirm that excavation causes surface settlement, and reveal that the rock mass around the area between vault 1N and vault 1S has the greatest instability risk, which can be used as an early warning for stability analysis and risk assessment, and provide a reference for the construction and implementation of large-scale microseismic monitoring system for underground water-sealed oil storage caverns.

References

1. Lindblom UE (1994) History and present status of hydrocarbon storage in excavated rock caverns. J Rock Mech Pet Eng 2(4):29–31
2. Wang ZK, Gu XW, Xie QK et al (2020) Controlled blasting of underground water-sealed oil storage cavern with super large section in hardrock. China Mining Magazine 29(2):138–142
3. Sun JP, Zhao ZY, Zhang Y (2011) Determination of three-dimensional hydraulic conductivities using a combined analytical/neural network model. Tunn Undergr Space Technol 26(2):310–319
4. Yu C, Deng SC, Li HB et al (2013) The anisotropic seepage analysis of water-sealed underground oil storage caverns. Tunn Undergr Space Technol 38:26–37
5. Song SJ (2010) Research of the duality classification on surrounding RVCK of storage cavern eased on the improved AHP fuzzy comprehensive evaluation. China University of Geosciences, Beijing
6. Song K (2012) Permeability of granitic gneiss rock mass and surrounding rock stability of the underground oil storage caverns underwater sealed condition. China University of Geosciences, Beijing
7. Wang ZQ, Yan EC, Lu GD et al (2014) Statistical analysis of in-situ stress field for underground water-sealed storage cavern in Chinese mainland. Rock Soil Mech 35(Supp. 1):251–256

8. Aberg B (1997) Prevention of gas leakage from unlined reservoirs in rock. In: Storage in Excavated Rock Caverns, Rockstore, vol 77, pp 399–413
9. Yang DW, Kim DS (1998) Preliminary study for determining water curtain design factor by optimization technique in underground energy storage. Int J Rock Mech Min Sci 35(4):409–514
10. Yamamoto H, Pruess K (2004) Numerical simulations of leakage from underground LPG storage caverns. Ernest Orlando Lawrence Berkeley National Laboratory, Berkeley
11. Benardos AG, Kaliampakos DG (2005) Hydrocarbon storage in unlined rock caverns in Greek limestone. Tunn Undergr Space Technol 20(2):175–182
12. Sun JP, Zhao ZY (2010) Effects of anisotropic permeability of fractured rock masses on underground oil storage caverns. Tunn Undergr Space Technol 25(5):629–637
13. Goel RK, Singh B, Zhao J (2012) Underground infrastructures: planning, design, and construction. Butterworth-Heinemann, Oxford
14. Xie HP, Ju Y, Li LY (2005) Criteria for strength and structural failure of rocks based on energy dissipation and energy release principles. Chin J Rock Mech Eng 24(17):3003–3010
15. Mccreary R, Mcgaughey J, Potvin Y et al (1992) Results from microseismic monitoring, conventional instrumentation, and tomography surveys in the creation and thinning of a burst-prone sill pillar. Pure Appl Geophys 139(3–4):349–373
16. Theodore IU, Trifu CI (2000) Recent advances in seismic monitoring technology at Canadian mines. J Appl Geophys 45(4):225–237
17. Wang HL, Ge MC (2008) Acoustic emission/microseismic source location analysis for a limestone mine exhibiting high horizontal stresses. Int J Rock Mech Min Sci 45(5):720–728
18. Kaiser PK (2009) Seismic hazard evaluation in underground construction. In: Proceedings of the 7th International Symposium on Rockburst and Seismicity in Mines (RaSiM7). Rinton Press, New York, pp 1–26
19. Ma K, Tang CA, Li LC et al (2013) 3D modeling of stratified and irregularly jointed rock slope and its progressive failure. Arab J Geosci 6:2147–2163
20. Hirata A, Kameokay Y, Hirano T (2007) Safety management based on detection of possible rockbursts by AE monitoring during tunnel excavation. Rock Mech Rock Eng 40(6):563–576
21. Tezuka K, Niitsuma H (2000) Stress estimated using microseismic clusters and its relationship to the fracture system of the Hijiori hot dry rock reservoir. Dev Geotech Eng 84:55–70
22. Li B, Dai F, Xu NW et al (2014) Microseismic monitoring system and its engineering applications of deep-buried underground powerhouse. Chin J Rock Mech Eng 33(Supp. 1):3375–3383
23. Hong JS, Lee HS, Lee DH et al (2006) Microseismic event monitoring of highly stressed rock mass around underground oil storage caverns. Tunn Undergr Space Technol 21:292–297
24. Ma K, Tang CA, Wang LX et al (2016) Stability analysis of underground oil storage caverns by integrated numerical simulation and microseismic monitoring. Tunn Undergr Space Technol 54:81–91
25. Ma K, Tang CA, Liang ZZ et al (2016) Stability analysis of the surrounding rock of underground water-sealed oil storage caverns based on microseismic monitoring during construction. Chin J Rock Mech Eng 35(7):1353–1365
26. Gnirk PF, Fossum AF (1979) On the formulation of stability and design criteria for compressed air energy storage in hard rock caverns. Proc Intersoc Energy Convers Eng Conf 1:429–440
27. Yang MJ, Guan BS (2001) Theoretical and numerical simulation study of underground gas-storage caverns with water curtain. Chin J Rock Mech Eng 20(3):301–305
28. Liu GQ, Han M, Song T et al (2007) Seepage simulation of underground petroleum storage caverns and necessity analysis of a water curtain system. Period Ocean Univ China 37(5):819–824
29. Zhang ZG, Tan ZS, Wan JL et al (2003) Three-dimensional seepage analysis of underground LPG storage with water curtain. Chin J Geotech Eng 25(3):331–335
30. Wu RJ, Li GM, Dong YH et al (2009) Analysis of seepage simulation of underground water sealed caverns in Jinzhou. J Yangtze River Sci Res Inst 26(10):87–91
31. Lu M (1998) Finite element analysis of a pilot gas storage in rock cavern under high pressure. Eng Geol 49(3–4):353–361

32. Lee CI, Song JJ (2003) Rock engineering in underground energy storage in Korea. Tunnel Undergr Space Technol 18(5):467–483
33. Wang ZY, Li YP, Guo ST et al (2005) Viscoelastic stability analysis of underground petroleum storage caverns. Rock Soil Mech 26(11):1705–1710
34. Chen X (2007) Evaluation on quality and analysis on stability of adjacent rock mass of the water-sealed underground oil tank in Huangdao. China University of Geosciences, Beijing
35. Shi HB (2010) Analysis and evaluation of water seal condition and surrounding rock stability for Huangdao water sealed underground petroleum storage caverns in rock. Beijing Jiaotong University, Beijing
36. Wang ZC, Li SC, Qiao LP et al (2013) Assessment of natural containment properties of an underground crude oil storage cavern using fluid flow-stress coupling method. Chin J Geotech Eng 35(8):1535–1543
37. Xu JC, Guo ST (2010) Seepage discharge calculation of surrounding rock groundwater in rock caverns sealed oil storage. Rock Soil Mech 31(4):1295–1302
38. Kim J, Cho W, Chung I et al (2007) On the stochastic simulation procedure of estimating critical hydraulic gradient for gas storage in unlined rock caverns. Geosci J 11(3):249–258
39. Tezuka M, Seoka T (2003) Latest technology of underground rock cavern excavation in Japan. Tunn Undergr Space Technol 18(2–3):127–144
40. Lian JF (2004) Evaluation of rock integrity parameters and surrounding rock stability of large underground water-sealed LPG caverns in Jinzhou. China University of Geosciences, Beijing
41. Yu C, Li HB, Zhou QS (2012) Analyses of surrounding rock stability of cavern group and seepage field at dalian underground oil storage. Chin J Rock Mech Eng 31(Supp. 1):2704–2710
42. Yang TH, Zhang FC, Yu QL (2011) Research situation of open-pit mining high and steep slope stability and its developing trend. Rock Soil Mech 32(5):1437–1472
43. Tang BL, Zhang ZL (2000) Introduction of safety monitoring of underground building powerhouse caverns of Ertan hydropower project. Dam Observ Geotech Test 24(3):25–28
44. Wang KN, Cao CR, Xia YQ, et al (2000) Monitoring and analysis of surrounding rock mass of underground powerhouse chambers of Jiangya Power Station. Yangtze River. 2:35–36, 50
45. Huang XY (1998) Prototype monitoring and analysis of underground powerhouse caverns in Tianhuangping Power Station. Hydropower 8:42–43
46. Dong JX, Xu GL, Li ZP et al (2014) Classification of failure modes and controlling measures for surrounding rock of large-scale caverns with high geostress. Chin J Rock Mech Eng 33(11):2161–2170
47. Huang RQ, Huang D, Duan SH et al (2011) Geomechanics mechanism and characteristics of surrounding rock mass deformation failure in construction phase for underground powerhouse of Jinping I hydropower station. Chin J Rock Mech Eng 30(1):23–35
48. Shi GB, Li N (2009) Analyses of block deformation characteristics of large underground cavern and its stability under high in-situ stress. Chin J Rock Mech Eng 28(Supp 1): 2884–2890
49. Xia YY, Zhu RG (1995) A monitoring method for surrounding rock of large caves and its successful application. Chin J Rock Mech Eng 14(4):329–335
50. Tan KY (2008) Design and implementation of the safety monitoring of underground cavern construction. Hydropower 34(9):81–84
51. Zhao XG, Qiu HT (2007) Application of fiberbragg grating sensing technology to tunnel monitoring. Chin J Rock Mech Eng 26(3):587–593
52. Zhang LG (2014) Study on deformation characteristics and control measures of super-large-section slate tunnel. China University of Geosciences, Beijing
53. Tang CA, Wang JM, Zhang JJ (2010) Preliminary engineering application of microseismic monitoring technique to rockburst prediction in tunneling of Jinping II project. J Rock Mech Geotech Eng 2(3):193–208
54. Ma K, Tang CA, Li LC et al (2013) Reinforcement effects of anti-shear gallery of dagangshan right bank slope based on microseismic monitoring and numerical simulations. Chin J Rock Mech Eng 32(6):1239–1247
55. Jiang FX, Xun L, Yang SH (2003) Study on microseismic monitoring for spati al structure of overlying strata and mining pressure field in longwall face. Chin J Geotech Eng 25(1):23–25

56. Gao MS, Dou LM, Zhang N et al (2007) Experimental study on earthquake tremor for transmitting law of rockburst in geomaterials. Chin J Rock Mech Eng 26(7):1365–1371
57. Pan YS, Zhao YF, Guan FH et al (2007) Study on rockburst monitoring and orientation system and its application. Chin J Rock Mech Eng 26(5):1002–1011
58. Jiang FX, Ye GX, Wang CW et al (2008) Application of high-precision microseismic monitoring technique to water inrush monitoring in coal mine. Chin J Rock Mech Eng 27(9):1932–1938
59. Kaiser PK, Vasak P, Suorineni FT, et al (2005) New dimensionals in seismic data interpretation with 3D virtual reality visualtion for burst-prone ground. In: The 6th International Symposium on Rockburst and Seismicity in Mines: Controlling Seismic Risk, Nedlands: Australian Centre for Geomechanics
60. Xu NW, Zhou Z (2017) Microseismic monitoring and stablity evaluation of rock slope in hydropower engineering. Science Press, Beijing
61. Tang LZ, Yang CX, Pan CL (2006) Optimization of microseismic monitoring network for large-scale deep well mining. Chin J Rock Mech Eng 25(10):2036–2042
62. Zhuang DY, Ma K, Tang CA et al (2019) Study on crack formation and propagation in the galleries of the Dagangshan high arch dam in Southwest China based on microseismic monitoring and numerical simulation. Int J Rock Mech Min Sci 115:157–172
63. Xu NW, Tang CA, Sha C et al (2010) Microseismic monitoring system establishment and its engineering applications to left bank slope of Jinping I Hydropower Station. Chin J Rock Mech Eng 29(5):915–925
64. Lu CP, Dou LM, Wu XR et al (2008) Experimental and empirical research on frequency-spectrum evolvement rule of rockburst precursory microseismic signals of coal-rock. Chin J Rock Mech Eng 27(3):519–525
65. Chen BR, Feng XT, Zeng XH et al (2011) Real-time microseismic monitoring and its characteristic analysis during TBM tunneling in deep-buried tunnel. Chin J Rock Mech Eng 30(2):275–283
66. Mo HH, Yang LD (1991) Failure mechanism of surrounding rock of hard rock underground cavern. Geotech Eng 3(2):1–7
67. Zhang B (1998) Stability analysis of surrounding rock of Ertan underground powerhouse system. Hydropower Station Des 14(3):72–76
68. Shi HG (1999) Evaluation on stability factors of surrounding rock of underground powerhouse of Ertan Hydropower Station. Hydropower Station Des 15(2):75–78
69. Ding WQ, Yang LD, Bao DB (2000) The surrounding rockstability analysis of underground plant in complication geological condition. Exploration and development of Rock Mechanics and Engineering in the new century. In: Proceedings of the Sixth Academic Conference of the Chinese Society of Rock Mechanics and Engineering. Beijing: Science and Technology Press of China, pp 632–637
70. Chen SY, Zhou WY, Yang Q et al (2003) Analysis of stability of surrounding rocks of Shuibuya underground plant is carried out by three-dimensional fast LaGrange method. Chin J Rock Mech Eng 22(7):1047–1053
71. Zhang QH, Wu AQ, Shi GH (2004) Application of key block theory to analysis of rock stability for underground plant in baize hydraulic project. Chin J Rock Mech Eng 23(15):2609–2614
72. Yang DS, Chen WZ, Yang WM et al (2004) Stability analysis of surrounding rock mass of Longtan underground caverns. Rock Soil Mech 25(3):391–395
73. Wang WY, Zhang SH (2005) Study on the stability of surrounding rocks of the underground chamber group for the left- bank powerhouse of Nuozhadu hydropower station. Water Power Generation 31(5):30–32
74. Yu YT, Huang SC (1984) Staged excavation of underground chambers in hard and incomplete rock masses. Undergr Eng 11:31–35
75. Zhu W, Wang P (1992) Application of dynamic programming in construction mechanics of cavern groups. Chin J Rock Mech Eng 11(4):330–331
76. Xiao M (2000) Three-dimensional numerical model of construction process for underground opening. Chin J Geotech Eng 22(4):421–425

77. Wang YS, Li XQ (2001) Elastic plastic finite element analysis and construction optimization of surrounding rock in underground cavern groups. Hydroelectr Power (6):35–38
78. Zhu WS, Li SC, Cheng F (2001) Application of energy dissipation model to optimization of construction order for large underground caverns. Chin J Geotech Eng 23(3):333–336
79. Chen WZ, Li SC, Zhu WS et al (2004) Excavation and optimization theory for giant underground caverns constructed in high dipping laminar strata. Chin J Rock Mech Eng 23(19):3281–3287
80. An HG (2003) Study of stability and optimization of large cavern group using synthetic integrated intelligent method. Chin J Rock Mech Eng 22(10):1760
81. Jiang AN (2005) Study of feedback optimizing and analyzing the schemes of excavation and supporting of large cavern group using integrated intelligent method. Northeastern University, Shenyang
82. Feng XT, Zhou H, Li SJ et al (2008) System of intelligent evaluation and prediction in space-time for safety of rock engineering under hazardous environment. Chin J Rock Mech Eng 27(9):1741–1756
83. Zhang CB, Zhang YQ, Hu MP et al (2019) Stability simulation on the oil storage caverns of large-section underground water-sealed oil storages. Oil Gas Stor Transp 38(7):827–832
84. China National Petroleum Pipeline Engineering Co., Ltd (2010) Geotechnical investigation report of Jinzhou National Petroleum Reserve Project. China Petroleum Pipeline Bureau, Langfang
85. Shi HB, Liu BG (2010) Design and seepage discharge analysis of artificial water curtains for water sealed underground petroleum storage caverns in rock. Chin J Geotech Eng 32(1):130–137
86. Gibowicz SJ, Kilko A (1994) An introduction to mining seismology. Academic Press Inc., San Diego, CA, pp 84–85
87. Lv JG, Jiang YD, Zhao YX et al (2013) Study of microseismic positioning based on steady simulated annealing-simplex hybrid algorithm. Rock Soil Mech 34(8):2195–2203
88. Li N, Wang EY, Sun ZY et al (2014) Simplex microseismic source location method based on L1 norm statistical standard. J China Coal Soc 39(12):2431–2438
89. Dong LJ, Li XB, Tang LZ et al (2011) Mathematical functions and parameters for microseismic source location without pre-measuring speed. Chin J Rock Mech Eng 30(10):2057–2067
90. Wang J (2012) Research on the source location accuracy and optimal geophone array designs for the oil well microseismic fracture monitoring. Jilin University, Changchun
91. Ma TH, Tang CA, Tang LX et al (2015) Rockburst characteristics and microseismic monitoring of deep-buried tunnels for Jinping II hydropower station. Tunn Undergr Space Technol 49:345–368
92. Cete A (1977) Seismic source location in the Ruhr District. In: Proceedings First Conference on Acoustic Emission Microseismic Activity in Geologic. Trans Tech Publications, Clausthal, pp 231–242
93. China National Petroleum Pipeline Engineering Co., Ltd. (2010) Geotechnical investigation report on Jinzhou Underground Water Sealing Oil Cave Storage. China National Petroleum Pipeline Engineering Co., Ltd., Beijing
94. Hustrulid W, Lu W (2002) Some general design concepts regarding the control of blast-induced damage during rock slope excavation. In: 7th Rock Fragmentation by Blasting, Beijing
95. Ma K, Tang CA, Xu NW et al (2013) Failure precursor analysis of surrounding rock mass around cross tunnel in high-steep rock slope. J Cent South Univ Technol 20:207–217
96. Laveov AV, Shkuratnik VL (2005) Deformation and fracture-induced acoustic emission in rocks. Acoust Phys 51(S1):2–11
97. Lindblom UE (1997) Design criteria for the Brooklyn Union gas storage caverns at JFK airport. Int J Rock Mech Min Sci 34(3–4):179
98. Ren WM, Cui W, Zhang A et al (2013) Application study of underground storage cavern stability analysis base on distinct element method. Chin J Undergr Space Eng 9(Supplement 2):1916–1921

99. Li Z, Ping Y, King C et al (2012) Assessments of containment and stability of underground crude oil storage caverns based on fluid-solid coupling theory for discrete medium. Chin J Rock Mech Eng 31(11):2161–2170
100. Yang TH, Jia P, Shi WH et al (2014) Seepage–stress coupled analysis on anisotropic characteristics of the fractured rock mass around roadway. Tunn Undergr Space Technol 43:11–19
101. Munjiza A, Owen DRJ, Bicanic N (1995) A combined finite-discrete element method in transient dynamics of fracturing solids. Eng Comput 12:145–174
102. Li SH, Tang DH, Wang J (2015) A two-scale contact model for collision between blocks in CDEM. Sci China Technol Sci 58(9):1596–1603
103. Wang LX, Li SH, Zhang GX et al (2013) A GPU-based parallel procedure for nonlinear analysis of complex structures using a coupled FEM/DEM approach. Math Probl Eng 15(2):1–15
104. Cai M, Kaisera PK, Moriokab H et al (2007) FLAC/PFC coupled numerical simulation of AE in large-scale underground excavations. Int J Rock Mech Min Sci 44:550–564
105. Lisjak A, Garitte B, Grasselli G et al (2015) The excavation of a circular tunnel in a bedded argillaceous rock (Opalinus Clay): short-term rock mass response and FDEM numerical analysis. Tunn Undergr Space Technol 45:227–248
106. Jing L, Stephansson O (2007) Fundamentals of discrete element methods for rock engineering: theory and applications. Elsevier Science, Oxford, pp 13–14
107. Li SH, Wang JG, Liu BS et al (2007) Analysis of critical excavation depth for a jointed rock slope using a face-to-face discrete element method. Rock Mech Rock Eng 40(4):331–348
108. Wang LX, Li SH, Ma ZS (2013) A finite volume simulator for singlephase flow in fractured porous media. In: Proceedings of the 6th International Conference on Discrete Element Methods and Related Techniques, pp 130–135
109. Zhao L, Wang JA, Li L et al (2015) A case study integrating numerical simulation and GB-InSAR monitoring to analyze flexural toppling of an anti-dip slope in Fushun open pit. Eng Geol 197:20–32
110. Li SH, Zhao MH, Wang YN et al (2004) A new numerical method for DEM block and particle model. Int J Rock Mech Min Sci 41(3):414–418
111. Zhang L, Wei ZA, Liu XY et al (2005) Application of three-dimensional discrete element face-to-face contact model with fissure water pressure to stability analysis of landslide in Panluo iron mine. Sci China Ser E Eng Mater Sci 48(1):146–156
112. Li SH, Liu XY, Liu TP et al (2008) Continuum-based discrete element method and its applications. In: UK-China Summer School/International Symposium on DEM, Beijing, pp 147–170
113. Ma ZS, Feng C, Liu TP et al (2011) A GPU accelerated continuous-based discrete element method for elastodynamics analysis. Adv Mater Res 320:329–334
114. Zhuang DY, Tang CA, Liang ZZ et al (2017) Effects of excavation unloading on the energy-release patterns and stability of underground water-sealed oil storage caverns. Tunn Undergr Space Technol 61:122–133

Chapter 3
Study on Stability of High Arch Dam During Impoundment Period Based on Microseismic Monitoring

Abstract This chapter investigates the stability of high arch dams during the initial impoundment period, combining microseismic monitoring and numerical simulation techniques. Centered on the Dagangshan high arch dam, the study begins with an overview of the research background, significance, and a review of existing methods, including numerical analysis, safety monitoring, and microseismic applications for high arch dams. Key challenges in current research are identified, forming the basis of the study. The deformation mechanism at the dam heel is analyzed using microseismic monitoring to reveal the spatial and temporal characteristics of seismic events and their role in initial impoundment-induced deformation. Crack formation and propagation in dam galleries are also studied through integrated microseismic monitoring and numerical simulation, offering insights into crack initiation and growth. Additionally, the chapter evaluates the overall working performance of the high arch dam during the impoundment period by synthesizing microseismic data and simulation results. This comprehensive analysis provides a deeper understanding of the stability mechanisms and offers theoretical and practical guidance for ensuring safety and optimizing design and monitoring strategies for high arch dams during critical operational phases.

3.1 Research Background and Significance

China is presently in a critical phase of constructing a modern and robust nation. The rapid development of the socio-economy has triggered a growing demand for energy in the country. Therefore, there is an urgent need to adjust the energy structure; fossil energy and other non renewable energy sources should be restricted and used sparingly, while clean energy sources such as hydroelectricity, wind energy and solar energy should be vigorously and rapidly developed. As early as the 2012 report from the 18th National Congress of the Communist Party of China, there was an emphasis on expediting water conservancy construction within the framework of building an ecological civilization [1]. The results from pertinent investigative data reveal that China possesses the most abundant hydropower resources globally.

K. Ma and C. Tang, *Microseismic Monitoring and Stability Analysis of Large Rock Mass Engineering*, https://doi.org/10.1007/978-981-96-6483-2_3

However, the distribution of water resources in China is highly disparate, with 70% of hydropower resources concentrated in the southwest. This region exhibits extremely intricate geological conditions, predominantly characterized by narrow and steep valleys. The high arch dam is a dam type known for its superior economy and safety, featuring high bearing capacity, exceptional seismic performance, and economic efficiency. It is particularly well-suited for construction in U or V shaped narrow river valleys. Consequently, the high arch dam type has found extensive application in water conservancy projects across China, notably in the southwest region. At the close of the twentieth century, China successfully completed its inaugural high arch dam, the Ertan arch dam (240 m). Over the subsequent nearly three decades, a succession of high arch dams gradually arose in the valleys of southwestern China. Dams of note include Xiaowan (294.5 m), Xiluodu (285.5 m), Jinping I (305 m), and Dagangshan (210 m), all of which were constructed and operationalized. Presently, all reservoirs have been filled to their standard water levels. Over years of hydraulic engineering construction practice, China has amassed substantial construction expertise, garnering worldwide recognition for notable achievements. Significant strides have been taken in pivotal research domains, encompassing high ground stress [2, 3], high slope excavation [4, 5], foundation surface optimization [6], concrete temperature control [7, 8], and intricate stress analysis of dam [9, 10].

High arch dams, in contrast to conventional arch dam structures, endure substantial water loads on the dam face during the reservoir filling period. For example, the total water thrust on the Xiluodu arch dam is approximately 147,150 MN, and for the Jinping I arch dam, it is around 127,530 MN [11]. Furthermore, these dams require a significant volume of concrete, with the concrete volume for the Ertan arch dam amounting to 4 million m^3, Jinping I arch dam utilizing 4.76 million m^3, and Xiaowan arch dam boasting an impressive 7.62 million m^3. In other words, the self-weight load of an arch dam and the water load on the upstream dam surface serve as significant tests for the bearing capacity of the foundation and the overall stability of the dam. Therefore, ensuring the safety and stability of the high arch dam during the impoundment period is crucial for guaranteeing its safety during normal operation. Damages incurred during the water storage process can lead to catastrophic consequences. Given that the arch dam is a high-order statically indeterminate structure, non-uniform deformation of the foundation or dam can easily transpire, creating localized stress concentration areas that lead to specific damage in the dam. Notably, tensile stress areas are prone to emerge in the dam heel region of high arch dams during water storage. Since concrete is a typical compressive and non-tensile material, the dam heel region is susceptible to cracking under the influence of substantial tensile stress. Consequently, ensuring the safety of the heel of the high arch dam is pivotal for the overall stability of the dam. The Kolnbrein arch dam in Austria experienced severe cracking and curtain detachment during the initial impingement stage, leading to significant leakage of the dam. Despite repeated reinforcement attempts, the reservoir had to be emptied and permanent support piers were installed on the downstream side for reinforcement [12]. In December 1959, as the storage level of the Malpasset arch dam in France approached its crest, movement occurred in both the dam and downstream foundation, resulting in extensive cracks at the dam heel and ultimately causing a sudden burst. This incident marked the complete destruction of an arch dam almost instantaneously. Similarly, when water storage levels near normal capacity at Schillergrund and Schleges arch dams in Austria, cracks appeared at upper sections of their respective heels, leading to gallery leakage that necessitated large-scale reinforcements. Being the country with the largest number of high arch dams worldwide, China has encountered the issue of arch dam cracking during impoundment periods for many times [13]. An illustrative instance is the first water storage of Shuanghekou dam, where the dam heel developed cracks when the water level approached 3 m near the dam crest. This resulted in significant water leakage on the downstream surface and multiple jets. Similarly, following the water storage of Ertan arch dam, the stiffness and strength of the foundation's weak zone diminished, leading to adjustments in the stress of the arch dam and the emergence of several cracks in the downstream area near the foundation [14]. The aforementioned engineering accident cases highlight that the impoundment period is particularly susceptible to accidents for high arch dams. Once the dam body develops cracks, reinforcing and consolidating become challenging, incurring substantial costs. In severe instances, the arch dam may collapse, resulting in irreparable losses. Consequently, ensuring the stability of high arch dams during the impoundment period has been a longstanding focus in the engineering field. Understanding the working behavior of high arch dams during the impoundment period is immensely beneficial for guaranteeing the safety of the dam body and promptly identifying potential accident risks.

Currently, microseismic monitoring technology has been widely used in mines, high and steep slopes, tunnel excavation, and has been recognized by the engineering field. As the water storage level increases, stress transfer and local stress concentration occur in the dam, leading to microcracks in the concrete material. Subsequently, the accumulated elastic–plastic energy is released into the surrounding area, resulting in microseismic phenomena. Based on a fundamental understanding of stress changes and microcrack phenomena of high arch dams, this chapter challenges the conventional approach to monitoring stress and displacement in high arch dams. Departing from the microcrack phenomena in the dam body during water storage, the chapter employs microseismic monitoring technology to establish the first set of microseismic monitoring system for the dam heel and gallery of a concrete high arch dam in China. It investigates the deformation mechanism of the dam heel in the Dagangshan arch dam during the impoundment period, unveiling the intrinsic relationship between the working behavior of the dam heel and microseismic events. Simultaneously, the chapter integrates microseismic monitoring and numerical simulation methods to investigate the causes of cracks in the gallery of the arch dam. It aims to identify and define potentially dangerous areas of the high arch dam and analyze the overall working behavior of the dam during the impoundment period. With a focus on engineering practice, this chapter introduces a new method for analyzing the working behavior of high arch dams during the impoundment period, providing valuable insights for those already built, under construction, or planned for the future.

3.2 Review of Research

3.2.1 Numerical Analysis Methods for High Arch Dams

Zhang et al. [15] conducted a comprehensive simulation of the entire process from construction to operation of the high arch dam, and analysis focused on the impact of cracks on the deformation and stress of the dam. Cheng et al. [16] based on the theory of deformation strengthening, conducted a comprehensive simulation analysis of the stress and displacement during the normal operation of the Jinping I arch dam during the reservoir filling period using the three-dimensional nonlinear finite element program TFINE. This analysis aimed to comprehensively evaluate the ultimate bearing capacity of the dam. Yang et al. [17] utilized the finite element method to establish a numerical model of a high arch dam, and analyzed the impact of slope and foundation deformation on the high arch dam during the initial stage of impounding. Lin et al. [18] took Xiluodu high arch dam as the research object, established a three-dimensional precise numerical model of the high arch dam by using nonlinear finite element method, and analyzed the working behavior of Xiluodu high arch dam under the coupling of stress and seepage. Lin et al. [19] developed a nonlinear finite element model for Mengdigou high arch dam. Employing the deformation reinforcement theory, they analyzed the variations in displacement and the plastic zone of the arch dam under normal water storage conditions or overload conditions. Additionally, they conducted an analysis of the overall stability of the arch dam. Shen et al. [20] established a three-dimensional nonlinear finite element model of Wudongde arch dam by using the three-dimensional finite difference program FLAC^{3D}, and analyzed the anti-sliding stability of dam abutment under normal water storage conditions. Zhou et al. [21] used the three-dimensional elastic–plastic finite element method to study the progressive failure process of Xiluodu high arch dam under the condition of overwater level, and analyzed the stability and safety of the dam. Ren et al. [22] assessed the safety level of stability for Xiaowan arch dam along the foundation base under normal water storage conditions using the strength reduction method. Linsbauer et al. [23] conducted an analysis of the mechanical mechanism of dam heel cracking during the initial filling of Kolnbrein arch dam by establishing a finite element numerical model.

3.2.2 Safety Monitoring Methods for High Arch Dams

Ensuring the safety of high arch dams during the impoundment period is a complex and interdisciplinary scientific challenge. Several factors crucial to dam safety remain unclear, including natural disasters (floods, earthquakes, etc.), the structural mechanics mechanism (arch dam load, etc.), material properties attenuation, construction, and various human factors. This lack of clarity poses potential risks for high arch dams during the impoundment period, emphasizing the need for real time safety monitoring to ensure the secure operation of these structures. Currently, the safety monitoring of high arch dams primarily relies on conventional monitoring equipment to conduct real time monitoring and data analysis of the main structure, dam foundation, bank slope, and other associated structures. Monitoring of arch dams primarily encompasses deformation, seepage, pressure, stress and strain, environmental parameters, and hydraulic monitoring. Safety monitoring of arch dams commenced in the 1890s. In 1908, Australia conducted observations on the deformation of an arch dam, marking the earliest dam deformation observation experiment worldwide [24]. In 1925 and 1926, the United States conducted uplift pressure observation and stress–strain monitoring on a test arch dam in the Stephenson Mountain area [25]. Since 1982, the United States has implemented a distributed data acquisition system on four arch dams, including Flaming Gorge, to monitor the safety of these structures [26]. As science and technology continually advance, and arch dam monitoring technology undergoes ongoing development, China has made a qualitative leap in arch dam safety monitoring. This transition spans from the initial use of optical and mechanical instruments for data collection to the current utilization of GPS, three-dimensional laser scanning, and other state-of-the-art technologies. Currently, the monitoring technology for arch dams is progressing towards automation, intelligence, high reliability, precision, continuity, real time capabilities, and networking [27]. Liu et al. [28] performed in-situ measurements of the volume deformation of concrete in the Jinping I high arch dam using the zero stress–strain meter. Yang et al. [29] employed fiber Bragg grating sensors to monitor the stress and strain of the geomechanical model of the Lizhou arch dam. They analyzed issues related to dam surface cracking and confirmed the feasibility of applying fiber Bragg grating sensors to arch dam safety monitoring. Hu et al. [30] investigated the backfilling effect of the Xiaowan high arch dam abutment based on prototype monitoring. This study utilized a combination of measured values and theoretical derivation. Through reviewing the results of arch dam safety monitoring, it is evident that the traditional monitoring method primarily focuses on macro deformation and stress monitoring of arch dams. However, it falls short in obtaining internal damage information of arch dams, hindering the ability to quickly and accurately identify potentially hazardous areas. Microseismic monitoring technology has the capability to capture real time information about microcracks within the dam. It can also infer the development trend of these microcracks, enabling accurate assessments of the stable state of the arch dam.

3.2.3 Microseismic Monitoring and Stability Analysis of Arch Dam

Dai et al. [31] investigated the damage mechanism of the surrounding rock in the underground powerhouse of Houziyan hydropower station using both conventional testing and microseismic monitoring techniques. They examined the progressive process and corresponding characteristics of deformation and damage in microcracks of the surrounding rock during excavation and unloading. This study offers valuable insights for the excavation and support of the underground powerhouse of Houziyan hydropower station. Xu [32] implemented a microseismic monitoring system on the steep and high rocky slope on the left bank of the Jinping I high arch dam. This system successfully conducted real time monitoring and analysis of the stability of the left bank slope. Zhang et al. [33] implemented the ISS microseismic monitoring system in South Africa with high accuracy. This system facilitated real time monitoring and early warning for the stability of the underground powerhouse at the fault-controlled Dagangshan power station. Building upon this, they established a safety evaluation method for the hydropower underground powerhouse, concluding that the overall stability of the rock mass in this region is higher, and the underground powerhouse tends to be stable after reinforcement. Based on the high-precision microseismic monitoring system, Li et al. [34] investigated the relationship between the accumulation degree of microseismic events and the deformation and damage of surrounding rock during the substantial deformation of the underground powerhouse cave on the right bank of Wudongde hydropower station. As a result, they proposed a warning method for large deformation in surrounding rock, offering guidance for the subsequent excavation and reinforcement of the underground powerhouse cave group at Wudongde hydropower station. Ma et al. [35] employed the ESG microseismic monitoring system to investigate and analyze the patterns of timing, spatial distribution, and intensity of rock bursts in the diversion tunnels of the Jinping II hydropower station. According to the 3S principle in seismology, they delved into the disaster mechanism and precursory patterns of rockbursts in deep-buried tunnels, applying four criteria for rockburst determination. The study revealed that microseismic events precede rockburst disasters in time and exhibit similar spatial distributions. It is evident that the use of microseismic monitoring in hydraulic engineering is primarily focused on the excavation of underground powerhouse and bank slopes. Currently, there are limited instances of microseismic monitoring for the dam body of high arch dams. The existing criteria for assessing the stability of arch dams mostly rely on conventional monitoring techniques that observe apparent phenomena like stress and displacement deformation. Little attention has been given to understanding the internal microcrack damage of arch dams.

Regarding the stability analysis of high arch dams, Kou et al. [36] utilized the three-dimensional fast Lagrange program FLAC^{3D} to perform stress-deformation analysis and parameter sensitivity analysis on Xiaowan arch dam. Their conclusion indicates that the dam foundation area is less secure under water load, experiencing both tensile stress and large compressive stress in this part. Zhou et al. [37] using three-dimensional nonlinear finite element modeling, analyzed the overall stability of the Jinping I arch dam. The feasibility of dam construction was validated through overload simulations. Huang et al. [38] created a hyperbolic arch dam model using the TFINE three-dimensional nonlinear finite element program, and analyzed the stress and displacement of the arch dam under normal water load conditions, investigated the overall stability of the arch dam, and put forward a reinforcement scheme for the arch dam. Xu et al. [39] examined the stress in the dam and the stability of the abutments of Dahuashui arch dam by combining the three-dimensional elastic-viscous plastic block method with the girder load split method. They formulated an overall balance equation to solve the displacement of block elements and internal nodes of the dam. The calculation process verified the overall abutment stability of the arch dam. Pan et al. [40] constructed a three-dimensional nonlinear finite element model of Baihetan arch dam. They conducted an analysis of the displacement and dam heel cracking of the arch dam, highlighting that the unbalanced force can forecast the location and direction of crack expansion. He et al. [41] employed a viscoplastic damage model to assess the long-term stability of high arch dams. The stress state of the dam undergoes constant adjustments during the storage of high arch dams, making it challenging to promptly apprehend changes in the stability state through conventional monitoring methods and finite element analysis. Hence, the integration of microseismic monitoring technology for real time observation of the mechanical behavior of concrete within the dam, coupled with numerical simulation, becomes essential for a comprehensive evaluation of the stability of high arch dams during the impoundment period. Based on the current research findings, this chapter posits that analyzing the stability of high arch dams through microseismic monitoring technology is both feasible and necessary. Consequently, the high arch dam at Dagangshan hydropower station was chosen as the research object, a 24 h all-weather microseismic monitoring and analysis system for dam heel and gallery of high arch dam was established for the first time in China. This system, combined with numerical analysis methods, was employed to analyze the stability of the arch dam at Dagangshan hydropower station. This initiative provides a novel approach for the safety and stability analysis and evaluation methods of high arch dams.

3.3 The Existing Problems

(1) Traditional monitoring methods for high arch dams primarily concentrate on observable phenomena such as stress, displacement, and deformation, lacking an understanding of the relationship between the authentic working behavior of arch dams and the microcrack of dam concrete. The occurrence of microcracks in dam concrete signifies the potential existence of macro cracks in the dam, potentially leading to irreversible damage. When significant cracks appear inside the arch dam, local stress is rapidly released, triggering stress transfer. However, conventional monitoring technologies detect stress changes only after microcracks or severe cracks have already emerged inside the dam. Therefore, traditional monitoring methods exhibit limitations and lag in the stability study of arch dams.

(2) Currently, the finite element method is predominantly employed for the numerical computation of high arch dams. Despite the gradual enhancements in the finite element method, the numerical model of high arch dams is unavoidably simplified due to the limitations of numerical calculation, deviating from the actual situation. Simultaneously, the majority of the analysis of the working

behavior of high arch dams remains in the static or quasi-static stage, making it challenging to simulate the complex and variable load conditions during the impoundment period. Consequently, on this basis, the stability analysis of high arch dams cannot be entirely accurate. The effective integration of numerical simulation and microseismic monitoring enables the exploration of the working behavior of the arch dam and the rapid identification of potential high-risk areas in high arch dams. This is crucial for studying the stability of high arch dams during the impoundment period.

3.4 Research Contents

(1) Based on the construction conditions of the arch dam at Dagangshan hydropower station, a microseismic monitoring system for the dam heel and gallery is established. The sensor layout scheme is determined based on the attenuation law of concrete microseismic signals and the relationship curve of microseismic monitoring distance and frequency. This ensures the formation of an effective spatial array of sensors and guarantees the spatial uniqueness of the microseismic event location solution. To address significant signal interference during construction and water storage, the wave velocity model of the microseismic monitoring system is established using a manual tapping method. Subsequently, the wave velocity is optimized to ensure the precision of microseismic monitoring.

(2) The mechanism of dam heel deformation and the arch crown of the gallery cracking is studied based on microseismic monitoring technology. The study delves into the relationship between dam heel deformation and microseismic events at Dagangshan hydropower station during the early stage of impoundment. Additionally, it identifies the causation of the cracking observed in the arch crown of the gallery.

(3) Based on the data of high arch dam of Dagangshan hydropower station, the three-dimensional finite element numerical model of the dam is established by using Abaqus and Ansys finite element software. The model simulates the stress, displacement, and deformation of the arch dam under various storage conditions. Through integration with microseismic monitoring and routine monitoring, we unveil the internal relationship between the working behavior and microseismic activity of the arch dam, delineating potentially hazardous areas.

3.5 Deformation Mechanism of the Dam Heel of Dagangshan High Arch Dam Based on Microseismic Monitoring During Initial Impoundment

In the whole life cycle of high concrete arch dams, the initial stage of impoundment is a stage with relatively large accident risk, and it is also a period with high incidence of various accidents [42–45]. The extensive scale of the project, elevated water levels, and substantial disturbances during the impoundment process result in the synergistic effects of factors such as creep, damage, and seepage in both concrete and rock mass. These factors contribute to diminished resistance in the dam and foundation, gradually revealing various potential defects not accounted for in the early stages [46–49]. Consequently, following impoundment, the arch dam may lead to damage in the dam foundation or localized harm to the dam. This can result in failure and the creation of a leakage channel, culminating in a potential disaster for the arch dam. An illustrative case is the sudden collapse of the Malpasset arch dam in France in 1959, occurring five years into impoundment. The collapse was attributed to the degradation of the dam foundation rock mass resulting from impoundment seepage [50]. In 1963, three years after the impoundment of the Vajoint arch dam in Italy, rock mass damage induced by seepage triggered landslides in the reservoir area. Approximately 300 million m^3 of rock mass rapidly slid down, displacing 50 million m^3 of reservoir water and ultimately resulting in the failure of the reservoir [51]. When the water level of the Kolnbrein arch dam in Austria increased from 1860 to 1890 m, a horizontal tensile crack, spanning a length of 100 m, emerged in the dam heel area due to water pressure. The grouting curtain was damaged, and the restoration of the dam took over ten years, incurring substantial costs [52]. In addition, the Sayano-Shushenskaya arch dam in Russia [53] and the Tolla arch dam in France [54] exhibited multiple cracks in the dam heel area during the initial storage phase, leading to substantial seepage. This occurrence garnered significant attention from the global dam construction community. During the initial impoundment period, the deformation mechanism and stability of both the dam and foundation have emerged as significant topics of public concern. Scholars have undertaken relevant studies, yielding a series of valuable results [55–58]. Based on the study of Shimen arch dam, Li [59] brought to light the occurrence of cracks in the dam heel area during the initial impoundment period, coinciding with the continuous rise of the reservoir water level. In light of these findings, Li Zan recommended a heightened focus on addressing the challenge of dam heel cracks specifically in wide valley arch dams. Shu [60] employed statistical and deterministic model theories to formulate a mathematical relationship between the external load on the dam and the force effect experienced by the Ertan arch dam. The determination of mathematical model parameters was based on measured values, and subsequent analysis encompassed the assessment of both displacement and seepage flow characteristics of the arch dam. Utilizing monitoring data on stress and deformation, Li et al. [61] conducted an analysis of the deformation patterns observed in both the arch dam and foundation throughout the initial impounding period of the Laxiwa hydropower station. Zhao [62] investigated the diverse impacts of reservoir water level variations on the displacement of the dam structure. This analysis was grounded in monitoring results obtained during the initial impoundment period of the Dachaoshan gravity dam. Wei et al. [63] conducted a spatiotemporal and causal analysis of stream displacement and dam deflection at Lijiaxia dam, thereby evaluating the overall safety of the dam. Han et al. [64] investigated the correlation between reservoir water level, and the deformation of both the dam and foundation of the Xiaowan arch dam, employing vertical line monitoring technology. The existing research findings serve as a crucial reference for the seamless water storage and regular operation of the project. Nevertheless, the conventional monitoring methods solely capture the deformation and stress patterns of the dam during the initial filling stage, deviating from the designed operational behavior description. For instance, during the initial filling stage of the Xiluodu hydropower station, even with the reservoir reaching its normal water level, the heel of the arch dam experiences pressure, and the upper valley width of the arch dam continues to contract under the influence of water pressure [65]. These characteristics contribute to a certain level of ambiguity among industry experts regarding the normal working behavior of arch dams. Presently, there is a lack of unified and widely accepted evaluation opinions. The existing evaluation criteria primarily rely on conventional monitoring techniques, centered on observable phenomena such as stress, deformation, and displacement. However, there is a deficiency in acknowledging the internal microcrack damage within the concrete of arch dams.

Given the continuous fluctuations in the initial loading during impoundment and the real time adjustments to the stress state of both the dam and foundation, relying solely on conventional monitoring analysis proves challenging in evaluating the microcrack distribution and comprehending the actual working behavior of the arch dam. This chapter introduces the pioneering application of microseismic monitoring technology to observe microcracks in the dam heel area of the Dagangshan hydropower station. The aim is to elucidate the initiation and evolution patterns of microcracks in both bedrock and concrete within the dam heel area during the initial water storage phase. By integrating these findings with conventional monitoring results, the chapter delves into the deformation mechanism of the dam heel throughout the initial filling stage, exploring its correlation with microseismic activity. This comprehensive approach aids in defining potential high-risk areas in concrete high arch dams, offering valuable technical support for understanding the real working behavior of high arch dam heels in China.

3.5.1 Overview of High Arch Dam of Dagangshan Hydropower Station

The dam site of the Dagangshan hydropower station is situated in Shimian County, in the upper reaches of the middle section of the Dadu river in Sichuan Province. Positioned as the 14th hydropower station among the planned 22 steps for the main stream of the Dadu river, it holds a significant role in the regional hydroelectric development plan. The hydro project primarily comprises structures such as the water retaining structure, left and right bank slopes, water inlet, and other related components (Fig. 3.1). The normal water level of the reservoir is set at 1130 m, with a dead water level of 1120 m. The rated water head is specified as 160 m, and the total installed capacity reaches 2600 MW [66]. The water retaining structure is a concrete double-curvature arch dam with the crest elevation of 1135 m, and the maximum dam height is 210 m. The width of crown cantilever is 10 m, and the width of dam base is 52 m. The concrete arch dam is divided into 29 sections, and the sections are gradually increased from left bank to right bank. The galleries in the dam are interlaced, including the upper inspection gallery, the lower inspection gallery, the access gallery, the foundation gallery, the drainage gallery and the climbing gallery. The dam is outfitted with four flood discharge deep holes at elevations of 1043 m and 1046 m, along with three diversion bottom holes positioned at elevations of 1011 and 1100 m (Fig. 3.2). The valley at the dam site exhibits a V shaped configuration, featuring abundant mountains and exposed bedrock on both sides, as illustrated in Fig. 3.2. The bedrock outcrops mainly include hoar and reddish medium-grained bio-adamellite of the Jinning-Chengjiang period in massive or sub-massive structure. Crisscrossing diabase dikes of class III_2–V are developed in granite at the dam foundation, such as the β_{21} and β_{28} in left bank dike, the β_{43} and β_8 in right bank dike, and the β_{88} in riverbed, etc. There are no regional fault crossings at the dam site. The structural configuration is marked by faults, compressional fracture zones, and joint fractures along the dikes. Typically, fracture development within the dikes is characterized as being discontinuous. In view of the complex engineering geological conditions at the dam site, a lot of foundation reinforcement measures have been applied, such as concrete pedestals, tunnel replacements and pre-stressed anchor cables, etc [67]. Concrete pouring for the arch dam was completed in November 2014, and the reservoir level was EL 975 m. On December 30, 2014, reservoir impounding was started by locking the diversion tunnel, and the reservoir level rose from EL 975 m to EL 1005 m. On May 29, 2015, the bottom outlets were locked, raising the reservoir level from EL 1015 m to EL 1120 m by July 4, 2014, and the average rate of water storage is approximately 2.9 m/d. The initial filling stage of the arch dam was from November 22, 2014 to July 31, 2015.

3.5.2 *Microseismic Monitoring System in the Dam Heel*

The phenomenon wherein materials, such as concrete, undergo deformation or fracture under load, releasing and propagating strain energy in the form of elastic waves, is termed acoustic emission (AE) [68]. AE is induced by non-uniform stress distribution within the material, serving as a function of deformation variables and demonstrating a relatively stable stage of crack propagation. This phenomenon is applicable for testing changes in the internal state of concrete and investigating the developmental

Fig. 3.1 Geomorphic photograph of dam site

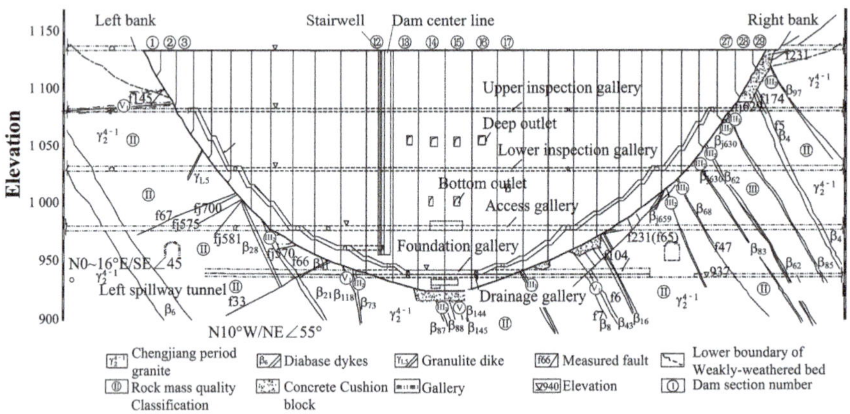

Fig. 3.2 A typical geological cross-section of the left and right ban k slope of Dagangshan hydropower station nearby the dam axis

process of internal defects in concrete [69–71]. This passive non-destructive monitoring method, which rapidly releases local strain energy in materials, is commonly referred to as microseismic monitoring in engineering. By monitoring microcrack signals during the deformation process of materials, it determines the timing and location of microcracks, deduces the development trend of its macro fracture, and identifies potential disaster activities. This approach finds widespread application in the fields of mines [72], rock slopes [73], underground oil storages [74], etc. Currently, AE in concrete is predominantly employed for laboratory test research, with limited reporting on the practical application of microseismic monitoring in large-scale concrete engineering projects. Due to the different application scope of AE and microseismic monitoring, there are obvious differences in monitoring instruments and data processing strategies. Using AE and microseismic monitoring of rock masses as examples, the AE tests in laboratory of rock masses specifically emphasize high-frequency signals throughout the entire process of rock sample failure. The signal quality is excellent, and the emphasis lies in interpreting source properties to investigate the precursory patterns of rock failure. To achieve a broader monitoring range, it is imperative to ensure a robust low-frequency response, with a particular focus on swiftly locating microcracks, calculating source strength, and studying the response patterns of microcracks in disturbed engineering rock masses. Therefore, in the construction of a microseismic monitoring system for mass concrete engineering, careful consideration should be given to selecting a system with a broad frequency response, high sensitivity, and fast, stable monitoring capabilities. In this chapter, the first microseismic monitoring system (MMS) for dam heel of concrete high arch dam is constructed.

1. MMS for the dam heel

The use of the Canadian ESG MMS involves a system composed primarily of a uniaxial accelerometer sensor, Paladin signal acquisition system, Hyperion data processing system, and the independently developed MMS-View remote wireless transmission and visualization system by MechSoft Technology (Dalian) Co., Ltd. The single-component acceleration sensor is constructed from stainless steel with a diameter of 32 mm. It possesses a frequency response range of 50–5000 Hz and a sensitivity of 30 V/g. The Paladin signal acquisition system has a sampling frequency of 20,000 Hz and 24-bit analog–digital conversion. Events were triggered by threshold and STA/LTA methods. Ji et al. [75] determined the dominant frequency of acoustic emission signals by conducting tests on concrete of various strength grades. In these tests, the acoustic emission signals for all strength grades were predominantly low-frequency signals, especially at lower stress levels (below 50% of the peak stress). For concrete below the strength grade of C40, the dominant frequency of acoustic emission signals was approximately 5000 Hz. In a manner analogous to large-scale concrete microseismic signals, and recognizing that the microseismic signal frequency tends to decrease in the low-frequency direction with an increase in concrete size, opting for a high-frequency response uniaxial accelerometer of 5000 Hz can effectively capture the dominant frequency of microseismic signals in arch dams. Base on the concrete microseismic signal attenuation law and the microseismic monitoring distance-frequency relationship curve (Fig. 3.3) [76], it was established that the distance from the microcracking sources to the sensor should be maintained at less than 50 m. Combined with the construction conditions of the arch dam heel area of Dagangshan hydropower station, the topological graph of the microseismic monitoring network of the arch dam heel is established, as shown in Fig. 3.4. To investigate the real time evolution of microcracks in the dam heel during the water storage process and to study the actual working state of the dam heel, the six accelerometers were arranged inside the foundation gallery at EL 940 m near the dam heel and the drainage gallery at EL 937 m close to the dam toe.

The average sensor spacing was about 45 m. Sensors S1, S3, and S5 were vertically loaded into the gallery base plate, while sensors S2, S4, and S6 were loaded into the side wall of the gallery at different angles. The accelerometers formed an effective spatial array ensuring the spatial uniqueness for positioning of the microseismic event. The MMS for dam heel can cover the dam sections #13–#17 below EL 980 m. It conducts all-weather continuous monitoring for microcracks in both the foundation and the concrete of the arch dam during the early stages of water storage. This system acquires real time dynamic parameters of microseismic events. To enhance accuracy, it employs methods such as setting thresholds and filtering to exclude background noise events. Additionally, the system recognizes, stores, and archives the complete waveforms and spectral analysis diagrams. The signal waveforms detected by the MMS for the arch dam heel are illustrated in Fig. 3.5. It can be seen from Fig. 3.5a that the knock signal waveform exhibits two peaks, primarily influenced by the reflection and refraction of elastic waves in the dam galleries. The signal amplitude is approximately 50 mV, and the main frequency distribution is around 2000 Hz. The signal from the hand-held drilling rig construction is the main construction noise signal in the dam heel area during the early stages of impoundment. The signal waveform exhibits a clearly periodic pattern. The typical microcracking signal displays an amplitude of 0.68 mV, characterized by a more developed signal coda and a main frequency distribution around 3000 Hz.

2. Wave velocity optimization

The selection of elastic wave propagation velocity in the rock directly affects the localization accuracy of microseismic events. After the sensor array has been determined, it is necessary to conduct on-site wave velocity tests to obtain an overall wave velocity model to attenuate the influence of the internal galleries of the dam on the propagation of elastic waves. On November 24, 2014, knock test waveforms were recorded by conducting knock tests at 11 points within the spatial interior and at the spatial boundary of the microseismic monitoring sensor array at the dam heel

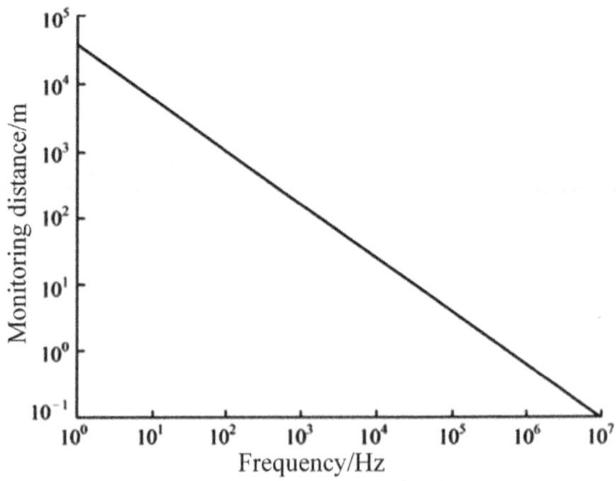

Fig. 3.3 Distance-frequency curve of microseismic monitoring

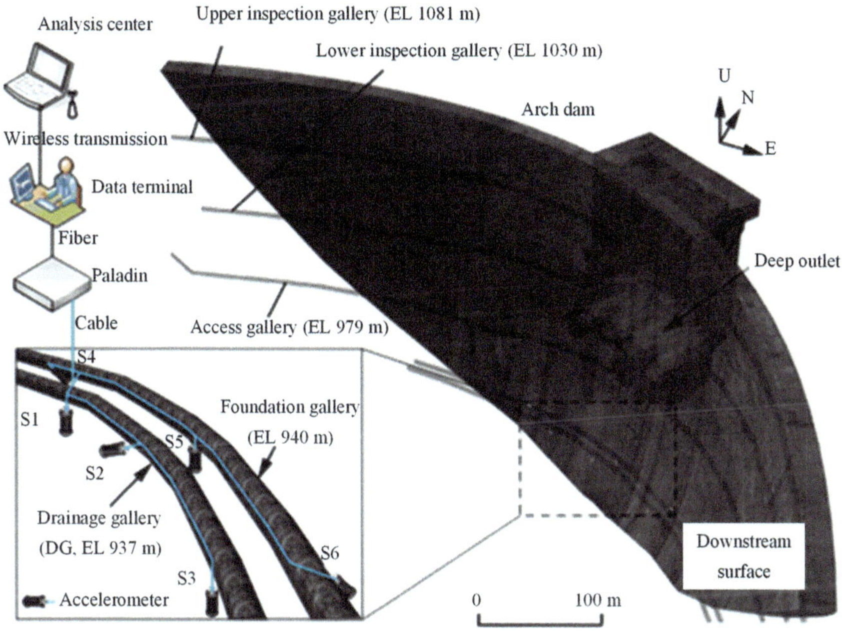

Fig. 3.4 Topological graph of the microseismic monitoring network of the arch dam heel

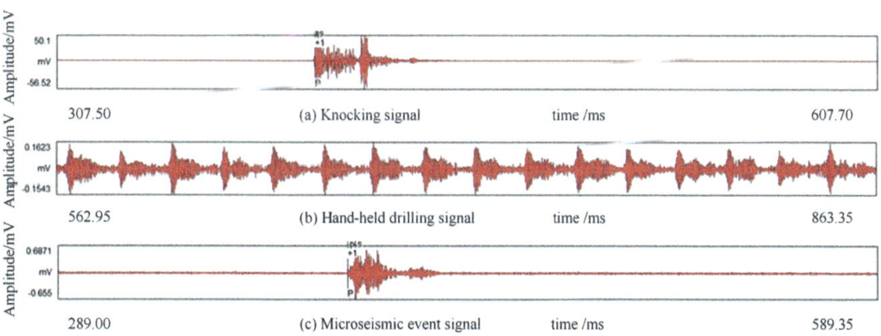

Fig. 3.5 Waveforms detected by monitoring system

(Fig. 3.5a). The system completely recorded the waveforms of each knocking test and verified the feasibility of the system. Based on the results of the dam concrete wave velocity measurement test, the P-wave wave velocity was set at 3700–4800 m/s across 12 different wave speeds, and the positioning errors of the knocking test under different wave velocities were calculated through traversal, as shown in Fig. 3.6. When the P-wave velocity is 4300 m/s, the knocking position has the least error with the positioning of the MMS, and the corresponding information of each knocking test is shown in Table 3.1. At this time, the positioning errors of the microseismic

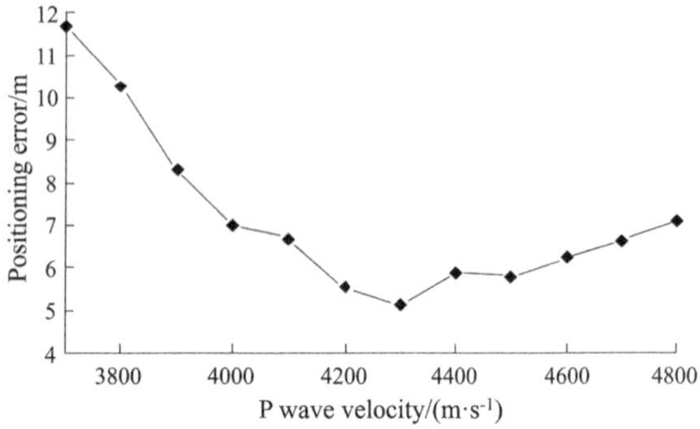

Fig. 3.6 The relationship between the knocking points and microseismic monitoring errors under different P-wave velocity conditions

monitoring system for the arch dam heel of Dagangshan hydropower station were all less than 8 m, and the average error reached 5.1 m.

3.5.3 Spatial and Temporal Distribution Characteristic of the Microseismic Events During Initial Impoundment Period

Since the establishment of the MMS for the dam heel of Dagangshan hydropower station on November 22, 2014, the system has operated effectively. The only disruption occurred when construction machinery severed the communication cables of the MMS from March to May 2015, leading to signal loss. Despite this, the system has successfully achieved all-weather monitoring of microseismic activity in the rock and concrete of the dam heel. As of July 31, 2015, a total of 323 valid events were monitored, including 11 knock events and 312 microseismic events. The distribution pattern of microseismic events with time illustrates the activity rate of microseismic events in the dam heel area during initial impoundment period (Fig. 3.7). Before the locking of the diversion tunnel, the microseismic activity rate in the dam heel was elevated, averaging about 7 microseismic events per day, and exhibited an increasing trend. During this period, the reservoir water level was EL 975 m, representing essentially an empty reservoir state. This suggests that microcracks in the dam heel were more pronounced under the influence of the gravity stress. On 30 December 2014, the microseismic activity in the dam heel remained strong during the locking of the diversion tunnel to store water to EL 1015 m. With the reservoir water level essentially stabilized at EL 1015 m, the microseismic activity in the dam heel decreased obviously. However, during the process of water level rise after the bottom outlets was locked, the microseismic activity exhibited a significant increase. This suggests a close correlation between microseismic activity in the dam heel during the initial impoundment period and the fluctuation of the reservoir water level. The water loads induced adjustments of the dam internal stresses during the high reservoir level state, resulting in an increase of the microseismic activity in the dam heel.

Table 3.1 The deviation of the locations of knocking under different P-wave velocities (Nov. 148, 2015)

No.	Time (h:m:s)	Knocking point position/m			System positioning coordinate/m			Absolute space error/m	Mean absolute space error/m
		N	E	U	N	E	U		
1	15:11:09	3,259,167.7	521,204.0	937.9	3,259,169.0	521,207.1	931.9	6.9	5.1
2	15:14:02	3,259,160.7	521,208.4	938.4	3,259,159.5	521,216.4	939.2	8.0	
3	15:16:02	3,259,158.6	521,213.6	937.8	3,259,153.1	521,211.0	934.9	6.8	
4	15:21:02	3,259,159.4	521,215.4	938.3	3,259,157.8	521,214.7	934.4	4.1	
5	15:23:01	3,259,153.0	521,221.7	938.8	3,259,151.8	521,223.3	937.1	2.6	
6	15:47:01	3,259,150.7	521,268.9	940.0	3,259,153.3	521,269.2	939.3	2.4	
7	16:00:00	3,259,170.0	521,231.0	941.2	3,259,170.5	521,231.2	940.5	1.2	
8	16:01:02	3,259,169.8	521,230.5	939.7	3,259,167.0	521,229.8	945.9	6.9	
9	16:04:00	3,259,168.8	521,228.3	941.2	3,259,169.3	521,232.7	938.1	5.4	
10	16:06:05	3,259,168.8	521,228.3	941.2	3,259,166.0	521,230.3	938.0	4.7	
11	16:08:02	3,259,173.1	521,220.3	939.7	3,259,168.3	521,225.2	937.9	7.3	

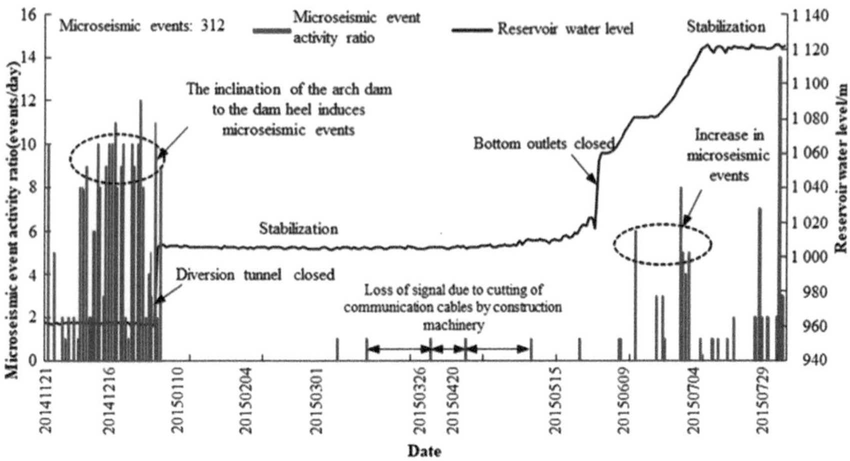

Fig. 3.7 Temporal distribution of microseismic activity of arch dam heel during initial impoundment period

As of 31 July 2015, the spatial distribution of microseismic events in the dam heel during initial impoundment period is shown in Fig. 3.8a, each ball in the figure represents a microseismic event, and its color represents the moment magnitude. The moment magnitude is mainly distributed in the range of − 4.0 to − 2.3. From Fig. 3.8b, it can be seen that the microseismic events in the dam heel during initial impoundment period are mainly distributed below the 979 m elevation, in the region of #13–#16 dam sections, and are clustered in the I, II, III, and IV regions shown in the figure. Area I is situated near the access gallery at EL 979 m within the sections #14–#16, displaying a horizontal strip distribution. Area II is located near the foundation surface of section #16, and is the interface of the high arch dam, the right bank foundation, and the concrete pedestal. Area III is situated close to the foundation surface of section #13 and extends toward the right bank along the foundation gallery at EL 940 m. Area IV is located in the dam toe area and concrete pedestal of the section #14, extending towards the left bank.

Combined with the analysis of the initial storage process of Dagangshan hydropower station, the reason why microseismic events are gathered in the above four regions is mainly due to the unbalanced distribution of local stress in the above regions under the joint action of self-weight of the high arch dam and upstream lateral water load, which leads to the concentration of stress in the above regions, which exceeds the corresponding strength of the concrete and the rock, leading to the initiation and accumulation of microcracks. The distribution of microseismic events is closely related to the impoundment process, and the evolution pattern of microseismic events effectively reveals the response of the high arch dam during impoundment.

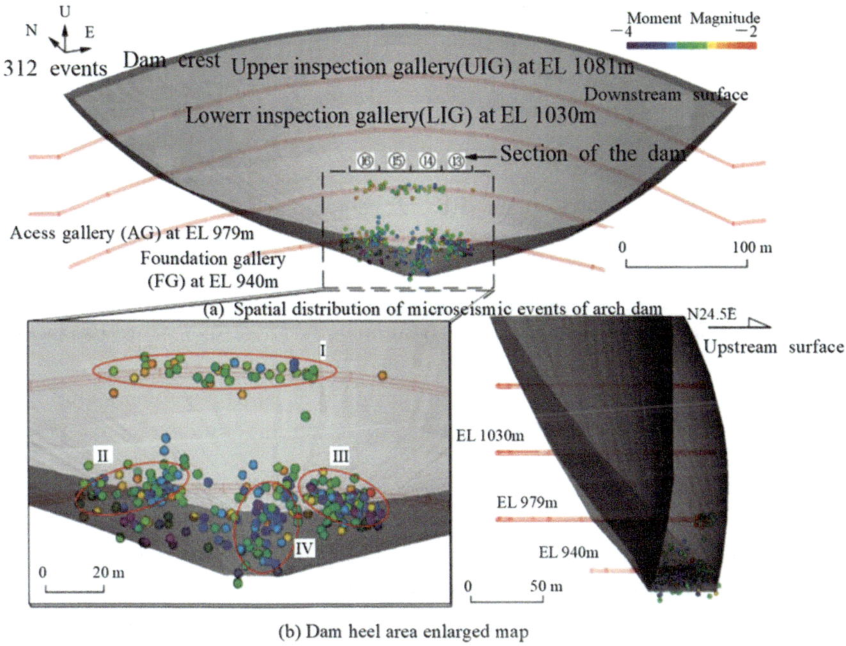

Fig. 3.8 Spatial distribution of microseismic events of arch dam heel during initial impoundment period

3.5.4 *Deformation Mechanism of the Dam Heel During Initial Impoundment Period*

The dam heel is a critical focus in ensuring the safety of high arch dam. The deformation analysis of dam heel during initial impoundment period and the investigation of its real working state forms the foundation for guaranteeing the long-term and reliable operation of high arch dam. Therefore, the MMS of bedrock and concrete in the dam heel area during the initial impoundment period was selected to plot the microseismic deformation distribution pattern in the dam heel area before the locking of the diversion tunnel for water storage, before the locking of the diversion tunnel to the locking of the bottom outlets, and after the locking of the bottom outlets (Fig. 3.9), which correspond to the reservoir levels of 975 m, 975–1015 m, and 1015–1120 m, respectively. It should be noted that microseismic deformation refers to the average microseismic strain (dimensionless) induced by concrete or bedrock under load within a certain period of time, which can be obtained by solving the cumulative seismic moment density of microseism [77, 78], and can be used to evaluate the deformation degree of concrete and rock mass in the dam heel area. As can be seen from Fig. 3.9a, before the locking of the diversion tunnel, the microseismic events in the dam heel area are mainly clustered at the junction of the left and right bank dam foundations and the concrete replacement block in the middle of the dam, presenting the characteristics of distribution along the dam heel towards the higher elevation, which is consistent with the areas II and III in Fig. 3.8b. At this time, the high arch dam is mainly subjected to gravity stress, and the dam is tilted towards the dam heel, resulting in unbalanced stress distribution in the above region and larger microseismic deformation. After the locking of the diversion tunnel, the water level was stabilized at EL 1015 m, and the microseismic deformation zone in the dam heel transferred to the downstream side and gathered at the foundation surface of section #16 (Fig. 3.9b). With the completion of the locking of the bottom outlets, the reservoir water level was stored from 1015 m to 1120 m, and the microseismic deformation in the toe area of dam sections #14 and #15 increased, indicating that the deformation zone of the dam base was transferred from the heel area to the toe area (Fig. 3.9c).

Figure 3.10a and b show the process lines of joint meters opening and closing at the dam foundation of sections #13 and #16 during initial impoundment period, respectively. Joint meters $JD_{13\text{-}1}$ ($JD_{16\text{-}1}$), $JD_{13\text{-}2}$ ($JD_{16\text{-}2}$), $JD_{13\text{-}3}$ ($JD_{16\text{-}3}$) and $JD_{13\text{-}3}$ ($JD_{16\text{-}3}$) are located at the upstream side, middle side and downstream side of dam foundation, respectively. The positive opening degree of the joint meter indicates the opening of the joint, while the negative value indicates the closing of the joint. Before the locking of the diversion tunnel, the joints of dam foundation at sections #13 and #16 were under compression, and the maximum compression of the two sections was 5.64 mm and 2.71 mm, respectively, which were located in the heel area of the section, corresponding to the high microseismic deformation area in Fig. 3.9a. After the locking of the diversion tunnel, the compression of the joints in the toe area of dam section #16 increased, while the other joints did not change significantly. This observation aligns with the high microseismic deformation area in Fig. 3.9b. With the storage of water under the bottom outlets to EL 1120 m, the compression of the joints in the heel area of the two dam sections decreased to different degrees, and the compression of the joints in the toe area of the section #16 increased further, which indicates that the dam foundation is still mainly subjected to compressive stress during initial impoundment period, with the compressive stress in the heel area decreasing and the compressive stress in the toe area increasing. The change process is in good agreement with the microseismic deformation law of the dam heel during initial impoundment period. This alignment reveals the deformation mechanism of the dam heel during initial impoundment period and also verifies the appropriateness of MMS for studying the working states of high arch dams during the impoundment period.

Before the locking of the diversion tunnel, the arch dam is basically in an empty reservoir state, and the dam shows a tendency of tilting deformation towards the dam heel. Continuous cross-stream cracks appeared at the arch crown of sections #13–#15 of the foundation gallery at EL 940 m in the dam heel area. The MMS of the dam heel captured the microseismic information of the concrete of the gallery at this level, drew a cloud map of the microseismic deformation process of the concrete at EL 940 m, and projected the dam onto the microseismic deformation cloud map of the same elevation (Fig. 3.11). On November 24, 2014, microcracks sprouted at section #13 of the foundation gallery at EL 940 m, and local microseismic deformation occurred

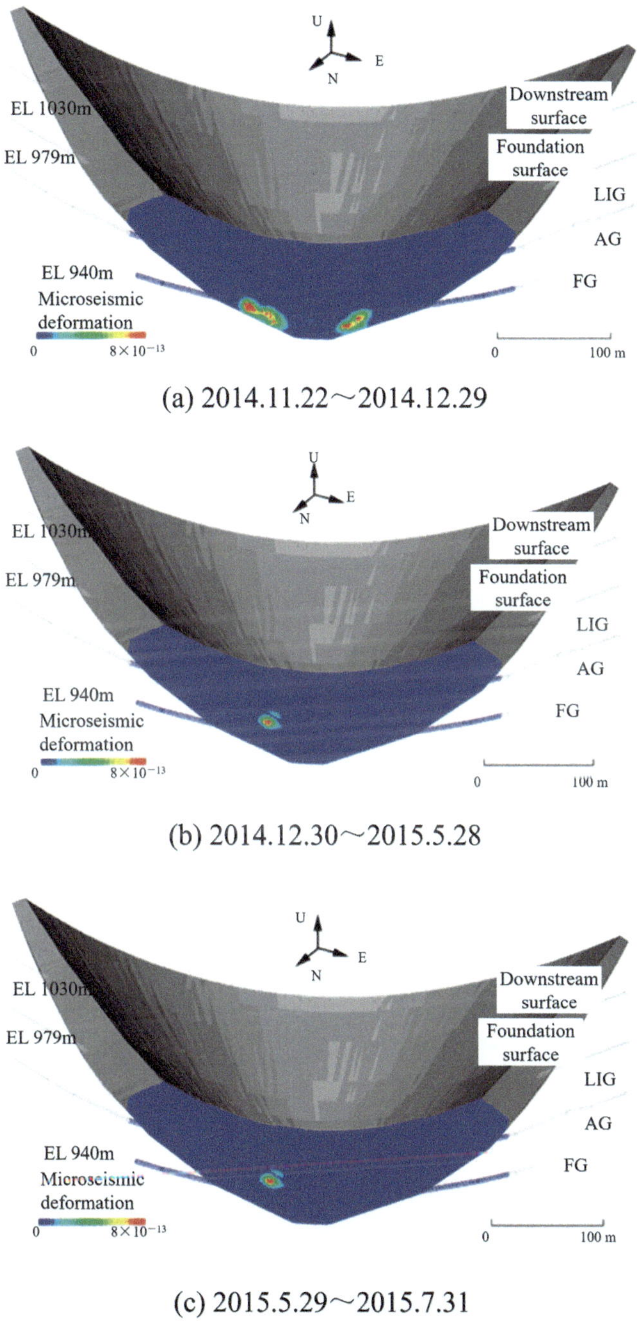

Fig. 3.9 Nephogram of microseismic deformation in dam heel during initial impoundment period

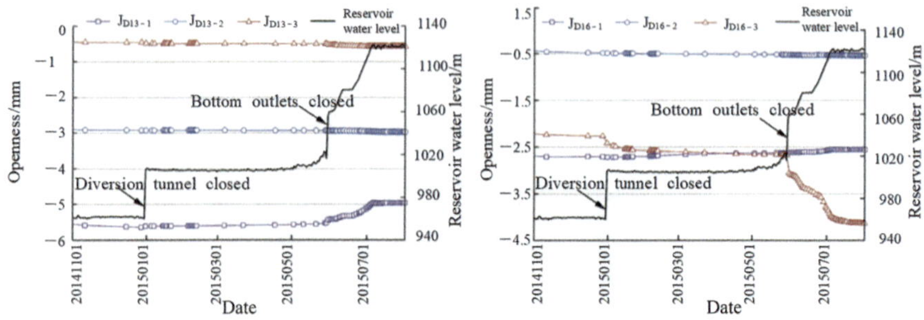

(a) Section #13

(b) Section #16

Fig. 3.10 The time curve of the opening width of the joint meters opening in dam foundation of sections #13 and #16

(Fig. 3.11a). On December 8, 2014, microcracks also occurred at section #16 of the foundation gallery (Fig. 3.11b), and the microseismic deformation continued to increase at both locations, forming a concentrated area of microseismic deformation along the axis of the foundation gallery at sections #13, #14 and #16 (Fig. 3.11c). On December 24, 2014, an inspection of the foundation gallery at the EL 940 m of the dam revealed a crack at the arch crown as shown in Fig. 3.12. It was analyzed that under the influence of the overlying self-weight load on the concrete of section #13 at EL 940 m, the circumferential stress on the arch crown increased, resulting in significant microseismic deformation of the arch crown (Fig. 3.12a), inducing the formation of cracks on the arch crown (Fig. 3.12c), and extending to the arch crown at the gallery of section #14 (Fig. 3.12b). Although no macrocracks were visually observed at section #16, the microseismic deformation cloud map of the area indicates a risk of cracking in the concrete of sections #16–#17. Therefore, site inspections of the concrete at this section should be strengthened. The arch dam of Dagangshan hydropower station experiences significant compressive stress in the heel area under empty reservoir conditions. The dam heel MMS can capture the initiation and deformation evolution law of concrete microcracks under self-weight load in real time, help to identify the potentially hazardous area of the concrete arch dam, and provide a reference to the analysis of the real working state of the high arch dam.

3.5.5 Chapter Conclusion

In this chapter, microseismic monitoring technology is employed for the first time to conduct real time monitoring of the evolution of microcracks in the heel area of the high arch dam. The goal is to analyze the deformation mechanism of the dam heel during initial impoundment period at Dagangshan hydropower station. Additionally,

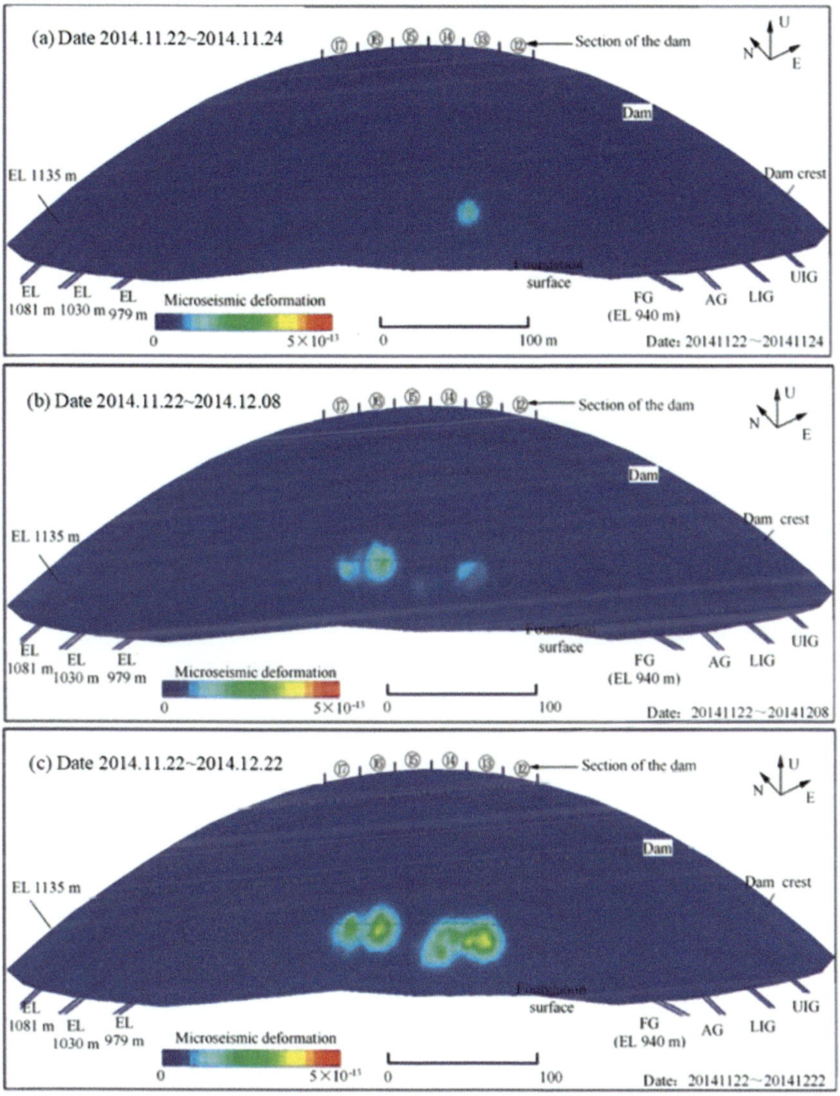

Fig. 3.11 Seismic deformation process of base gallery at EL 940 m

an attempt is made to reveal the real working state of the dam heel during initial impoundment period, leading to the following main conclusions.

(1) Successfully constructed the first MMS for the heel of the high arch dam in China, and verified the feasibility of microseismic monitoring for large volume concrete projects. The equivalent P-wave velocity of the heel is determined to be 4300 m/s, and the positioning error of the system is controlled within 8 m,

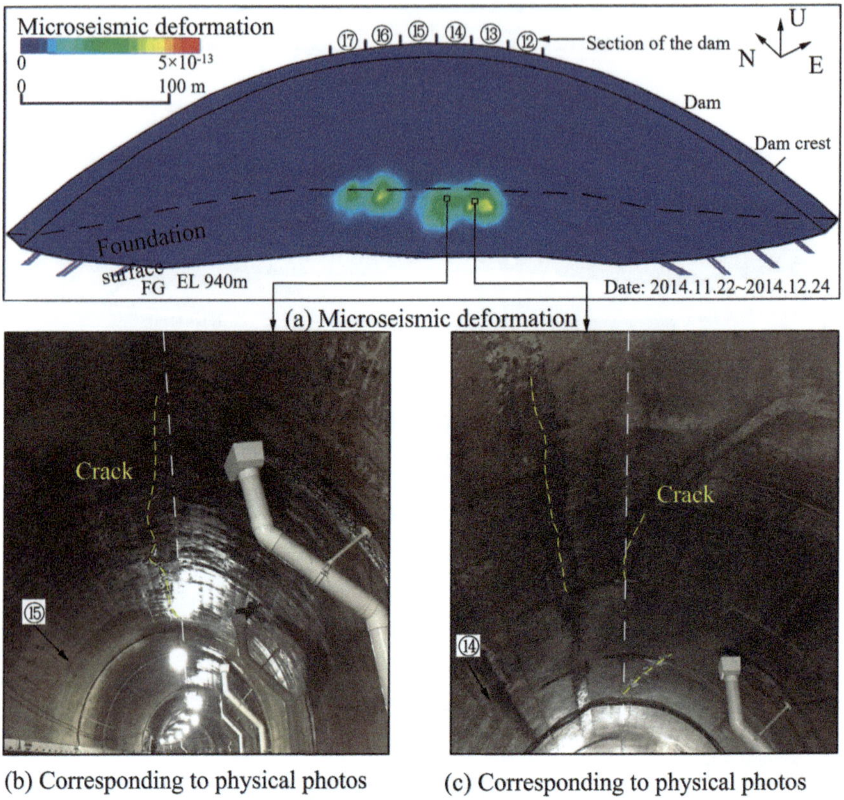

Fig. 3.12 Microseismic deformation and cracks of base gallery at EL 940 m (Dec. 24, 2014)

so as to realize the all-weather monitoring of the microcracks evolution in the dam heel during initial impoundment period.

(2) The microseismic activity in the dam heel of Dagangshan high arch dam is closely related to the reservoir level during initial impoundment period. Under high reservoir level, the microseismic activity in the dam heel increases significantly. Under the joint action of the self-weight of the arch dam and the upstream water load, the stress distribution in the junction area of the arch dam, the left and right bank foundations and the concrete pedestal is uneven, which is the main distribution area of microcracks in the dam heel.

(3) After the diversion tunnel is locked, the microseismic aggregation area of the arch dam of Dagangshan hydropower station moves from the upstream to the downstream of the foundation surface, and realizes the transfer from the heel to the toe of the dam after the bottom outlets is locked. The compressive deformation of the dam heel decreases, and the deformation of the dam toe increases.

(4) The MMS for the dam heel of the high arch dam at Dagangshan hydropower station reproduces the phenomenon of microcracks aggregation induced by the self-weight load at the arch crown of section #13–#14 of the foundation gallery at EL 940 m before locking of the diversion tunnel. It defines the potentially dangerous area of the arch dam concrete, offering a reference for the analysis of the real working behavior of the high arch dam.

3.6 Study on Crack Formation and Propagation in the Galleries of the Dagangshan High Arch Dam in Southwest China Based on Microseismic Monitoring and Numerical Simulation

The high concrete arch dam is widely used around the world benefitting from its advantages of high bearing capacity, outstanding seismic performance and high economic efficiency, etc [79–81]. Because of the rapid development of hydropower energy, more arch dams have been constructed in mountains and canyons. The dam height has exceeded 300 m, indicating that the construction technology of high arch dams has become mature [82–84]. However, due to the complex factors, such as drying shrinkage, thermal stress, hydraulic thrust, deterioration of concrete, uneven settlement of the dam and foundation, nearly every high arch dam has suffered from unexpected cracking during the construction stage and in the service conditions [85, 86]. Various cracks appear in many sections of the dam-foundation system, affecting the integrity and safety of the dams, which has been one of the most challenging problems [87–89]. The text discusses extensive research conducted by scholars both domestically and internationally. For instance, Sheibany et al. [90]. used three-dimensional finite element simulations to analyze the distribution of temperature-induced stress during the operational period of the Karaj arch dam under the influence of air temperature, reservoir water temperature, and solar radiation. They found that the downstream face of the arch dam experienced a significant temperature rise due to the absorption of summer solar radiation, leading to increased tensile stress and widespread cracking on the downstream face and the dam crest. Jin et al. [91] applied a simplified MgO concrete autogenous strain model to nonlinear finite element simulations of the temperature and stress fields of the Changsha arch dam. They suggested that the early hydration heat during arch dam construction, high concrete temperature gradient, and rapid autogenous swelling rate induced concrete cracking. Another study by Jin et al. [92] utilized the ASHRAE Clear Sky model and ray-tracing algorithm to simulate the uneven temperature distribution caused by solar radiation on the arch dam. They investigated the temperature stress on the downstream face, along with the characteristics of crack initiation and propagation. Abdulrazeg et al. [93] observed that high-temperature zones could form in the concrete near the upstream side of the dam body, close to the dam abutment and foundation, with the temperature gradient formed by environmental temperature being a critical factor for inducing tensile stress distribution and cracking. Li et al.

[94] established a regression estimation method for online monitoring of arch dam cracks based on a crack safety monitoring statistical model and mutation theory. They pointed out that crack abnormal behavior is mainly influenced by the concrete age and temperature stress. Santillan et al. [95] studied the impact of future climate change on the temperature, deformation, and cracking of arch dams using a finite element model. Li et al. [96] conducted wedge splitting tests on high arch dams' cast-in-place concrete, establishing a relationship between equivalent maturity and fracture parameters, describing the cracking characteristics with temperature and age variations. Zhang et al. [97] proposed strict temperature control and intelligent means for overall temperature control and construction quality monitoring of high arch dams, aiming to improve work efficiency and prevent temperature-related cracks. However, despite effective temperature control measures [98], the Dagangshan high arch dam still experienced multiple continuous transverse cracks on arch crown of the low elevation gallery before impoundment, posing a serious threat to the safety of the high arch dam during water storage and operation. Therefore, it is necessary to study the mechanism of cracks in the high arch dam gallery.

In recent years, scholars both domestically and internationally have conducted research on the factors causing cracks in high arch dam galleries. Malla et al. [99] used a three-dimensional finite element model to analyze the stress state of the gallery during the dam's operation, discussing the influence of alkali-aggregate reaction-induced concrete volume expansion on the formation of gallery cracks. Wieland et al. [100] installed crack gauges on already cracked galleries of the arch dam, monitoring the changes in gallery crack width over time and analyzing the correlation between gallery crack width and seasonal temperature. Chen et al. [101] used finite element numerical simulation to study the stress state of the gallery under the self-weight and temperature effects during arch dam construction. They believed that the top arch of the high arch dam's gallery is more affected by self-weight stress than temperature, and the gallery located in the inverted part of the high arch dam is prone to surface cracks due to circumferential tensile stress exceeding the limit. Although the aforementioned studies revealed critical factors causing galleries to crack under different conditions, they lack a detailed explanation of the formation and evolution process of gallery cracks. Without establishing the initiation threshold of gallery cracks, the inherent mechanism and evolution mechanism of gallery cracks cannot be comprehensively revealed.

Therefore, this chapter takes the high arch dam of the Dagangshan hydropower station as an example. It utilizes microseismic monitoring technology to conduct real-time monitoring of gallery microcracks before and after impoundment, analyzes the spatiotemporal distribution patterns of microcracks in the low-elevation gallery of the arch dam before and after impoundment, reproduces the process of gallery crack formation, and proposes a microseismic threshold for top arch cracking. Simultaneously, using ABAQUS software, the chapter simulates the background stress field of the high arch dam gallery under both empty and full reservoir conditions. It explores the inherent driving forces behind the formation and evolution of gallery cracks, combines microseismic monitoring results to reveal the mechanism and process of gallery crack formation, and evaluates the stability of gallery cracks.

3.6.1 Characteristics of the High Arch Dam Gallery

The Dagangshan arch dam is located in the Dadu River Basin in Sichuan Province. The total reservoir capacity is 742 million m^3 and the installed capacity is 2600 MW. The basic seismic intensity at the dam site is Grade VIII. The corresponding fortification criterion is a peak horizontal acceleration of 0.5575 g in the base rock with the exceedance probability of 2% in 100 years, which is the most stringent among similar projects in the world. The Dagangshan arch dam is a concrete double-curvature arch dam (Fig. 3.13a). The elevation of the dam crest is EL 1135 m, and the width is 10 m. The elevation of the lowest foundation surface is EL 925 m, and the width of dam base is 52 m (Fig. 3.13b). The maximum dam height is 210 m and the normal water level is EL 1130 m. The arch dam is casted in 29 sections with a section length of 22 m, which are numbered from #1 to #29 in sequence from the left at EL 937 m. Multi-layer galleries are set inside the high arch dam of Dagangshan hydropower station. Multiple galleries are built inside the Dagangshan high arch dam (Fig. 3.13c), including the upper inspection gallery, the lower inspection gallery, the access gallery, the foundation gallery, the drainage gallery, the climbing gallery, etc.

The galleries are inter-lacing, so as to meet the requirements of traffic, inspection, grouting, monitoring, etc. Table 3.2 is a summary of main galleries of the Dagangshan high arch dam. 5 layers of galleries are located at different elevations, namely, the

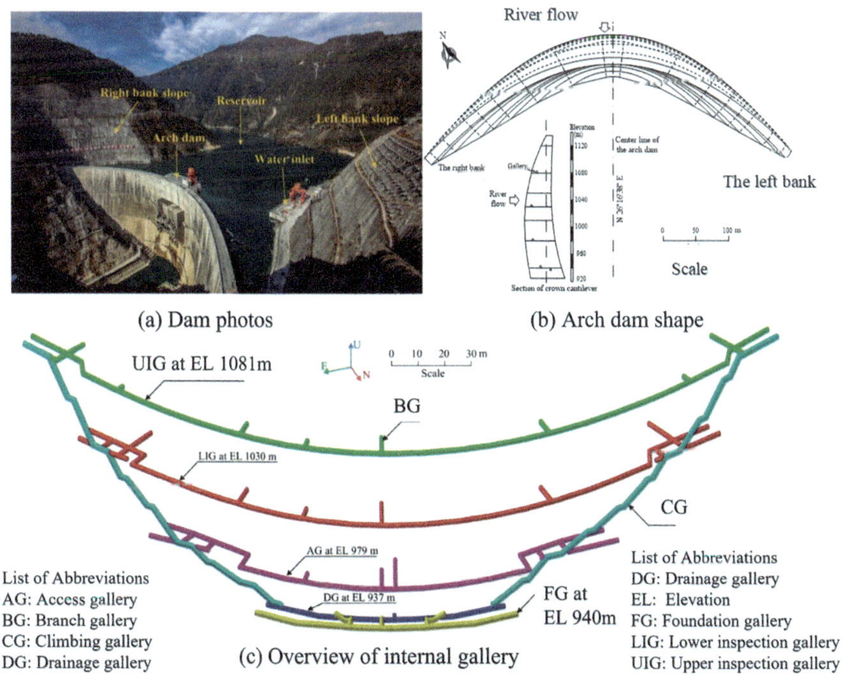

Fig. 3.13 Overview of the Dagangshan high arch dam and internal galleries

upper inspection gallery at EL 1081 m, the lower inspection gallery at EL 1030 m, the access gallery at EL 979 m, the foundation gallery at an elevation of 940 m and the drainage gallery at EL 937 m. The foundation gallery and the drainage gallery are constructed in the riverbed concrete replacement. All the gallery axes are transverse to the river direction, and the cross sections of the galleries are city-gate shaped. The cross-sectional dimensions are 3.0 m × 3.5 m (width × height) for the foundation gallery, 2.0 m × 2.5 m for the drainage gallery, and 2.5 m × 3.0 m for the other three galleries. Concrete pouring for the arch dam was started in September 2011 and completed in November 2014. On December 30, 2014, reservoir impounding was started by locking the diversion tunnel, and the reservoir level rose from EL 975 m to EL 1005 m. On May 29, 2015, the bottom outlets were locked, raising the reservoir level from EL 1015 m to EL 1120 m by July 4, 2014. On October 30, 2015, the impounded water level reached the normal water level at EL 1130 m, i.e., the full reservoir condition.

The mountain massifs on both banks of the Dagangshan dam are tremendous with a steep, narrow and symmetrical V-shaped valley. The natural slope angle is generally 35°–65°. The bedrock outcrops mainly include hoar and reddish medium-grained bio-adamellite of the Jinning-Chengjiang period in massive or sub-massive structure. The in-situ stresses are high, many microcracks exist in the bedrock, and the rock mass is subjected to strong unloading toward the free surface of the valley along the existing structural planes. Crisscrossing diabase dikes are developed in granite at the dam foundation (Fig. 3.14), such as the dike β21 in the left bank (110°∠70°), the dike β43 in the right bank (280°∠50°), and the dike β88 in the riverbed (300°∠65°), etc. Though no regional faults are revealed at the dam site, the dam site is characterized by the faults, the crushed and fractured zones and the joints and fractures developed along the fractured dikes. In view of the complex engineering geological conditions at the dam site, a lot of foundation reinforcement measures have been applied, such as concrete pedestals, adit replacements and prestressed anchor cables, etc., which help to improve the bearing capacity of the foundation surface and therefore ensure the stability of the arch dam foundation during construction and operation [102].

Table 3.2 Specifications of galleries

Name	Cross-sectional dimension (width × height) (m × m)	Bottom elevation (m)
Upper inspection gallery	2.5 * 3.0	1081
Lower inspection gallery	2.5 * 3.0	1030
Access gallery	2.5 * 3.0	979
Foundation gallery	3.0 * 3.5	940
Drainage gallery	2.0 * 2.5	937

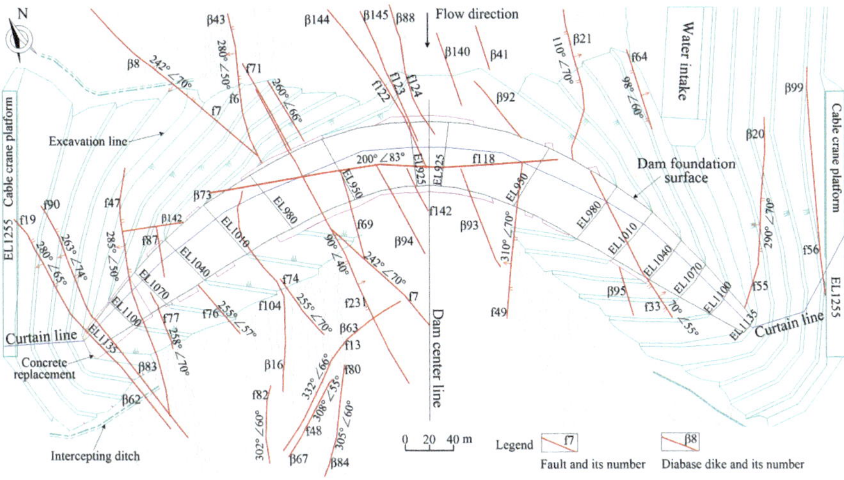

Fig. 3.14 Geological map of the Dagangshan dam site

3.6.2 *Characteristics of Cracking in the Galleries*

During site inspection on 23rd and 24th of December 2014, macrocracks were spotted at the arch crowns of the access gallery sections #11–#19 at EL 979 m and the foundation gallery sections #13–#15 at EL 940 m. The statistical characteristics of the cracks (Table 3.3 and Fig. 3.15) were obtained through field measurement and analysis. The crack depth was primarily determined by the sonic wave testing method. As a supplement, core sampling perforation vertical to the gallery axis was carried out when the crack depth cannot be determined by sonic wave testing at some locations. A typical core sample containing cracks in the access gallery section #17 is shown in Fig. 3.16. The cracks in the access gallery sections #11–#19 were distributed continuously at the center of the arch crown along the direction transverse to the river. The cracks at sections #11 and #19 were 5.0 m and 4.2 m long, respectively, while the cracks ran through the entire section for other sections. Further-more, 2 cracks transverse to the river direction appeared at the arch crown of sections #13-#15 (Fig. 3.15a). The crack width was less than 0.1 mm and the crack depth varied from 0.1 to 3.7 m. The crack depth at the arch crowns of sections #14 and #15 was the greatest, reaching 3.6 m. At the arch crown of the foundation gallery sections #13 and #15, two cracks occurred separately in the position adjoining section #14, with the length of 2.2 m and 2.5 m, respectively. The crack at section #14 ran through the arch crown of the entire section (Fig. 3.15b). The crack width was within 0.15 mm and the depth was between 1.6 and 2.4 m. Such apparent deep cracks in the galleries of high concrete arch dam after casting would certainly affect the structural stresses in the galleries during impoundment, and eventually affect the bearing capacity and safety of the arch dam. Such case is relatively rare in the construction of high arch dams in China.

Table 3.3 Characteristics of cracks in the galleries

Elevation (m)	Section	Crack length (m)	Crack width (mm)	Crack depth (m)	Inspection method
979	#11	5.0	< 0.1	2.2	Sonic wave testing
979	#12	24.1	< 0.1	2.2–3.0	Sonic wave testing
979	#13	22.3	< 0.1	3.0	Sonic wave testing
979	#14	22.7	< 0.1	3.7	Sonic wave testing
979	#15	22.6	< 0.1	3.6	Sonic wave testing
979	#16	23.2	< 0.1	1.2–3.0	Sonic wave testing
979	#17	23.0	< 0.1	0.15	Core sampling
979	#18	23.4	< 0.1	2.2	Sonic wave testing
979	#19	4.2	< 0.1	0.1	Core sampling
940	#13	2.2	< 0.15	1.6–2.4	Sonic wave testing
940	#14	20.7	< 0.15	1.6	Sonic wave testing
940	#15	2.5	< 0.15	1.6–2.4	Sonic wave testing

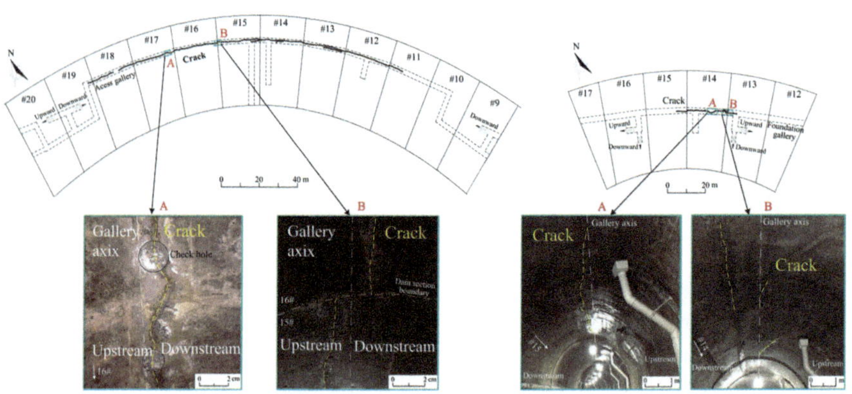

(a) Access gallery at EL 979m (b) Access gallery at EL 940m

Fig. 3.15 Schematic distribution of cracks in the vault of the gallery

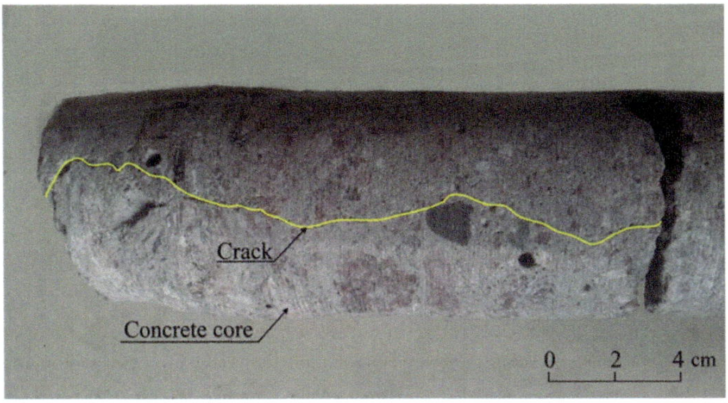

Fig. 3.16 Fracture riding seam core sample diagram

3.6.3 MMS Configuration

Materials such as concrete deform or crack under loads, releasing strain energy in the form of elastic waves. Such phenomenon is called acoustic emission. Passive acoustic emission detection technology, which takes advantage of the local strain energy quickly released in engineering materials, is also known as the microseismic monitoring technology. It can capture the microseismic signals, which can be used to resolve the time and location of microseismic events by the back analysis methods, and further extrapolate the evolution of macroscopic structural failure and the patterns of potential disasters. MMS has been widely adopted in the fields of mines [103, 104], rock slopes [105], underground oil storages [106], etc. In acoustic emission tests on small-scale concrete samples, the high-frequency signals during the failure process are collected for analysis of the acoustic emission source parameters, identification of the defects inside the concrete structure and safety evaluation of the concrete structure. In order to cover larger monitoring scope for large-scale concrete projects, sensors with high sensitivity and good low-frequency response are required to capture more valid microseismic signals [107]. Besides, high-performance and reliable data acquisition instruments and processors are equipped to work together for signal recording, data processing and data storage. On this basis, microcracking is located quickly and microseismic intensity is calculated efficiently for the timely hazard zone delineation of massive concrete engineering. Therefore, for microseismic monitoring of high concrete arch dams, the MMS with wide frequency response, high sensitivity, quick reaction and good reliability shall be chosen.

Analysis of acoustic emissions in concrete under uniaxial compression indicates that, during the early stage of damage, the dominant acoustic emission signals are mostly high-frequency signals in the range of 0.1–1 MHz. With further damage, the frequencies of dominant signals are shifted to kHz level, or even lower [108, 109]. The spectrum analysis of acoustic emission signals during uniaxial compression shows that the dominant frequency of acoustic emissions in concrete samples below

C40 is about 5000 Hz [110]. This also applies to microseismic signals in large-scale concrete structures. Considering that the dominant frequency of microseismic signals tends to decrease with increasing scales of concrete structures, a sensor with high-frequency response of 5000 Hz is capable of capturing the microseismic signals with the dominant frequency in the arch dam concrete. The MMS for galleries in concrete arch dam was established based on this principle (Fig. 3.17a). The sensors adopted in the MMS were uniaxial accelerometers with a frequency response range of 50–5000 Hz and a sensitivity of 30 v/g. The data acquisition instrument is produced by ESG, Canada, and has a sampling frequency of 20,000 Hz and 24-bit analog–digital conversion. Events were triggered by threshold and STA/LTA methods. Specifically, the thresholds in different channels varied with their background noise levels, and the threshold for single channel was adjusted from time to time. The STA window was 30 ms while the LTA window was 100 ms with a STA/LTA threshold of 3. As for noise filtering, a band-pass filter that removes frequencies higher than 5000 Hz and lower than 50 Hz, a baseline filter that removes a constant DC offset and a pole filter for suppressing relatively low frequencies were configured. Moreover, the MMS was equipped with the MMS-view remote wireless transmission and visualization system. Thus, real-time acquisition and remote transmission of microseismic signals can be achieved. According to the distance-frequency relationship for microseismic monitoring technique, the detectable distance corresponding to the dominant frequency of 5000 Hz is around 50 m (Fig. 3.18) Considering the attenuation law of microseismic signals in concrete, the distance between the emission source and the sensor should be less than 50 m. Therefore, in view of the construction conditions for the low-elevation galleries in the Dagangshan arch dam, the sensor array was designed as shown in Fig. 3.17b. Six accelerometers were arranged inside the foundation gallery at EL 940 m near the dam heel and the drainage gallery at EL 937 m close to the dam toe. The average sensor spacing was about 45 m. Sensors S1, S3 and S5 were installed vertically downwards into the gallery baseplate. Sensor S2 was installed horizontally into the side wall at the downstream side of the drainage gallery at EL 937 m. Sensors S4 and S6 were installed in the side wall at the upstream side of the foundation gallery at EL 940 m with an inclination angle of 45°. The accelerometers formed an effective spatial array ensuring the spatial uniqueness for positioning of the microseismic event. The MMS in the galleries can cover the dam sections #13 – #17 below EL 980 m, continuously monitor the access gallery, the foundation gallery and the drainage gallery, and capture the source parameters of microcracks in real time.

3.6.4 Positioning Accuracy Test

For verification of the MMS in the galleries of the Dagangshan arch dam, a knocking test was carried out at 11 points within or outside the scope of the sensor array on November 24, 2014 and the time, location and test waveform of the knocking tests were recorded. The P-wave velocities range from 4198 m/s to 4440 m/s with a mean

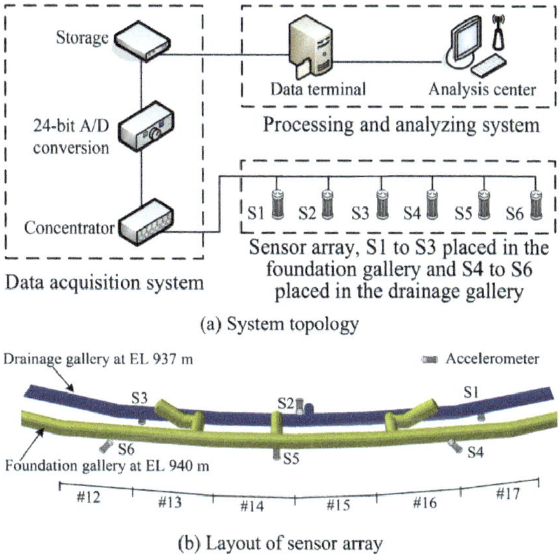

Fig. 3.17 MMS for the galleries

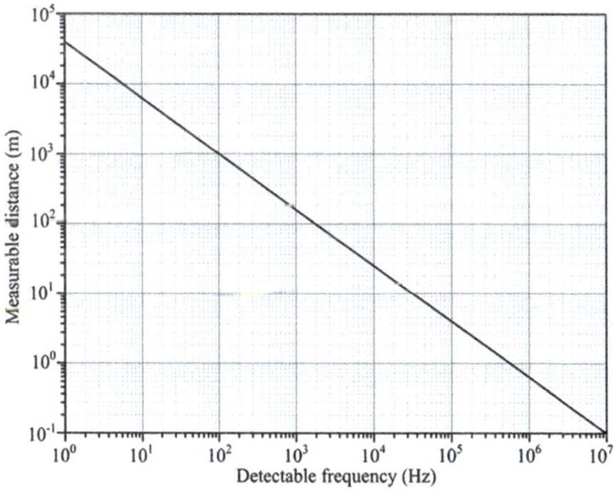

Fig. 3.18 Relationship curve between measurable frequency and monitoring distance [76]

value of 4311 m/s in different knocking tests. Besides, according to the wave test of concrete samples from the Dagangshan arch dam, the velocity has a range of 3700–4800 m/s. By traversing the velocities within the range with an interval of 100 m/s, MMS gives the positioning errors of the knocking tests. When the P-wave velocity is 4300 m/s, the average errors be-tween the actual knocking positions and the source locations reach the minimum value of 5.1 m. It is noticeable that the overall P-wave velocities acquired by two different methods are almost the same. Therefore, the P-wave velocity adopted by the MMS in the galleries of the Dagangshan arch dam is 4300 m/s and the corresponding positioning errors of 11 knocking tests are controlled within 8 m [110]. The time–frequency characteristics of signals from knocking and

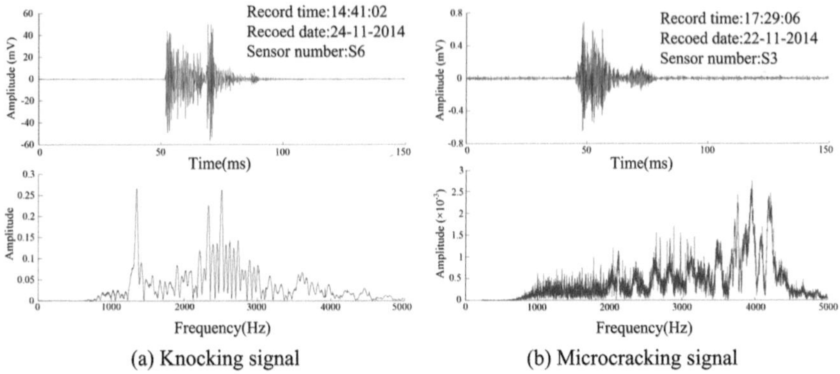

Fig. 3.19 Time–frequency characteristics of signals collected by the MMS

microcracking captured by the MMS are shown in Fig. 3.19. It can be seen from Fig. 3.19a that the knocking test signal has two peaks with an amplitude of about 50 mV. The frequency distribution is relatively dispersed and the main frequency is lower than 2500 Hz. The amplitude of typical microcracking signals is 0.6 mV and the signal coda is more developed. The signal energy is mainly concentrated in the high-frequency area, and the main frequency is distributed at about 4000 Hz (Fig. 3.19b). By analyzing the time–frequency characteristics of different types of signals in the arch dam, the event triggering threshold and filtering parameters of the MMS can be determined, which have been introduced above. This allows automatic filtering of noise signals from construction or background, and effective collection and storage of the microcracking signals.

3.6.5 Microseismic Identification of Gallery Crack Formation in High Arch Dams

Monitoring of microseismic events was started from November 22, 2014 till October 31, 2015. During this period, the arch dam pouring was completed, and the impounded water level rose from EL 975 m to EL 1130 m gradually. The spatial distributions of microseismic events within 8 m away from the access gallery, the foundation gallery and the drainage gallery are shown in Fig. 3.20. Each ball in the figure represents a microseismic event and its color represents the moment magnitude. During this period, 287 microseismic events were detected around three galleries, mainly distributed near the foundation gallery and the drainage gallery. The microseismic events near the access gallery were mainly concentrated at sections #14–#16, with some sparsely distributed at sections #13 and #17. The moment magnitudes were between − 3.5 and − 2.8. The microseismic events near the foundation gallery were concentrated at sections #13–#16, and the moment magnitudes were mostly smaller than − 3.5. Sections #14–#15 of the drainage gallery were also a concentrated zone of microcracks with moment magnitude smaller than − 3.5, which is also the case at the foundation gallery. By comparing the spatial distributions of microseismic events in the galleries at different elevations, it can be found that microcracks initiated in sections. The moment magnitude of microcracking in the access gallery at EL 979 m is significantly greater than two galleries at lower elevations. However, the count of microseismic events (MS count) is smaller. The accelerometers were installed in the galleries at EL 940 m and EL 937 m, which allowed more microcrack signals to be detected near the two galleries. The microcrack signals with low moment magnitudes near the access gallery at EL 979 m, however, were not collected due to signal attenuation. This resulted in more microseismic events with small moment magnitude in the two lower galleries than the upper one.

Seismic moment is a basic parameter representing the source strength of a microseismic event and the mechanical state at the source. It is directly related to the physical essence of the microcracking process, and is often used for the study of the intensity and evolution of microcracks in an engineering structure [111]. The cumulative moments of microseismic events over time in the access gallery, the foundation gallery and the drainage gallery are presented in Fig. 3.21. Here, the volume of interest is a cubic area within 8 m away from the corresponding gallery and the period of

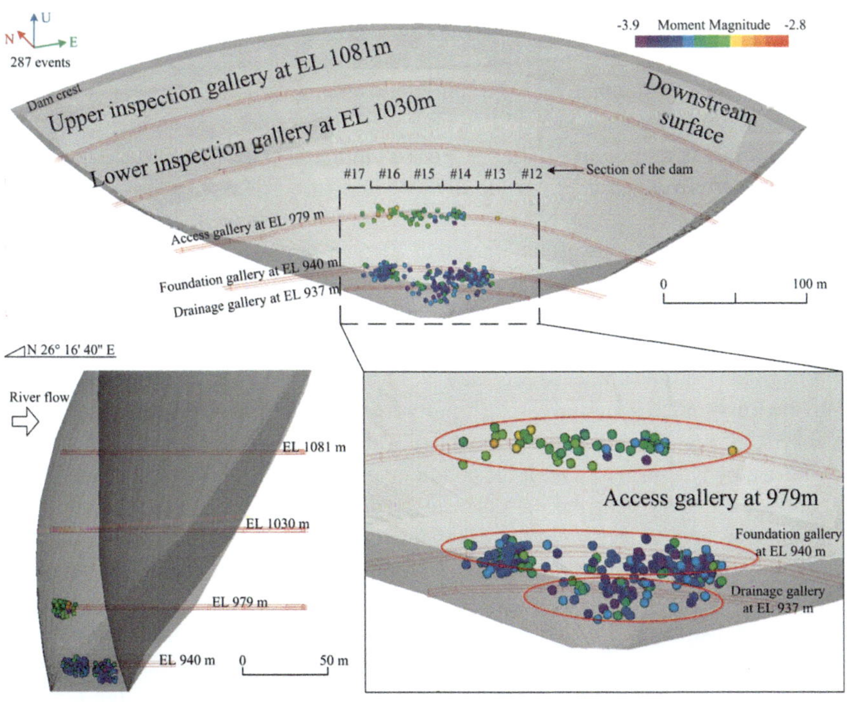

Fig. 3.20 Spatial distribution of microseismic events in the galleries of the Dagangshan arch dam

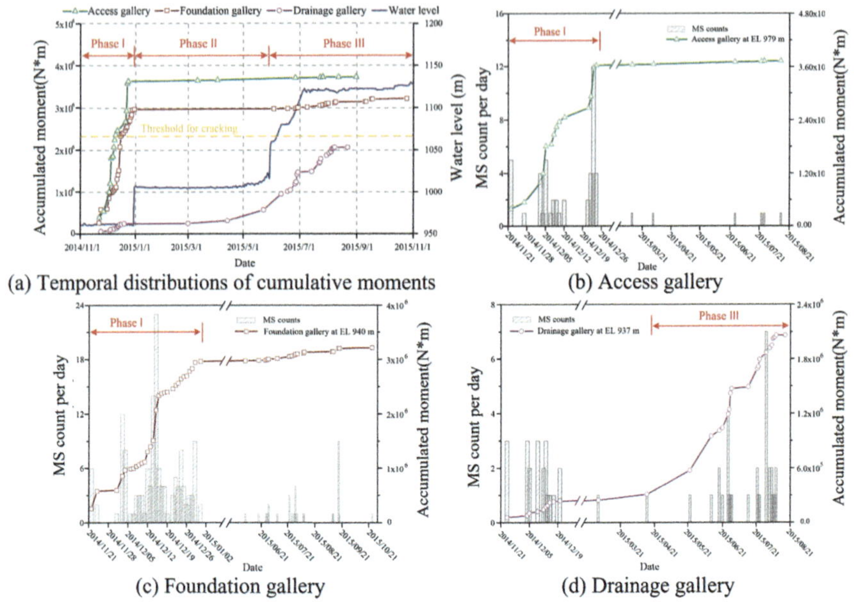

Fig. 3.21 Temporal distributions of MS counts and cumulative moments in galleries

time is from the startup of the MMS to the origin time of microcracking. Figure 3.21a shows the relationship between the cumulative moments in the three galleries and the impounded water levels. From November 22, 2014 to October 31, 2015, the cumulative moments in the three galleries were significantly correlated to the impounded water levels. In general, it can be divided into three phases, namely phases I, II and III. Phase I corresponds to the time before December 30, 2014, during which the impounded water level maintained at EL 975 m, basically corresponding to an empty reservoir. The arch dam was mainly subjected to the gravitational stress. Phase II corresponds to the time between December 31, 2014 and May 29, 2015, during which the impounded water level was between EL 1005 m and EL 1015 m. Phase III corresponds to the time between May 30, 2015 and October 31, 2015. Before July 4th the impounded water level rose rapidly from EL 1015 m to EL 1120 m at a rate of 2.74 m per day. After that, the impounded water level rose to the normal level at EL 1130 m slowly. During Phase I, the cumulative moments in the access gallery and the foundation gallery near the dam heel increased significantly, while the cumulative moment of the drainage gallery close to the dam toe increased slightly. The changes of cumulative moments in the three galleries during Phase II were insignificant, with the exception for the drainage gallery where a minor increase was observed in the later stage. During Phase III, the cumulative moments of the two upper galleries were almost the same as that of Phase II. However, the cumulative moment in the drainage gallery followed the increasing trend in the previous phase and rose significantly. It finally became stable at 2.06e6 N m, which is lower than the cumulative moments in the two upper galleries, as shown in Fig. 3.21a. The main reasons for the phenomena happened in three phases are as follows. The daily MS counts and the cumulative moments in the three galleries were distributed in phases. To be specific, the increase of the daily MS count and the cumulative moment in the access gallery at EL 979 m mainly occurred in Phase I. On December 5, 2014 the daily MS count reached 5 and then decreased to 2 from December 12 to December 22. As to the cumulative moment, two periods of high growth in the microseismicity were observed in Phase I. The first period is from December 5–12 and the second is from December 23–24. In the first period, the access gallery suffered a significant increase of cumulative moment to the value of 2.3e6 N m. After Phase I, the microseismicity restored to a low level. The microseismicity in the foundation gallery has similar characteristics. The daily MS count significantly increased on December 3 and 15, and remained at a high-level during Phase I. The cumulative moment underwent a sharp increase to the value of 2.3e6 N m on December 15 and then the microseismicity significantly decreased, followed by a minor increase of the cumulative moment as the impounded water level rose. In contrast, microseismicity in the drainage gallery was concentrated before December 20, 2014 and after May 30, 2015, accompanied by the slow increase of cumulative moment before December 20, 2014. As the impounded water level rose, the cumulative moment increased significantly in this gallery.

Deriving from seismic deformation in quantitative seismology, microseismic deformation refers to the average microseismic strain of microseism induced by loads. It is an effective parameter for comprehensively evaluating the characteristics of microcracks in concrete or rock mass providing an insight to the cracking process [112]. The distributions of microseismic deformation in the section of the arch crown cantilever of the Dagangshan arch dam before and after impoundment are shown in Fig. 3.21. "Before impoundment" corresponds to the days before December 30, 2014 or Phase I in Fig. 3.21a. "After impoundment" corresponds to the days between December 31, 2014 and October 31, 2015, or Phase II and III in Fig. 3.21a, during which the impounded water level rose from EL 975 m to EL 1130 m gradually. Before impoundment, large microseismic deformation was distributed around the access gallery and the foundation gallery near the arch dam heel. The microseismic deformation of the upper gallery was significantly greater than that of the lower gallery (Fig. 3.22a). After impoundment, the microseismic deformation of the access gallery in the section of the crown cantilever decreased significantly. The large microseismic deformation zone transferred to some locations around the drainage gallery close to the arch dam toe. The microseismic deformation was notably smaller than that before impoundment (Fig. 3.22b). It is worth mentioning that, before impoundment (Phase I), there were varying degrees of cracking observed in the crown arches at elevations of 979 and 940 m. In contrast, the gallery at an elevation of 937 m did not exhibit macroscopic cracks before or after impoundment. This observation corresponds well with the distribution and evolution characteristics of cumulative seismic moment and microseismic deformation in the arch dam of the Dagangshan hydropower station. Based on this, it can be inferred that under the self-weight of the arch dam, the galleries at elevations of 979 and 940 m, near the dam heel, experience higher stress, leading to the continuous initiation and accumulation of microcracks. This results in the formation of a zone with high microseismic deformation, ultimately triggering macroscopic cracks in the crown arches of these galleries. The microseismic deformation in the access gallery and the foundation gallery reached the threshold for cracking before impoundment in the section of the arch crown cantilever, leading to crack formation at the arch crown of galleries. As the impounded water level rose, the microseismic deformation zone transferred to the drainage gallery close to the arch dam toe. However, the microseismic deformation in the drainage gallery consistently remained lower than the threshold. Consequently, no macrocrack was formed. By correlating the time points of cumulative seismic moment and microseismic deformation in Figs. 3.21 and 3.22 with the occurrence of macroscopic cracks, an attempt can be made to infer the threshold values of cumulative seismic moment and microseismic deformation corresponding to the crown arch cracking in the galleries of the Dagangshan hydropower station's arch dam before and after impoundment. Assuming that the galleries at elevations of 979 and 940 m share the same structural measures and construction methods, and that the mechanisms of microcrack initiation are consistent in both levels of galleries, the threshold for the formation of crown arch cracks in both galleries should also be the same. Based on this assumption, the threshold for the cumulative seismic moment leading to crown arch cracks is determined to be 2.3×10^6 N m (as indicated by the dashed line in Fig. 3.21a), and the threshold for microseismic deformation is set at 8×10^{-13}. Consequently, the cracking of the crown arch in the gallery at an elevation of 979 m on December 12th indicates a rapid increase in cumulative seismic moment before this date, followed by a slowing growth rate in the subsequent week. This suggests that the formation of crown arch cracks in the gallery at 979 m involves the release of local high stress in the crown arch. Similarly, the appearance of macroscopic cracks in the gallery at an elevation of 940 m on December 15th is followed by a slowdown in the growth of cumulative seismic moment. The evolution process of cumulative seismic moment in both galleries, reflecting stress accumulation and release trends, to some extent, validates the speculated microseismic threshold for crown arch cracking in the galleries. It should be noteworthy that the threshold of microseismic deformation may vary significantly in different MMS due to the differences in frequency range, system configuration and inversion algorithm for microseismic source. Thus, attention should be paid to the settings of the MMS when applying to similar cases.

1. 979 m Access gallery

Based on the cumulative moments, microseismic deformation and their thresholds for cracking in the galleries, the cracking processes of the access gallery, the foundation gallery and the drainage gallery can be analyzed. The distributions of microseismic deformation in the three galleries of the Dagangshan arch dam before and after impoundment are shown in Fig. 3.23. In this figure, the horizontal sections of dam body at EL 979 m, EL 940 m and EL 937 m are projected on the microseismic deformation nephogram. As is shown in Fig. 3.23a, the microseismic deformation was relatively high at sections #14–#16 of the access gallery, and a large microseismic deformation zone was generally distributed along the axis of the gallery and tended to extend toward the section #17 before impoundment (Fig. 3.15a). Accordingly, a

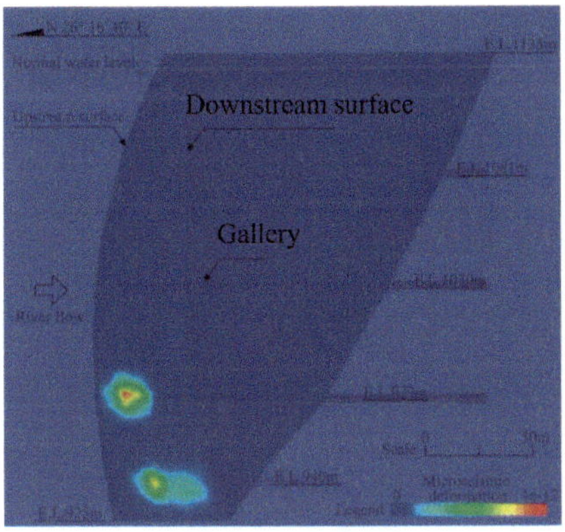

(a) Before impoundment

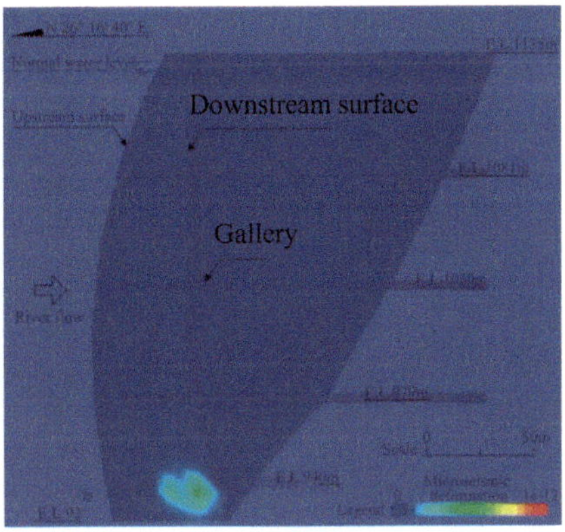

(b) After impoundment

Fig. 3.22 Distribution of microseismic deformation in the section of arch crown cantilever

potential danger area of the access gallery can be delineated based on the distribution of high microseismic deformation. Referring to Fig. 3.21b, the cumulative moment of the access gallery rose rapidly on December 12, 2014. The microseismic deformation of the access gallery also accumulated dramatically and the concentration of microseismicity exceeded the threshold for cracking in the galleries. Hence, a macrocrack was formed at the arch crown of the access gallery on December 12, 2014 with the cracks at the dam sections #14–#16 being the most severe. It can be found that the cracking process in the access gallery included microcrack initiation, concentration and coalescence until the formation of macrocracks. Moreover, the distribution of high microseismic deformation was basically consistent with the crack distribution in the access gallery (Fig. 3.15a), and the large microseismic deformation zone may spread toward the section #17 of the gallery. It can be inferred that the cracking of the access gallery would be deteriorated and the cracks may propagate if the arch dam continues inclining upstream. After impoundment on December 30, 2014, the microseismic deformation of the access gallery decreased significantly (Fig. 3.21b). At the same time, chiseling, drilling, pipe laying and grouting treatment were performed for the access gallery. A joint meter along the river direction was installed at the middle of arch crown at section #14. The joint meter record is shown in Fig. 3.24, in which a positive value represents opening of the joint and a negative value represents closing of the joint. After gallery treatment measures were applied, the cracks at the arch crown of section #14 were continuously closing under compression as the impounded water level rose, except small fluctuation during the initial stage of installation. It can be concluded that the cracks in the access gallery would not propagate further but become closed.

2. 940 m Foundation gallery

The cracking process in the foundation gallery at EL 940 m was similar to that of the access gallery. Before impoundment, the microcracks in the foundation gallery were

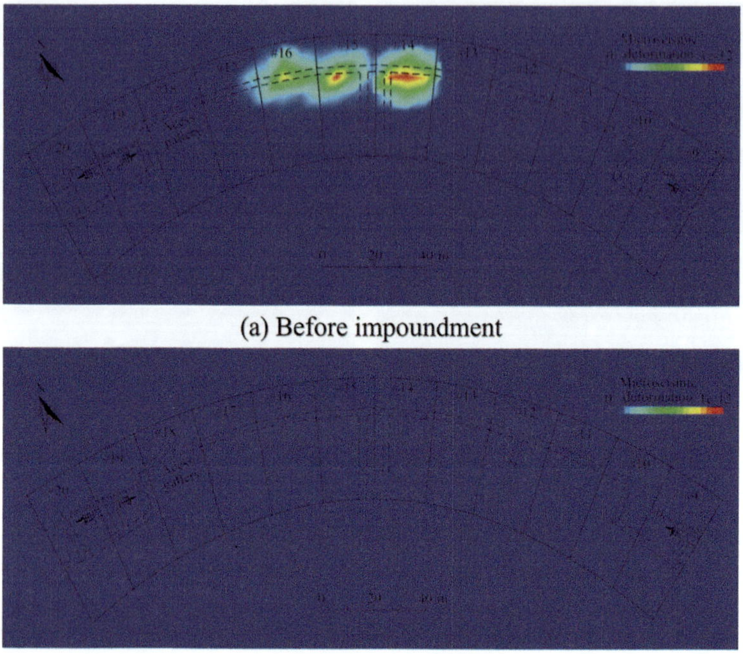

(a) Before impoundment

(b) After impoundment

Fig. 3.23 Distribution of microseismic deformation at the horizontal sections of the galleries

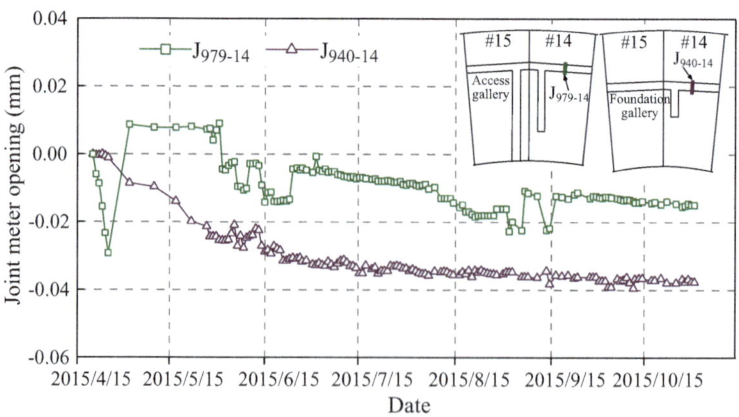

Fig. 3.24 Joint meter records at the arch crown of the galleries

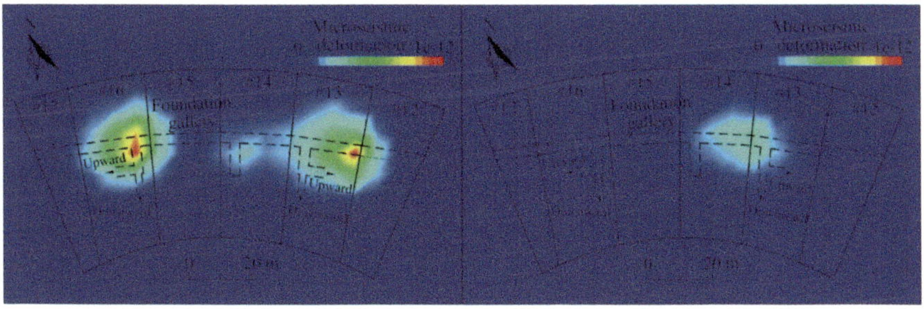

(a) Before impoundment (b) After impoundment

Fig. 3.25 The distribution map of microseismic deformation along the 940.00 m elevation profile

concentrated at sections #13, #14 and #16, among which the intersection between #13 and #14, and the middle part of section #16 were both high microseismic deformation zones (Fig. 3.25a). The foundation gallery was also subjected to the gravitational stress before impoundment and microcracks initiated and accumulated continuously. As a result, cracking occurred at sections #13 and #14 on December 15, 2014 and the cracks may propagate towards the dam section #15. Furthermore, the microseismic deformation at section #16 of the foundation gallery was relatively high before impoundment. It was probably caused by the stress concentration in the connection of the climbing gallery at EL 940 m. After impoundment, the microseismicity in concrete at the foundation gallery decreased significantly. The microseismic deformation was distributed solely at section #14 adjacent to section #13 (Fig. 3.25b), and the values, which were lower than the threshold, decreased remarkably. Hence, no crack growth occurred at the arch crown of the foundation gallery. According to Fig. 3.24, the joint meter at the middle of arch crown of section #14 of the foundation gallery always showed a negative value, indicating that the cracks at the arch crown were closing, which was consistent with the microseismic deformation analysis result.

3. 937 m Drainage gallery

The distribution of the microseismic deformation in the drainage gallery at EL 937 m was very different from those of the access gallery and the foundation gallery. Before impoundment, only some low magnitude microcracks at the intersection between sections #14 and #15 were detected, and the microseismic deformation was relatively small (Fig. 3.26a). After impoundment, more microcracks initiated and transferred towards section #14 with a relatively large microseismic deformation zone distributed at section #14 of the drainage gallery (Fig. 3.26b). The microseismic deformation of the drainage gallery was smaller than the cracking threshold before and after impoundment. According, no macrocracks appeared in the gallery. However, the microseismicity accumulated and transferred continuously during the impounding process, indicating a potential danger zone which needs caution during operation. Therefore, the monitoring and measurement at sections #14 and #15 of the drainage

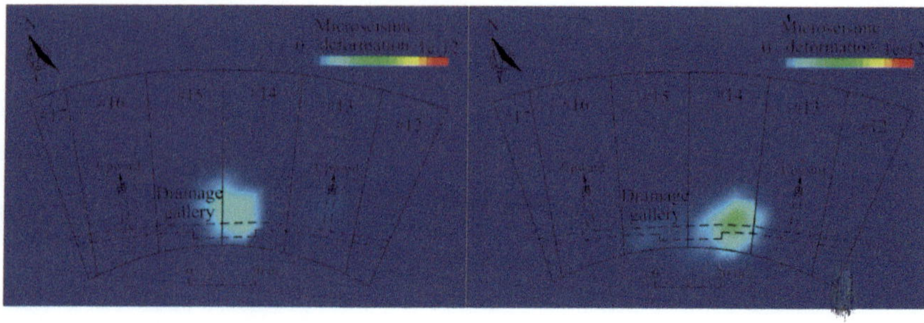

(a) Before impoundment (b) After impoundment

Fig. 3.26 Microseismic deformation distribution along the 937.00 m elevation profile

galleries should be enhanced. The evolution law of microcracks in this gallery shall be closely analyzed in order to guide the gallery maintenance during the impoundment and operation of the arch dam.

3.6.6 Numerical Simulation of Gallery Crack Formation in High Arch Dams

1. Numerical model

In order to obtain the background stress field of the galleries in the Dagangshan high arch dam and explore the internal driving force for crack formation and evolution in the galleries, a 3D finite element model of the Dagangshan arch dam was constructed using ABAQUS (Fig. 3.27). The model covers 1200 m along the transverse direction, 610 m in the vertical direction from 100 m above the dam crest to 300 m below the arch dam bottom, and 630 m along the river direction. The total number of tetrahedron meshes is 835,802, among which 684,832 meshes are for the dam body, including the gallery model [113]. The ideal Drucker-Prager constitutive model was chosen due to its ability to reflect the mechanical parameters of concrete as well as its convergence. The constitutive parameters of Drucker-Prager were obtained by converting the cohesive strength c and the friction angle of the Mohr–Coulomb model [114]. The model parameters were obtained based on laboratory tests, as shown in Table 3.4, where $f = \tan \phi$.

2. Model verification

The displacement curves of the arch crown cantilever along the river direction after the reservoir impounding to the normal water level at EL 1130 m are shown in Fig. 3.28, including the numerical simulation results, the geomechanical model results [115], and the in-situ monitoring results. The displacements obtained by numerical simulation increased with increasing elevation, with the maximum displacement appearing at the top crest. The trend was consistent with the geomechanical model results. Meanwhile, the numerical simulation results and the in-situ

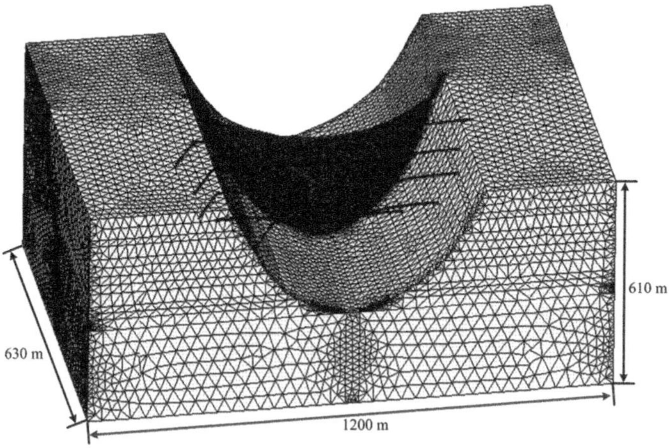

(a) Mesh for the entire model

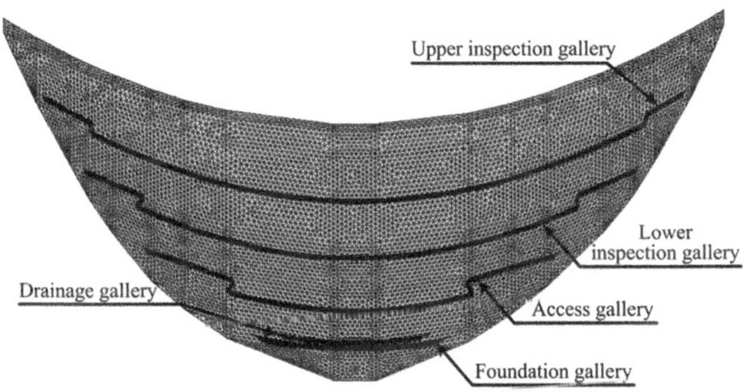

(b) Mesh for the dam body

Fig. 3.27 Numerical model of the Dagangshan arch dam

Table 3.4 Material parameters of numerical model of Dagangshan arch dam

Material	ρ/(kg/m^3)	E/GPa	μ	c/MPa	f
Dam body	2,400	25.2	0.16	2.5	1.7
Rock mass of dam foundation	2,700	16.7	0.25	1.6	1.4

monitoring results of the crown cantilever at EL 1081 m were very lose, but the in-situ monitoring results indicate that the displacements of the arch crown cantilever at EL 1081 m and EL 1135 m almost remained the same. This is probably related to the reservoir deformation, such as valley contraction [14], which was not considered in the numerical simulation in this study.

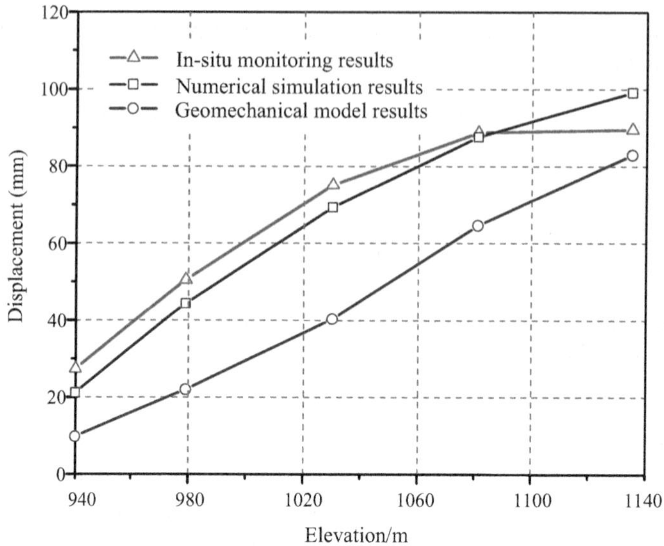

Fig. 3.28 Displacements of the arch crown cantilever along the river direction after impoundment

The distribution of minor principal stresses in the galleries of the Dagangshan arch dam in the case of empty reservoir is shown in Fig. 3.29. In this case, the arch dam was subjected to gravitational stress, and varying levels of tensile stresses occurred in the low-elevation galleries, especially in the access gallery and the foundation gallery (Fig. 3.29a). Relatively high tensile stresses appeared at the top and bottom of the access gallery at EL 979 m. At some locations, the tensile stresses exceeded the control standard of 1.5 MPa in Design specification for concrete arch dams [116]. The area with tensile stress exceeding the control standard was continuously distributed along the gallery and reached a depth of 1.6 m in the section of the crown cantilever (A in Fig. 3.29b). The distribution area of tensile stress in the foundation gallery at EL 940 m was smaller than that of the access gallery. The tensile stresses were mainly distributed transverse to the river direction at sections #13–#16 (Fig. 3.29a). The depth of tensile stress zone was smaller than that of the access gallery (B in Fig. 3.29b). The locations of tensile stress zone in the two galleries are basically consistent with the crack distribution, which indicates that the root cause of the crack formation at the arch crowns of the galleries is the circumferential tensile stresses which exceed the concrete tensile strength. To be specific, as the access gallery is located at the overhanging part of the dam, the overlying gravitational stress results in the continuous circumferential tensile stress transverse to the river direction at the arch crown, and therefore cracking at the arch crown of the gallery. The height of the dam body over the foundation at EL 940 m is the greatest. However, it is further away from the overhanging part of the dam compared to the access gallery. Thus, the degree of cracking in the foundation gallery is less severe than the access gallery. The analysis results of microseismic deformation have also revealed the similar phenomena.

In the case of full reservoir, the distribution of minor principal stresses in the galleries is significantly different. Under self-weight and hydraulic thrust, the tensile

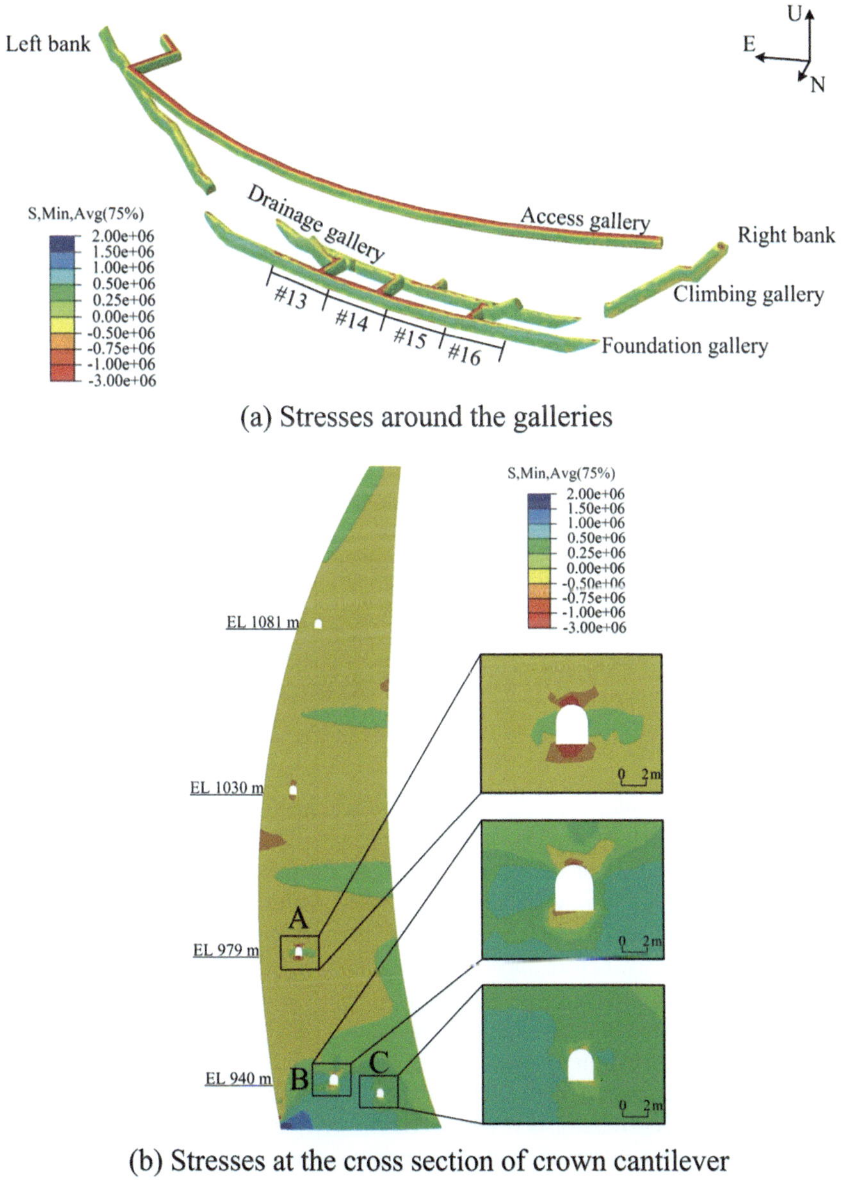

Fig. 3.29 Distribution of minor principal stresses of the arch dam under empty-reservoir condition

stress area in the access gallery vanished (A in Fig. 3.30b). The tensile stress in the foundation gallery near the dam heel decreased significantly at sections #14–#15, except that some low tensile stresses appeared in the upstream abutment (B in Fig. 3.30b). Apparent tensile stresses in the two galleries before impoundment led to cracking. However, the tensile stresses in the galleries significantly decreased after impoundment, indicating that the cracks due to self-weight would not propagate further but be closed by themselves. It can be found that apparent tensile stresses were also distributed in the drainage gallery close to the dam toe. Compared with the stress distribution before impoundment, the tensile stresses deviated from the centers of gallery arch crown and gallery bottom to the upstream spandrel and the downstream side wall bottom, respectively (C in Fig. 3.30b). Besides, tensile stresses were distributed continuously along the gallery axis. The results indicate that cracking may occur in this gallery after impoundment. The tensile stress distribution and the large microseismic deformation zone obtained by numerical simulation were basically consistent. They together reveal the crack formation and possible propagation in the galleries of the high arch dam.

3.6.7 Chapter Conclusion

In this study, a microseismic monitoring system was established in the Dagangshan high arch dam to study the spatial and temporal distributions of microcracks in the galleries before and after impoundment. The crack formation and propagation in the galleries were investigated. Furthermore, the background stress field, the crack formation and propagation in the galleries of the Dagangshan high arch dam was analyzed by 3D numerical simulation. The cause of crack formation before impoundment and whether the crack may propagate after impoundment were revealed by microseismic monitoring and numerical simulation. The main conclusions are as follows:

(1) The microseismicity in the access gallery, the foundation gallery and the drainage gallery of the Dagangshan high arch dam was highly correlated to the impounded water levels, and presented a phased distribution. The microseismicity of the access gallery and the foundation gallery near the dam heel increased significantly while that of the drainage gallery close to the dam toe increased slightly before impoundment. However, as the impounded water level rose, the microseismicity of the upper two galleries decreased while that of the drainage gallery showed a rapid growth.
(2) Before the formation of macrocracks in the galleries, there was a process of microcrack initiation, concentration and coalescence. According to the spatial and temporal correlation between microseismicity and cracks in the galleries, the cracking thresholds of the cumulative moment and microseismic deformation at the gallery arch crowns can be identified (2.3×10^6 N m and 8×10^{-13}). By the derived thresholds, it is inferred that macrocracks in the access gallery

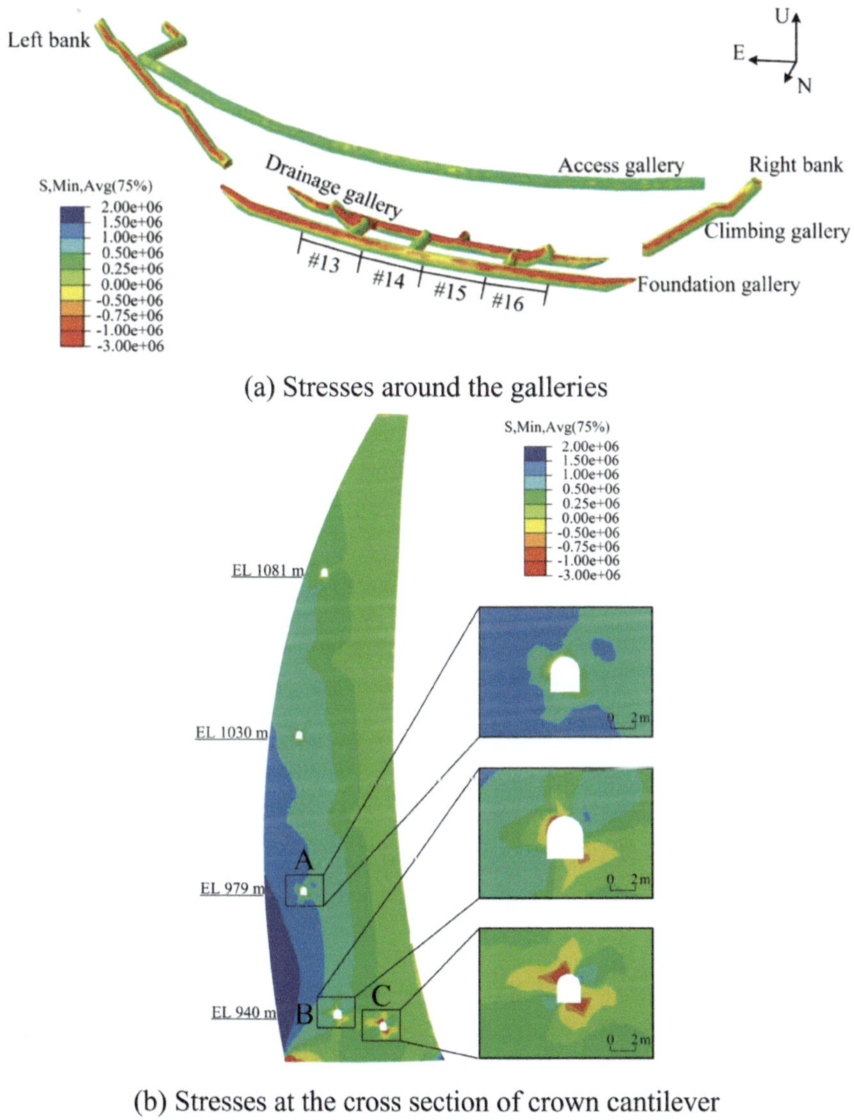

Fig. 3.30 Distribution of minor principal stresses of the arch dam under full-reservoir condition

and the foundation gallery occurred on December 12, 2014 and December 15, 2014, respectively.

(3) During impoundment, the large microseismic deformation zones transferred toward the drainage gallery close to the dam toe. The cracks in the access gallery and the foundation gallery would not propagate further but be closed

as the reservoir water level rose. Although no cracks appeared in the drainage gallery, it is still a potential danger zone after impoundment.

(4) Continuous circumferential tensile stresses appeared in the access gallery and the foundation gallery due to the gravitational stress from the overlying dam body. The tensile stresses exceeded the control standard for concrete arch dam within a depth of 1.6 m, which is the internal cause of deep cracks. The tensile stresses in the galleries at EL 979 m and EL 940 m decreased significantly after impoundment. However, there were noticeable tensile stresses in the spandrel of the drainage gallery at EL 937 m and cracking may occur in this gallery during operation of the arch dam.

3.7 Study on Working Performance Evaluation of the High Arch Dam of Dagangshan Hydropower Station During the Initial Impoundment Period

Ensuring the stability of arch dams is the key to the design and construction. In the full cycle of a concrete high arch dam, the impoundment period is often the stage of an arch dam with a higher probability of hazardous accidents [41, 43, 44]. Due to the large scale and high-water level of the arch dam project, the impoundment process is affected by the water level change, and the water level difference between upstream and downstream increases dramatically, which may lead to local damage to the dam foundation and dam body, resulting in cracking, and easily inducing an arch dam catastrophe, causing serious losses. For example, the Malpasset arch dam in France suddenly broke on December 2, 1959, after five years of impinging, because the seepage caused the rock mass of the left dam foundation to slide under the full head pressure, resulting in its collapse [50]. The Kolnbrein arch dam in Austria suffered deformation of the riverbed foundation during initial impoundment period, and cracking at the heel dam in upstream extended to the bottom gallery, leading to a seepage surge that resulted in large-scale rehabilitation of the arch dam [117]. The Vajoint arch dam in Italy caused a catastrophic landslide event on the upstream left bank slope during the impoundment period due to heavy rainfall and reservoir seepage [51]. The Shuanghe arch dam in China was found to have multiple through cracks in the dam body before impoundment due to weakened foundations in the middle and upper elevations of the left and right bank slopes, and after impoundment there were areas of loosening of the upper dam abutments [118]. The Tolla arch dam in France showed multiple cracks in the dam during initial impoundment period and had to be reinforced with additional arch ribs and arch-ring on the downstream of the dam [87]. The Sayano-Shushenskaya arch dam in Russia showed multiple horizontal cracks at the dam heel during the initial impoundment period, leading to severe seepage in the gallery. The safety and stability of arch dams during the impoundment period has been a key research issue in the engineering community around the world and has attracted extensive attention from many scholars [53]. Huang et al. [119] investigated the stability of the Laxiwa arch dam by using a comprehensive analysis of finite element modelling and physical tests to determine the value of the water load safety reserve for this arch dam. Zhang et al. [120] analyzed the overall stability of Yangfanggou arch dam using a combination of three-dimensional nonlinear finite element (TFINE program) and the laboratory model test, and summarized the displacement and stress distribution law of the arch dam. Li [59] studied the actual occurrence of dam heel cracks in Shimen arch dam, and discussed the judgement of the occurrence of cracks, the range of cracking, the special seepage reflections and the safety of fissure dams. Lin et al. [121] established a three-dimensional elemental confinement model and used the Drucker- Prager yield criterion to analyze the influence of the structural face of the abutment of the Wudongde high arch dam on the overall stability of the arch dam, and put forward effective measures for the reinforcement of the structural face of the dam abutment. Zhou et al. [122] discussed the influence of the actual measurement method and calculation method on the evaluation of dam heel stress, and proposed an improved combination of the actual measurement method and numerical calculation to provide a reliable basis for evaluating the heel stress of high arch dams. Zhang et al. [65] evaluated the working condition of Xiluodu special high arch dam during the initial impoundment period, and pointed out that the dam was under pressure at the heel during the period of self-weight and impoundment, and the whole impoundment process was safe and controllable, and the dam was in normal working condition. Song et al. [123] used Baihetan high arch dam as an engineering example, and focused on the characteristics of stress, deformation and damage mode of high arch dam with asymmetric foundation through geomechanical model test and establishment of three-dimensional numerical model. Feng et al. [124] proposed an inverse analysis hybrid model based on the simulated stresses during the construction period of the Xiluodu extra-high arch dam as a background, combined with various deformation monitoring data during the construction period, to obtain elastic and deformation moduli close to the actual ones. Yang et al. [125] carried out a study on the overall stability of Jinping I high arch dam after reinforcement treatment by comprehensively analysing the results of model tests and numerical calculations, and obtained the working behavior of the arch dam under normal working conditions.

3.7.1 Microseismic Monitoring of the High Arch Dam During the Initial Impoundment Period

The arch dam of Dagangshan hydropower station adopts the ESG MMS, Canada, which provides all-weather real time monitoring of the arch dam during the initial impoundment period. Figure 3.31 shows the spatial distribution of microseismic events of the arch dam obtained using the MMS during the initial impoundment from November 22, 2014 to July 31, 2015, with the ball representing a single microseismic event and the color representing the moment magnitude of a single microseismic event (− 4 to − 2.3). From the figure, it can be seen that most of the microseismic

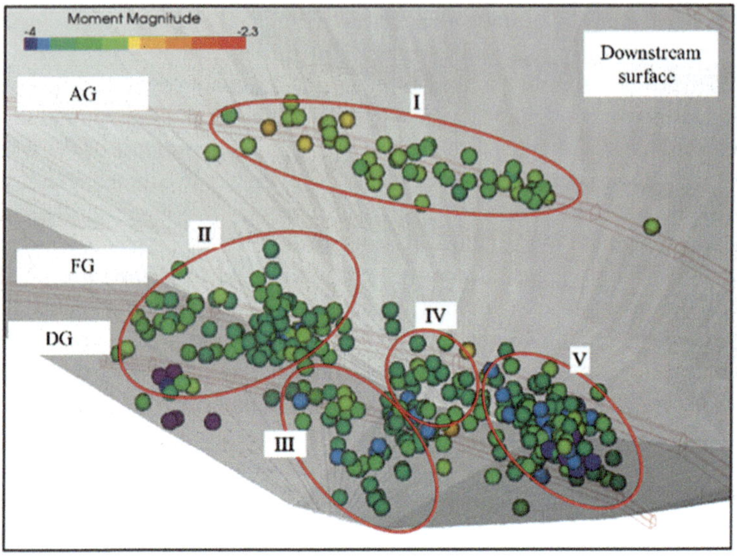

Fig. 3.31 Spatial distribution of MS events of the arch dam during the initial impoundment period

events are concentrated in the 979 m access gallery and the following areas, of which area I is located around the access gallery, area II and V are located in the dam near the foundation surface, area III is located in the dam toe, and area IV is located in the dam heel. When the rock or concrete material is subjected to external loading, the phenomenon of stress transfer or stress concentration will occur within it, in the process of stress change, elastic–plastic energy aggregation will occur in the region of high stress concentration, when the energy is aggregated to the critical value, microcracking phenomenon will occur within the material. Once microcracks are generated or developed, they release stored energy in the form of elastic or stress waves to the surrounding area. The vibration waves generated by each microseismic event include S wave and P wave, and the E_S/E_P ratio of energy released by S wave and P wave is an important basis to reflect the stress state of rock mass or concrete material at the source. By analysing the cumulative frequency distribution characteristics of E_S/E_P ratio, it can roughly reflect whether the material is tensile damage or compression shear damage. When the material is tensile damaged, the P-wave energy released is larger, and the E_S/E_P value is smaller. When the material is compression shear damaged, the S-wave energy released is much larger than the P-wave energy, and the E_S/E_P value is larger. In this section, the MMS is used to obtain the energy released during the occurrence of each microseismic event inside the dam, and by analysing the cumulative frequency distribution characteristics of the E_S/E_P ratio of microseismic events, it is possible to quickly determine the stress state of the arch dam during the initial impoundment period.

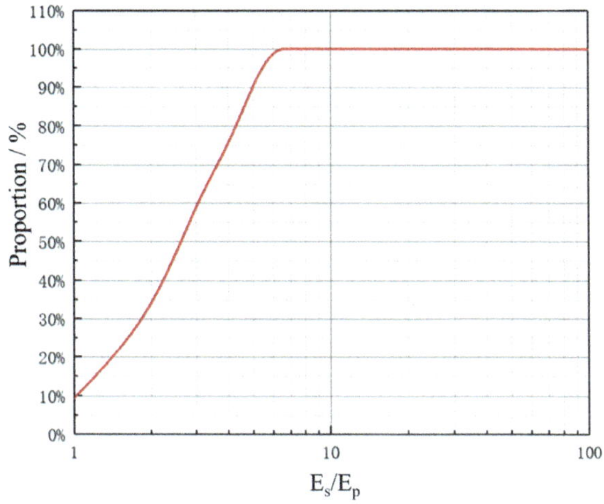

Fig. 3.32 Accumulated frequency distribution of E_S/E_P of microseismic events at the access gallery before impounding

Microseismic events within the range below the EL 979 m of the arch dam were selected to plot the cumulative frequency distribution of the E_S/E_P ratio of microseismic events at the arch dam access gallery at the water level of EL 975 m (prior to impoundment) (Fig. 3.32). When the water storage level reaches EL 975 m, the tensile stress in the arch crown of the access gallery exceeds the cracking limit under the overriding self-weight load of the arch dam, leading to the occurrence of cracks in the arch crown of the gallery along the axis [126]. At this time, the microseismic events are typically induced by the tensile stress. Therefore, it can be determined that the E_S/E_P ratio under the dominant action of tensile stress are all less than a certain value. Figure 3.32 shows that almost all microseismic events have $E_S/E_P \leq 6$. Therefore, it is assumed that tension damage occurs in the arch dam when $E_S/E_P \leq 6$, and compression-shear damage occurs in the arch dam when $E_S/E_P > 6$.

The arch dam of Dagangshan hydropower station, as a typical double-curvature arch dam, has its toe under pressure all the time during the initial impoundment period, so the cumulative frequency distribution of E_S/E_P of microseismic events at the dam toe during the initial impoundment period was plotted (Fig. 3.33). At the dam toe there are 40% of $E_S/E_P>6$ and 60% of $E_S/E_P \leq 6$. Obviously in the compression region of the dam toe, there is a larger proportion of tensile damage, accompanied by the occurrence of a certain proportion of compressive shear damage, the reason is that the rock and concrete brittle materials have the basic characteristics of compressive but not tensile resistance, even if deformation damage occurs under compressive stress, but often need to go through the cumulative development of the tensile damage to the process of macro-shear damage [127].

Combined with the above analyses of the cumulative frequency distribution characteristics of the E_S/E_P ratio of microseismic events, the cumulative frequency distribution of the E_S/E_P ratio of microseismic events at the dam heel before and after impoundment was plotted (Fig. 3.34a and b). Before impoundment, as shown in Fig. 3.34a, the percentage of $E_S/E_P \leq 6$ is 59% at the dam heel and $E_S/E_P > 6$ is

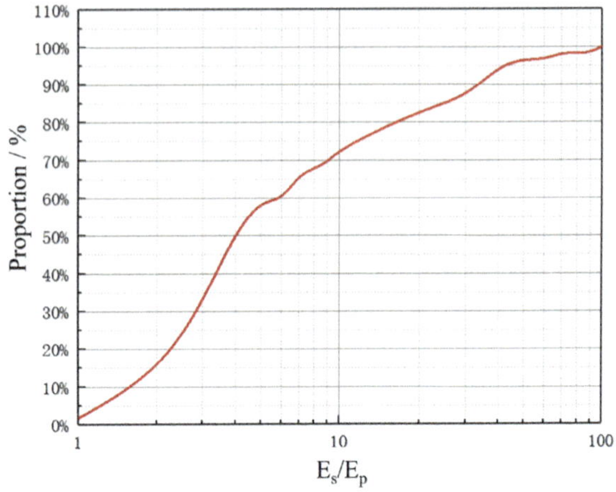

Fig. 3.33 Accumulated frequency distribution of E_S/E_P of microseismic events at the dam toe during the initial impoundment period

41%. After impoundment, as shown in Fig. 3.34b, the percentage of $E_S/E_P \leq 6$ is 57% at dam heel and $E_S/E_P > 6$ is 43%. It was deduced that the arch dam heel was consistently under pressure during the initial impoundment period. The feasibility of using the cumulative frequency distribution characteristics of microseismic events to judge the stress state of the arch dam is verified.

Figure 3.35 presents the cloud map of microseismic deformation distribution at the dam heel during the initial impoundment period. By observing the microseismic deformation distribution location at different impoundment levels and combining it with the value, it is possible to reveal the law of stress change in the heel of the arch dam during the initial impoundment period [110]. The microseismic deformation of the arch dam before impoundment is primarily concentrated at the dam heel. As

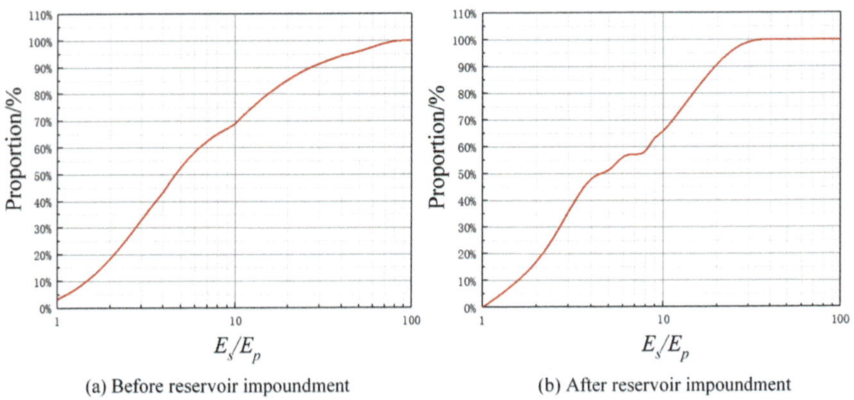

Fig. 3.34 Accumulated frequency distribution of E_S/E_P of microseismic events at the dam heel during the initial impoundment period

shown in Fig. 3.34a, the microcracking phenomenon of the dam heel before impoundment is attributed to compression-shear damage, indicating that the compressive stress concentration phenomenon occurs at the dam heel before impoundment. After locking the diversion tunnel and storing water, when the storage level reaches EL 1015 m, the microseismic deformation distribution area transfers to the downstream and is primarily concentrated at the foundation surface of the section #16 (Fig. 3.35b). This transfer occurs because the foundation surface is now in a compressed state, and the microseismic events occurring at this time are also caused by the concentration of compressive stresses. At the time of impoundment to EL 1120 m, as shown in Fig. 3.35c, the microseismic deformation of the arch dam mainly occurred at the dam toe of sections #14–#15. By observing the change in the microseismic deformation region, combined with the consistency of the spatial and temporal distribution of the microseismic deformation region and the high-stress concentration region, it is preliminarily judged that the transfer of the compressive stress aggregation region from the dam heel to the dam toe was completed during the initial impoundment period. It is evident that the high compressive stress concentration and the transfer phenomenon of arch dams are the intrinsic driving forces for the generation and transfer of microseismic events.

In order to support the rationality of microseismic monitoring to analyze the force state of the arch dam, the measured results of the opening and closing of the joint meter at the dam base of section #16 were selected, and the process line of the opening and closing of the joint meter at the dam base of dam section #16 was plotted (Fig. 3.36). Joint meters $JD_{16\text{-}1}$, $JD_{16\text{-}2}$, and $JD_{16\text{-}3}$ were arranged in the upstream, midstream, and downstream of the dam foundation, respectively. The positive and negative values of the joint meters indicate the opening and closing of the joints at the dam foundation, respectively, and the change of the measured openness value reflects the real stress

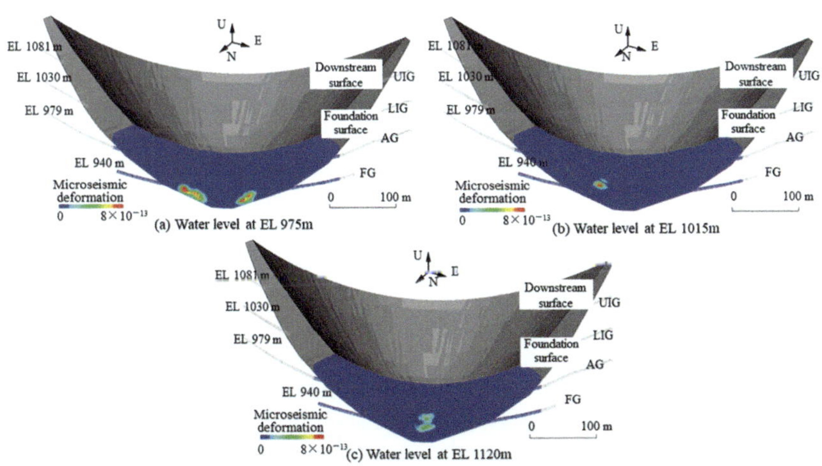

Fig. 3.35 Nephogram of microseismic deformation in the dam heel during the initial impoundment period

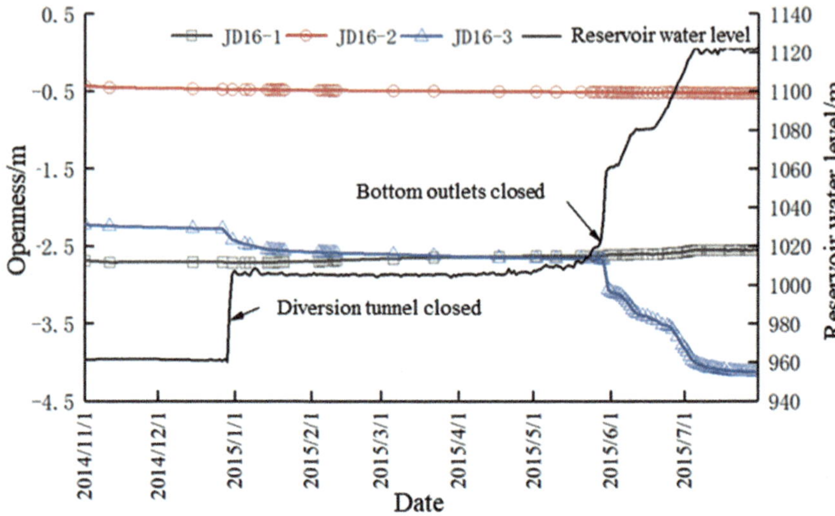

Fig. 3.36 Time curves of the opening width of the joint meter opening in the dam foundation of section #16

condition of the dam. As can be seen in Fig. 3.36, the dam foundation is in a closed state before the diversion tunnel is locked, and the maximum compression measured by the $JD_{16\text{-}1}$ on the upstream is 2.71 mm at this time, which is located in the dam heel, and the maximum compression measured by the $JD_{16\text{-}3}$ on the downstream at the dam toe in this period is 2.42 mm, which is slightly smaller than that at the dam heel, which indicates that before impoundment, due to the effect of the self-weight load of the arch dam, the dam is in a compressed state. As the water level rises and the bottom outlets are locked for storage to EL 1120 mm, the compression of the joints in the dam heel area measured by $JD_{16\text{-}1}$ decreases slightly, whereas the compression at the toe measured by $JD_{16\text{-}3}$ increases to 4.13 mm, which is analysed in conjunction with the initial storage process, due to the increase in the upstream water load, which leads to an increase in the level of compressive stresses at the dam toe, resulting in an increase in the amount of compression at the dam toe. Figure 3.36 further clarifies that the dam foundation has been under pressure during the initial impoundment period, which is consistent with Fig. 3.34 that the dam heel is under pressure during the initial impoundment period, and verifies the rationality of the judgement of the force state through the cumulative frequency distribution characteristics of the E_S/E_P ratio of microseismic event. The alterations in compression at the dam heel and toe illustrated in Fig. 3.36, coupled with the relocation of the microseismic deformation accumulation zones depicted in Fig. 3.35, collectively underscore that monitoring the evolving pattern of microseismic deformation in the dam body can unveil the actual stress state of the dam. In addition, the above results also indicate that the arch dam is affected by the upstream lateral water load during the initial impoundment period, and the stress changes in the dam base region are frequent, which should be paid special attention to the safety of this region during the operation of the dam.

3.7.2 Numerical Simulation of the High Arch Dam During the Initial Impoundment Period

In order to further understand the stress and displacement deformation law of the high arch dam of Dagangshan hydropower station, and to investigate the correlation between microseismic events and the working state of the arch dam, the three-dimensional finite element numerical model of the Dagangshan arch dam is established by using ANSYS. The range of model establishment is based on the center point of the upstream face on the dam crest, and extends 700 m to the left and right bank slopes along the axis of the dam, totaling 1400 m, and the boundaries of the left and right sides of the model are about 1.89 times the height of the dam from the cross-stream width of the dam abutments on both sides. Along the stream direction to extend upstream 402 m, downstream extension of 448 m, a total of 850 m. The model upstream side of the boundary from the base point of the stream direction to the length of about 1.91 times the height of the dam, the length of the downstream side of the length of about 2.13 times the height of the dam, the straight direction to build to the bottom of the surface of the 715 m elevation, the two sides of the top of the rock body elevation of 1,185 m. The thickness of the rock body below the EL 925 m at the foundation level is 210 m, which is 1 time the height of the dam. The three-dimensional mesh model is shown in Fig. 3.37a, and the internal structure of the dam body is depicted in Fig. 3.37b. The numerical model comprises 1,760,000 tetrahedral elements, with 941,000 elements dedicated to the dam. The model accuracy meets the requirements. Solid 65 elements were chosen to model the dam body, while the shell elements were selected to model the gallery inside the dam (Fig. 3.37). This section primarily addresses the stress state of the arch dam body. To minimize computational costs, the dam foundation is assumed to be an elastic material, and solid 45 elements are employed with introduced initial ground stresses. The model takes the upstream direction and the slope direction of the left bank as the positive direction. The mechanical parameters of materials are presented in Table 3.5. The calculation working conditions are detailed in Table 3.6. An ideal elastic–plastic Drucker-Prager constitutive model was utilized for the concrete arch dams as:

$$f = \alpha I_1 + \sqrt{J_2} - k \leq 0 \tag{3.1}$$

where I_1 is the first invariant of the stress; J_2 is the second invariant of the bias tensor S_{ij}; α and k are parameters related to the internal friction angle ϕ and cohesion c of the material:

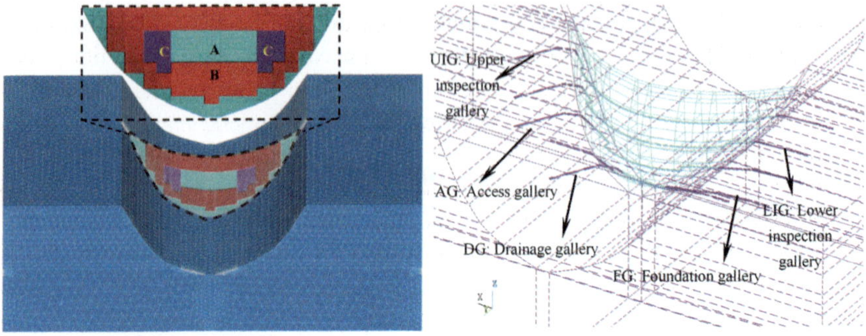

(a) Mesh for the entire model (b) FEM model for arch dam

Fig. 3.37 3D finite element model of the dam

Table 3.5 Mechanical parameters of materials

Structures	Materials	Compressive strength/ MPa	Tensile strength/ MPa	Deformation modulus E/ GPa	Poisson's ratio μ	Cohesive c/MPa	Internal friction angle $\phi/(^\circ)$
Dam (A zone)	$C_{180}36$	34.2	2.78	26.8	0.16	9.75	58
Dam (B zone)	$C_{180}30$	32.7	2.80	25.2	0.15	9.57	57.5
Dam (C zone)	$C_{180}25$	20.5	2.57	23.9	0.15	8.85	57.5
Dam (D zone)	Rock	–	–	15	0.25	–	–

Table 3.6 Calculating conditions and load combinations

Case	Load combination					
	Upstream reservoir water level				Gravity	Downstream reservoir water level
	EL 975 m	EL 1015 m	EL 1120 m	EL 1130 m		EL 960 m
1	Δ				Δ	Δ
2		Δ			Δ	Δ
3			Δ		Δ	Δ
4				Δ	Δ	Δ

Note Δ represents the calculated selection of the case

$$\alpha = \frac{2\sin\phi}{\sqrt{3}(3+\sin\phi)} \tag{3.2}$$

$$k=\frac{6\cos\phi}{\sqrt{3}(3+\sin\phi)} \tag{3.3}$$

1. Stress analysis of the arch dam

Figure 3.38 illustrates the distribution of the principal tensile stress in the arch dam during the initial impoundment period. Before the locking of the diversion tunnel and the start of reservoir filling, the upstream reservoir area is essentially in an empty state. Due to the slightly upstream-biased center of gravity of the hyperbolic arch dam, the dam tilts towards the upstream side under the action of self-weight stress. The region of principal tensile stress (0.5–2.89 MPa) in the arch dam is located around the left and right dam abutments (Fig. 3.38a), with the maximum principal tensile stress of 2.89 MPa occurring on the upstream side of the left dam abutment. As the reservoir filling progresses to EL 1015 m after the diversion tunnel is locked, the region of principal tensile stress (0.5–2.64 MPa) in the arch dam remains concentrated around the left and right dam abutments (Fig. 3.38b). However, influenced by the water load on the upstream side, the distribution range of the principal tensile stress region decreases, and the maximum principal tensile stress slightly reduces to 2.64 MPa. From Fig. 3.38c and d, it can be observed that as the reservoir water level continues to rise, the tensile stress at the junction of the dam heel and foundation increases continuously. As the reservoir filling progresses to EL 1015 m after the bottom outlets is locked, the main tensile stress area (0.5–2.48 MPa) transfers to the upstream dam heel. This indicates that under the influence of the water load on the upstream side, the main tensile stress zone in the dam transfers from the dam abutments to the dam heel. When the water level reaches the normal storage level of 1130 m, the main tensile stress area (0.5–3.43 MPa) expands. The rise in the upstream water level leads to a significant increase in the maximum principal tensile stress in the dam, reaching 3.43 MPa. The distribution area is mainly at the dam heel and the junction between the upstream side of the dam and the foundation surface. The increase in the maximum principal tensile stress may lead to cracking in the vicinity of the dam heel, and special attention should be paid to this. In actual monitoring results, the dam foundation has been under compression, but in numerical calculations, there is a significant tensile stress at the dam heel, which is inconsistent with reality. This is mainly because the arch dam itself has a significant pre-stress, and the tensile stress at the dam foundation is not sufficient to completely offset this portion of the compressive stress. This is also the reason why, in Fig. 3.36, the compression at the dam heel during the initial impoundment period decreases but still remains in a compressed state.

Figure 3.39 illustrates the distribution of the principal compressive stress in the arch dam of the Dagangshan hydropower station during the initial impoundment period. From Fig. 3.39a, it is evident that before the locking of the diversion tunnel, the arch dam experiences significant compressive stress at the dam heel due to the self-weight load. Consequently, the primary compressive stress area (10–17.1 MPa)

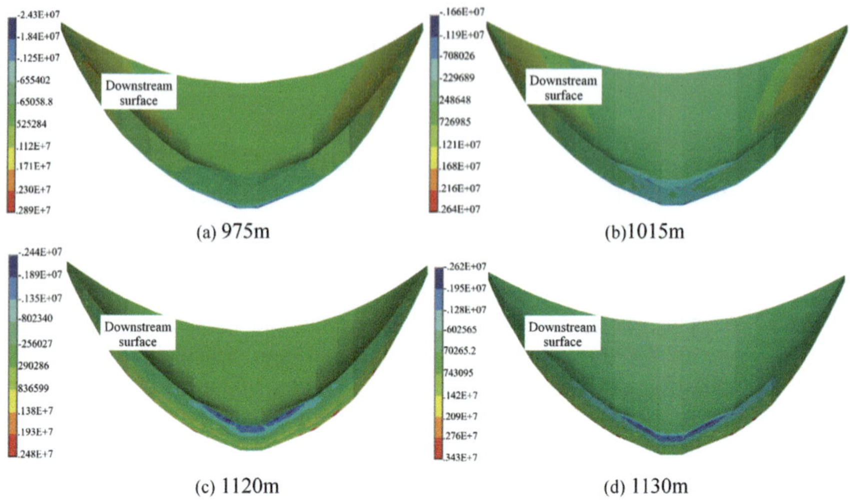

Fig. 3.38 Nephogram of the main tensional area of Dagangshan dam (unit: Pa)

in the arch dam is located near the dam heel, with a maximum principal compressive stress of 17.1 MPa. Figure 3.35a shows that at this time, microseismic deformations primarily occur in the dam heel area, confirming that the accumulation of compressive stress is an intrinsic factor for microseismic events and supporting the rationality of determining the compressive state in the dam heel area using the cumulative frequency distribution characteristics of the E_S/E_P ratio of microseismic events before reservoir filling, as shown in Fig. 3.34. During the reservoir filling to EL 1015 m after the locking of the diversion tunnel (Fig. 3.39b), although the primary compressive stress area (10.0–13.5 MPa) in the arch dam is still near the dam heel, its range significantly reduces under the influence of the upstream water load. The maximum principal compressive stress decreases to 13.5 MPa, and the high compressive stress concentration area begins to transfers toward the dam toe. The microseismic deformation at the dam heel before reservoir filling have released the accumulated elastic–plastic energy in the dam heel region. However, the gradual accumulation of high compressive stress near the dam toe causes the microseismic deformation at this time to primarily occur at the foundation face of section #16 rather than in the dam heel. During the reservoir filling to EL 1120 m after the locking of the diversion tunnel, as observed in Fig. 3.39c, the primary compressive stress area (10–21.5 MPa) in the arch dam is clearly located at the dam toe. This indicates a transfer of compressive stress from the dam heel to the dam toe. The main compressive stress significantly increases under the influence of the upstream water load, reaching a maximum principal compressive stress of 21.5 MPa. At this time, the localized elastic–plastic behavior of the dam toe concrete further concentrates and ultimately releases, leading to the generation and development of microseismic deformation in the dam toe (Fig. 3.35c). This validates the close correlation between the development of microseismic events in the dam heel and toe and the transfer and

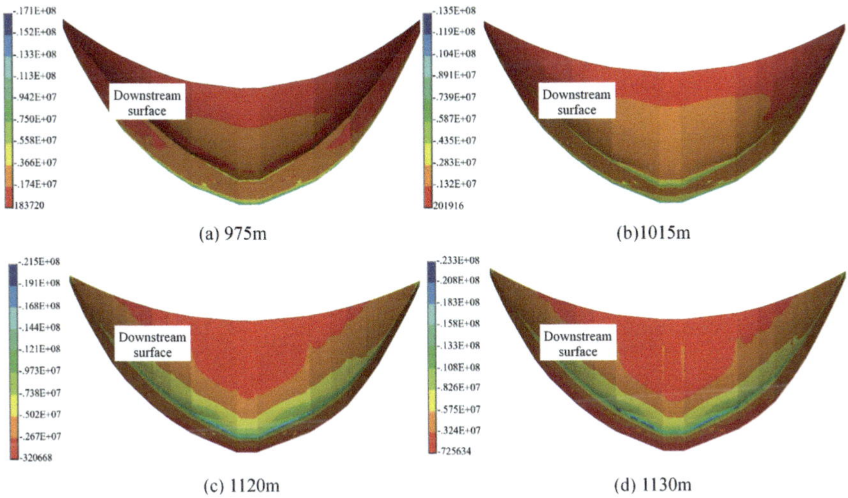

Fig. 3.39 Nephogram of main compression area of Dagangshan dam (unit: Pa)

concentration of compressive stress. As the water level reaches the normal reservoir level of EL 1130 m, the main compressive stress area (10–23.3 MPa) in the dam body is observed to be in the dam toe, with the maximum principal compressive stress increasing to 23.3 MPa due to the further rise in water level. Comparing the stress values at the dam toe under different reservoir level, it is evident that the compressive stress values at the dam toe are consistently greater than the tensile stress values. This confirms that the dam toe remains under compression during the initial impoundment period, and the microseismic events in both the dam heel and toe are attributed to compressive-shear failure. The results are consistent with those obtained from the cumulative frequency distribution of the E_S/E_P ratio of microseismic events shown in Figs. 3.33 and 3.34. The findings indicate that the variation in upstream water loading is the primary cause of the transfer of the main compressive stress area from the dam heel to toe. The accumulation and transfer of high compressive stress are the actual causes of microseismic events. Through analyzing the spatial and temporal distribution of microseismic deformation and the cumulative frequency distribution characteristics of the E_S/E_P ratio, combined with numerical simulations, it is possible to comprehensively reflect the stress state of the arch dam. This approach allows for a focused investigation of areas where stress concentration in the dam may occur, facilitating early detection of potential hazards that could impact the overall stability and safety of the arch dam.

2. Deformation analysis of the arch dam

The radial displacement monitoring values at various elevations of the #14 dam section were selected at the reservoir water levels of EL 1120 m and 1130 m, and the radial displacement monitoring curves for section #14 were plotted (Fig. 3.40, Due

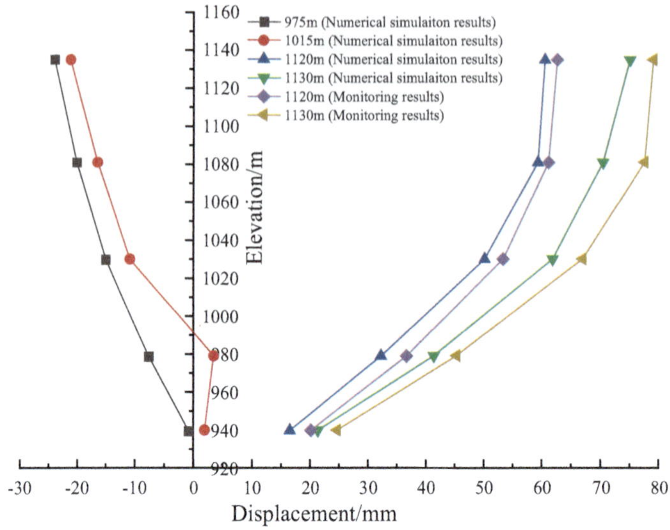

Fig. 3.40 Comparison between the radial displacements of section #14 obtained from site monitoring and FEM simulation

to the lack of measured data for radial displacement monitoring values at reservoir water levels of EL 975 m and 1015 m, the corresponding curves are not provided). Simultaneously, calculated radial displacement values for section #14 at elevations corresponding to reservoir water levels of EL 975 m, 1015 m, 1120 m, and 1130 m were selected and plotted in Fig. 3.40. In the numerical simulation, before the locking of the diversion tunnel, when the reservoir water level reached EL 975 m, the arch dam exhibited upward displacement deformation, gradually increasing from lower to higher elevations. At this time, the maximum radial displacement of the dam was 24.10 mm, located at the top of the crown cantilever. The phenomenon of stress concentration occurred in the dam heel due to the self-weight of the dam, leading to the initiation of microseismic events. When the diversion tunnel was locked and impounded water to the elevation of 1015 m, under the influence of the upstream water load, the arch dam exhibited downward displacement deformation in the mid-low elevations, while the high elevations of the dam still experienced upward displacement under the effect of self-weight. The maximum radial displacement was 21.22 mm, located at the top of the crown cantilever. When the bottom outlets were locked and impounded water to the elevation of 1120 m, under the action of the water load on the upstream side, the arch dam as a whole is displaced and deformed to the downstream side, and the maximum radial displacement value is 60.38 mm, which is located in the top of the crown cantilever, and the direction of the radial displacement of the arch dam changes under the influence of the water load on the upstream side, and it also causes the main tensile stress area of the arch dam to be transfered to the dam heel, and the main compressive stress area is transfered from the dam heel to the toe, and the area of the microseismic deformation is also transfered from dam heel to the toe. When the water was stored to the elevation of 1130 m, the arch dam was displaced to the downstream side, and the maximum radial displacement was 74.91 mm, which was located in the area near the top of the crown cantilever. It was found that the numerically calculated radial displacement value was slightly smaller than the actual monitoring value, but the trend was the same, the radial displacement gradually increased from the low elevation to the high elevation. The maximum radial displacements of the arch dam at the beginning of impoundment occurred at the top of the crown cantilever. The results show that the displacement and deformation occurred under the influence of self-weight of the dam body and upstream side water load at the early stage of impoundment, which led to the continuous change and adjustment of the dam body stress. The microseismic deformation aggregation area was also transferred, and it is obvious that the upstream side water load is the key factor for the change of the microseismic deformation area of the arch dam.

The tangential displacement of the arch dam is shown in Table 3.7. Before the locking of the diversion tunnel, when the storage level is at EL 975 m, the maximum displacement of the right dam abutment is 9.0 mm, and the maximum displacement of the left dam abutment is − 9.6 mm, and the left and right deformations are basically symmetrical. The maximum displacement of the right dam abutment was 8.5 mm and the maximum displacement of the left dam abutment was − 9.3 mm at the storage level of EL 1015 m. At the storage level of EL 1120 m, the maximum displacement of the right dam abutment was − 21.0 mm and the maximum displacement of the left dam abutment was 19.9 mm, and the direction of the tangential displacement of the left and right dam abutments was changed. This is consistent in time with the changes in the location of the main tensile stress area and the main compressive stress area of the arch dam, and the left and right deformations are coordinated. When the reservoir water level reached EL 1130 m, the maximum displacement of the left and right dam abutments were both near the dam crest. The maximum displacement at the right dam abutment was − 19.7 mm, while at the left dam abutment, it was 19.1 mm, and the displacement changes were generally symmetrical on the left and right sides. In the initial stages of reservoir filling at a low water level, the arch dam exhibited tangential deformation towards the riverbed. As the reservoir water level continued to rise, reaching 1120 m, the arch dam began to undergo tangential deformation

Table 3.7 Tangential displacements of the dam

Reservoir water level/m	Maximum displacement of left abutment/mm	Deformation direction of left abutment	Maximum displacement of right abutment/mm	Deformation direction of right abutment
975	− 9.6	Riverbed	9.0	Riverbed
1015	− 9.3	Riverbed	8.5	Riverbed
1120	19.9	Slope	− 21.0	Slope
1130	19.1	Slope	− 19.7	Slope

Note The tangential displacement is positive in the direction of the left slope towards both sides, conforming to the typical deformation pattern of arch dams. This indicates that the arch dam possesses good deformation coordination capability.

3.7.3 Chapter Conclusion

By employing a combined approach of microseismic monitoring and numerical computations, this study investigates the intrinsic correlation between microseismic events and dam stress. The analysis reveals the factors triggering microseismic events in arch dams and explores the patterns of stress variations, as well as displacement deformations in the high arch dam of Dagangshan hydropower station. The main conclusions are as follows.

(1) The stress transfer is consistent with the emergence of microseismic events in arch dam and the changes in compression at the dam foundation in terms of temporal and spatial changes, and the high compressive stress aggregation and migration are the intrinsic driving force for the occurrence and development of microseismic events. Most of the microseismic events in the arch dam of Dagangshan hydropower station are caused by tensile stresses, and it is determined that microseismic events is caused by tensile stresses while $E_S/E_P \le 6$; the method of determining the region of the dam high compressive stress concentration is that most of the microseismic events have $E_S/E_P \le 6$, and at the same time, there is a certain proportion of microseismic events $E_S/E_P > 6$. The cumulative frequency distribution characteristics of microseismic events can be used to judge the real stress state of arch dams.
(2) Arch dams undergo varying and adjusting stresses throughout the impoundment process. During the impoundment period, the primary compressive stress area in the arch dam of Dagangshan hydropower station transfers from the dam heel to the toe, while the primary tensile stress area transfers from the dam abutment to the dam heel. The migration process of the primary stress-acting area is accompanied by changes in the microseismic deformation gathering area. Arch dams are predominantly influenced by compressive stresses, and under normal water storage conditions, the compressive stress level in the dam foundation is significantly higher than the tensile stress level, resulting in compression at the dam foundation.
(3) The radial displacement of the arch dam under low water level conditions is minimal, and the direction of radial displacement changes from the upstream to the downstream with an increase during the impoundment period. Under the water load on the upstream, the radial displacement value reaches 74.91 mm when the normal storage level is reached. This indicates that the self-weight of the dam and the change in water load on the upstream during the impoundment period are the primary reasons for the continuous variation in the displacement of the arch dam. The tangential displacement deformation of the high arch dam at Dagangshan hydropower station is symmetrical and coordinated,

aligning with the general pattern of tangential displacement changes observed in arch dams. Furthermore, the foundation of the arch dam has consistently been under compression, leading to the initial belief that the arch dam at Dagangshan hydropower station exhibits a high degree of stability.

3.8 Conclusion

This chapter employs microseismic monitoring technology, focusing on the Dagangshan high arch dam, to successfully establish the first set of MMS for the dam heel and gallery in China. The feasibility of microseismic monitoring for large scale concrete structures is validated, enabling real time, all-weather monitoring of microcrack evolution in the dam heel and gallery during the initial impoundment period. The microseismic monitoring study on the dam heel and gallery of the Dagangshan arch dam reveals the patterns of microcracks generation and evolution, elucidates the relationship between microseismic activity in the dam heel and gallery, and the operational state of the arch dam, and identifies and delineates potential danger zones in the high arch dam during the impoundment period. This research can serve as a reference for the construction and construction practices of high arch dams. The main conclusions of this chapter are as follows:

(1) Based on the analysis of the structural characteristics of the Dagangshan high arch dam, considering the tendency of microseismic signals in massive concrete to attenuate from high frequency to low frequency, the high frequency response of the microseismic monitoring uniaxial velocity sensor is determined as 5000 Hz. The microseismic monitoring system for the dam heel and gallery of the Dagangshan arch dam is successfully constructed, capturing microseismic signals during the impoundment period.
(2) By employing threshold setting, filtering techniques, and other methods, signal interference from the construction site is successfully eliminated. Additionally, manual tapping optimization of the sensor wave velocity is conducted on-site, resulting in an equivalent P-wave velocity of 4300 m/s with an average error of 5.1 m.
(3) Analysis of the temporal distribution patterns of microseismic events during the impoundment period of the Dagangshan high arch dam reveals a close correlation between the microseismic event activity rate in the dam heel region and the reservoir water level. During the locking of the diversion tunnel for storage, the water level rises rapidly, leading to a significant increase in microseismic events in the dam heel. When the reservoir water level stabilizes, microseismic activity decreases significantly. Under high reservoir water level conditions, microseismic activity in the dam heel increases significantly. Microcracks in the dam heel mainly occurs at the junction of the arch dam, left and right bank dam foundations, and concrete pedestal.

(4) The study of the spatial distribution of microseismic events in the dam heel reveals distinct clustering areas that are not fixed but transfer during the impoundment process. Throughout the impoundment process, microseismic clustering areas transfer from the dam heel towards the dam toe. During the impoundment period, the stress distribution in the high arch dam is adjusting, indicating a correspondence between microseismic events and stress transfer in the arch dam.
(5) Examining the spatial distribution of microseismic events in the gallery reveals the middle section of the 940 m elevation foundation gallery as a cluster area for microseismic events. Analyzing the cumulative seismic moment in this gallery over time indicates continuous microseismic event generation during the empty reservoir period. It suggests that, under the self-weight of the arch dam, this elevation gallery experiences high stress, is in a stress concentration zone, leading to increased microcracking that eventually exceeds the concrete cracking threshold, resulting in cracks. The high stress zone aligns with the high microseismic deformation area. Microseismic monitoring identifies the #13–#14 section of the 940 m elevation foundation gallery as a potential danger zone for the high arch dam. As the reservoir water level rises, microseismic activity significantly increases in the 937 m elevation drainage gallery near the downstream side of the dam body, confirming the correlation between microseismic activity at the gallery and reservoir water level.
(6) Combining microseismic monitoring with numerical simulation methods, the study investigates the inherent connection between the working state of the arch dam and microseismic events. The internal stress changes in the arch dam during the impoundment period exhibit high consistency with microseismic events in terms of both time and space. Numerical simulations show that, at low water levels, the dam heel is a zone of concentrated compressive stress, mainly caused by the upward tilt of the arch dam under its self-weight. As the water level rises, the compressive stress concentration zone transfers from the dam heel to the dam toe. The microseismic deformation clustering area corresponds to the compressive stress concentration area, revealing the intrinsic reasons for the transfer of the microseismic deformation area and reconfirming the potential danger zone of the arch dam during the impoundment period. This provides a reference for stability analysis and danger warnings for high arch dams.

References

1. Feng F (2013) Simulation of working performance of super-high arch dam during construction period based on entire dam and whole process, vol 6. China Institute of Water Resources & Hydropower Research (IWHR), p 100038
2. Yuan FB, Liu J, Li PJ et al (2007) Back analysis and multiple-factor influencing mechanism of high geostress field for river valley region of Laxiwa Hydropower Engineering. Rock Soil Mech 4:836–842

3. Qi SW, Wu FQ (2011) Stresses field characteristics of a valley slope in high geo-stress area. Rock Soil Mech 32(5):1460–1464
4. Song SW, Gong MF, Lei CD (2006) Study on stability of high rock slope in hydropower engineering [J]. Chin J Rock Mech Eng 2:226–234
5. Zhong W (2009) Study on excavation stability of complex rock slope at high field stress areas. Southwest Jiaotong University
6. Wang RK (2007) Analysis and evaluation on optimization of design for foundation interface of super high arch dam. Tsinghua University
7. Lin P, Li QB, Zhou SW et al (2013) Intelligent cooling control method and system for mass concrete. J Hydraul Eng 44(8):950–957
8. Huang JW, Yuan H, Zhou YH et al (2019) Evaluation of concrete temperature control effects for high arch dams based on cloud model. Water Power 45(4):65–69
9. Zhu BF, Xu P (2002) Simulation analysis of the whole process of concrete high dam. Water Resources Hydropower Eng 12:11–14
10. Zhou WY, Yang RQ, Yan GR (1997) Stability analysis system for large dams. Chin J Rock Mech Eng 5:26–32
11. Wang RK (2015) Review of construction achievements and technological development of super-high arch dam in China. Adv Sci Technol Water Resour 35(5):13–19
12. Baustadter K, Widmann R (1985) The behavior of the Kolnbrein arch dam. In: Proceedings of the 15th ICOLD, Lausanne, Switzerland, pp 633–651
13. Lin C (2016) Evaluation of high arch dam heel cracking and reinforcement effect of foundation treatment and its application. Tsinghua University
14. Pan YW (2015) Reservoir deformation mechanism during impoundment and operation and its influence to safety of high arch dam. Tsinghua University
15. Zhang G, Liu Y, Zheng C et al (2011) Simulation of influence of multi-defects on long-term working performance of high arch dam. Sci China Technol Sci 54(1):1–8
16. Cheng L, Liu YR, Pan YW et al (2016) Research on ultimate bearing capacity of Jinping-I Arch Dam based on impoundment period inversion. Rock Soil Mech 37(5):1388–1398
17. Yang Q, Pan YW, Cheng L et al (2015) Impounding influence of slope and foundation deformation on high arch dam. Chin J Rock Mech Eng 34:3979–3986
18. Lin P, Liu XL, Hu Y et al (2013) Deformation stability analysis of Xiluodu arch dam under stress-seepage coupling condition. Chin J Rock Mech Eng 32(6):1145–1156
19. Lin C, Yang Q, Wang HB et al (2016) Numerical simulation of Mengdigou arch dam based on nonlinear finite elements. Rock Soil Mech 37(9):2624–2630
20. Shen H, Luo XQ, Li Y et al (2012) Three-Dimensional stability analysis of dam abutment of Wudongde arch dam. Chin J Rock Mech Eng 31(5):1026–1033
21. Zhou W, Chang XL, Tang ZM et al (2002) Research of gradually destruction process simulation analysis and stabilization safety on Xiluodu high arch dam. J Sichuan Univ Eng Sci Ed 4:46–50
22. Ren QW, Qian XD, Zhao Y et al (2002) Methods for analyzing sliding resistance stability along the base surface of high arch dam. J Hydraul Eng 2:1–7
23. Linsbauer HN, Ingraffea AR, Rossmanith HP et al (1989) Simulation of cracking in large arch dam: part I. J Struct Eng 115(7):1599–1615
24. Jeon J, Lee J, Shin D et al (2009) Development of dam safety management system. Adv Eng Softw 40(8):554–563
25. Huang HN, Zhou Q, Hua XS (2005) Actuality on the theory and technology of monitoring and controlling for dam safety. Dam Saf 2:54–57
26. Zhao ZR, Xu R (2010) Advance and prospect of monitoring technology for dam safety [J]. Hydropower Autom Dam Monitor 34(5):52–57
27. Wang J (2015) Research on dam dynamic three-dimensional deformation monitoring based on laser scanning technology. Zhengzhou University
28. Liu MJ, Wang HJ (2016) Analysis of monitoring results of non-stress meters of arch dam of Jinping Hydropower Station I. Yangtze River 47(1):91–94

29. Yang GX, Zhang L, Zhu HH et al (2012) Application of fiber Bragg grating sensor in geomechanical model test of arch dam. J Yangtze River Sci Res Inst 29(8):52–57
30. Hu B, Guan B, Wang SJ et al (2011) Evaluation and research on replacement reinforcement quality at abutment of ultra-high arch dam based on prototype monitoring. J Hydraul Eng 42(7):876–882
31. Dai F, Li B, Xu NW et al (2015) Characteristics of damaged zones due to excavation in deep underground powerhouse at Houziyan hydropower station. Chin J Rock Mech Eng 4:735–746
32. Xu NW (2011) Study on microseismic monitoring and stability analysis of high steep rock slope. Dalian University of Technology
33. Zhang BH, Deng JH, Gao MZ et al (2012) Safety evaluation research based on microseismic monitoring in underground powerhouse of hydropower station. Chin J Rock Mech Eng 5:937–944
34. Li B, Xu NW, Dai F et al (2017) Microseismic monitoring and large deformation forecasting research during excavation of underground powerhouse at Wudongde hydropower station. Chin J Rock Mech Eng 2:4102–4112
35. Ma TH, Tang CA, Tang LX et al (2016) Mechanism of rock burst forcasting based on microseismic monitoring technology. Chin J Rock Mech Eng 3:470–483
36. Kou XD, Zhou WY, Yang RQ (2000) Stability analysis of arch dam using 3-D fast Lagrangian method. J Hydraul Eng 9:8–14
37. Zhou WY, Chen X (2001) Stability analysis of Jinping arch dam. J North China Inst Water Conserv Hydroelectr Power 3:31–34
38. Huang YS, Zhou WY, Chen X et al (2002) Analysis on stability in the large of Laxiwa hyperbolic arch dam. Chin J Rock Mech Eng 2:2413–2417
39. Xu MY, Xue LL, Wang WM et al (2005) Coupled analysis of Dahuashui arch dam with trial load method and block element method. Chin J Rock Mech Eng 2:5281–5286
40. Pan YW, Liu YR, Zhang L et al (2014) Optimization research on Baihetan arch dam foundation and pedestal. Chin J Rock Mech Eng 33(1):2641–2648
41. He Z, Liu YR, Deng JQ et al (2015) Evaluation of long-term stability for high arch dam based on the viscoplasticity damage model, vol 45(10). Science China Press, pp 1105–1110
42. Xing LS, Zhou JB (1986) Accident analysis of dam in initial impoundment period. Water Power 41(1):31–35
43. Ru NH, Jiang ZS (1995) Dam accidents and safety. China Water and Power Press
44. Okamoto S, Mizukoshi T, Miyata Y (1986) On the observations of microearthquake activity before and after impounding at the Takase Dam. A special report. Phys Earth Planet Interiors 44(2):115–133
45. Luo DN, Lin P, Li QB et al (2014) Performance analysis of Xiluodu super-high arch dam in impounding period. J Hydraul Eng 45(1):18–26
46. Yang Q, Pan YW, Cheng L et al (2015) Mechanism of valley deformation of high arch dam and effective stress principle for unsaturated fractured rock mass. Chin J Rock Mech Eng 34(11):2258–2269
47. Zhou QJ, Zhang GX, Liu Y (2013) A regression analysis and simulation feedback-prediction method for behavior of super high arch dam during first impounding period. J Hydraul Eng 43(1):73–79
48. Aleksandrovskay ÉK (1986) Stress-strain state of the Sayano-Shushenskoe dam during filling of the reservoir. Hydrotech Constr 20(3):123–129
49. Pemyakova LS, Epifanov AP (2011) Formation of the stress-strain state of the dam at the Sayano-Shushenskaya HPP during 2010 filling of reservoir. Power Technol Eng 45(3):188–192
50. Londe P (1987) The Malpasset dam failure. Eng Geol 24(1–4):295–329
51. Kiersch GA (1964) Vaiont reservoir disaster. Civ Eng 34(2):32–39
52. Pan JW, Jin F, Xu YJ et al (2010) Analysis of dam heel cracking and ultimate bearing capacity of Kolnbrein arch dam. J Hydroelectr Eng 29(3):148–153
53. Chen SH, Wang WM, Xu MY et al (2003) Finite element analysis of the crack propagation in high arch dam heel of Xiaowan project. J Hydraul Eng 1:66–71

54. Zhu BF (2006) On the feasibility of building high quality arch dams without cracking and the relevant techniques. J Hydraul Eng 37(10):1155–1161
55. Sun SA, Xiang AM, Ping Z et al (2006) Gravity change and its mechanism after the first water impoundment in three gorges project. Acta Seismol Sin 28(5):485–492
56. Tomida N, Sato N, Soda H et al (2011) Estimation of rockfill dam behavior during impounding by elasto-plastic model. In: Dams and reservoirs under changing challenges. Taylor and Francis Group, London, pp 27–34
57. Yuan H, Li DD, Zhang K (2013) Study on security risk of impounding safety in China. Adv Mater Res 798–799:1119–1122
58. Zhou QJ, Zhang GX, Liu Y (2013) Prediction of and early warning for deformation and stress in the Xiaowan arch dam during the first impounding stage. Appl Mech Mater, pp 2463–2472
59. Li Z (1999) Study on cracks at the heel of the arch dam of Shimen Project. J Hydraul Eng 11:61–65
60. Shu Y (2006) Back Analysis of displacement monitoring results during Ertan arch dam first impoundment. Des Hydroelectr Power Station 22(2):34–41
61. Li PJ, Wei P, Zhang Q (2009) Behavior of Laxiwa ach dam in the period of first impounding. Water Power 35(11):12–15
62. Zhao Y (2004) Study on level displacement observed during the reservoir in Dachaoshan project. Dam Saf 6:48–51
63. Wei C, Tian ZH, Zhang B (2010) Statistical model analysis of observed data of horizontal displacement of Lijiaxia Arch Dam. J Hohai Univ Nat Sci 38(6):651–654
64. Han SD, Zhao B, Liao ZY et al (2010) Analysis and evaluation of monitoring data by plumb lines in early impoundment period of Xiaowan arch dam. Dam Saf (3):38–41, 210003
65. Zhang C, Wang RK, Tang XJ (2016) Safety evaluation of Xiluodu ultra-high arch dam during the initial impoundment period. J Hydraul Eng 1:85–93
66. Dong JH, Xie HP, Zhang L et al (2007) Experimental study of 3D geomechanical model on global stability of Dagangshan double curvature arch dam. Chin J Rock Mech Eng 26(10):2027–2033
67. Zhang L, Liu YR, Yang Q et al (2014) Geomechanical model tests on lobal stability of an arch dam at Dagangshan. Chin J Rock Mech Eng 33(5):971–982
68. Cheng B, Yao W, Wu KR et al (2000) Application of acoustic emission technique to the study of concrete. Nondestruct Test 22(9):387–390
69. Ji HG, Jia LH, Li ZD (1996) Study on the AE-model of concrete damage. Acta Acust 2(4):601–608
70. Tang CA, Zhu WC (2003) Damage and fracture of concrete-numerical test. Science Press, Beijing
71. Hu SW, Lu J, Fan XQ (2011) Study on acoustic emission technique for normal concrete fracture test. J Hydroelectr Eng 30(6):16–19
72. Liu C, Tang CA, Li LC et al (2009) Analysis of probability of water inrush from grout curtain based on background stress field and microseismicity. Chin J Rock Mech Eng 28(2):366–372
73. Xu NW, Tang CA, Li LC et al (2011) Microseismic monitoring and stability analysis of the left bank slope in Jinping first stage hydropower station in Southwestern China. Int J Rock Mech Min Sci 48(6):950–963
74. Ma K, Tang CA, Liang ZZ et al (2016) Stability analysis of the surrounding rock of underground water-sealed oil storage caverns based on microseismic monitoring during construction. Chin J Rock Mech Eng 35(7):1353–1365
75. Ji HG, Hou ZF, Zhang L et al (2011) Experimental studies on the frequency characteristics of acoustic emissions in concrete material and its dependences on strength parameters. J Appl Acoust 30(2):112–117
76. Grosse CU, Ohtsu M (2008) Acoustic emission testing. Springer, New York
77. Ekstrom G, England P (1989) Seismic strain rates in regions of distributed continental deformation. J Geophys Res Solid Earth. 94(B8):10231–10257
78. Axwell SC, Urbancic TI, Prince M et al (2003) Passive imaging of seismic deformation associated with steam injection in Western Canada. In: SPE Annual Technical Conference and Exhibition. Society of Petroleum Engineers, SPE84572

79. Novak P, Moffat AIB, Nalluri C et al (2007) Hydraulic structure, 4th edn. Taylor & Francis Group, London
80. Ren Q, Xu L, Wan Y (2007) Research advance in safety analysis methods for high concrete dam. Sci China Ser E Technol Sci 50:62–78
81. Haftani M, Gheshmipour AA, Mehinrad A et al (2014) Geotechnical characteristics of Bakhtiary dam site, SW Iran: the highest double-curvature dam in the world. Bull Eng Geol Env 73:479–492
82. Wu SY, Shen MB, Wang J (2010) Jinping hydropower project: main technical issues on engineering geology and rock mechanics. Bull Eng Geol Env 69:325–332
83. Chen Y, Zhang L, Yang B et al (2015) Geomechanical model test on dam stability and application to Jinping high arch dam. Int J Rock Mech Min Sci 76:1–9
84. Shen X, Niu X, Lu W et al (2017) Rock mass utilization for the foundation surfaces of high arch dams in medium or high geo-stress regions: a review. Bull Eng Geol Env 76(2):795–813
85. Léger P, Leclerc M (2007) Hydrostatic, temperature, time-displacement model for concrete dams. J Eng Mech 133(3):267–277
86. Fu S, He T, Wang G et al (2011) Evaluation of cracking potential for concrete arch dam based on simulation feedback analysis. Sci China Technol Sci 54(3):565–572
87. Li TC, Wang RK, You QS, et al (2007) Study on safety evaluation methods for high arch dams. J Hydraul Eng 1:78–83, 105
88. Lin P, Zhou W, Liu H (2015) Experimental study on cracking, reinforcement, and overall stability of the Xiaowan super-high arch dam. Rock Mech Rock Eng 48(2):819–841
89. Wang R (2016) Key technologies in the design and construction of 300m ultra-high arch dams. Engineering 2(3):350–359
90. Sheibany F, Ghaemian M (2006) Effects of environmental action on thermal stress analysis of Karaj concrete arch dam. J Eng Mech-Asce 132(5):532–544
91. Jin F, Zhang GX, Luo XQ et al (2008) Modelling autogenous expansion for magnesia concrete in arch dams. Front Architect Civil Eng China 2(3):211–218
92. Jin F, Chen Z, Wang J (2010) Practical procedure for predicting non-uniform temperature on the exposed face of arch dams[J]. Appl Therm Eng 30(14–15):2146–2156
93. Abdulrazeg AA, Noorzaei J, Jaafar MS et al (2014) Thermal and structural analysis of RCC double-curvature arch dam. J Civ Eng Manag 20(3):434–445
94. Li Z, Gu C, Wang Z et al (2015) On-line diagnosis method of crack behavior abnormality in concrete dams based on fluctuation of sequential parameter estimates. Sci China Technol Sci 58(3):415–424
95. Santillan D, Salete E, Toledo MA (2015) A methodology for the assessment of the effect of climate change on the thermal-strain-stress behaviour of structures. Eng Struct 92:123–141
96. Li Q, Guan J, Wu Z et al (2016) Equivalent maturity for ambient temperature effect on fracture parameters of site-casting dam concrete. Constr Build Mater 120:293–308
97. Zhang GX, Liu YZ (2016) Liu Y (2016) Temperature control and crack prevention of super-high arch dam. J Hydraul Eng 47(3):382–389
98. Liu X, Zhang C, Chang X et al (2015) Precise simulation analysis of the thermal field in mass concrete with a pipe water cooling system. Appl Therm Eng 78:449–459
99. Malla S, Wieland M (1999) Analysis of an arch-gravity dam with a horizontal crack. Comput Struct 72(1):267–278
100. Wieland M, Kirchen GF (2012) Long-term dam safety monitoring of punt dal gall arch dam in Switzerland. Front Struct Civ Eng 6(1):76–83
101. Chen PP, Zhang GX (2014) Study on the prevention measures and causes of cracks in arch crests of galleries in high concrete dams. Adv Mater Res 919:813–819
102. Zhang L, Liu YR, Yang Q (2015) Evaluation of reinforcement and analysis of stability of a high-arch dam based on geomechanical model testing. Rock Mech Rock Eng 48(2):803–818
103. Ohtsu M (1996) The history and development of acoustic emission in concrete engineering. Mag Concr Res 48(177):321–330
104. Snelling PE, Godin L, Mckinnon SD (2013) The role of geologic structure and stress in triggering remote seismicity in Creighton mine, Sudbury, Canada. Int J Rock Mech Min Sci 58:166–179

105. Ma K, Tang CA, Liang ZZ et al (2017) Stability analysis and reinforcement evaluation of high-steep rock slope by microseismic monitoring. Eng Geol 218:22–38
106. Zhuang DY, Tang CA, Liang ZZ et al (2017) Effects of excavation unloading on the energy-release patterns and stability of underground water-sealed oil storage caverns. Tunn Undergr Space Technol 61:122–133
107. Behnia A, Chai HK, Shiotani T (2014) Advanced structural health monitoring of concrete structures with the aid of acoustic emission. Constr Build Mater 65:282–302
108. Schiavi A, Niccolini G, Tarizzo P et al (2011) Acoustic emissions at high and low frequencies during compression tests in brittle materials. Strain 47(S2):105–110
109. Krajcinovic D (1996) Damage mechanics. Elsevier, North-Holland
110. Ma K, Jin F, Tang CA et al (2017) Deformation mechanism of the dam heel of Dagangshan high arch dam based on microseismic monitoring during initial impoundment. Chin J Rock Mech Eng 36(5):1111–1121
111. Gibowicz SJ, Kijko A (1994) An introduction to mining seismology. Academic Press Inc., San Diego, California
112. Maxwell SC, Urbancic TI, Prince M et al (2003) Passive imaging of seismic deformation associated with steam injection in Western Canada. In: SPE Annual Technical Conference and Exhibition. Society of Petroleum Engineers, SPE84572
113. Liu XZ, Tang CA, Li LC et al (2017) Microseismic monitoring and 3D finite element analysis of the right bank slope, Dagangshan hydropower station, during reservoir impounding. Rock Mech Rock Eng 50(7):1901–1917
114. Jin F, Hu W, Pan J et al (2011) Comparative study procedure for the safety evaluation of high arch dams. Comput Geotech 38(3):306–317
115. Luo D, Lin P, Li Q et al (2015) Effect of the impounding process on the overall stability of a high arch dam: a case study of the Xiluodu dam, China. Arab J Geosci 8(11):9023–9041
116. Development N, Commission R (2007) Design specification for concrete arch dams: DL/T 5346-2006. China Electric Power Press, Beijing
117. Baustaedter K, Widmann R (1985) The behavior of the Kolnbrein arch dam. In: Proceeding of the 15th International Commission on Large Dams, Lausanne, pp 633–651
118. Lin P, Cheng X, Zhou WY et al (2003) Simulation on back analysis of Shuanghe arch dam cracking. Rock Soil Mech 24(2):53–56
119. Huang YS, Zhou WY, Yang RQ et al (2006) Stability analysis and evaluation of Laxiwa arch dam. Chin J Rock Mech Eng 25(5):901–905
120. Zhang L, Liu YR, Yang Q et al (2013) Global stability of Yangfanggou Arch Dam by 3D nonlinear FEM analysis and geomechanical model tests. Chin J Geotech Eng 35(1):239–246
121. Lin P, Shi J, Zhou H et al (2016) Analysis on the influence of structural plane and the stability of coordinated reinforcement of Wudongde dam abutment. Chin J Rock Mech Eng 35(2):3937–3946
122. Zhou JP, Du XK, Zhang LB et al (2017) Determination of high arch dam heel stress: a comparison between monitoring and calculating methods. J Hydroelectr Eng 36(12):87–94
123. Song ZH, Liu YR, Yang Q et al (2017) Research on asymmetry of high arch dam foundation and its improvement effect. Rock Soil Mech 38(2):507–516
124. Feng F, Qiu XJ, Zhang GX et al (2017) Inversion of mechanical parameters of super-high arch dam based on deformation monitoring during construction period. Rock Soil Mech 38(1):237–246
125. Yang BQ, Zhang L, Chen Y et al (2017) Experimental and numerical comprehensive analysis on overall stability of Jinping I high arch dam. J Hydraul Eng 48(2):175–183
126. Ma K, Zhuang DY, Tang CA et al (2018) Study on formation causes of gallery cracks in Dagangshan high arch dam based on microseismic monitoring. Chin J Rock Mech Eng 37(7):1608–1616
127. Tang CA, Liu H, Lee PKK et al (2002) Numerical studies of the influence of microstructure on rock failure in uniaxial compression—Part I: effect of heterogeneity. Int J Rock Mech Min Sci 37:555–569

Chapter 4
Research on Water Inrush Mechanism of Coal Mine Floor Based on Microseismic Monitoring

Abstract This chapter delves into the mechanism of water inrush in coal mine floors, employing microseismic monitoring, theoretical analysis, and numerical simulation, with the Dongjiahe Coal Mine as a representative case study. It begins by establishing the research background and significance, followed by a detailed literature review that examines global and domestic studies on water inrush mechanisms and monitoring methods, thereby identifying critical gaps in current knowledge. Building on this foundation, the chapter explores the stress field and failure characteristics of the coal mine floor, integrating elastic mechanics theory, numerical simulations, and microseismic monitoring to uncover the interactions between stress distribution and structural failure. Attention is then directed to the development of water inrush channels, analyzing stress variations and the evolution of stress fields in fault-influenced surrounding rock. Further, the study investigates the failure characteristics of floor faults above confined aquifers through mechanical modeling, stress field solutions, and numerical simulations, supported by microseismic observations of fault damage. The integration of these approaches provides a coherent understanding of the processes driving water inrush and offers actionable insights for predicting and mitigating such hazards in coal mining, ensuring both safety and operational efficiency.

4.1 Research Background and Significance

Coal is the second largest energy resource after petroleum in the world's primary energy consumption structure. China has the third largest coal geological reserves in the world, and the annual coal mining volume has been around 3.5 billion tons in recent years, making it the largest producer and consumer of coal in the world. The endowment of energy resources in China determines that coal will remain the main energy source for a considerable period of time. Currently, coal accounts for over 65% of China's primary energy consumption structure, and it is expected to account for 50% in 2050 [1]. In recent years, as the national economy has entered a period of structural adjustment, the dependence on energy for economic growth has

K. Ma and C. Tang, *Microseismic Monitoring and Stability Analysis of Large Rock Mass Engineering*, https://doi.org/10.1007/978-981-96-6483-2_4

decreased, the intensity and growth rate of energy demand have declined, and the demand for coal has slowed down at a low speed. However, coal will still maintain its dominant position as the main energy source in China, and coal still has a large space for development in the medium and long term [2].

Due to the complex hydrogeological and tectonic geological conditions in China, as well as the fact that 95% of coal resources are mined using underground mining methods, mine water inrush disasters have always been a prominent problem affecting the safety production of coal mines in China. According to incomplete statistics, in the past 20 years until 2005, water inrush accidents have occurred in more than 250 coal mines, with direct economic losses of over 35 billion yuan [3], resulting in significant property loss, serious casualties, and destruction of underground water resources and the ecological environment. In the early twenty-first century, a major water inrush accident occurred in the Dongpang mine of Hebei Jinniu Energy Group Co. Ltd., with a peak water inrush rate of about 75,000 m^3/h, causing economic losses of about 800 million yuan. In 2010, the driving face of the return main airway of the No.16 coal seam in Luotuoshan Coal Mine was close to a hidden small Ordovician limestone water-conducting karst collapse column, causing the confined water to break through the aquiclude and resulting in floor water inrush. The accident caused 32 deaths, 7 injuries, and direct economic losses of 48.53 million yuan. In the "3.28" water inrush accident in Wangjialing, Shanxi, in 2010, the accidental drilling into old goaf water zone of small kilns caused a water inrush accident, trapping 153 people underground. Despite nearly 100 million yuan spent on rescue efforts, 38 people unfortunately lost their lives, causing serious losses to the country and the people. On January 30, 2015, a water inrush accident occurred in the Zhuxianzhuang Coal Mine of Anhui Huaibei Coal Mining Group, flooding the mine and causing 7 deaths and 7 injuries. On May 25, 2017, a water inrush accident occurred in the 12,123-working face floor connection roadways of Pan Er Coal Mine of Huainan Mining Group, with a maximum water inrush rate of 14,520 m^3/h, causing the mine to be flooded. Numerous water inrush cases indicate that most of the major water inrush accidents and mine flooding accidents are caused by the confined water in the thick Ordovician limestone aquifer and Taiyuan Formation aquifer below the coal seam breaking through the floor aquiclude and entering the working face. According to data statistics [4], as of 2001, in the North China region, the main coal seams of more than 230 mines were threatened to varying degrees by the floor karst confined water, and the threatened coal reserves reached 14.97 billion tons, accounting for about 39% of the total reserves.

In recent years, with the depletion of shallow resources in China, major mining areas have successively entered the stage of deep (800–1500 m) coal mining. In the deep mining environment, the rock mass is often in a complex geological and mechanical environment of "three highs and one disturbance" (high ground stress, high temperature, high karst water pressure, and intense mining disturbance), making the threat of floor water inrush more severe. Therefore, floor water inrush accidents have become an important factor restricting the development of China's coal resources.

The key issue that makes it difficult to control floor water inrush accidents lies in the insufficient understanding of the mechanism of floor water inrush and the lack of a comprehensive theory and method for predicting and preventing floor water inrush. Essentially, floor water inrush in coal mines refers to the phenomenon where confined water in the underlying aquifers below the coal seam breaks through the floor rock layer barrier and enter the working face in various forms, with a sudden, gradual or delayed onset, along the floor fractures. Mining activities inevitably disrupt the balance of the original rock stress field, leading to damage and fractures in the floor rock mass. This damage and fracture increase the permeability of the rock mass, resulting in the seepage of confined water in the floor. The changes in the seepage field of the floor will then interact with the rock mass through the seepage force and hydro-physical and chemical actions (such as flushing, softening, etc.), further promoting the development of floor damage and fractures. This water–rock coupling continues to evolve and accelerate. Finally, when the floor formed a "water inrush channel" that communicates with the mining face, it caused the water inrush disaster of the coal seam floor. Rock damage and fractures are the fundamental causes of water inrush disasters and the main topic of the investigation on "safe coal mining under confined water conditions." Especially for some geological structures in the floor rock mass, it is easier to undergo damage and form water inrush channels in the multi-field interaction environment of water–rock coupling. Therefore, to deeply reveal the mechanism of floor water inrush in coal seams, it is necessary to comprehensively consider the original rock stress, mining-induced stress, water pressure, and geological structures, and focus on the initiation and development of rock mass microfractures, studying the evolution laws of deformation, damage, and fracture of the floor rock mass.

In the field of energy science, "control and prevention of engineering geological disasters of rock mass" is a priority topic in the national medium- to long-term scientific and technological development plan and a key supported object in the 13th Five-Year Plan for scientific and technological development. In this context, in response to the current problems faced by China's coal resources, it is of great strategic significance to conduct research on the mechanism, early warning, control, and other related theoretical issues of floor water inrush disasters. In this context, in response to the current problems faced by China's coal resources, it is of great strategic significance to conduct research on the mechanism, early warning, control, and other related theoretical issues of floor water inrush disasters. Therefore, in order to develop coal resources threatened by water inburst and achieve safe production, this chapter focuses on the 22517 working face of Dongjiahe Coal Mine of Shaanxi Coal Chenghe Mining Co., Ltd. By introducing high-precision microseismic monitoring technology into the mining process of the working face and simplifying the mechanical model from the specific engineering situation, this study comprehensively applies elastic mechanics and structural mechanics, combined with microseismic data and numerical simulation, to study the characteristics of the floor stress field distribution and damage depth during the initial pressure-bearing of the 22517-working face, as well as the stress, displacement, and seepage conditions at the fault before and after the working face crosses the fault. The causes of microfractures are theoretically analyzed, and the potentially hazardous areas of the floor rock mass are delineated, in order to provide theoretical support for achieving "safe coal mining with geological structures in the floor under confined water conditions".

4.2 Literature Review

Floor water inrush is a complex geological and mining-induced phenomenon. It refers to a natural disaster where confined water below the coal seam breaks through the floor aquiclude and enters the working face in the form of delayed, gradual, or sudden water inflow, causing an increase in water inflow or mine flooding [5]. Over the years, many scholars at home and abroad have conducted extensive research on the floor mining-induced stress field, floor rock mass damage, and confined water seepage field characteristics through theoretical analysis, similar model tests, numerical simulations, and field monitoring. They have also made continuous efforts and innovations in predicting and forecasting methods for floor water inrush and have achieved many beneficial results in understanding the mechanism of the formation of water inrush channels in the floor and the occurrence of water inrush disasters.

4.2.1 Literature Review of Water Inrush Mechanism in Abroad

The study of the structure, deformation, and failure characteristics of the floor rock mass induced by floor water inrush disasters in coal mines was carried out earlier and has a history of over a hundred years [6, 7]. However, due to the differences in hydrogeological conditions of coal-bearing strata, only major coal-producing countries in Europe have been threatened to varying degrees by floor water inrush [8]. Therefore, research on floor water inrush has mainly focused on these major coal-producing countries in Europe.

In the 1940s and 1950s, Hungarian engineer Weg Frence first proposed the concept of floor "relative aquiclude" and established the relationship between water pressure, aquiclude thickness, and floor water inrush. Later, it was adopted directly or indirectly by many countries [9, 10].

In the 1950s and 1960s, Soviet scholar Sresalev [11] first regarded the floor of the coal seam as a beam structure fixed at both ends and subjected to uniformly distributed loads. He began to study the deformation and failure mechanisms of coal seam floor under the action of confined water from the view of statics, and derived theoretical formulas for calculating the theoretical safe water pressure for avoiding floor water inrush based on rock mass strength theory. This provided a reference for floor water control work. Hungarian scholars focused on intact rock layers in the floor and paid particular attention to changes in floor stress during the mining process and the characteristics of floor damage.

In the 1970s and 1980s, scholars from the Soviet Union, Yugoslavia, and other countries considered the effect of mining-induced stress and the flow of confined water on the destruction of the relative aquiclude [12]. Soviet scholar Mironenko et al. [13] conducted research using theoretical analysis and field measurements, pointing out that floor water inrush is the result of the combined effects of rock mass stress field, fracture network, and hydrostatic pressure changes under mining disturbance. Italian scholar Sammarco [14, 15] studied water inrush accidents in underground coal mines and found that water inrush is accompanied by changes in water level and gas outburst, revealing the phenomenon of multi-hazard coupling in coal mines and proposing the use of such precursor information to achieve effective prediction of mine water inrush.

In the 1990s, Polish scholar Motyka et al. [16] clarified the development law of karst in the region by investigating a large amount of drilling data from mine floors, realizing that the most direct cause of mine water inrush disasters is the connection of mining-induced fractures with karst aquifers. Soviet scholar Kuznetsov et al. [17] recognized that mine water inrush is related to the structure of the underground rock mass through the establishment of a water–rock coupled geological mechanical model. Soviet scholar Mironenko et al. [13] studied the hydrogeological mechanism of mine water inrush and believed that the process of mine water inrush is a complex interaction between underground water and underground rock mass structures under the influence of coal mining. They also proposed methods for preventing and controlling mine water inrush and rock mass failure.

Since the beginning of the twenty-first century, coal production in aboard has gradually decreased, and there are fewer coal mines with complex hydrogeological conditions. In addition, developed countries have strict regulations on groundwater protection. Currently, there are very few occurrences of floor water inrush in producing mines, and research on floor water inrush in these countries is basically stagnant. On the contrary, due to China's rapid economic development and increasing demand for coal resources, research on mine water inrush in China continues to develop.

4.2.2 Literature Review of Water Inrush Mechanism in China

Research on coal mine water inrush in China started relatively late. This is because in the early stages of mining, the coal seam was shallow and the hydrogeological conditions of the coal-bearing strata were relatively simple, resulting in few coal mine water inrush accidents. Since the 1980s, with the increase in mining intensity and depth, floor water inrush accidents have become more serious, attracting the attention of researchers to the prevention and control of floor water inrush and its theoretical research. Through theoretical analysis, numerical simulation, similar model tests, and field monitoring, China has gone through a growth stage from learning and drawing on foreign research experience and achievements to independent innovation. Various distinctive classical theories and hypotheses have been proposed.

1. Water Inrush Coefficient

In the 1960s, water inrush coefficient was proposed by summariing empirical formulas based on water inrush data documented in the Fengfeng (Hebei province), Jiaozuo (Henan province), Zibo (Shangdong province), and Jingxing (Hebei province) coalfields. During the hydrogeological battle in Jiaozuo coalfield in 1964, some scholars referred to the concept of floor relative aquiclude from abroad and defined the water pressure borne by the unit thickness of the floor aquiclude as the water inrush coefficient [18]. Since then, through the analysis of factors such as the depth of floor damage, the height of confined water uplift, the different lithological composition of the floor aquiclude, and the head loss during the uplift of confined water, the water inrush coefficient has been revised multiple times [19–21], with the main focus of revision being the determination of the effective aquiclude thickness.

2. Down Three Zones

In 1988, Li et al. [22] from Shandong University of Science and Technology conducted extensive field measurements and experimental research on floor damage. They believed that there are three different zones with different aquiclude properties in the mined coal seam floor. They proposed the "Down Three Zones" theory for floor water inrush. They divided the rock layers of the floor affected by mining into the water-conducting fractured zone, the intact rock layer zone (or the rock strata protection zone) and the zone of water rises from the confined aquifer (or concealed head zone). Shi et al. [23] further subdivided the intact rock layer zone based on damage mechanics, fracture mechanics, and mining pressure theory, into the newly damaged zone and the original damaged zone. They derived the calculation formulas for the thickness of each zone and provided a method for determining floor water inrush.

3. "Thin Plate Structure" Theory

In 1989, Jincai Zhang et al. from the China Coal Research Institute propose a thin plate model for floor water inrush used the theory of elastic–plastic mechanics. They considered the floor aquiclude as a thin plate supported on all sides and subjected to uniformly distributed loads. Based on the Mohr–Coulomb strength theory and Griffith strength theory, they used the elastic solution for a semi-infinite body subjected to a uniform vertical load on a certain length to calculate the maximum depth of floor fracturing due to mining [24–26].

4. Floor "Key Layer" Theory

Since 1996, Xu et al. [27] and Li et al. [28] have borrowed the concept of key layer from the roof to identify the strongest load-bearing rock layer above the confined aquifer as the key layer of the floor. They consider the key layer as the main controlling factor for floor water inrush and establish the theory of key layer for floor water inrush. Based on the characteristics of layered rock structure in the floor, this theory identifies the roles of hard rock layers and aquicludes in the floor water inrush prevention process and reveals the mechanisms of structural damage and failure in the floor rock layers under the coupling of mining pressure and water pressure.

5. Vulnerability index method

Wu et al. and Song et al. [29–31] believe that floor water inrush is a nonlinear dynamic phenomenon controlled by various factors and complex formation mechanisms. Based on the integrated theory of multi-source information and the "nested theory," they propose a vulnerability index method for floor water inrush using the analytic hierarchy process and geographic information system. They also introduce a variable weight model to reflect the internal differences in indicator values of the main controlling to study floor water inrush in coal seams [29–31].

In addition, domestic scholars have conducted a lot of work on theoretical analysis, numerical simulation, and laboratory tests of floor water inrush. In terms of theoretical analysis, Song et al. [31] theoretically analyzed the impact of mining-induced stress on floor water inrush. They establish a predictive control mechanical model for water inrush during cross-fault mining and formulate relevant control criteria. Guo et al. [32] analyzed the floor water inrush induced by complex geological structures under high-pressure water conditions in deep mines. They divide the deep floor water inrush into three types: the water inrush disaster mode in deep mining was divided into three types, floor fracture spreading type, primary channel conduction type and buried structure shear type, and deduced the corresponding water inrush criterion. Lu et al. [33] use the limit equilibrium theory to derive the analytical formula for hydraulic pressure of floor fault water inrush and analyze the influence of fault dip angle and working face advancement direction on the critical water inrush hydraulic pressure. Zhou et al. [34] use the extended finite element and fracture mechanics methods to numerically simulate floor water inrush induced by activated concealed faults. They quantitatively evaluate the stress concentration intensity of the fault tip stress field and analyze the influence of mining depth, water pressure, fault dip angle, and length on fault activation. Hua et al. [35] establish a mathematical model based on catastrophe theory to evaluate the hazard of water inrush caused by fault activation through the analysis of geological factors, fault parameters, and mining factors, combined with a sample database of fault activation-induced water inrush. Hu et al. [36] study the distribution of stress fields and the influence of normal stress and shear stress on fault activity and the distribution of effective shear stress on faults based on the distribution law of stress fields in the floor under supporting pressure.

In terms of numerical simulation, Li et al. [37, 38] use RFPA software to numerically simulate the damage process of the floor during floor water inrush, analyze the response characteristics of the floor stress field and seepage field during the development of water inrush channels, and reveal the mechanism of floor water inrush caused by mining-induced floor and fault-containing floors. Liu et al. [39] use $FLAC^{3D}$ software to simulate the process of water inrush induced by concealed faults and point out the uplift height of confined water and the development height of plastic zones in the fault damage zone. Cheng et al. [40] and Liu et al. [41] conduct a large number of simulation studies on delayed water inrush in floors containing fractured structures using FLAC^{3D}. Yang et al. [42] and Lu et al. [43] use the COMSOL to study the water inrush mechanism in coal seam floors.

In terms of laboratory experiments, Li et al. [44] used similarity simulation experiments to study the dynamic failure process of floor aquicludes under mining influence and propose precursor information for floor water inrush induced by normal faults. Zhang et al. and Sun et al. have developed a similar simulation test system for high-water-pressure floor water inrush in deep mining. The system can effectively simulate the phenomena of water inrush in deep mining and the coupling of stress and seepage [45–47]. Xu et al. [48] conducted similarity simulation tests on floor heave water inrush disasters in deep roadways. Combined with field pressurized water injection tests, they discovered the sudden change pattern of roadway floor heave water inrush.

4.2.3 Literature Review of Water Inrush Monitoring

Floor water inrush is a complex process of rock mass damage influenced by multiple factors, and the dynamic damage of the coal rock mass occurs during the advancement of the working face. It is difficult to accurately reflect the engineering reality solely through theoretical analysis or laboratory tests. Therefore, on-site monitoring of floor rock mass damage during mining is necessary, and currently, it is mainly achieved through methods such as borehole water injection (grouting) and geophysical exploration.

Geophysical exploration techniques that are widely used include geological radar detection, seismic CT imaging, and the direct current resistivity method. Cheng et al. [49] used rock mass velocity tests to assess the state of the rock mass and determine the development range of water-conducting fractures in the floor. In addition to acoustic CT observations, dynamic process tests were conducted using single-hole acoustic waves and segmented water injection measurements to obtain the maximum depth of floor damage. Zhang et al. [50] used seismic CT detection technology to collect seismic CT data at different stages, combined with detection profiles formed between boreholes and galleries in the coal seam working face, and performed inversion and data processing to obtain the maximum depth of fracture zone development. Liu et al. [51] designed an electromagnetic method floor water inrush monitoring system around the working face and summarized the response characteristics of geological anomalies at different spatial positions, proposing the concept of a water inrush coefficient threshold. Liu et al. [52] studied the spatial transient variation characteristics of electric field parameters during rock mass water conduction. They conducted water inrush monitoring using parameters such as potential, current over time, and apparent resistivity over time surface, and carried out corresponding field verifications.

4.3 Unresolved Issues

At present, traditional monitoring methods for rock engineering mainly focus on macroscopic information such as deformation, deformation rate, and acceleration. For specific floor water inrush problems in coal mines, displacement monitoring methods that primarily focus on deformation cannot detect the characteristics of microfracture activity inside the rock mass before macroscopic damage occurs in the floor. They also cannot provide real-time monitoring of the "dynamic" rupture of water channels during coal seam mining. However, microfractures in the rock mass are the precursors to floor water inrush accidents, and traditional monitoring methods cannot perceive the characteristics of microfracture activity inside the rock mass or reveal the essence of floor water inrush mechanisms.

Due to the heterogeneous, discontinuous, and nonlinear characteristics of the surrounding rock mass of the mining face, as well as the complex effects of loading and boundary conditions, it is currently impossible to accurately solve the stress distribution within the rock mass using mathematical mechanics. However, approximating the stress state around the working face is beneficial and necessary for a deeper understanding of floor deformation and water inrush disasters. However, the discussion on selecting appropriate mechanical models for floor water inrush is lacking, and the application of multiple mechanics in studying the mechanism of floor water inrush is not sufficiently in-depth, especially in the study of the mechanism of delayed water inrush caused by faults and other structures. This is a major reason for the limited effectiveness of water control in coal mines at present.

4.4 Research Content

The essence of floor water inrush in coal seams is the problem of water breaking through the aquiclude of the floor rock mass under the coupling of mining-induced stress and seepage, leading to the disaster of increased water inflow or mine flooding. In this chapter, real-time dynamic monitoring of the process of floor water inrush channel development is conducted through the establishment of an on-site microseismic monitoring system. Combined with different mechanical models, the dynamic evolution characteristics of the formation process are studied to reveal the mechanism of floor water inrush disasters and the precursor information of microfractures. The content of this chapter includes the following parts.

(1) Analysis of microseismic activity characteristics of coal rock mass during the development of floor water inrush channels. Real-time dynamic monitoring of the process of coal rock mass damage is achieved through the microseismic monitoring system during mining. By processing and analyzing the monitoring data, a large number of progressive fractures, instability, and disseminated microfractures are obtained, fully reproducing the entire process of the formation, expansion, and connection of water inrush channels. The spatial

distribution, magnitude, and energy distribution of microseismic events are analyzed to analyze the temporal and spatial evolution characteristics of the floor damage process and its range during the advancement of the working face. The E_S/E_P value is used to analyze the forms of damage and failure in the floor coal rock mass, and the cumulative laws of tensile and shear failure during the development of water inrush channels are analyzed.

(2) Mechanical analysis and numerical simulation of the development process of floor water inrush channels. Based on the above research, combined with elastic mechanics and structural mechanics, the distribution characteristics of floor stress fields when the working face is initially under pressure and the stress and displacement at the fault before and after the working face crosses the fault are studied. At the same time, numerical simulation methods are used to reproduce the process of the development of floor water inrush channels. The evolution laws of microfractures and damage in different types of floor water inrush disasters are studied. The focus is on the response characteristics of coal rock mass microfractures and damage before the occurrence of water inrush disasters, comparing the formation, evolution, and connection laws of different types of water inrush disasters, revealing the common characteristics and precursor laws of coal rock mass mining-induced fractures, faults, and collapsed pillars forming water inrush channels, exploring the formation mechanism of water inrush channels, and revealing the static mechanical mechanism of mine water inrush disasters.

Studying the mechanism and precursor laws of mine water inrush disasters is most important for tracking the development process of water inrush channels and finding precursor information during the formation of water inrush channels. This chapter captures microfracture information in the process of coal rock mass damage through microseismic monitoring technology, fully reproducing the entire process of the formation, expansion, and connection of floor water inrush channels. By combining microseismic monitoring, mechanical analysis, and numerical simulation, the static mechanical mechanism of water inrush channel formation in the floor mining-induced fractures is studied, revealing the mechanism of delayed fault water inrush, evaluating the danger of floor water inrush, and establishing a research method for the mechanism and precursor laws of floor water inrush disasters based on microseismic monitoring. The content of this chapter is of great scientific significance and engineering practical value for revealing the essence of floor water inrush, clarifying the conditions for water inrush formation, and dynamic early warning and prediction.

4.5 Research on Stress Field and Failure Characteristics of Floor Based on Elastic Mechanics

Abundant Ordovician limestone water in the floor is the main water inrush source in north China coal fields. Mining-induced Ordovician limestone water inrush in the floor poses a great threat to safe and efficient mining [26, 53]. The key to realizing safe mining under confined water pressure is to understand the failure mechanism of the coal seam floor under the combined effects of mining pressure and water pressure [37, 54], which attracts scholars' attention to the failure mechanism, early warning, and reinforcement of the coal seam floor.

At present, the methods for investigating the floor failure characteristics mainly include theoretical analysis, numerical simulation, and field monitoring. Theoretical analysis guides engineering practice. Many scholars have proposed a variety of floor failure analysis theories. Baiying Li put forward the "Down Three Zones" hypothesis according to the rock mass failure characteristics of different layers of the floor during mining operations, and divided the rock strata into the water-conducting fractured zone, the intact rock stratum zone and the zone of water rises from the confined aquifer from top to bottom [55, 56]. Liu et al. [57] established a mechanical model for floor failure along the inclined coal seam and calculated the maximum failure depth in the inclined coal floor undermining and revealed the distribution characteristics of plastic zone of floor under the combined action of mine pressure and water pressure. Lu et al. [58] obtained performance function that calculates the reliability probability by applying the theory of elastic thin plate, analyzed the water-resistant floor failure mode, and calculated the reliability probability of different failure modes by Monte Carlo method. It is found that the above scholars consider water pressure as a uniform load with a certain length in their theoretical analysis, but the reliability of this assumption remains to be discussed.

Numerical simulation is a powerful tool for solving problems related to groundwater in mines. Lu et al. [59] proposed a A micromechanics-based coupled damage and flow model, and applied finite element method to study the progressive development of fractures and the associated water flow in the floor strata during mining above a confined aquifer, reproduce the stress re-distribution, acoustic emission (AE) evolution, fracture development, permeability changes and water inrush channel formation in the floor strata during the mining process. Liu et al. [39] used $FLAC^{3D}$ to simulate the insidious fault lagging water inrush process induced by floor failure under the creep-seepage coupling, and proposed the concept of the potential water inrush channel with time effect based on the differential flow deformation theory and the subcritical crack propagation theory. Wang et al. [60] simulated the deformation and failure characteristics of deep stope floor due to working face mining, and explored the distribution characteristics of horizontal stress, vertical stress, shear stress, horizontal displacement, vertical displacement and plastic zone of floor. However, these numerical studies can rarely fully reflect the progressive failure process of the surrounding rock masses. The initiation, expansion, penetration and interaction of microcracks in coal and rock mass are the key to study the failure characteristics of floor.

Field monitoring and experiment are direct means for understanding the damage and failure of coal and rock masses. Zhou et al. [61] monitored the failure zones of the extra-thick coal seam floor in deep mining by using the distributed optical fiber sensing technology (BOTDR) and obtained the failure characteristics and the failure depth of the coal seam floor. Li et al. [62] developed a set of comprehensive methods, including transient electromagnetic method, borehole drilling, water injection test, and numerical simulation, to obtain the groundwater distribution, the aquifer thickness, and the floor failure depth. Xu et al. [63] observed the rock angular displacement to reflect whether the floor of coal seam was bent and failed, obtained the variation law between the rock mass failure depth and angular displacement, and designed a water inrush warning system on this basis. These methods can effectively monitor and evaluate the macroscopic deformation and the process of water inrush that occurred in the coal seam floor, however, microfracture and its evolution law in coal seam before the formation of water inrush channels can hardly be known.

Scholars have made a lot of achievements in studies on failure characteristics of the coal seam floor. However, the deformation and failure process of the confining floor is affected by the heterogeneity of coal and rock, the mining-induced stress redistribution and the fluid–solid coupling, etc. So, failure characteristics and the evolution law of coal seam floor can hardly be studied comprehensively by utilizing only one or two methods. Therefore, further crossing and integration of various methods are necessary so as to put forward a comprehensive and efficient method for studying mining-induced failure characteristics of the coal seam floor.

In this paper, taking a typical working face of Dongjiahe Coal Mine during initial weighing as an example, the floor stress components due to confined water are derived based on the characteristics of the confined aquifer. The floor stress distribution law and failure characteristics are obtained according to the mining stress evolution and the failure law. Besides, the evolution process of microfractures from initiation, propagation, and coalescence in the coal seam floor is reproduced by combining numerical simulation and microseismic monitoring. Finally, based on comparative analyses, a comprehensive method integrating multiple means is proposed, which is expected to provide some reference for control of inrush hazards and studies on water inrush mechanism of the coal seam floor.

4.5.1 Project Overview

Dongjiahe Coal Mine, which belongs to the Chenghe mining area, is located at about 3.5 km southwest of Chengcheng County, Weinan City, Shannxi Province, China, as shown in Fig. 4.1. The strata are Quaternary, Permian, Carboniferous, and Ordovician strata in descending order. The coal seam No. 5 of Shanxi Formation with a stable occurrence and an average thickness of 3.3 m is the main coal seam undermining. The comprehensive mechanized longwall mining technology is employed as the main mining method. The full-caving method is adopted for the roof. The typical working face is high in the west and south, and low in the east and north. The maximum elevation difference is 50 m along the east–west direction and 20 m along the north–south direction. The working face is 1217 m long along the strike direction and 185 m along the dipping direction. The dip angle is about 3°, which can be approximately regarded as a horizontal coal seam. The floor elevation is in the range of + 225 m to + 275 m, with an average of + 250 m. The open-off cut is located at 1217 m from the entrance of the track roadway and the stop line is at 100 m from the entrance of the track roadway. The progress of working face advancing is shown in Table 4.1.

No large-scale tectonic structures, such as faults and folds, were revealed. According to geological investigation and microseismic monitoring, an anticline exists at 480–720 m from the entrance of the track roadway. The maximum height difference between its core and wings is 4 m. There is a hidden fault at 200–350 m away from the entrance of the track roadway, with a dip angle of about 73° and a fault throw of 3 m. The coal seam floor water includes the Taiyuan Formation aquifer and the Ordovician limestone aquifer. The Taiyuan Formation aquifer is located between EL + 381.3 to + 384.8 m, with relatively high water pressure of about 1.5 MPa, which presents a threat to normal mining at the typical working face.

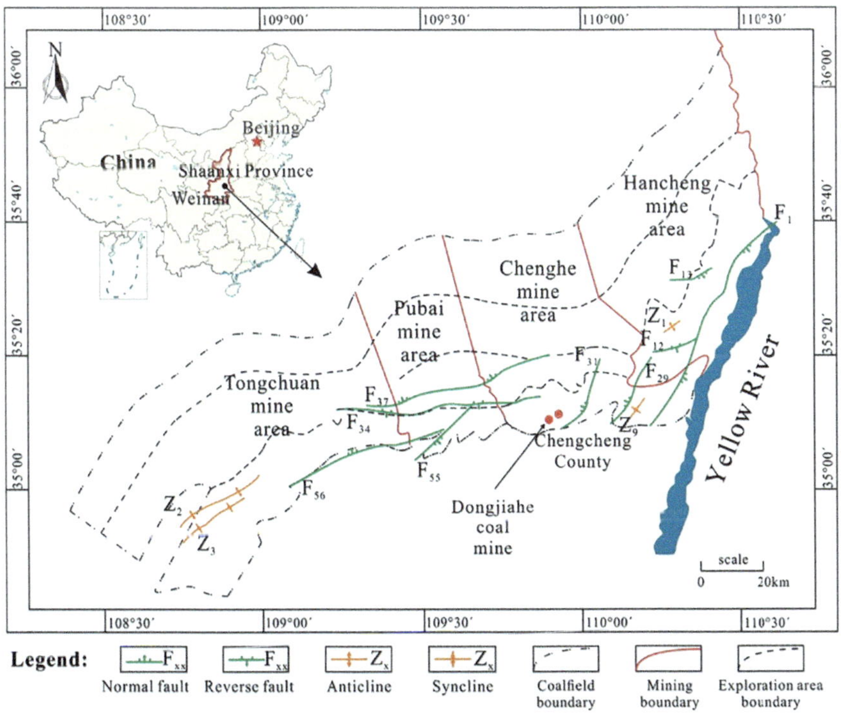

Fig. 4.1 Location and structural geological map of the Dongjiahe Coal Mine

Table 4.1 Some part of advancing progress of working face #22517

Time	Distance from the entrance of the track roadway/m	Time	Distance from the entrance of the track roadway/m
2014/12/23	1005	2015/09/08	615
2015/01/09	981	2015/09/19	603
2015/02/09	885	2015/10/09	578
2015/03/09	779	2015/11/09	514
2015/04/09	750	2015/12/09	457
2015/05/08	730	2015/12/18	443
2015/06/09	700	2016/01/08	410
2015/07/09	681	2016/01/30	379
2015/08/08	657	2016/02/05	369

4.5.2 *Stress Components Solution and Failure Analysis of the Floor*

1. Model establishment

The rock masses in the middle part along the mining direction are taken as the research object and considered as a plane strain problem in elastic mechanics. The floor rock masses are assumed to satisfy the linear elastic constitutive model. As the probability of water inrush from the coal seam floor during initial weighting is higher than that during periodic weighting [64], the analysis starts from the condition under initial weighting. The schematic diagram of loading is shown in Fig. 4.2. The load distribution in the floor is based on the "Down Three Zones" hypothesis proposed by Li [55]. The floor is subjected to the advanced abutment pressure and vertical strata stress at the top and evenly distributed water pressure at the bottom. The gravitational stress of the floor is also considered. The total thickness of the floor is h, the thicknesses of the water-conducting fractured zone, the rock strata protection zone and the zone of water rises from the confined aquifer are h_1, h_2 and h_3, respectively. The dynamic load due to falling gangue in goaf is not considered in the analysis.

2. Floor stress components caused by advanced abutment pressure

Firstly, the floor stress components due to the advanced abutment pressure are calculated. According to elastic mechanics, for a semi-infinite solid subjected to a normal distributed force on the boundary (Fig. 4.3), the stress components at Point M are [65]:

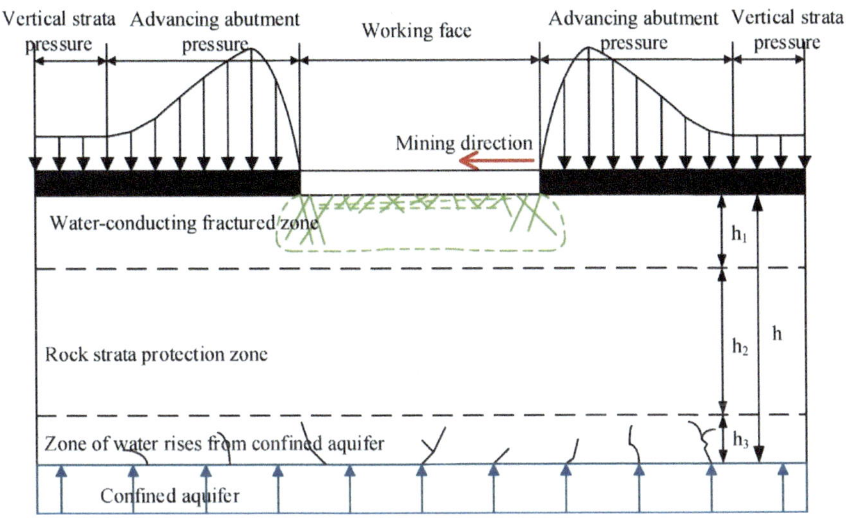

Fig. 4.2 Schematic diagram of load on coal seam floor above the confined aquifer along the working face during first weighting

$$\left.\begin{aligned}\sigma_x &= -\frac{2}{\pi}\int\limits_{-b}^{a}\frac{q(\xi)x^3\mathrm{d}\xi}{\left(x^2+(y-\xi)^2\right)^2}\\ \sigma_y &= -\frac{2}{\pi}\int\limits_{-b}^{a}\frac{q(\xi)x(y-\xi)^2\mathrm{d}\xi}{\left(x^2+(y-\xi)^2\right)^2}\\ \tau_{xy} &= -\frac{2}{\pi}\int\limits_{-b}^{a}\frac{q(\xi)x^2(y-\xi)\mathrm{d}\xi}{\left(x^2+(y-\xi)^2\right)^2}\end{aligned}\right\} \tag{4.1}$$

where the load q is a function of ξ, and the stress components can be obtained by integration.

The mechanical model for the coal seam floor subjected to the advanced abutment pressure is established as Fig. 4.4, where 2l represents the initial weighting interval, a is the length of the plastic zone due to the advanced abutment pressure, b is the length of the elastic zone. The plastic zone caused by the advanced abutment pressure is simplified into two triangular loads q_1 and q_3, and the elastic zone due to the advanced abutment pressure is simplified into two triangular loads q_2 and q_4. Their maximum value is $n\gamma h$, where n is the coefficient of stress concentration (generally in the range of 1–3), γ is the average bulk density of the overlying rock and soil, h is the depth of the coal seam. In situ stress relief occurs in the goaf as a free face, which is considered as an evenly distributed rectangular load q_5 with a magnitude of γh. Taking the middle of distance of the initial weighting as the origin, the x-axis is parallel to the floor and pointing horizontally to the boundary between coal ad rock,

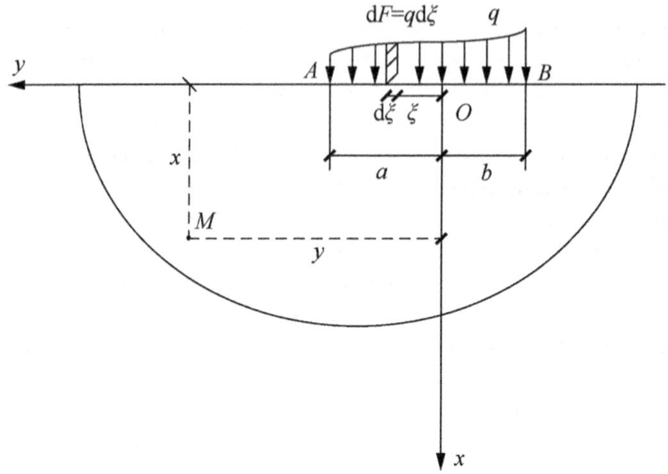

Fig. 4.3 Sketch of a semi-infinite solid subjected to normal distributed force on the boundary

and the z-axis is perpendicular to the floor and pointing downward. θ_1, θ_2, θ_3, θ_4, θ_5 and θ_6 are the intersection angles between the line connecting an arbitrary point M below the floor and the edge of loading.

The floor stress components generated by the triangular load q_1 are:

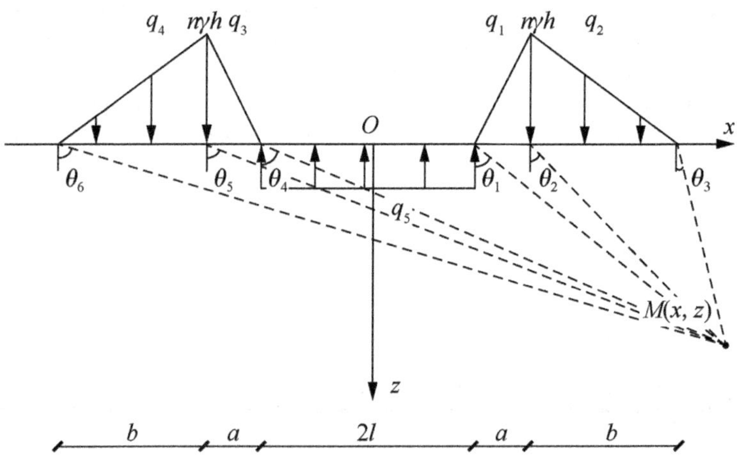

Fig. 4.4 Mechanical model of advanced abutment pressure

$$\left.\begin{aligned}\sigma_{z1} &= \frac{n\gamma hz}{a\pi}\left[\tan\theta_1(\theta_1-\theta_2)-\frac{1}{2}\tan\theta_1\sin 2\theta_2+\sin^2\theta_2\right]\\ \sigma_{x1} &= \frac{n\gamma hz}{a\pi}\left[\tan\theta_1(\theta_1-\theta_2)+\frac{1}{2}\tan\theta_1\sin 2\theta_2+2\ln\frac{\cos\theta_1}{\cos\theta_2}-\sin^2\theta_2\right]\\ \tau_{xz1} &= \frac{n\gamma hz}{a\pi}\left[\tan\theta_1(\cos^2\theta_2-\cos^2\theta_1)+\theta_2-\theta_1-\frac{1}{2}(\sin 2\theta_2-\sin 2\theta_1)\right]\end{aligned}\right\} \tag{4.2}$$

where $\theta_1 = \arctan\frac{x-l}{z}$, $\theta_2 = \arctan\frac{x-l-a}{z}$. For stress components caused by the triangular load q_2, θ_1 and θ_2 in Eq. (4.2) shall be changed to θ_3 and θ_2, respectively. Similarly, the stress components corresponding to the triangular load q_3 and q_4 can be obtained. In these equations, $\theta_3 = \arctan\frac{x-l-a-b}{z}$, $\theta_4 = \arctan\frac{x+l}{z}$, $\theta_5 = \arctan\frac{x+l+a}{z}$ and $\theta_6 = \arctan\frac{x+l+a+b}{z}$.

The floor stress component due to the rectangular load q_5 is:

$$\left.\begin{aligned}\sigma_{z5} &= -\frac{n\gamma h}{2\pi}[2(\theta_4-\theta_1)+(\sin 2\theta_4-\sin 2\theta_1)]\\ \sigma_{x5} &= -\frac{n\gamma h}{2\pi}[2(\theta_4-\theta_1)-(\sin 2\theta_4-\sin 2\theta_1)]\\ \tau_{xz5} &= \frac{n\gamma h}{2\pi}(\cos 2\theta_1-\cos 2\theta_4)\end{aligned}\right\} \tag{4.3}$$

3. Floor stress components caused by confined aquifer

For the derivation of the floor stress components contributed by the water pressure, the water pressure can be simplified into an evenly distributed load with an uncertain length. In previous studies, most scholars only considered the water pressure within the range of advanced abutment pressure [37, 54, 56]. However, the water pressure outside this range also affects the state of stress of the floor. In this section, the water pressure is first considered as an equivalent infinitely wide uniform load. The stress components corresponding to the overall water pressure is calculated based on a simplified mechanical model shown in Fig. 4.5. In this figure, the x-axis is parallel to the floor and pointing horizontally to the boundary between coal and rock, and the z-axis is perpendicular to the floor and pointing downward. The floor depth is H. The water pressure is equivalent to an infinitely wide uniform load on the floor, with a magnitude of p.

When analyzing the floor stress components contributed by the water pressure separately, the confining water only affects the zone of water rises from the confined aquifer. In addition, during the period of initial weighting, the measurement of footwall heaving is small compared to the floor thickness. Therefore, the boundary between coal and rock is considered as a fixed boundary. In view of symmetry in this problem, the displacements $u_x = 0$, $u_y = 0$, $u_z = w(z)$ and the volumetric strain $\varepsilon_v = \frac{\partial u_x}{\partial x} + \frac{\partial u_y}{\partial y} + \frac{\partial u_z}{\partial z} = \frac{\mathrm{d}w}{\mathrm{d}z}$. By substituting the displacements and strain into

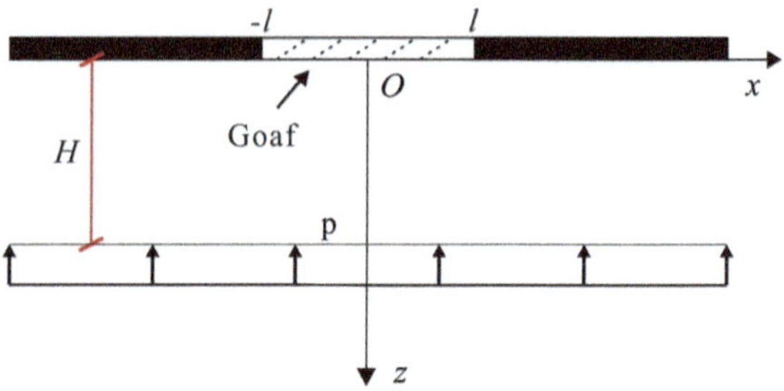

Fig. 4.5 Mechanical model of water pressure

the equilibrium equation, which expresses the tensor in terms of displacement, we have:

$$G\nabla^2 u_i + (\lambda + G)u_{j,ji} + f_i = 0 \tag{4.4}$$

where ∇^2 is the Laplace operator, λ is the Lame constant, G is the shear modulus, f_i is the body force in the i direction, which is 0 in Eq. (4.4). The equation is then simplified into $w = Bz + A$, where A and B are arbitrary constants. The stress can be calculated from displacements as:

$$\left.\begin{aligned} \sigma_x &= 2G\frac{\partial u_x}{\partial x} + \lambda\varepsilon_v = B\lambda \\ \sigma_y &= 2G\frac{\partial u_y}{\partial y} + \lambda\varepsilon_v = B\lambda \\ \sigma_z &= 2G\frac{\partial u_z}{\partial z} + \lambda\varepsilon_v = 2GB + B\lambda \\ \tau_{zx} &= \tau_{zy} = \tau_{xy} = 0 \end{aligned}\right\} \tag{4.5}$$

Given the boundary condition $[\sigma_z]_{z=H} = -p$ and constant σ_z in Eq. (4.5), $\sigma_z = -p$. B can be solved and substituted into Eq. (4.6), $\sigma_x = \sigma_y = -[\lambda/(2G+\lambda)]p$. A can be obtained by $[w]_{z-0} = 0$ We assume that the compressive stress is positive. The floor stress components caused by the confined aquifer are $\sigma_z = p$, $\sigma_x = \frac{\lambda}{2G+\lambda}p$, $\tau_{zx} = 0$. In this method, the floor heave is ignored and the water pressure is considered as an infinitely wide uniform load. The vertical and horizontal stresses at an arbitrary point are fixed and no shear stress is generated. It is consistent with the actual condition and simple in form.

4. Distribution law of floor stress

The total stress in the floor is the summation of the stress components caused by the advanced abutment pressure, water pressure and in situ stresses. According to the above derivation, the vertical stress, the horizontal stress and the shear stress at an arbitrary point M (x, z) under the floor during initial weighting can be obtained based on stress superposition principle in the elastic theory:

$$\left.\begin{aligned}\sigma_z &= \sigma_{z1} + \sigma_{z2} + \sigma_{z3} + \sigma_{z4} + \sigma_{z5} + \sigma_{zs} + \gamma(h+z)\\ \sigma_x &= \sigma_{x1} + \sigma_{x2} + \sigma_{x3} + \sigma_{x4} + \sigma_{x5} + \sigma_{xs} + \lambda\gamma(h+z)\\ \tau_{xz} &= \tau_{xz1} + \tau_{xz2} + \tau_{xz3} + \tau_{xz4} + \tau_{xz5} + \tau_{xzs}\end{aligned}\right\} \tag{4.6}$$

where σ_{zs}, σ_{xs}, τ_{xzs} are the floor stress components caused by the confined aquifer, z is the vertical distance between a point in the floor and the bottom of the working face. The geological investigation report and relevant data indicate that the depth of the coal seam is 355 m, the initial weighting interval is 40 m, the length of elastic zone is 15 m, the length of plastic zone is 5 m, the floor thickness is 30 m, the average bulk density of the overlying rock and soil is 24.5 kN/m^3, and n is taken as 1.5. The stress contour can be obtained by using MATLAB.

Under the gravitational stress from the overlying rock masses, the vertical stress on the coal seam close to the advanced abutment pressure is relatively high. The vertical stress spreads outward and downward, forming a bulb of pressure. The rock masses below the goaf are less affected by the advanced abutment pressure. The vertical stress gradually increases with increasing depth. This is mainly due to the effect of gravitational stress, as shown in Fig. 4.6a. The variation of horizontal stress is approximately the same as the vertical stress contour. However, tensile stress occurs in the floor near the goaf. This is caused by the relatively high compressive stresses in the two directions under the advanced abutment pressure. It is known from the generalized Hooke's law that $\varepsilon_x = (\sigma_x - \nu\sigma_z)/E$, horizontal compressive stress occurs locally in the rock masses below the advanced abutment pressure. The vertical stress in the goaf is relatively low. In order to maintain the compatibility of horizontal deformation, tensile stress occurs in the rock masses below the goaf. With increasing depth, the stress in the floor below the abutment pressure gradually spreads and decreases, and no more compressive stress is generated in the rock masses below the goaf, as shown in Fig. 4.6b. Under the action of symmetric loading, two bulbs of shear stress with the same magnitude and opposite directions are generated and spread outward under the advanced abutment pressures zones, as shown in Fig. 4.6c.

5. Failure depth and characteristics

As can be seen from Fig. 4.6b, tensile stress exists in the floor. The modified Mohr–Coulomb failure criterion with tension cut-off is employed to reflect rock failure under tensile stress and shear stress. The formula is:

$$\left.\begin{aligned}&\sigma_3 > \sigma_t\\ &\sigma_3 > \sigma_1 \tan^2(45^\circ - \frac{\phi}{2}) - 2c\tan\left(45^\circ - \frac{\phi}{2}\right)\end{aligned}\right\} \tag{4.7}$$

When Eq. (4.7) is satisfied, no failure occurs in the floor. In this formula, σ_t is the uniaxial tensile strength, c is the cohesion and ϕ is the internal friction angle. According to Table 4.2, the average friction angle φ and cohesion c are determined by the weighted average method. The contour of σ_3 which can satisfy the Eq. (4.7) is obtained by using MATLAB, i.e., the floor failure shown in Fig. 4.7. As can be seen

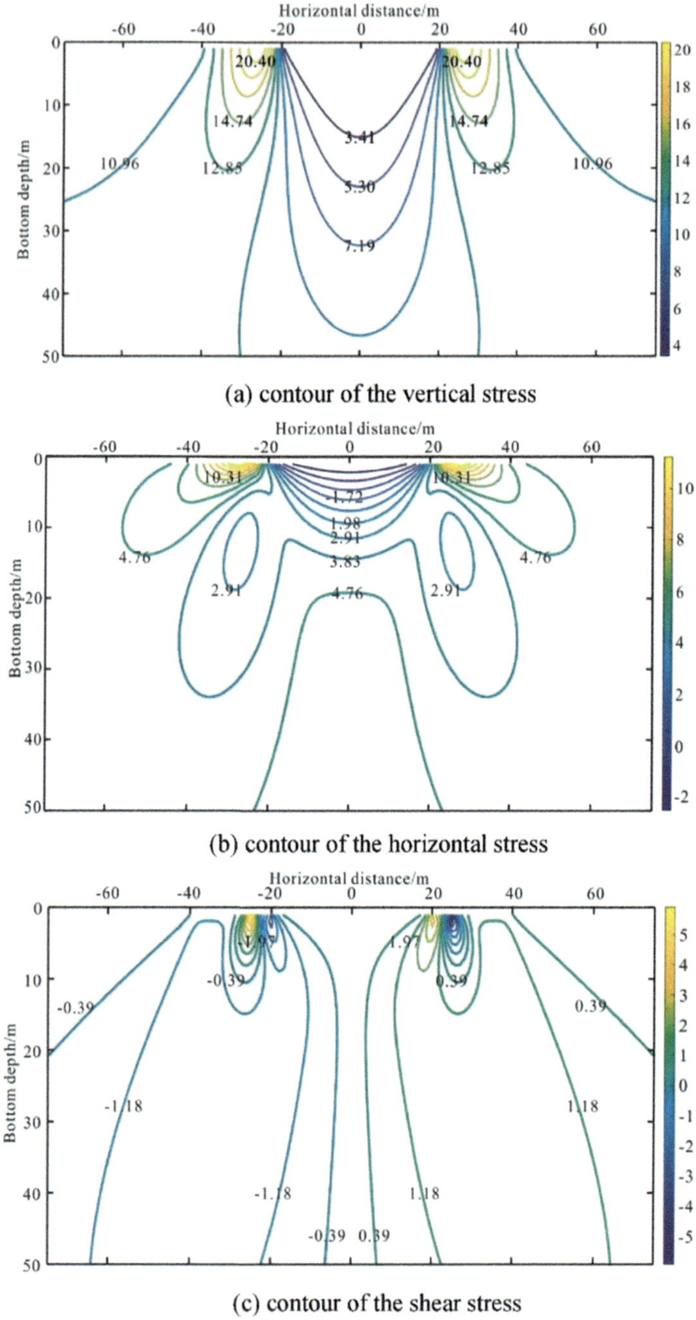

Fig. 4.6 Contour of stress components and failure zones (unit: MPa)

from this figure, the failure zone is in a shape of an inverted horse saddle. The failure depth is the greatest in the floor under the advanced abutment pressure, reaching 11.0 m. The rock masses within 1.7 m below the goaf are subjected to tensile failure, leading to floor heave. Shear failure occurs in 1.7–3 m below the goaf. A shear failure zone is formed at the two sides 5–11 m below the floor. With increasing floor failure depth, the confined water can rise up to the floor bottom easily from the shear failure zone, subsequently leading to water inrush events.

Table 4.2 Physical and mechanical parameters of rock strata

Stratum No	Rock stratum	Elastic modulus/MPa	Compressive strength/MPa	Friction angle/(°)	Poisson's ratio	Density/(kg/m^3)	Thickness/m
1	Coarse sandstone	8800	45	26	0.24	2450	23
2	Fine sandstone	7800	24	29	0.28	2450	10
3	Coarse sandstone	8800	45	26	0.24	2450	4
4	Coal seam 5#	1650	8	36	0.32	1500	4
5	Fine sandstone	8200	35	28	0.26	2450	13
6	Quartz sandstone	8850	48	25	0.24	2650	17
7	Aluminous mudstone	8800	20	30	0.25	2660	4
8	Ordovician limestone	5000	68	37	0.24	2770	25

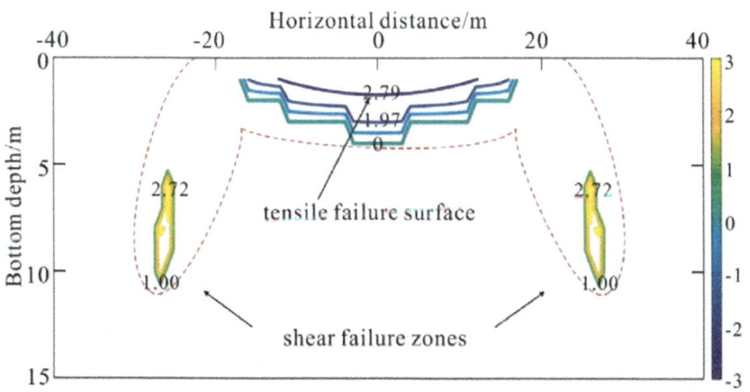

Fig. 4.7 Contour of tensile and shear failure zones (unit: MPa)

The effects of various loads on the floor failure depth are analyzed below. As can be seen from Fig. 4.8, 2c represents the distributed length of the water pressure; *L* represents the advancing length of the working face; inf represents infinity. When the coefficient of stress concentration is constant, the floor failure depth decreases with increasing water pressure, indicating that the water pressure constrains floor failure. It shows that the water inrush coefficient method is not applicable for the floor with a complete confining bed, which is consistent with the conclusion by Guo and Liu [66]. In addition, the coefficient of stress concentration indicates the magnitude of the advanced abutment pressure. In case of constant water pressure, the floor failure depth increases with increasing advanced abutment pressure. It shows that the floor failure depth is proportional to the advanced abutment pressure and the advanced abutment pressure is a key load affecting the failure depth. When the coefficient of stress concentration is 1.5 and the water pressure is 2.4 MPa, the floor failure depth decreases rapidly. This is because the maximum failure depth is mainly controlled by the shear failure zone. No shear failure zone appears under the given condition. Instead, tensile failure only occur below the goaf. When the shear failure zone occurs, the floor failure depth decreases approximately linearly. The gradient is independent of the advanced abutment pressure. In engineering practices, for mining area with larger advanced abutment pressure, the floor can be reinforced by grouting or rockbolts to reduce the impact of advanced abutment pressure and the risk of water inrush.

In this section, the analytic solutions of floor stress components are derived based on the semi-infinite solid in elastic theory. The stress figures are obtained by using MATLAB. The modified Mohr–Coulomb criterion with tension cut-off is employed to analyze the floor failure depth and failure characteristics. The advantage of theoretical analysis is that it can perform sensitivity analysis for various parameters

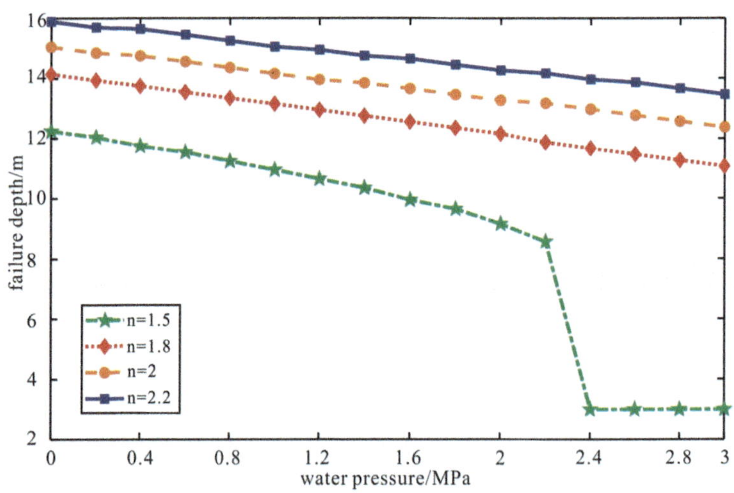

Fig. 4.8 Floor failure depth under different conditions

conveniently by using the derived formulas. The influencing factors and variation law of the floor failure depth can then be obtained. However, the theoretical analysis involves a number of assumptions and the model is simplified. A linear elastic constitutive model is assumed for the floor rock masses, the water pressure is considered as an evenly distributed rectangular load and the seepage of confined water is ignored.

4.5.3 Numerical Simulation

1. Basic principle of RFPA^{2D}-flow

In this section, RFPA^{2D}-flow is employed to simulate the progressive failure process of the coal seam floor above confined aquifer during mining. RFPA^{2D}-flow is a rock failure analysis program predicated on stress–strain-permeability test results and seepage-stress-damage coupling [67]. It can perform preliminary analysis on the coupling mechanism of seepage and damage during the rock failure process. It has been widely applied in hydraulic fracturing [68, 69], water inrush in coal mines [3], and other fields in recent years. The following assumptions are made in the RFPA^{2D}-flow program [67]: (1) Seepage in rock material conforms to the Biot seepage theory; (2) The elastic damage theory is satisfied during the loading and unloading processes; (3) The maximum tensile strength criterion and the Mohr–Coulomb criterion are the criteria of element damage threshold; (4) The relationship between stress and permeability coefficient can be described by a negative exponential equation and the permeability coefficient increases remarkably after material failure; (5) The elastic modulus and strength of elements are assigned according to a certain probabilistic distribution, such as the Weibull distribution.

The stress-seepage coupling equation adopted in RFPA^{2D}-flow is:

$$k(\sigma, p) = k_0 e^{-\beta(\sigma_{ii}/3-\alpha p)} \tag{4.8}$$

where k_0 is the initial value of rock mass permeability, β is the coupling coefficient determined by experiments, σ_{ii} is the sum of the three principal stresses, α is the pore pressure coefficient, p is the pore pressure.

The influence of stress on the seepage equation is mainly reflected in the variation of permeability. The permeability of rock mass changes with the evolution of rock damage according to the stress state of the element. When the stress in an element reaches its shear strength, the permeability after damage can be expressed as:

$$k = \xi k_0 e^{-\beta(\sigma_1-\alpha p)} \tag{4.9}$$

When the stress in an element reaches its tensile strength, the permeability after damage can be expressed as:

$$k = \xi k_0 e^{-\beta(\sigma_3-\alpha p)} \tag{4.10}$$

where ξ is the permeability jump coefficient, which is taken as 100 when elements are detached and 0.01 when elements are in contact. It can also be determined by experiments.

2. Model establishment

Figure 4.9 shows a two-dimensional numerical model for a typical working face in the coal mine. The model is 200 m long and 100 m high. It is divided into 400 * 200 = 80,000 elements. One element represents a small square with a side length of 0.5 in actual size. On the premise of satisfying the computation accuracy, the adjacent rock strata with similar physical and mechanical properties are combined. The model consists of 8 rock strata, including coarse sandstone, fine sandstone, coarse sandstone, coal seam, fine sandstone, quartz sandstone, aluminous mudstone, and Ordovician limestone. The coarse sandstone is the immediate roof and the fine sandstone is the main roof. The physical and mechanical properties of the rock strata are listed in Table 4.2. The boundary conditions are: the horizontal displacement is set to be 0 at the two side boundaries, the bottom boundary is fixed at two direction and the upper boundary is free. An evenly distributed vertical load (6.6 MPa) is applied on the upper boundary to simulate the gravitational force from the overlying rock strata. The model is simplified to a plane strain problem. In order to eliminate the boundary effect, excavation is carried out continuously and in steps toward the left starting at 80 m away from the right boundary. The coal seam is 4 m thick. Full-seam mining is adopted. The first step is to calculate the state of stress under gravitational force. From the second step, excavation is carried out in 4 steps, with each step being 10 m. 150 m water head is applied on the Ordovician limestone stratum to simulate the confined aquifer with a water pressure of 1.5 MPa.

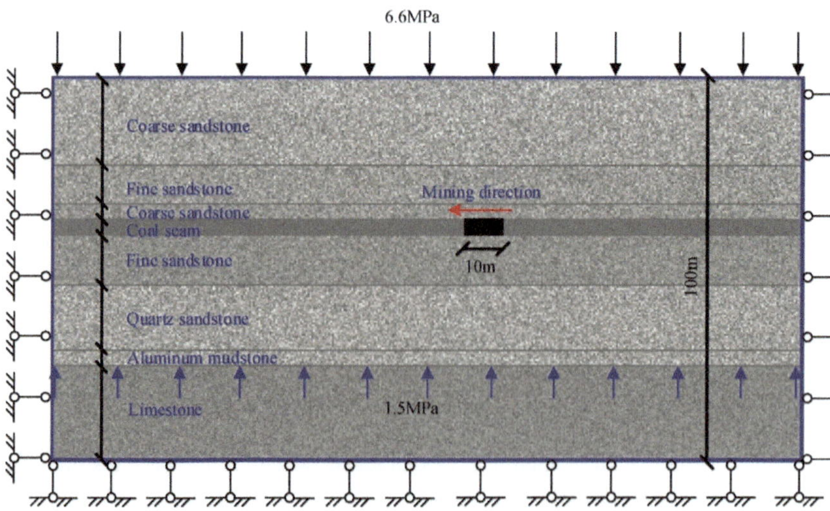

Fig. 4.9 Numerical model for analysis of floor failure

3. Analysis of numerical simulation results

In the modeling using RFPA^{2D}-flow, the element reaching the ultimate strength is weakened. The element strength and elastic modulus approach zero. Such elements are displayed in black in the shear stress plot to reflect rock mass failure. The shear stress and damage of elements (Fig. 4.10) near the coal seam during mining are analyzed. With the advance of the working face (advancing distance 0–30 m), obvious stress concentration occurs in the four corners and the short sides of the goaf. Only a small amount of shallow fractures are observed in the floor. When the advancing distance reaches 40 m, the main roof and the immediate roof are detached and the ultimate span is reached. Fracturing occurs and the initial weighting is formed. The fractures in the lower floor are coalescent and extend downward. The effect of advanced abutment appears.

A large number of experiments show that strain energy is accumulated in rock-like materials under loading. When the strain energy reaches a certain threshold, microfractures initiate and propagate in the material and the strain energy is released in the form of the stress wave. This process is called acoustic emission. It is assumed in RFPA^{2D}-flow that element damage is accompanied by acoustic emission [70] and different colors represent different types of failure. Red dots indicate that the stress state of the element reached the maximum tensile strength criterion and failed, white dots indicate that the stress state of the element reached the Mohr–Coulomb criterion and failed, and black dots mean that the failed elements occurred in the prior calculation steps. As shown in Fig. 4.11, The rock mass obliquely below the working face and open-off cut are mainly subjected to shear failure and the shallow rock mass below the floor is mainly subjected to tensile failure. This is consistent with the floor failure characteristics obtained by theoretical analysis in the previous section. The acoustic emission signals can be detected at a maximum depth of 10.5 m below the floor, indicating that the floor failure depth calculated by RFPA^{2D}-flow is 10.5 m.

The elements satisfy the displacement compatibility equation in RFPA^{2D}-flow simulation, i.e., the displacement is continuous between elements. The stress of the model can be obtained by multiplying the stiffness matrix and the displacement vector. As the elastic modulus conforms to the Weibull distribution, the stress distribution is not continuous or smooth. As shown in Fig. 4.12, taking the elements at 2 m below the floor for analysis, the horizontal stress below the floor in the goaf follows an approximately V-shaped distribution. The horizontal stress below the middle point of the goaf is the lowest, i.e., − 3 MPa. Together with the acoustic emission distribution in Fig. 4.11, tensile failure occurs in the floor under tensile stresses. The horizontal stress increases gradually with the distance away from the middle of the goaf and has a dramatically rise near the corners of the goaf. Finally, the horizontal stress changes from tensile stress to compressive stress. This is mainly due to the effect of advanced abutment pressure.

In this section, the floor failure depth and failure law are discussed based on acoustic emissions obtained by RFPA^{2D}-flow simulation. Numerical modeling is an important supplement to analytic analysis. It can provide numerical solutions for

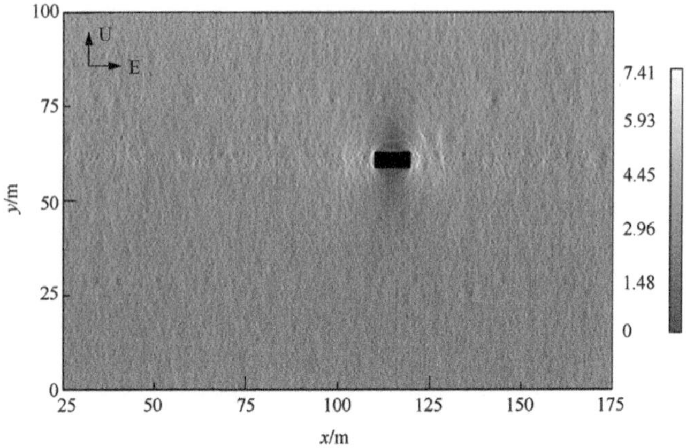

(a) Working face is 10 m away from open-off cut

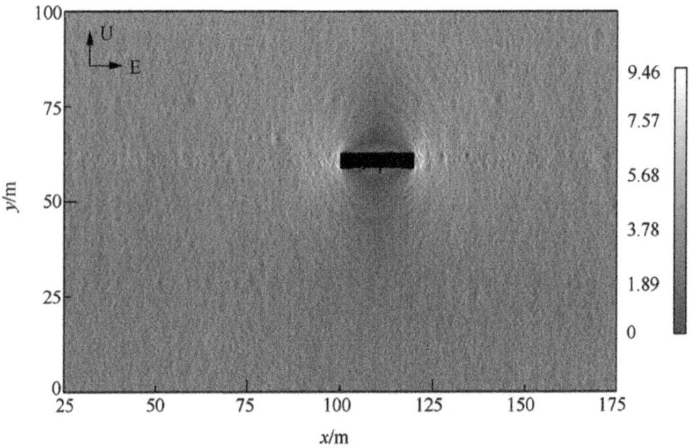

(b) Working face is 20 m away from open-off cut

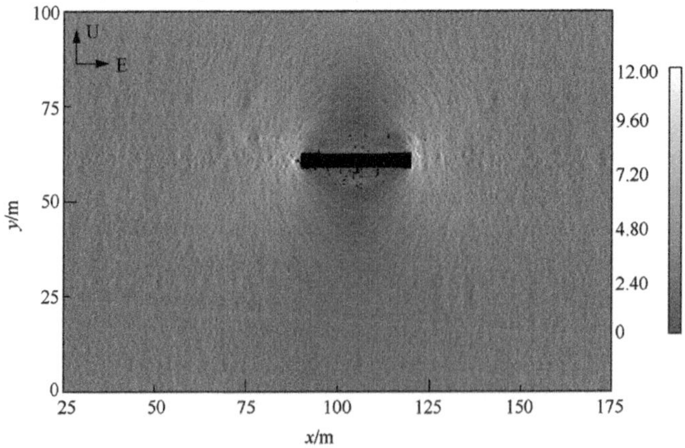

(c) Working face is 30 m away from open-off cut

Fig. 4.10 Evolution of shear stress in the coal seam floor (unit: MPa)

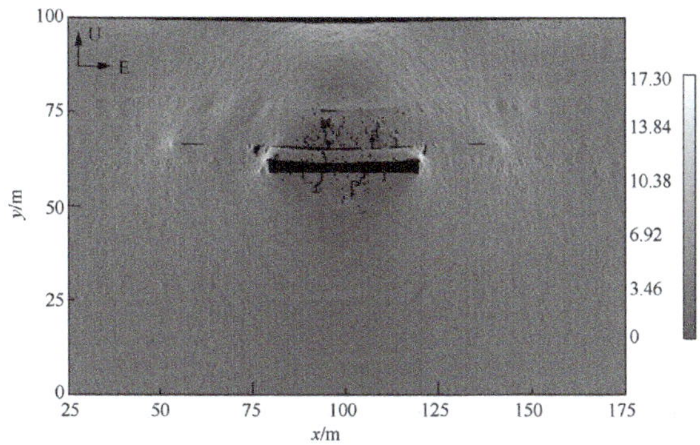

(d) Working face is 40 m away from open-off cut

Fig. 4.10 (continued)

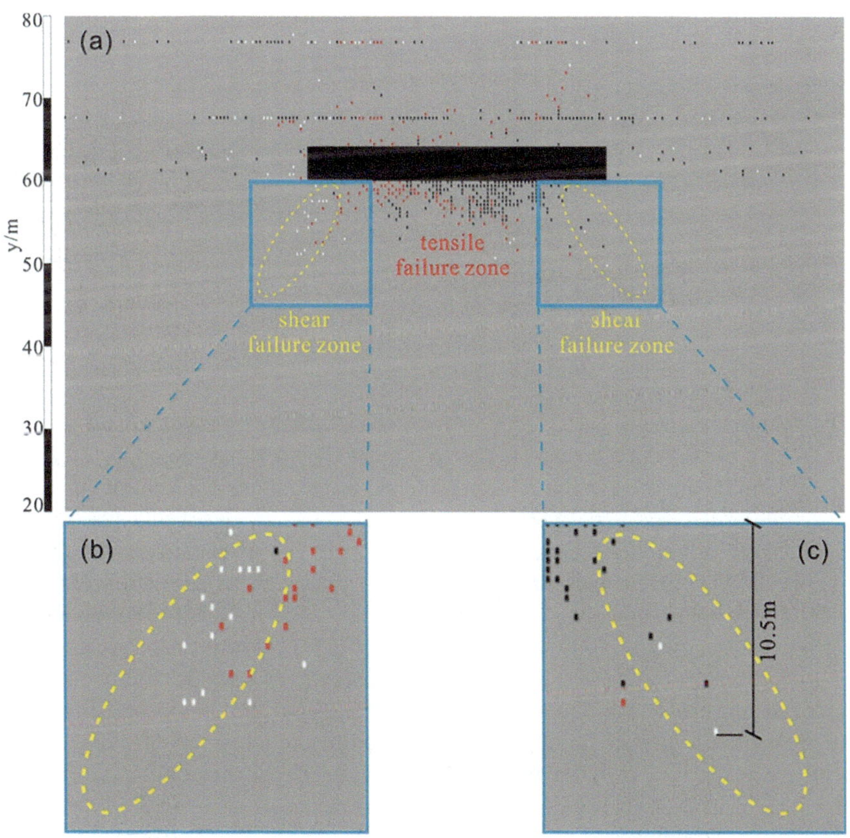

Fig. 4.11 Acoustic emissions in the floor (step 5)

Fig. 4.12 Variation of horizontal stress with distance from the open-off cut

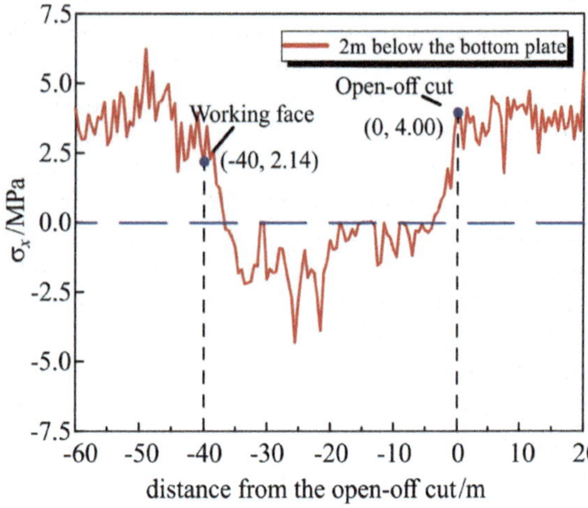

some partial differential equations which are hard to solve analytically and consider the influencing factors more comprehensively. However, it still has some deficiencies in the treatment of fractures and geological structures in the model and has difficulties in determining some parameters.

4.5.4 Microseismic Monitoring

1. Principle of microseismic monitoring

The essence of microseism in rock or rock mass, natural earthquake, and acoustic emission is the same, i.e., strain energy accumulated in rock-like materials is released in the form of stress wave during initiation and propagation of microfractures [71]. The scale of the geological body and the signal frequency received by sensors are significantly different for earthquakes, microseism, and acoustic emission. In the mining field, microseismic events are generally triggered by excavation activities, with energy approximately in the range of 0–10^{10} J, corresponding to the Richter earthquake magnitude of 0–4.5. Relevant researches [72] indicate that a great number of microfractures (microseism) are formed before macroscopic instability occurs. If microseismic events continue to increase and accumulate, failure occurs. Therefore, by analyzing the distribution characteristics and evolution trend of microseismic events received by sensors, the mechanical behaviors of rock material can be inferred, and the scale and properties of rock mass fracture can be determined. Early warning, prevention, and control can be carried out correspondingly. Microseismic monitoring has been widely applied in large-scale engineering projects such as mines [58], rock slopes [73], underground storage caverns [74], and underground powerhouses [75] in recent years.

2. Establishment of microseismic monitoring system

The high-precision microseismic system manufactured by ESG Canada was employed for the working face. The system consists of microseism sensors, the Paladin digital signal acquisition system, and the Hyperion data processing system. The sensors installed in the roadway floor received microseismic signals released in the form of elastic waves. The signals were transmitted to the Paladin signal acquisition system through cables and stored in the Hyperion data processing system for further analysis. Data transmission and interaction were implemented by the Internet. The microseismic sensors were seismometers with response frequency in the range of 15–1000 Hz and sensitivity of 43.3 V/(m s). The topographic structure of the monitoring system is shown in Fig. 4.13.

3. Analysis of microseismic monitoring results

The typical working face was put into production on October 5, 2014. In order to investigate the distribution of microseismic events and damage in the floor, the microseismic monitoring system started to operate on November 25, 2014 so as to monitor and analyze the rock masses in the coal seam floor in real-time. According to the excavation progress of the typical working face at Dongjiahe Coal Mine, the microseismic events detected at 600–1000 m from December 23, 2014 to May 19, 2015 were selected for analysis. This is because these microseismic events were largely caused by periodic weighting and the period assured that the working face was away from the tectonic structures, such as the fold and hidden fault. Hence, the

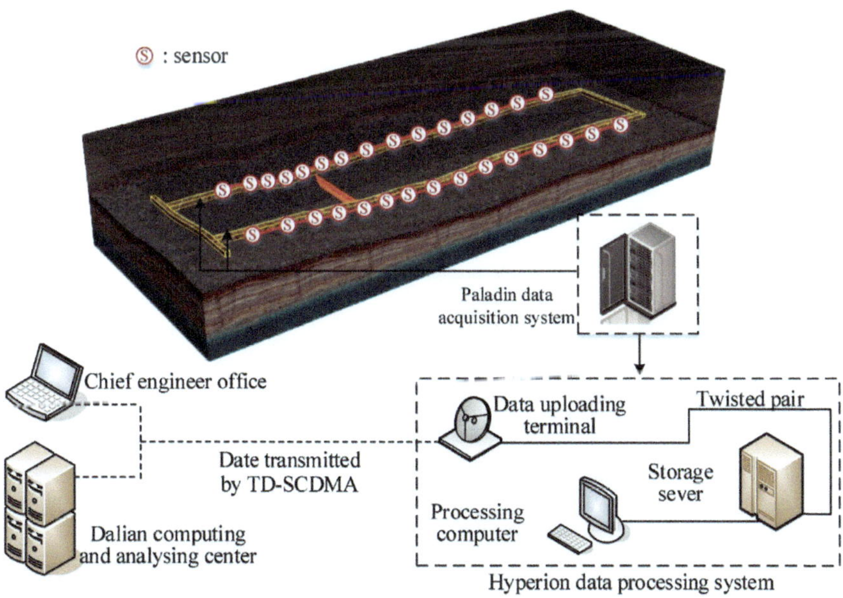

Fig. 4.13 Topographic map of microseismic monitoring system

effects of tectonic structures can be ignored. The microseismic events in the whole model and the magnified map for coal seam at 600–1000 m along the mining direction are shown in Fig. 4.14. This section mainly explores the location and distribution of microseismic events, the strata are simplified in the model, namely, coal seam roof, coal seam, coal seam floor, and Ordovician limestone from top to bottom. There are two roadways marked in red in Fig. 4.14. The north roadway is the transportation roadway, and the south roadway is the track roadway. The model is 1400 m long, 250 m high and the corresponding elevation is in the range of + 132 m to + 382 m. The small spheres in the figure represent microseismic events. The color indicates the magnitude of microseismic events. In this period, a total number of 29 microseismic events in the floor were recorded, among them, 12 were located within 600–1000 m and all these events occurred above the confined aquifer. According to the preliminary judgment, there was no water inrush event on the floor as of May 19, 2015.

Combined with the on-site working face water inflow data (Fig. 4.15), the water inflow of the working face during this period was relatively stable, with an average value of about 60 m^3/h. Moreover, the engineering report showed that the water inflow of the working face increased slightly from March 2015 to May 2015, because the working face passed through the thin coal belt, and the mining method changed from comprehensive mechanized mining to explosive mining. Moreover, according to the data offered by the field, there was not water inrush accident occurring as of May 24, 2015.

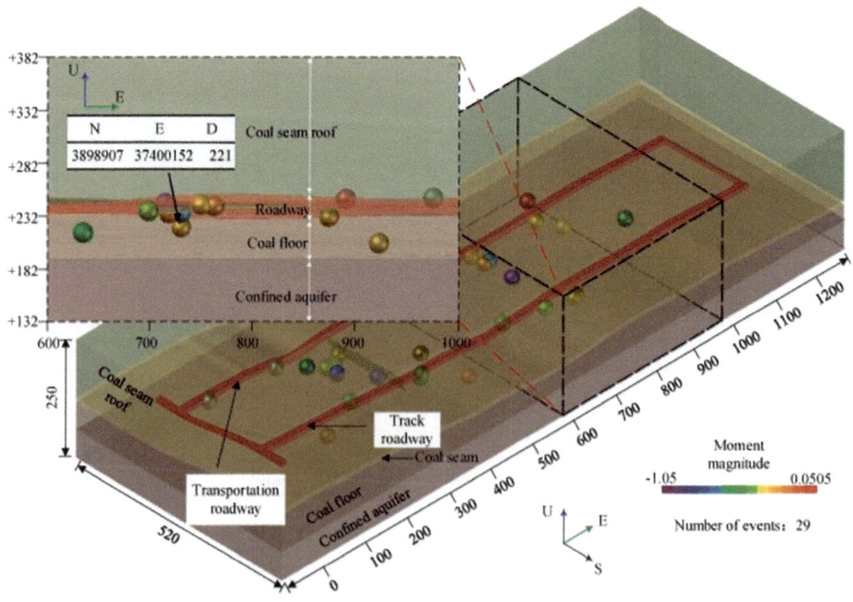

Fig. 4.14 Distribution of microseismic events (unit: m)

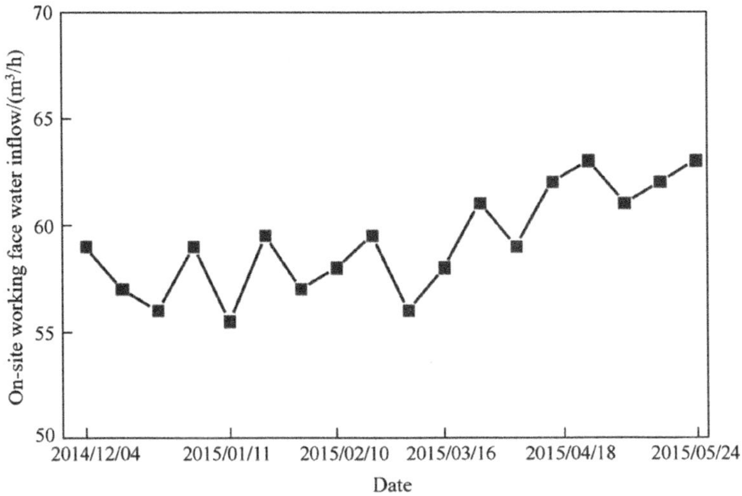

Fig. 4.15 Variation of water inflow in working face

The density of microseismic events in the floor at 600–1000 m is analyzed, as shown in Fig. 4.1, where the two white dashed lines stand for the boundaries between the coal seam and the upper and lower rock strata, the white solid line indicates the boundary between the confined aquifer and the coal seam floor. The areas with higher microseismic event density are marked in this figure. The microseismic events are relatively concentrated at 726 m, showing that microfracture of rock masses in this area is clustered and tends to form coalescent fractures. Together with the moment magnitude of microseismic events shown in Fig. 4.14, the magnitude magnitude is an absolute mechanical scale that can be measured for large-scale earthquakes, small-scale earthquakes, microseisms, even ultra-microseisms, and deep earthquakes. It can reflect the magnitude of the deformation scale [76] and can be expressed as:

$$M_w = \frac{2}{3} \lg M_0 - 6.033 \tag{4.11}$$

where M_w is the moment magnitude and M_0 is the earthquake moment. As the moment magnitude of event of the lower elevation is relatively high, it can be considered that through cracks have formed in this area. It is known from the positioning information of microseismic events provided by the ESG software, such as the microseismic event occurred at EL + 221 m. The elevation of the corresponding floor is + 234 m. Hence, it can be concluded that the depth of the floor disturbance caused by the excavation is 13 m utilizing MS monitoring (Fig. 4.16).

4. Failure mode analysis

In seismology, the propagation of seismic waves inside the earth can be divided into two basic types: P wave and S wave, and the ratio of energy released by P wave and S

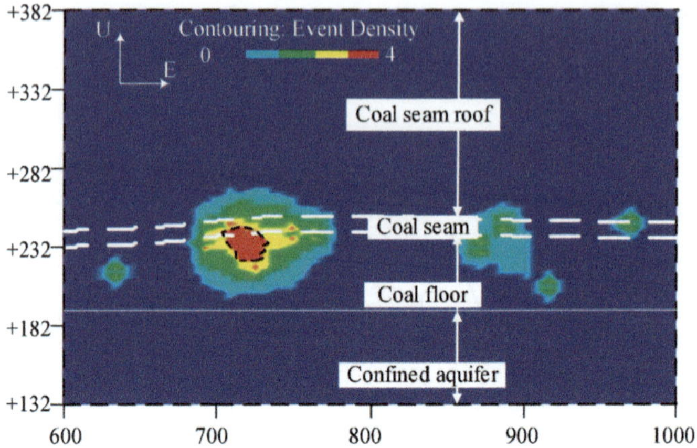

Fig. 4.16 Density map of microseismic events (unit: m)

wave is a powerful index for the analysis of earthquake focal mechanism. Generally, the P-wave velocity is larger than the S-wave velocity, so the P-wave and S-wave can be detected by this characteristic in the microseismic system. Gibowicz et al. [77] draws the following conclusions through the research and statistics of a large number of microseismic information: when the ratio $E_S/E_P \geq 10$, the S-wave energy is much larger than the P-wave energy, and microseismic events are induced by fault or shear failure;when the ratio $E_S/E_P \leq 3$, the microseismic event is induced by non-shear type failure, such as strain rockburst, tension failure, etc.;when $3 < E_S/E_P < 10$, microseismic events are induced by both shear and tension failure. Hudyma et al. [78] made statistics on E_S/E_P of microseismic events in hard rock mines and concluded that the failure modes in mines were consistent with the monitoring situation. Xu et al. [79] conducted microseismic monitoring on the left bank slope of the Jinping I hydropower station in China, statistically analyzed the distribution of E_S/E_P and found that the rock failure mode obtained from on-site monitoring is consistent with the numerical simulation results.

The E_S/E_P of microseismic events in the floor at 600–1000 m is analyzed. Figure 4.17 shows the elevation of microseismic events and distribution characteristics of E_S/E_P. It can be seen from the figure that most E_S/E_P of microseismic events are ≤ 3 when the elevation ranges from + 230 to + 250 m. The elevation of the coal seam floor in this area is about + 235 m, indicating that most of the microseismic events within 5 m below the coal seam are induced by tension failure. When the elevation ranges from + 205 to + 220 m, the E_S/E_P of microseismic events are ≥ 10, indicating that the microseismic events within 15 m below the coal seam are induced by shear failure. This failure law of the coal seam floor is similar to that in theoretical analysis and numerical simulation.

In this section, the maximum failure depth and failure mode of the floor are studied by means of microseismic monitoring and quantitative seismology. As a three-dimensional spatial monitoring method, microseismic monitoring overcomes

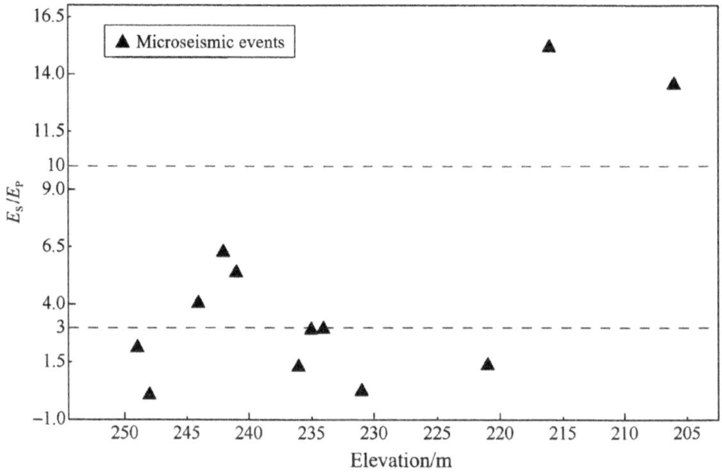

Fig. 4.17 The elevation of microseismic events and distribution characteristics of E_S/E_P

the limitations of traditional monitoring methods such as local monitoring at a "point" or on a "line" and the difficulty of capturing micro-fractures in rock mass. It can reflect the real state of the measured are. However, its disadvantages are also obvious: firstly, the rock mass is regarded as uniform and isotropic medium in determining P-wave velocity; Secondly, the measured spatio-temporal evolution of microseismic events cannot fully characterize the entire process of damage [80]. Finally, high precision positioning technology and data processing speed need to be improved.

4.5.5 Comprehensive Method

The comparison of the three methods in the above sections shows that, for analysis of the floor failure depth and failure characteristics, any single method has its limitations. Therefore, in this section, a comprehensive method integrating the features and utilizing the advantages of theoretical analysis, numerical simulation, and field monitoring is put forward.

This method is mainly applied for studies on the floor failure characteristics and failure depth. The preparatory work includes: to collect information on typical features in the mining area, such as engineering geological structures (faults, joints, lithology), physical and mechanical parameters of rock strata, hydrogeological condition, the layout of microseismic sensors. Secondly, combined with the advantages of different methods, the failure depth and failure characteristics of the floor are analyzed. The floor stress during initial weighting is solved with a simplified model and the sensitivity of loading is analyzed. The floor failure depth and failure characteristics can be obtained by applying the modified Mohr–Coulomb criterion with tension cut-off. The results can serve as theoretical guidance and powerful supplement to numerical simulation and microseismic monitoring. Numerical modeling is an important tool to remedy the shortcomings of the analytic method. By simulating the continuous excavation process, the stress contour can be used to study the stress variation trend and verify the theoretical results in combination with the acoustic emission plot. Meanwhile, it can be a tool for approximate examination of the positioning accuracy for microseismic events. By processing the microseismic waveforms received by sensors, the temporal and spatial distribution of microseismic events can be investigated. The distribution law of microseismic events can also help to identify the geological structures unrevealed in the earlier stage, which can be a supplement to theoretical analysis and numerical simulation. The maximum floor failure depth can be obtained according to the location of microseismic events. The floor failure depth and failure characteristics obtained by the three methods can be mutually verified, and the reliability of the calculated results can be verified by combining with the on-site water inflow data. In addition, for other important excavation points, this method can also be used to analyze the failure depth and failure characteristics of the floor, determine whether water inrush occurs in combination with specific working conditions, and provide reinforcement schemes according to the failure characteristics. The flow chart is shown in Fig. 4.18.

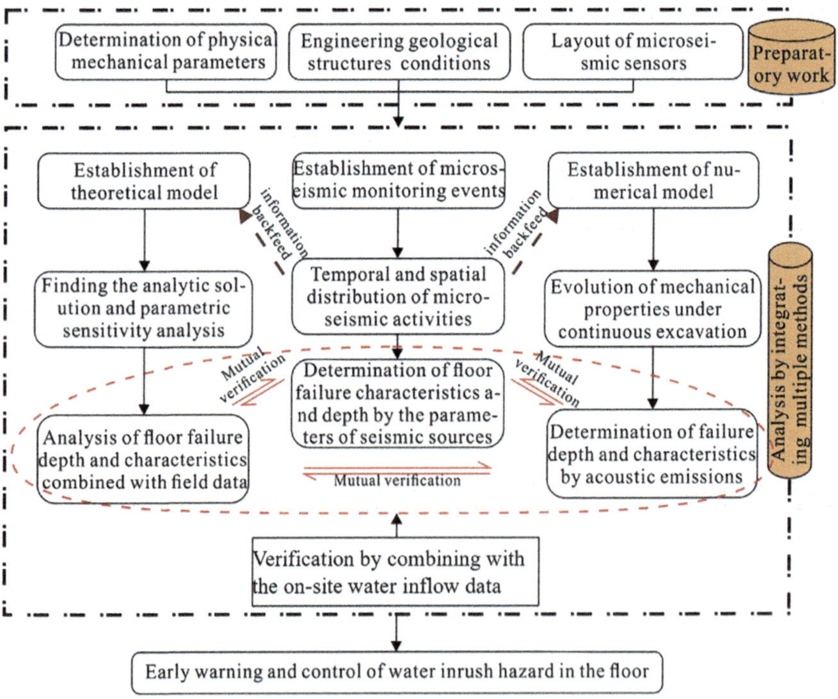

Fig. 4.18 Flow chart of analysis by integrating multiple methods

4.5.6 Summary

In this section, by establishing a mechanical model for the floor of a typical working face at Dongjiahe Coal Mine, the analytic solutions of floor stress components are derived based on the semi-infinite solid in elastic theory, and the state of stress of the coal seam floor undermining is analyzed. In combination with $RFPA^{2D}$-flow modeling and microseismic monitoring, the failure characteristics and failure depth of the coal seam floor above a confined aquifer is investigated. The following conclusions can be drawn:

(1) The failure characteristics of the coal seam floor at Dongjiahe Coal Mine are analyzed by using the elastic theory and the modified Mohr–Coulomb with tension cut-off. The rock masses within 1.7 m below the goaf are subjected to tensile failure during initial weighting. Meanwhile, shear failure zones are formed at the two sides within 5–11 m below the goaf. Due to the existence of a complete rock strata protection zone, the floor failure depth is positively proportional to the advanced abutment pressure and negatively proportional to the water pressure.

(2) The progressive failure process of coal during initial weighting on the typical working face at Dongjiahe Coal Mine is reproduced by $RFPA^{2D}$-flow. With the advancement of the working face, the shallow rock masses in the floor are mainly subjected to tensile failure, and the rock masses obliquely under the working face and the open-off cut are mainly subjected to shear failure. The failure zone is in the shape of an inverted horse saddle, with a failure depth of 10.5 m.

(3) By establishing a microseismic monitoring system at the typical working face, the density and moment magnitude of microseismic events in the floor are analyzed. It can be concluded that the depth of the floor disturbance caused by the excavation is 13 m. Most of the microseismic events at 5 m below the coal seam are induced by tension failure, while those below 15 m below the coal seam are induced by shear failure. The results are verified by combining with the water inflow data of the working face.

(4) The logic of theoretical analysis is clear but there are many assumptions, numerical simulation is comprehensive but cannot reproduce the actual engineering background, and microseismic monitoring can reflect the actual engineering background but has positioning errors A comprehensive method integrating the advantages of three methods is proposed to overcome the limitations of a single method. Mutual verification can be achieved through different methods to evaluate the floor failure characteristics and failure depth more comprehensively. This method also can provide a reference for tunnel water inrush problem and selection of proper reinforcement measures.

4.6 Research on Mechanism of Water Inrush Channel Development in Dongjiahe Coal Mine Floor Based on Microseismic Monitoring

As coal mines in North and East China shift towards deep mining, the problem of water inrush from the mine floor is becoming increasingly prominent. The process of water inrush from the coal mine floor is a damage and fracture process caused by the coupling of disturbance of the floor rock mass and seepage effect, its mechanism of action is complex, seriously affecting the safe mining of coal mines, attracting widespread attention from many scholars. Song et al. [31] theoretically analyzed the impact of mine pressure on floor water inrush, established a mechanical model for predicting and controlling water inrush in cross-fault mining, and formulated related control criteria. Wu et al. [81, 82] introduced a variable weight model to reflect the changes in the index values of the main controlling factors of water inrush under abrupt conditions and the combination relationship between the main controlling factors, proposed a method for determining the variable weight interval threshold and adjustment parameter of the main controlling factors, and constructed a variable weight model for predicting and evaluating the problem of floor water inrush. Shi et al. [83] proposed a model forecasting the water inrush risk based on principal component analysis, fuzzy, particle swarm optimization algorithm and support vector machine, which obtained accurate prediction results in practical application. Hu et al. [84] used the ratio of the actual strength of the floor rock mass and the strength when the rock mass was destroyed after reduction to evaluate the risk of water inrush and analyzed its influencing factors. He et al. [85] comprehensively considered factors such as the water pressure of the aquifer, the thickness of the aquifer, and the yield ratio of the aquifer from a mechanical perspective, established a function related to time to evaluate the floor water inrush, and expounded the water inrush mechanism of underground mining. Zhang et al. [86] used similar material experiments and numerical experiments to study the damage activation law of the floor fault under the influence of mining, revealing the formation process of the floor water inrush channel under the action of the mining stress field and seepage field. Gao et al. [87] established a mechanics model, analyzed the propagation of micro-fracture from a microscopic perspective, studied the effects of water pressure on the effective water-resisting layer from a macroscopic point of view through the thin plate theory, and revealed the influence mechanism of abutment pressure and floor water pressure on the expansion of visible fractures and propagation of micro-cracks and fracture damage of the effective water-resisting layer. Currently, most of the research on floor fault water inrush in coal mines uses theoretical analysis, numerical simulation, and similar material experiments, and some practical criteria for evaluating the risk of floor water inrush have been proposed. However, the diversity of the mechanical properties of the floor rock mass causes certain difficulties in the selection of its parameters, often unable to accurately obtain the relevant parameters of the rock mass, and it is difficult to effectively describe the evolution process of the coal rock mass water-conducting fracture zone under the influence of mining, causing the failure of the water inrush criteria. In recent years, microseismic monitoring technology has been widely used in the analysis of large-scale rock mass engineering such as coal mines [88–91], metal mines [92, 93], hydropower station diversion tunnels [94, 95], underground plants [96, 97], hydropower slopes [98], and petroleum caverns [99], and has achieved good results. A series of research results have been achieved in the aspects of coal mine microseismic waveform characteristics, source location, mine seismic activity law, ground pressure monitoring and early warning [100, 101], but the application of microseismic monitoring technology to the research of the evolution of floor water inrush channels is rarely reported.

This section constructs a microseismic monitoring system for the typical 22517-working face of Dongjiahe Coal Mine, conducts real-time monitoring of rock mass damage and fracture in the floor fault zone, and studies the law of micro-fracture generation and evolution inside the floor rock mass before and after passing through the fault. The on-site microseismic monitoring results are combined with the rock fracture process analysis system RFPA^{2D} to intuitively reflect the damage evolution process of the coal rock mass under mining disturbance through micro-fracture information, reproducing the whole process of the generation, expansion and development of the water-conducting fracture zone in the coal mine floor, and analyzing variation of the stress field during the breeding process of the water inrush channel.

4.6.1 General Situation of Typical Working Face in Dongjiahe Coal Mine

The working face #22517 of Dongjiahe Coal Mine is located in the east of the haulage-dip of galley level 2, mining area #2. Its north and south are both unmined coal seams. Coal seam #5 with Shanxi formation is the primary mining area due to stable coal deposit there. The thickness of this strata is 2.5–4.1 m with an average of 3.3 m. The seam is generally inclined 3 degrees to the northeast. The maximum differences in altitude are 50 m along the east–west direction and 20 m along the south-north direction. The gradient of the whole working face is about 3 degrees, which is an approximately flat seam. The working face is 1217 m long along the strike and 185 m wide along the dip. The excavation was implemented from east to west, corresponding to the margins from 1217 to 0 m. The open-off cut is located at the point distance of 1217 m, while the stopping line is at the point distance of 100 m. The false roof for mining this coal seam is made of mudstone with 0.6 m thickness. The direct roof is 0.65–1.66 m thick coarse siltstone, with thin strata or strips of fine sandstone inclusions on its top. The basic roof is 5.7–11.77 m thick medium grain and fine grain sandstones. The direct bottom which is dense and hard with well-developed fissures is 0.8–1.72 m thick quartz sandstone. The basic bottom is 0.89–3.23 m thick fine siltstone.

According to the results of microseismic monitoring and three-dimensional seismic exploration, there is a small reverse fault located 200–350 m in front of

Table 4.3 Some part of advancing progress of working face #22517

Date	Distance from the entrance of the track roadway/m	Date	Distance from the entrance of the track roadway/m
2015/10/19	553	2016/01/08	410
2015/10/30	530	2016/01/18	397
2015/11/09	514	2016/02/05	369
2015/11/18	499	2016/02/18	357
2015/11/29	476	2016/02/28	330
2015/12/09	457	2016/03/09	310
2015/12/18	443	2016/03/18	300
2015/12/30	426	2016/03/29	281.2

the track heading, with a dip angle of about 73° and a displacement of 3 m. There is a syncline located at a distance of 480–720 m in front of the working face, with a maximum height difference of 4 m between the core and the wings. The progress of the working face advancement is shown in Table 4.3. The water-bearing layers in the floor of the No. 5 coal seam mainly include the Taiyuan Formation limestone (sandstone) aquifer and the Ordovician limestone confined aquifer. The Taiyuan Formation limestone (sandstone) aquifer undergoes a phase transition to quartz sandstone in the floor of the working face. The second section of the Fengfeng Formation in the Middle Ordovician is the most important source of water inrush in the floor mining of the No. 5 coal seam. It is mainly composed of calcite, with dissolved fissures, developed karst, and a fissure ratio of 4%. The drilling unit inflow rate in the second section of the Fengfeng Formation is 0.2–1.51 L/(s m), with a salinity of 0.000563 g/L and a water temperature of 25.5 °C. The thickness of this water-bearing layer is generally 150 m, with a maximum of 180 m. The water level in the Ordovician limestone aquifer is stable at around + 375 m, with a maximum water pressure of about 1.5 MPa. It is a highly water-rich karst fissure confined aquifer and poses a risk of flooding the working face.

4.6.2 Microseismic Monitoring Analysis of Water Inrush Channel in Working Face

1. Construction of microseismic monitoring system

The advanced high-precision microseismic monitoring system produced from ESG Corporation, Canada, had been installed for working face used for the real-time monitoring, location, and analysis of microfracture of coal and rock mass. The system is mainly composed of microseismic sensors, Paladin mining data loggers, Hyperion ground data loggers, and 3D visualization software based on the remote network transmission developed by Mechsoft Co. Ltd., Dalian, China. Microseismic sensors

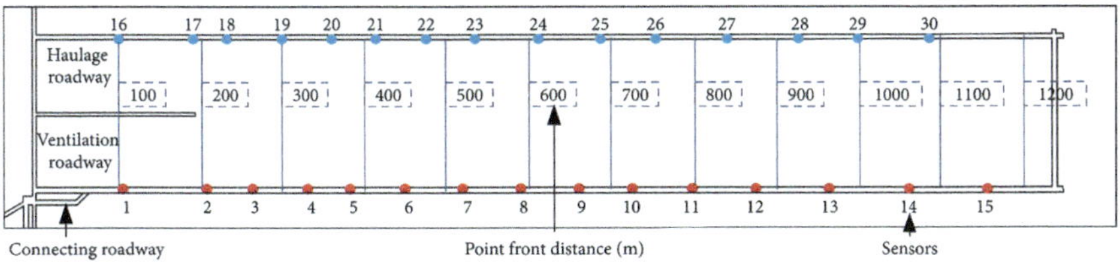

Fig. 4.19 The flat profile projection of the working face sensor layout

are geophones with a response frequency range of 15–1000 Hz and sensitivity of 43.3 V/(m s). Combined with the mining and geological conditions of the working face, the geophones are arranged under the detection area track lane and the coal seam floor of the transport roadway. Along the strike direction, sensors are installed at intervals of 100 m in non-mining slope areas. These sensors are staggered along the strike direction and vertical direction of the working face in the track lane and transport lane, as shown in Fig. 4.19.

To protect the goaf cable from casing and landfills and to ensure the signal transmission after the collapse of the mined out area, the sensors in the track lane and the transportation lane will transmit the signal through the cable to the workstations under the shaft, and the data collected by each sensor will be uploaded to the data processing server, the data storage, and the transmission server at the ground grouting station through the optical cable (via connection roadways No. 5 and No. 7). After being processed by field workers, they are sent to scientific research units and the party units. The topological structure of the microseismic monitoring system is shown in Fig. 4.20.

2. Distribution law of water inrush channel

The working face #22517 was put into production on October 5, 2014. The microseismic system was installed on November 25, 2014 and began monitoring. The monitoring was stopped on October 10, 2016. To study the distribution pattern of microseismic events in the floor and the damage and failure of the surrounding rock with fault when the working face approached the fault, the microseismic events in the floor from December 1, 2015 to March 31, 2016 were selected for analysis. During this period, the development of water inrush channels in the floor was mainly affected by the fault. The activity pattern of microseismic events before and after crossing the fault was analyzed from the strike direction of the working face, and the generation process of the macro water-conducting fracture zone in the floor was explored. n the microseismic event distribution diagram, a circular ball represents the microseismic event caused by the damage of rock mass, and the sphere colour represents the seismic event moment magnitude.

As shown in Fig. 4.21a, on December 24, 2015, the working face was about 85 m away from the fault, and microseismic events began to occur in the surrounding rock of the floor fault due to mining disturbances. On December 31, 2015, when the working face was about 76 m away from the fault, 8 microseismic events were

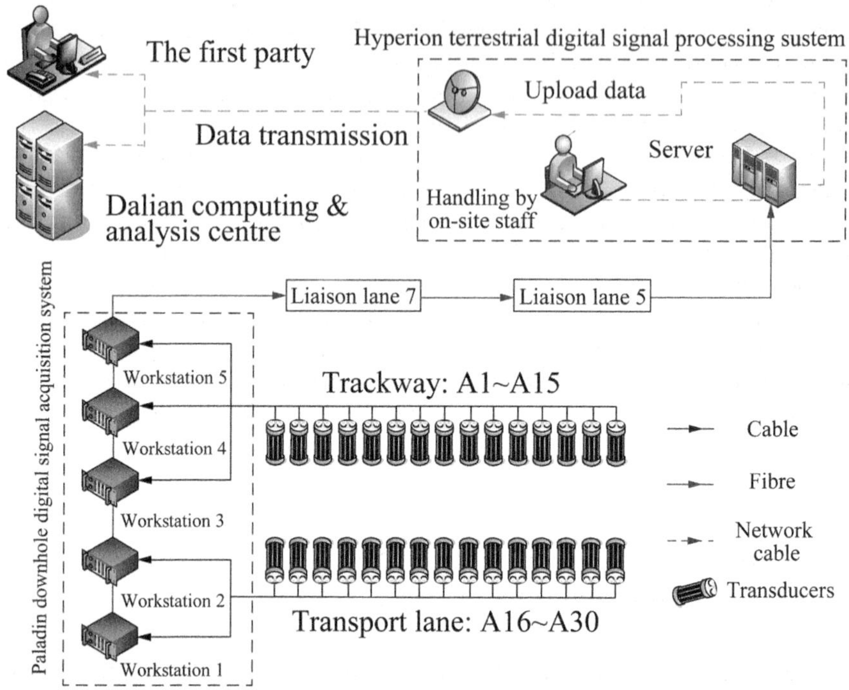

Fig. 4.20 The topological structure of the microseismic monitoring system

recorded in the floor within a range of about 25 m below the floor, with moment magnitudes ranging from − 1.03 to − 0.02. he microseismic events clustered in the area labeled as I in the figure, with a length of 50 m in the strike direction and concentrated below the floor in the vicinity of 25 m, with a relatively dispersed distribution and local damage to the surrounding rock of the fault. In addition, the damage and fracture of the floor strata were concentrated in an area labeled as II, which extended about 35 m × 15 m in front of the fault at a distance of about 120 m. As shown in Fig. 4.21b, sporadic microseismic events occurred in the floor surrounding rock from January 1, 2016, to February 29, 2016. The distribution range of the microfracture concentration zone I near the fault remained basically unchanged, but the internal distribution of events became denser. In the strike direction, the microfractures gradually extended towards the goaf, and as the working face advanced towards the fault, the floor microseismic events extended to the goaf about 200 m in front of the fault. The concentrated zone II extended about 150 m in front of the fault. As shown in Fig. 4.21c, during the period from March 1, 2016, to March 31, 2016, the floor surrounding rock was in an active phase of microseismic events. The working face had passed through the fault, and 32 microseismic events were recorded in the floor, extending further into the depth of about 35 m. The moment magnitudes ranged from − 0.73 to 0.49, and the cumulative seismic moment of microfractures increased significantly. In the strike direction, the events extended into the goaf area for about

220 m, and the area labeled as I expanded towards the goaf, with the strike extending to 80 m and the depth extending to 35 m. One microseismic event was recorded at a depth of 40 m. The damage degree and range of the floor in the goaf area further intensified.

As shown in Fig. 4.22a, the microdamage energy concentration area before crossing the fault is mainly distributed within about 30 m near the fault, among which the I area is a high energy density area with a strike length of about 15 m, located 5–25 m below the coal seam, and the rock layer 5 m below the coal seam is in the sub-high energy area. The fault damage is mainly concentrated in 5–25 m below the coal seam, while the fault within 5 m adjacent to the coal seam is still in a stable state. As shown in Fig. 4.22b, after crossing the fault, the high energy zone extends towards the goaf area and deeper regions. Along the strike direction of the working face, it extends to an area of about 80 m in the goaf area and extends to a depth

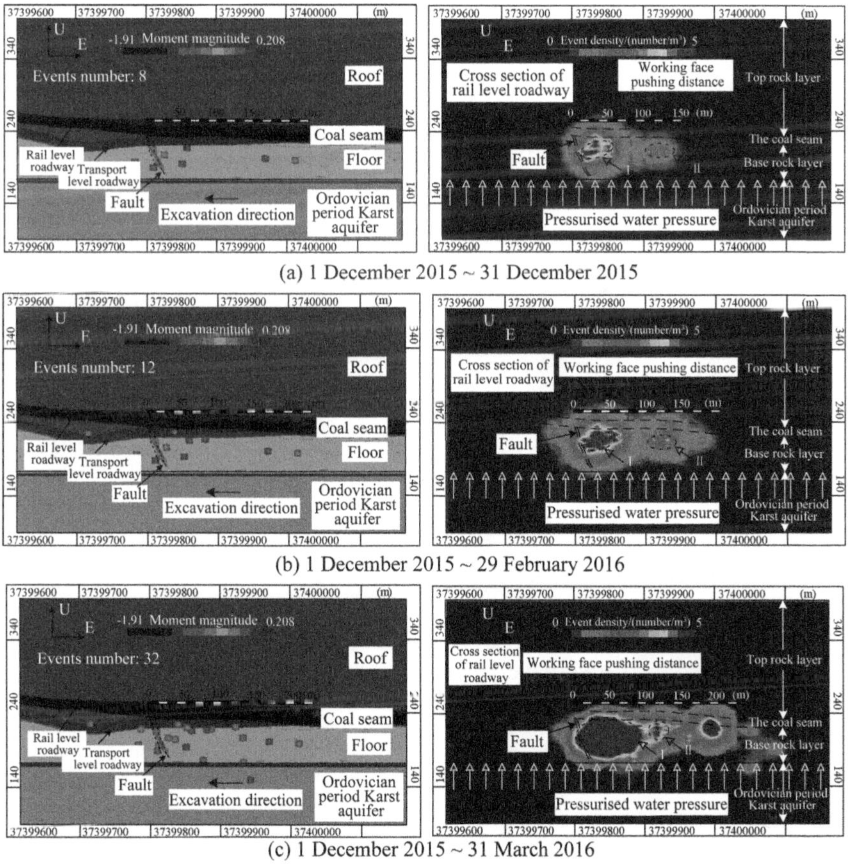

Fig. 4.21 Distribution of microseismic events in coal seam floor

of 35 m below the coal seam. The floor fractures develop and connect, potentially forming a pathway for floor water inrush.

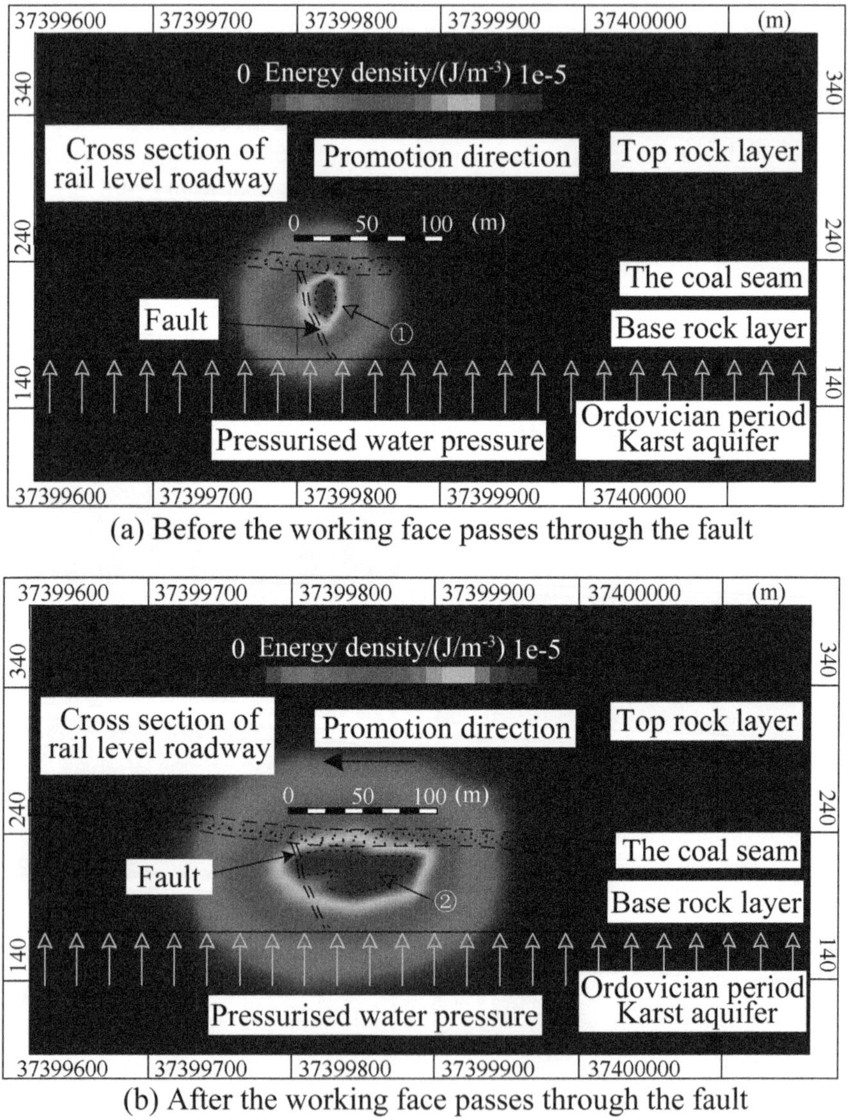

Fig. 4.22 The microdamage energy concentration area before and after crossing the fault

4.6.3 Study on Stress Variation Mechanism Near Water Inrush Channel

The progressive process of microfracture initiation, development, and formation of water inrush channels in the surrounding rock of the fault under mining disturbance was simulated using RFPA^{2D}. The model had a strike length of 500 m, a height of 200 m, and a coal seam thickness of 4 m, divided into 100,000 elements as shown in Fig. 4.23. The side boundaries of the model were subjected to horizontal displacement constraints, the bottom surface had vertical displacement constraints, and the upper boundary was subjected to a uniformly distributed load of 4 MPa to simulate the overlying rock strata self-weight, considering the weight of the model. The yield criterion was the Mohr–Coulomb criterion. The physical and mechanical parameters of each rock layer are shown in Table 4.4, where layers 1, 4, 6, 12, 14, and 16 are fine siltstone; layers 2, 7, and 17 are coarse siltstone; layers 3, 9, 18, and 22 are medium-grained sandstone; layers 5, 8, and 10 are fine sandstone; layer 11 is the coal seam; layers 13, 15, and 19 are quartz sandstone; layer 20 is aluminum shale; and layers 21 and 23 are limestone. Among them, layers 20, 21, and 23 are Ordovician limestone. In the model, weaker elements were used to simulate the fault with a dip angle of 73°, as shown by the dotted line in Fig. 4.23.

Then, a numerical model of stepwise excavation method was used to simulate the mining of the working face. The influence of boundary effects was fully considered. The excavation started 170 m away from the right boundary of the model with each step being 10 m. The top and bottom of the model were set as impermeable boundaries. The top hydraulic head is set as 0 m. 150 m water head is applied on the Ordovician limestone stratum to simulate the confined aquifer with a water pressure of 1.5 MPa. The water pressure was transmitted through the boundaries to the underlying aquifer beneath the coal seam.

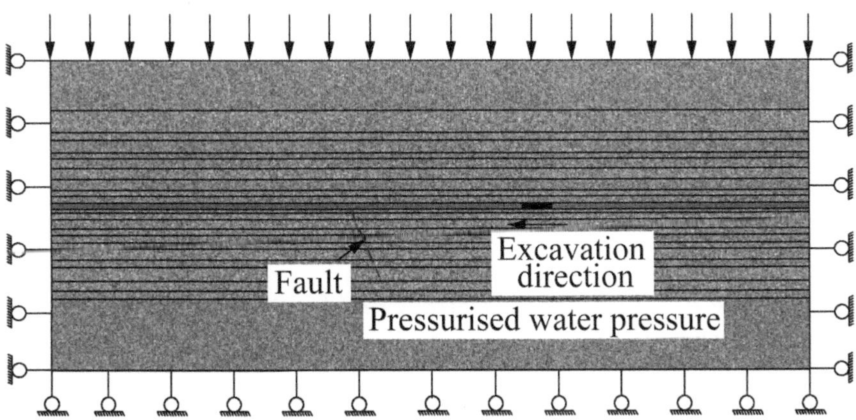

Fig. 4.23 Numerical model for analysis of floor water inrush

Table 4.4 Physical and mechanical parameters of rock strata

Rock stratum	Elastic modulus/MPa	Compressive strength/MPa	Friction angle/(°)	Poisson's ratio	Density/(kg/m^3)
Coarse siltstone	8800	45	26	0.24	2450
Fine siltstone	8200	35	28	0.26	2450
Medium-grained sandstone	8500	29	28	0.26	2450
Fine sandstone	7800	24	29	0.28	2450
Coal seam	1650	8	36	0.32	1500
Quartz sandstone	8850	48	25	0.24	2650
Ordovician limestone	5000	68	30	0.24	2770
Fault	1000	7	28	0.26	2000

In distribution of acoustic emission events from RFPA simulation, white, red, and black dots represent compression failure, tension failure, and accumulative acoustic emission failure, respectively. As shown in Fig. 4.24a, As the working face approached the fault from the hanging wall, acoustic emission events occurred first in the surrounding rock of the fault about 25 m below the floor, and gradually extended upward, producing few acoustic emission events along the strike, which are mainly compression failure events. The acoustic emission energy generated by each excavation was less than 2000 J, and the surrounding rock of the fault suffered local damage and fracture, with the characteristics of segmented initiation of microcracks. As shown in Fig. 4.24b, acoustic emission events gathered near the fault, and microcracks gradually penetrate to form macrocracks and extend to the deeper zone along the fault trend. When the working face advances 35 m past the fault, the acoustic emission events are mainly distributed within 35 m below the floor and dominated by tensile failure events, forming through cracks within 20 m. As shown in Fig. 4.25, when the working face advanced 15 m past the fault, the acoustic emission energy surged to 6305 J, and the accumulated number of acoustic emission events increased to 35. Moreover, the micro-fracture events gradually penetrated, the damage of the fault surrounding rock intensified, and a water inrush channel was likely to form, which was consistent with the microseismic monitoring results.

4.6.4 The Evolution Law of Stress Field in Surrounding Rock with Fault

As shown in Fig. 4.24, when the working face is located in the hanging wall, there are more compressive failure acoustic emission events at the fault, and there are sporadic tensile failure acoustic emission events. The surrounding rock of the fault

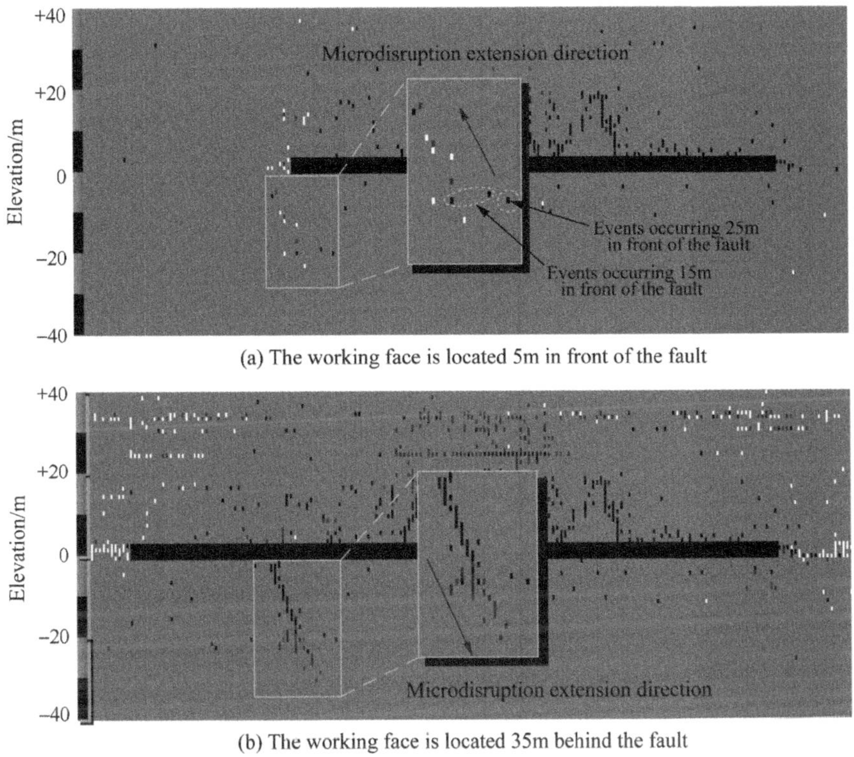

Fig. 4.24 Distribution of acoustic emission events in water Inrush channel

Fig. 4.25 Number and energy statistics of acoustic emission events in the rock of the floor fault

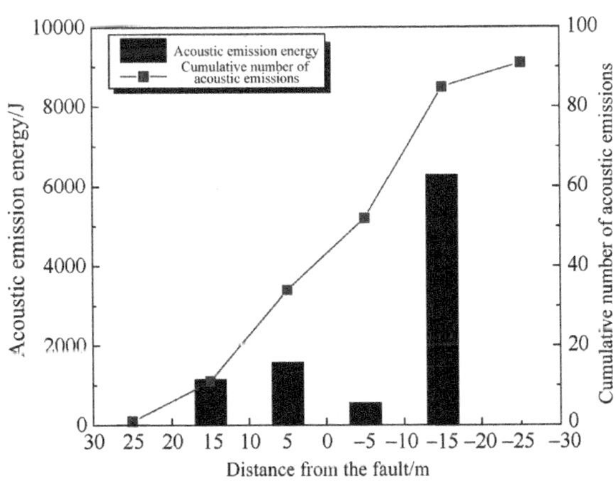

mainly undergoes compressive failure; when advancing to the foot wall, the acoustic emission events are mainly tensile failure. This indicates that when the working face is located in the hanging wall, the fault is in the compressive stress zone and mainly undergoes compressive shear failure. After crossing the fault, the stress is readjusted, and a tensile stress zone appears near the fault. The damage and failure of the surrounding rock are aggravated under the action of tensile stress.

As shown in Fig. 4.26a, the plastic zone first appeared below the coal seam at 15–35 m in the surrounding rock with fault under the mining disturbance. When the working face advanced to 25 m in front of the fault, the boundary of the minimum principal stress distribution ① locally extended to the fault at 15 m below the coal seam. As shown in Fig. 4.26b, when the working face advanced to 15 m in front of the fault, the deformation of the surrounding rock further developed, but the range of plastic zone distribution basically remained unchanged; the minimum principal stress concentrated in the local area of the fault at 15–30 m below the coal seam, and the minimum principal stress gradually extended to the fault within 15 m below the coal seam. As shown in Fig. 4.26c, when the working face advanced to 5 m in front of the fault, the deformation of the surrounding rock with fault further developed, and the plastic zone extended upward, with a tendency to communicate with the floor of the working face; the boundary of the minimum principal stress distribution ① extended to the foot wall with the working face advancement, and the stress concentration zone scattered along the fault appeared within 15 m below the coal seam. As shown in Fig. 4.26d, when the working face passed 5 m after the fault, the plastic zone of the fault surrounding rock overlapped with the plastic zone of the floor, both of which were in the pressure relief zone of the goaf floor. The fault surrounding rock at 15 m below the coal seam underwent local micro-fracture, causing stress release, forming the minimum principal stress concentration zone at 15 m, and having a tendency to extend downward.

As shown in Fig. 4.27a, when the working face advances 15 m past the fault, the minimum principal stress concentration zone is still distributed below the coal seam at 15 m, but its distribution range slightly increases, and the local damage of the above fault increases. As shown in Fig. 4.27b, when the working face advances 25 m past the fault, the stress concentration zone extends downward along the fault to 20 m, and the damage zone extends downward and gradually connects. As shown in Fig. 4.27c, when the working face advances 35 m past the fault, the stress concentration zone extends downward to 25 m, and cracks occur within 15 m below the coal seam, and there is a tendency to further extend downward and communicate with the aquifer, posing a risk of water inrush.

After the calculation is completed, the fault shear stress at the vertical distances of 2, 10, 20, and 30 m from the coal seam along the fault strike direction is quantitatively analyzed. As shown in Fig. 4.28, during the process of the working face advancing to 5 m in front of the fault, the fault shear stress at 2 m below the coal seam is negative and gradually decreases. The closer to the fault, the faster the shear stress changes, and the minimum value is − 4.37 MPa. When the working face passes 5 m after the fault, the stress value is + 1.49 MPa, and the shear stress direction reverses. As the working face advances, the fault damage and failure and the stress

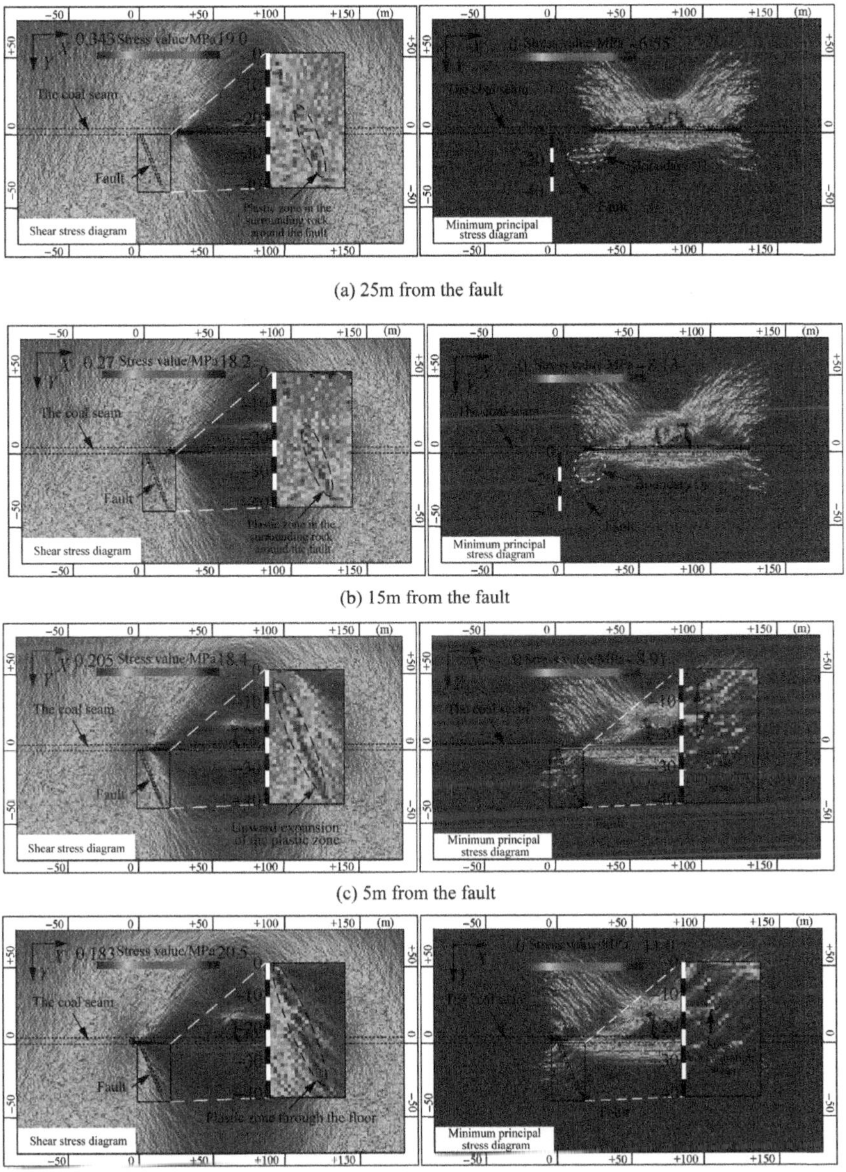

(a) 25m from the fault

(b) 15m from the fault

(c) 5m from the fault

(d) -5m from the fault

Fig. 4.26 Stress distribution diagram of water inrush channel. *Note* "–" indicates the working face crossed the fault

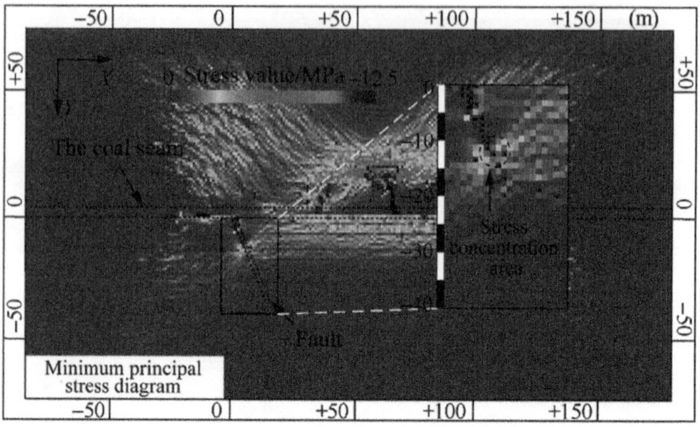

(a) -15m from the fault

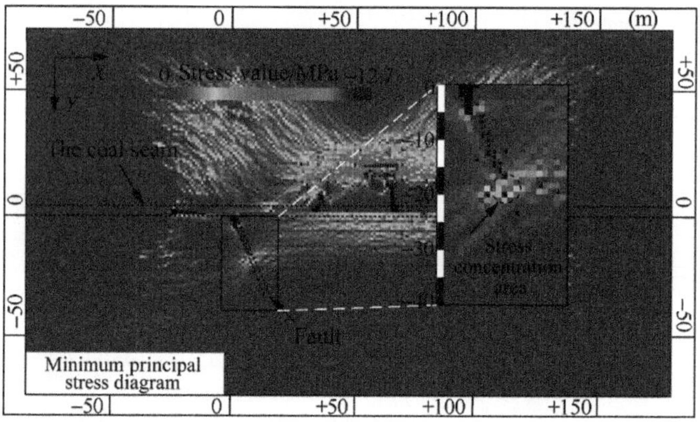

(b) -25m from the fault

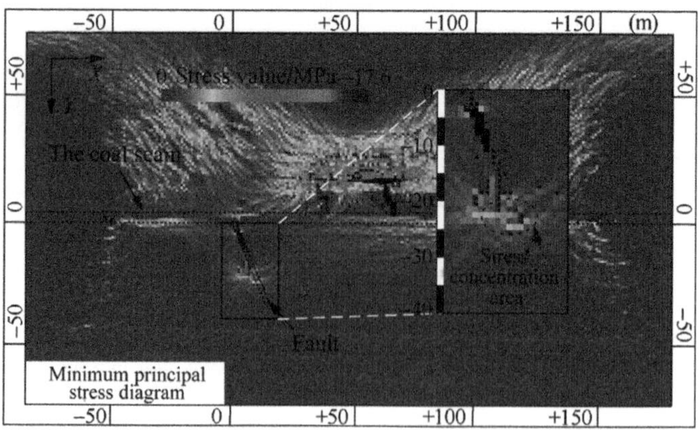

(c) -35m from the fault

Fig. 4.27 Minimum principal stress of water inrush channel after crossing fault value drop to 0 MPa. Along the fault strike direction, other measuring points have the same tendency. The fault shear stress at 10 m, 20 m, and 30 m below the coal seam decreases to the minimum when the working face advances to 25 m, 35 m, and 45 m away from the fault, respectively. The deep shear stress on the floor fault plane reaches the maximum first and exceeds the bearing capacity of the surrounding rock. The acoustic emission events extend from bottom to top. The fault shear stress at 10 m, 20 m and 30 m below the coal seam reverses the direction when it is 5 m, 5 m and 15 m away from the fault, respectively. The stress value drops to 0 MPa when it is − 15 m, − 25 m and − 35 m away from the fault, respectively. After the stress reversal, the shear stress in the shallow 20 m range on the floor fault plane decreases first, and the shallow fault cracks and fails. The direction of damage changes from top to bottom to extend to the deep.

In the process of the working face approaching the fault from the hanging wall, the shear stress is negative and gradually decreases, indicating a downward sliding tendency of the fault. During the process across the fault, the shear stress becomes positive, and the stress direction reverses, causing an upward movement of the hanging wall. Considering the floor as a beam structure, the shear stress value and its variation amplitude are the smallest in the 20 m below the coal seam, making it the neutral layer for the bending deformation of the floor. Considering the occurrence of the rock layers in the floor, the thick and strong quartz sandstone above the measurement point plays a key role in the floor deformation, which is consistent with the characteristics of the neutral layer in the floor deformation.

As a natural fault, the fault acts as a barrier for stress transmission between the hanging wall and foot wall. There are significant differences in deformation between

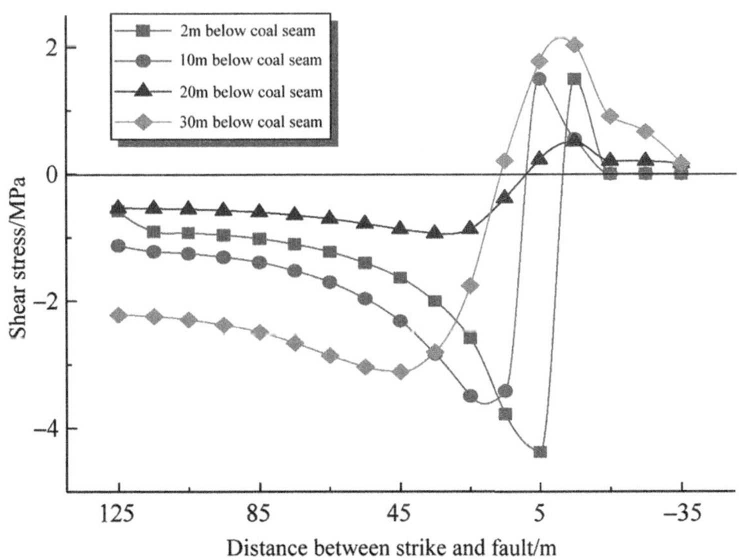

Fig. 4.28 Fault shear stress diagram

the surrounding rock of the hanging wall and foot wall, leading to fault damage and slip. In order to study the stress variation pattern of the fault cracking from top to bottom after the working face crosses the fault and forms a water inrush channel, stress variation curves of the floor along the working face are obtained. As shown in Fig. 4.29a, during normal advancement of the working face, the horizontal stress distribution in the goaf floor exhibits a "U" shape, with tensile stress of 3 MPa in the middle of the goaf floor and gradually decreasing within a 10 m range from the edges, transitioning to compressive stress in the solid coal support area outside the goaf and exhibiting an advance support effect. The surrounding rock with the fault is under the influence of compressive stress and undergoes compressive failure due to the influence of the advance support as the working face approaches. When the working face is in the foot wall, the fault is influenced by the tensile stress of the goaf floor, and a region of increased tensile stress is formed near the fault. When the working face advances 15 m past the fault, the tensile stress in the overlying floor layers near the fault at a distance of 25 m increases, reaching a maximum value of approximately 8 MPa. At this time, the surrounding rock with fault undergoes tensile failure, and the damage intensifies under high stress.

As the depth from the coal seam increases, there are significant differences in the horizontal stress state of the floor. As shown in Fig. 4.29b, when the working face advances 25 m past the fault, a region of increased tensile stress appears near the fault at a distance of 20 m below the coal seam, with a maximum value of approximately 1.6 MPa. As shown in Fig. 4.29c, when the working face advances 35 m past the fault, a region of increased tensile stress appears near the fault at a distance of 30 m below the coal seam, with a maximum value of approximately 1.1 MPa. This is because the fault near the coal seam first undergoes cracking and damage, reducing the load on the underlying rock layers, resulting in tensile stress near the deep fault surrounding rock, and the cracks gradually extend deeper. Due to the continued loading from the overlying strata, the tensile stress in the deep surrounding rock with fault is relatively small, resulting in less damage. Only a small amount of acoustic emission damage occurs, and microfractures in the rock layers below 30 m below the coal seam do not connect.

4.6.5 Summary

By constructing a microseismic monitoring system for water inrush channels in the 22517-working face of Dongjiahe Coal Mine and combining it with the numerical simulation using RFPA^{2D}, the micro-fracture information of the floor rock mass is studied, and the following conclusions are obtained.

(1) Using microseismic monitoring technology to continuously and dynamically monitor the failure characteristics of the surrounding rock in the floor fault zone, and track the generation process of the water inrush channel. By combining numerical simulation analysis of the stress changes during the progressive failure

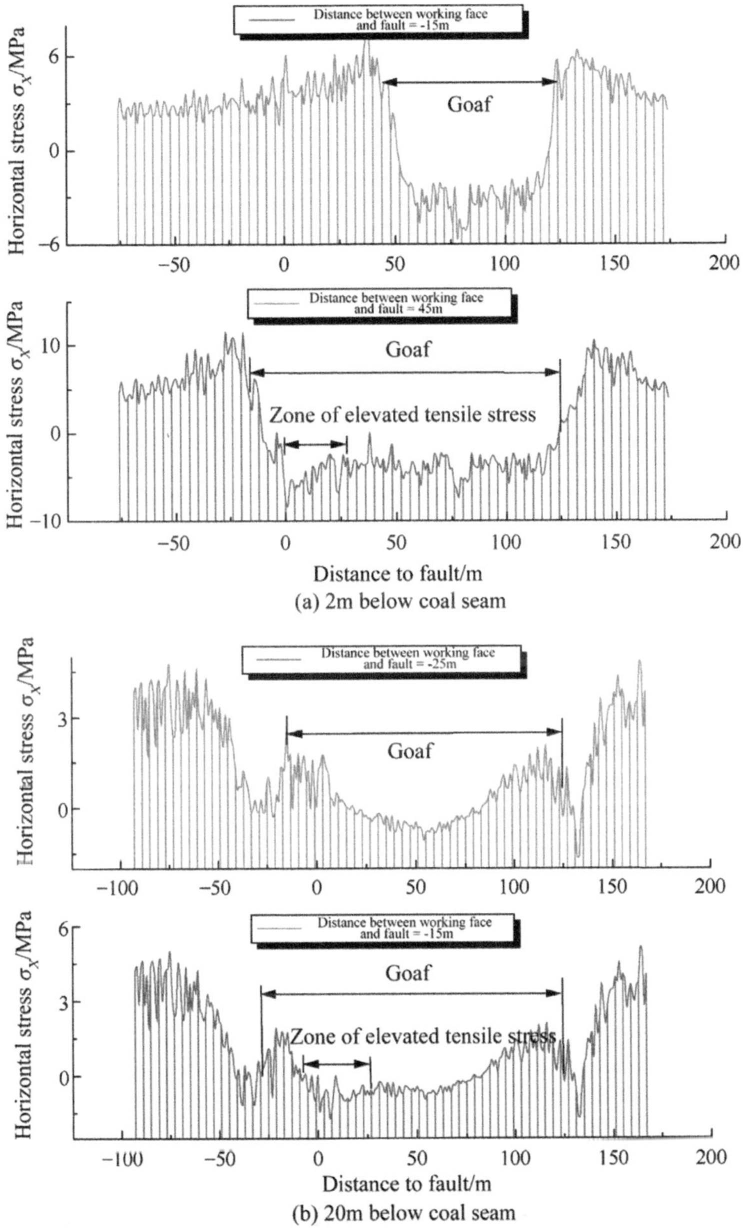

Fig. 4.29 Horizontal stress variation diagram of working face direction

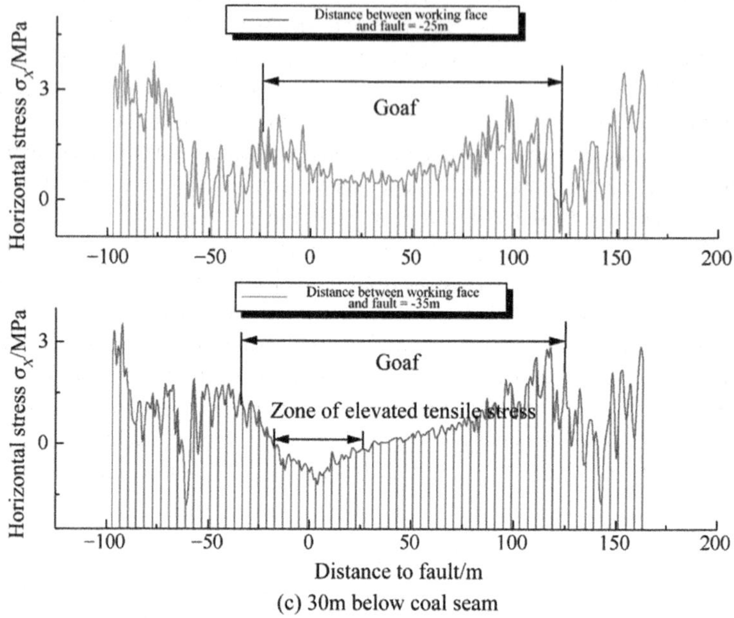

Fig. 4.29 (continued)

process of the surrounding rock, research methods for the formation of water inrush channels in the floor fault area have been enriched.

(2) Through the spatiotemporal distribution characteristics of microseismic activity, the damaged area of the surrounding rock of the floor with fault has been delineated. Microfractures in the floor fault began to occur when the working face was 85 m away from the fault. Before crossing the fault, the microfracture depth near the fault reached 25 m, showing segmented local damage characteristics. After crossing the fault, the maximum depth of microfracture reached 35 m, and the microfractures gradually connected, making it easier to form water inrush channels and induce water inrush in the goaf.

(3) Based on microseismic monitoring and numerical simulation analysis, the expansion process of the water inrush channel in the floor can be divided into two stages: before and after crossing the fault. Before crossing the fault, microfractures first occurred around 25 m below the coal seam in the surrounding rock with the fault, and extended upwards along the fault. Local microfractures occurred in the range of 5 ~ 25 m below the coal seam, but did not connect, and the floor showed segmented local damage characteristics. After crossing the fault, the microfractures extended from top to bottom and gradually connected, forming water inrush channels.

(4) Before crossing the fault, the direction of the fault shear stress was downward along the dip. The hanging wall had a tendency to slide downward under the pressure of advance support and the floor water pressure, with the absolute

value reaching a maximum from bottom to top. The surrounding rock mainly underwent compressive-shear failure. At the same time, scattered minimum principal stress concentration areas are formed near the fault, and local tensile failure occurs in surrounding rock.

(5) When the working face is near the fault, the shear stress direction quickly reverses and reaches the maximum, and the hanging wall tends to rise under the action of confined water. The minimum principal stress concentration zone appears in the surrounding rock with fault and extends deeper along the fault. The tensile stress concentration zone appears from top to bottom within 25 m near the fault, and the surrounding rock undergoes tensile failure.

4.7 Study on Fault Failure Characteristics of Floor Above Confined Aquifer

The process of water inrush through fault is essentially the deformation and destruction of the surrounding rock around the fault under the disturbance of mining, Specifically, the destruction zone of the surrounding rock gradually expands to the aquifer, resulting in the weakening of the mechanical properties for the rock body in the aquiclude as well as the influx of pressurised water into the fault under the hydraulic pressure. The stress field of the rock beneath the coal seam changes during the mining process, with the combination of newly developed fissures and natural cracks creating new flow paths for groundwater [102]. Between 2008 and 2018, 52 serious water inrush accidents occurred in state-owned key coal mines, most of which were related to faults [32], causing serious economic losses and many fatal accidents. Faults not only change the mechanical properties of the rock body, the distribution of the stress field and its evolution characteristics. As natural cracks, they are further damaged and destroyed under the coupling effect of mineral pressure and pressurised water pressure, gradually expanding the cracks to form a passage of water inrush. As for the on-site monitoring of floor water inrush, Wu et al. [81, 82] considered that the water inrush on the floor of the quarry is a kind of nonlinear dynamic phenomenon which is controlled by various factors and complex formation mechanism, proposed a vulnerability index method based on hierarchical analysis method and geographic information system, introduced a variable weight model reflecting the variability of the index value of the main controlling factors of water inrush, and predicted the water inrush on the floor of the coal seam. Zhang et al. [50] used seismic wave CT technology to collect the floor exploration data in different periods, obtaining the dynamic development law and characteristics of the floor damage in the process of mining back. Wu et al. [103] used ground boreholes in parallel with three-dimensional electrical methods to detect geological anomalies in the floor and locate the water-rich zones, which strongly guided the design of grouting boreholes and the construction of borehole grouting. Track the gestation process of the water inrush channel is the key to study the mechanism of coal mine water inrush disaster. In terms of the evolution of the water inrush process, Qian et al. [104] studied the evolution of water inrush accidents and divided the fault water inrush into a steady flow growth phase, a rapid growth phase, a decline phase and a stable phase, pointing out that the first two phases are caused by permeability increase due to infiltration damage, and the latter two phases are affected by the water level of the aquifer. Zhang et al. [45], physically simulated and pointed out that the floor water inrush channel is formed by connecting the tectonic rock zone and the floor damage zone on the basis of the research of fluid–solid coupling mechanics and solid materials, revealing the hysteresis water inrush mechanism of the floor under the action of stress field and seepage field. Liu et al. [39] used FLAC3D to simulate the water inrush process of concealed faults, pointing out the height of the uplift of pressurised water in concealed faults and the height of the development of plastic zones in fault fracture zones, revealing the temporal effect of water inrush in concealed faults. However, the above studies rarely pay attention to the influence of the deformation mode of fault-containing floor structure on fault damage. Before and after the working face crosses the fault, the over-supporting pressure crosses the fault, and the deformation mode of the floor structure changes significantly, affecting the damage location and damage nature of the surrounding rock on the fault. Besides, it is difficult to monitor the internal damage of the rock body in real time and identify its destructive nature by conventional monitoring means, which makes it difficult to verify the results of theoretical analyses.

This section takes the typical working face mining of Dongjiahe Coal Mine as the research background, establishes the mechanical model of the fault damage in floor, theoretically solves the fault stress distribution and its changing law before and after crossing the fault, and researches the influence of the deformation mode of the floor structure on the damage of the fault and its destructive nature. Numerical simulation was used to study the fault displacement, stress and seepage evolution characteristics under different structural deformation modes of the floor slab before and after crossing the fault. At the same time, the microseismic monitoring technology is used to capture the micro rupture signal of the floor, to determine the damage area of the fault before and after the fault, so as to identify the nature of surrounding rock damage, to verify the correctness and feasibility of the theoretical model and numerical simulation, providing a certain reference for the prevention and control of water in the mine.

4.7.1 Mechanical Analysis of Fault Failure

The tectonic stress field around the fault is complex, with the stress direction and magnitude changing at the fault plane during the advancement of the working face. Liu et al. [105] derived an analytical equation for the ultimate water pressure of the water-resisting strata in the floor by applying the theory of limiting equilibrium of the rock mass and considering the nature of the fault itself and the role of the stress reduction zone in the mine pressure. Lu et al. [33] regarded the floor of fault as a semi-infinite body to derive the elastic mechanical solution of the fault stress, which in turn derived the analytical equation of the water pressure of the water inrush in the floor under the influence of the fault. Chen et al. [106] established a simplified fracture mechanics model for water inrush on concealed faults in the floor, deducing the critical water pressure and influencing factors for splitting damage to occur on water-conducting faults. However, these methods neglect the influence of the deformation mode of the floor structure on the floor and fault stress. In order to study the influence of the structural evolution of the floor on the fault damage before and after crossing the fault, without considering the influence of tectonic stress, this section takes the surrounding rock in the central part of the quarry of unit width as the research object, simplifying the floor of the working face approximately into a beam structure to analyse the change of the fault stress and the damage process.

4.7.2 Model Setup

The open cut eye of the working face is about 900 m away from the fault area, where the support pressure is difficult to influence the fault activity, regarded as the fixed end. Neglecting the effect of the fault on the computational model until the fault is destroyed, the computational model can be simplified to a beam model that is solidly supported at both ends. The stress distribution in the strike profile of the working face is shown in Fig. 4.30. The over-support pressure on the floor is equivalent to the vertical uniform load $k\gamma H/2$ (H is the depth of the coal seam; γ is the average bulk weight of the overlying rock layer; k is the over-support stress concentration coefficient, generally taken as 3–4); the length m is twice the distance from the working face to the stress peak; the water pressure of the pressurised water is simplified to the uniform load q; l is the span of the solid support beam; x is the span of the upper plate beam; and m is the length of the over-support stress effect. The weight of the fallout gangue from the mining area is not taken into account, as well as the effect of the fallout caused by the small fallout of the fault.

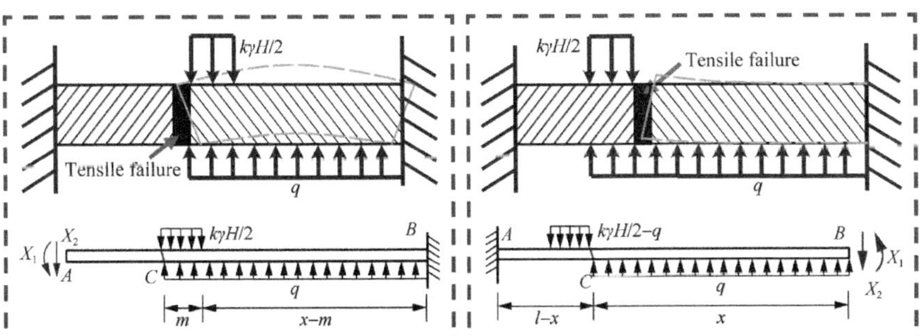

(a) Before passing through the fault (b) After passing through the fault

Fig. 4.30 Fault-bearing floor force and fault damage in mining on pressurised water

4.7.3 Fault Plane Stress Solution

Based on the mechanical model, the choice was made to apply the principle of force method without considering the axial deformation of the beam. According to the principle of superposition, in the basic structure shown in Fig. 4.30, X_1 represents the bending moment and X_2 represents the shear force applied at bearing A, with the basic structure being made to satisfy the corresponding deformation co-ordination conditions. Typical equations for the force method, the flexibility coefficients and free terms in the equations are as follows:

$$\begin{cases} \delta_{11}X_1 + \delta_{12}X_2 + \Delta_{1p} = 0 \\ \delta_{21}X_1 + \delta_{22}X_2 + \Delta_{2p} = 0 \end{cases} \tag{4.12}$$

where, $\delta_{11} = \frac{l}{EI}$; $\delta_{12} = \delta_{21} = \frac{l^2}{2EI}$; $\delta_{22} = \frac{l^3}{3EI}$;

$$\Delta_{1p} = \int \frac{\overline{M} \cdot M_p}{EI} \mathrm{d}s = \frac{q_1 - q}{12EI} \cdot m^3 + \frac{1}{EI}\left[\frac{mq_1}{2}(x^2 - xm) - \frac{q}{12}(x^3 - m^3)\right];$$

$$\begin{aligned} \Delta_{2p} &= \int \frac{\overline{M} \cdot M_p}{EI} \mathrm{d}s \\ &= \frac{1}{EI}\left\{\frac{q_1 - q}{16}\left[(l - x + m)^4 - (l - x)^4\right] + \frac{q_1 - q}{6}(x - l)\left[(l - x + m)^3 - (l - x)^3\right]\right. \\ &\quad + \frac{q_1 - q}{8}(x - l)^2\left[(l - x + m)^2 - (l - x)^2\right] + \frac{mq_1}{3}\left[l^3 - (l - x + m)^3\right] \\ &\quad + \left(\frac{mq_1 x}{2} - \frac{m^2 q_1}{4} - \frac{mq_1 l}{2}\right)\left[l^2 - (l - x + m)^2\right] - \frac{q}{16}\left[l^4 - (l - x + m)^4\right] \\ &\quad \left. - \frac{2(x - l)}{3}\left[l^3 - (l - x + m)^3\right] - \frac{(x - l)^2}{2}\left[l^2 - (l - x + m)^2\right]\right\}. \end{aligned}$$

where E represents the elasticity model of the coal rock body; I represents the moment of inertia of the coal rock body.

Solve for the bending moment by Cramer's law:

$$X_1 = \frac{\Delta_{2p}\delta_{12} - \Delta_{1p}\delta_{22}}{\delta_{11}\delta_{22} - \delta_{12}^2}$$

$$X_2 = \frac{\Delta_{1p}\delta_{12} - \Delta_{2p}\delta_{11}}{\delta_{11}\delta_{22} - \delta_{12}^2}$$

As shown in Fig. 4.30a, before crossing the fault, the fault section of the upper disc is in the solid coal bearing area. Affected by the function of over-supporting stress of the working face, bending deformation occurs upward on the upper disc, with bending moments in the counterclockwise direction at the fault. The fault above the neutral layer is squeezed while the fault below the neutral layer stretches. The surrounding rock of the fault below the neutral layer undergoes tensile damage first in case of insufficient strength of the surrounding rock. Taking the AC section as the object of study, it is known that the fault is under the bending moment $M_1 = X_1 + X_2(l - x)$. According to the material mechanics formula, the maximum positive stress suffered at the fault is $\sigma_{\max} = M_1/W_z$, where W_z is called the bending section coefficient. The damage criterion at the fault is $\sigma_{\max} \geq [\sigma]$, where $[\sigma]$ is the allowable tensile stress in the surrounding rock of the fault; $\sigma_{\max}$ is the actual maximum tensile stress on the fault surface. When $\sigma_{\max} \geq [\sigma]$, the strength damage occurs at the fault, and the greater the distance from the neutral layer, the greater the tensile stress at the fault, with the surrounding rock of the fault starting to be damaged from the depth and expanding upwards along the fault. Similarly, as shown in Fig. 4.30b, after crossing the fault, the working face over-supporting stress moves to the lower solid coal bearing area, while the upper beam structure is completely in the hollow area, losing the function of over-supporting stress, bending upwards under the function of pressurised water hydraulic pressure. The fault is affected by a clockwise bending moment, with tensile stresses at the shallow part of the fault and compressive stresses at the deep part of the fault. From the basic structure shown in Fig. 4.30b, the bending moment at the fault is M_2 and the maximum positive stress at the fault is $\sigma_{\max} = M_2/W_z$. When $\sigma_{\max} \geq [\sigma]$, strength damage occurs in the surrounding rock of the fault. At this time, the micro-rupture at the fault extends from top to bottom to the depth, which is opposite to the direction of micro-damage extension of the fault before the workface crosses the fault.

4.7.4 Numerical Simulation of Water Inrush Channel Formation Process

1. Seepage model of floor water inrush

Establish the RFPA^{2D} strain model of the floor water inrush, as shown in Fig. 4.31, set the model size as 500 m long in the horizontal direction and 200 m long in the vertical direction, totally divided into 500 × 200 = 100,000 cells, with each cell representing 1 m of the actual rock stratum. On the premise of satisfying the accuracy of the calculation model, rock layers with similar lithology or smaller thickness are merged, taking into account the simplicity of the calculation. In the model, beds 1, 4, 6, 12, 14, 16 are fine-flour sandstones; beds 2, 7, 17 are coarse-flour sandstones; beds 3, 9, 18, 22 are medium-grained sandstones; beds 5, 8, 10 are fine-grained sandstones; bed 11 is a coal seam; beds 13, 15, 19 are quartz sandstones; bed 20 is an aluminous mudstone; and beds 21, 23 are limestones. Of these, layers 20, 21 and 23 are Ordovician rock stratum. Faults are modelled using the weaker mechanical unit with a dip of 73°, without considering the effect of the drop due to the small drop of the faults in the modelling process. The physical and mechanical parameters of each formation are shown in Table 4.5.

Horizontal displacement constraints are applied to the left and right sides of the model, with the vertical displacement constraints at the bottom of the model. A uniform load of 4 MPa is applied to the top of the model to simulate the self-weight of the overlying rock layer in addition to the model. The top and bottom of the model are watertight boundaries. Set a fixed head boundary with a top of 0 m and a bottom of 150 m high to simulate pressurized water pressure (1.5 MPa). The water pressure is transmitted through the boundary into the overlying aquifer beneath the

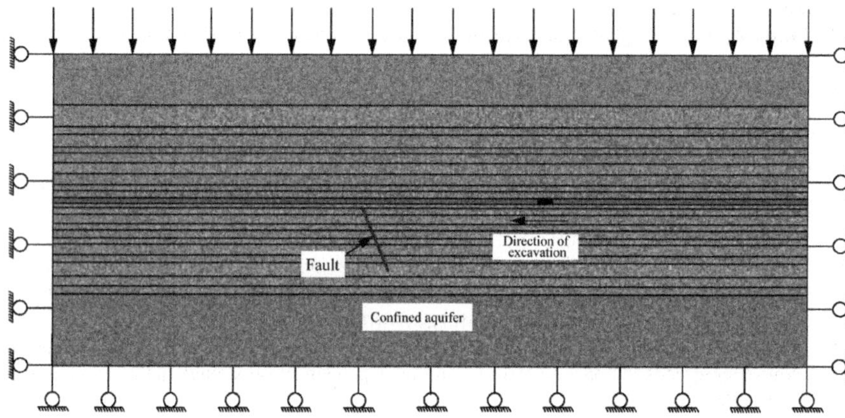

Fig. 4.31 Numerical modelling of floor water inrush

Table 4.5 Physical and mechanical parameters of the rock layers in the model

Rock stratum	Elastic modulus/MPa	Compressive strength/MPa	Friction angle/(°)	Poisson's ratio	Density/(kg/m^3)
Coarse siltstone	8800	45	26	0.24	2450
Fine siltstone	8200	35	28	0.26	2450
Medium grained sandstone	8500	29	28	0.26	2450
Fine siltstone	7800	24	29	0.28	2450
Coal seam	1650	8	36	0.32	1500
Quartz sandstone	8850	48	25	0.24	2650
Ordovician limestone	5000	68	30	0.24	2770
Fault	1000	7	28	0.26	2000

coal seam, with the rock mass being subjected only to self-gravitational stresses and water pressure.

The model is subjected to elastic–plastic assumptions with the Mohr–Coulomb criterion used as the yield criterion in the calculations. The second step starts to simulate the workface recovery by using stepwise continuous excavation with 10 m excavation steps. Excavation to the left at 170 m from the right-hand boundary of the model, taking full account of boundary effects.

2. Characteristics of displacement change at fault

After the model calculation, the displacement information of the fault-upper disc contact unit before and after crossing the fault is extracted. Horizontal and vertical displacements at calculation step 13-1 before crossing the fault and at calculation step 14-1, calculation step 14-30, and calculation step 14-33 after crossing the fault are selected for analysis. The data are plotted at intervals of one unit along the depth direction of the fault, as shown in Fig. 4.32 (*X*-direction displacement to the right is "+" and *Y*-direction displacement down is "+").

As shown in Fig. 4.32a, before the workface crosses the fault, the vertical displacement of the deep part of the fault is negative, while the shallow part of the fault is positive, with the absolute value increasing gradually as the workface approaches the fault. Calculations to Calculation Step 13-1 show a maximum vertical displacement of − 13 mm at the deep end and a maximum vertical displacement of 27 mm at the shallow end. The upper disc of the fault is bounded by 15 m below the coal seam, with vertical displacement downwards above and upwards below. Besides, the horizontal displacements are positive in the deeper part and negative in the shallower part, and their absolute values increase gradually as the working face approaches the fault. The maximum horizontal displacement in the deep part is 5 mm while the maximum horizontal displacement in the shallow part is − 20 mm by calculation to step 13-1 of the calculation. The upper disc of the fault is bounded by 17 m below the coal seam, with horizontal displacement to the left above and to the right below. Therefore, before the working face crosses the fault, the upper disc of the fault is bounded by 15–17 m below the coal seam. The angle between the maximum displacement in the deep part and the horizontal direction is 17° slanting to the upper right, while the angle between the maximum displacement in the shallow part and the horizontal direction is 197° slanting to the lower left, consistent with the deformation mode of the beam structure before crossing the fault in Sect. 4.7.3.

As shown in Fig. 4.32b, after the working face crosses the fault, the vertical displacement at the junction of the fault and the upper plate changes abruptly. When calculating step 14-1, the overall vertical displacement changes to negative value, with the absolute value gradually increasing with the increase of the displacement at the depth of burial. The minimum displacement in shallow part is − 7 mm, and the maximum displacement in deep part is − 21 mm. The absolute value of displacement in the shallow vertical direction continues to increase while the absolute value of vertical displacement in the deeper part of the calculation continues to increase to a lesser extent by calculation step 14-30. For calculations up to calculation step 14-33, the vertical displacements remain essentially unchanged. After crossing the fault,

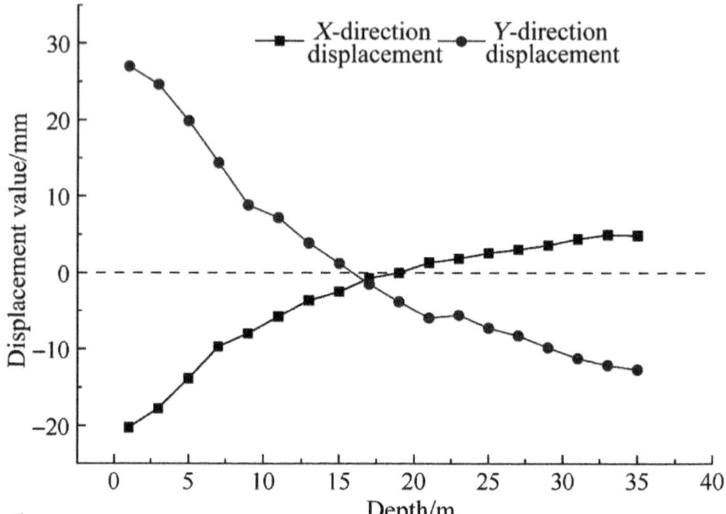

(a) Before working face passed through fault

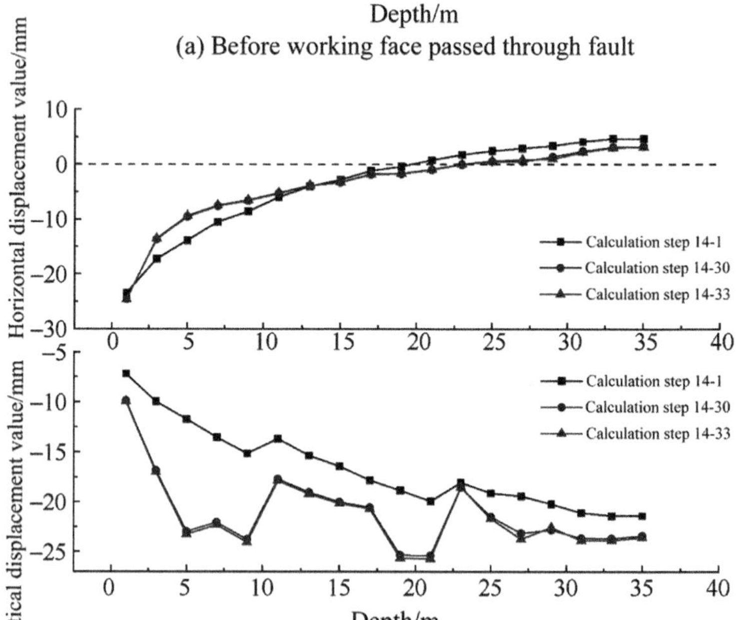

(b) Working face passed through fault

Fig. 4.32 Displacement variation curve of upper plate fault the change of horizontal displacement is small, and the horizontal displacement in step 14-1 and step 13-1 are basically equal. When calculating step 14-30, the absolute value of horizontal displacement in shallow part and the absolute value of horizontal displacement in deep part both decrease slightly, among which the maximum decrease in shallow part is 4 mm and the maximum decrease in deep part is 2 mm. Therefore, after crossing the fault, the deformation at the fault of the upper disc is mainly vertically upward, with the horizontal direction being opposite to that before crossing the fault, consistent with the way of deformation of the beam structure after crossing the fault in Sect. 4.7.3.

3. Fault stress variation characteristics

To study the effect of structural deformation of the floor beams on the fault stress before and after crossing the fault, the calculation results of the fault units in calculation step 1-1, 9-1, 11-1, 12-1, 13-1 before crossing the fault, and calculation step 14-1, 14-20, 14-25, 14-30, and 14-33 after crossing the fault were taken, extracting information of fault stresses at every interval of the units along the tendency to extract fault stress information at each cell interval along the tendency.

As shown in Fig. 4.33a, in calculation step 1-1, the model is not excavated, and the horizontal stress of the fault is not affected by mining, with the stress value of about 2 Mpa. Calculation to calculation step 9-1, the fault stress began to change. Horizontal stress increases in the range of 0–21 m below the coal seam, while horizontal stress decreases in the range of 25–35 m. The cut-off points for stress increase and decrease compared to the initial stress is located 21–27 m below the coal seam. Calculation step 11-1, the shallow horizontal stress continues to increase, with the maximum value of 4.3 MPa. The stress continues to decrease in the deep part, in which the horizontal stress of the fault near 31 m below the coal seam decreases to 0 MPa, compared with the initial stress, the cut-off point of stress increase and decrease is located 15–17 m below the coal seam. Calculation to the calculation step 12-1, the maximum horizontal stress in the shallow part is 5.4 Mpa. The deep horizontal stress continues to decrease, in which the horizontal stress of the fault in the local area within the range of 15–35 m below the coal seam is less than 0 MPa, implying that the fault may undergo tension damage, with the cut-off point of stress increase and decrease located at 9–13 m below the coal seam. Calculation to calculation step 13-1, the fault stress does not completely show the trend of decreasing in the deep part and increasing in the shallow part. The maximum value of shallow horizontal stress decreases to 4.5 MPa, and the local horizontal stress in the range of 5–20 m below the coal seam is less than 0 MPa, which may cause pull-up damage, while the horizontal stress in the deeper part of 20–35 m starts to increase. At this time, the working face is 5 m away from the fault, according to Sect. 4.7.3, the working face is located in the upper disc and lower disc when the deformation trend of the floor structure is different. By this time, the over-above support pressure has gradually moved to the lower disc, and the fault end of the upper disc is gradually unloaded, leading to a gradual change in the deformation trend of the floor structure, resulting in the fault stress complexity at this time, while having the characteristics of horizontal stress change before and after over-faulting. In summary, during the excavation process, the horizontal stress of the deep fault decreases while the horizontal stress of the shallow fault increases as the work advances towards the fault, with the demarcation point between the stress increase and decrease gradually moving upwards. This is consistent with the pattern of change in fault stress before crossing the fault in Sect. 4.7.3.

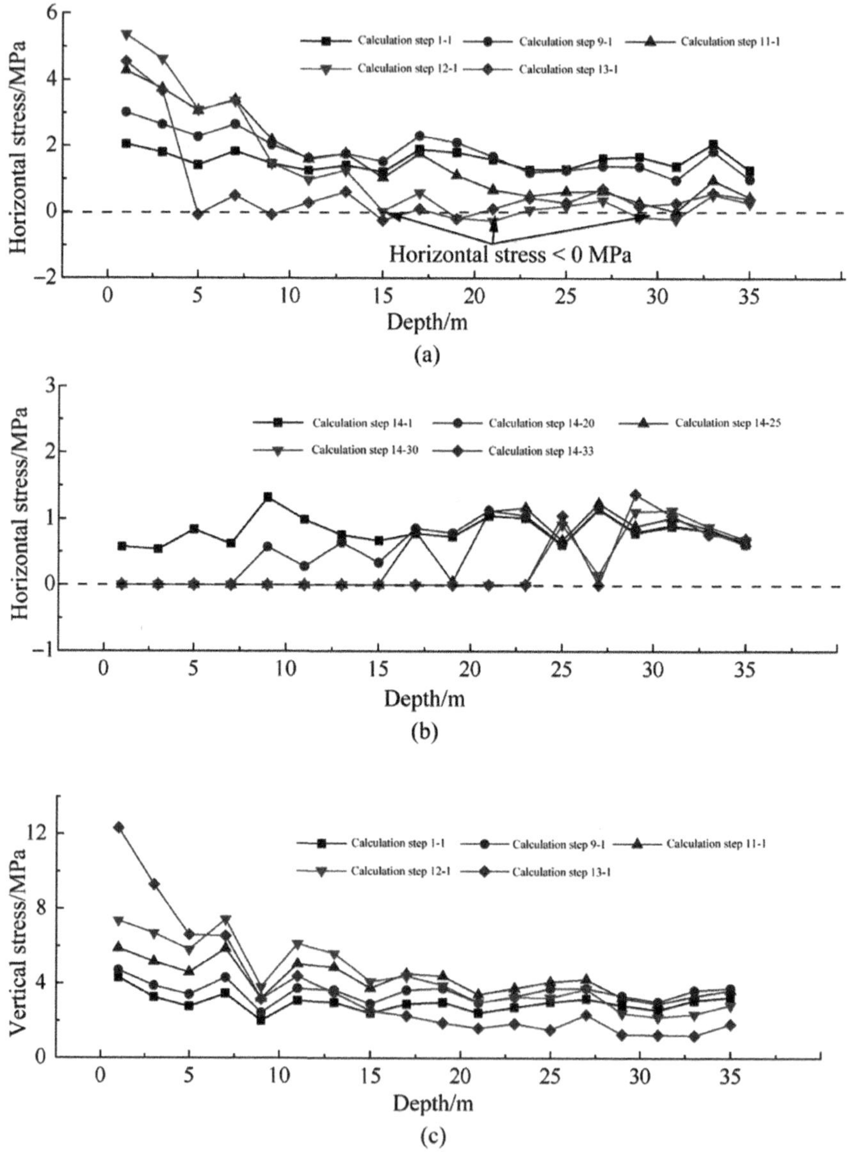

Fig. 4.33 Stress curves on the fault plane before and after crossing the fault

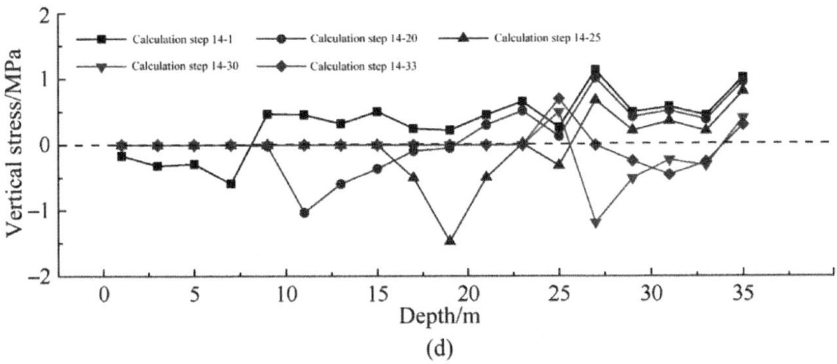

Fig. 4.33 (continued)

As shown in Fig. 4.33b, the workface pushes through the fault for 5 m with the overtopping support pressure moving fully into the lower disc. Calculation to calculate step 14-1, the fault horizontal stress value is approximated to 1 MPa, compared with the results of calculation step 1-1 fault stress reduction. This is the fault from the upward pressure and downward pull gradually changed to the upward pull and downward pressure process, in which deep and shallow horizontal stress values are equal for a short period of time. And before crossing the fault, the surrounding rock of the fault is damaged and destroyed locally, so the stress is reduced compared with the calculation step 1-1 fault. Calculation to calculation step 14-20, the fault breaks within 7 m below the coal seam with horizontal stress reduced to 0 MPa. Calculation to calculation step 14-25, the fault damage extends to 15 m below the coal seam, with the horizontal stress reduced to 0 MPa. Calculations continue to extend deeper to 23 m below the coal seam at calculation step 14-30, releasing the fault stress. The horizontal stresses on the fault are essentially unchanged up to calculation step 14-33. In summary, the horizontal stress of the fault gradually decreases to 0 MPa from shallow to deep, destroying the surrounding rock from top to bottom. This is consistent with the pattern of change in fault stress after crossing the fault in Sect. 4.7.3. As shown in Fig. 4.33c, the average vertical stress on the fault is about 3 MPa in calculation step 1-1. With the advancement of the workface, the calculation-to-calculation step 9-1 and Calculation Step 11-1 shows an overall increase in stress on the fault face, but the increase in vertical stress is greater at shallow than at deeper depths. Calculation to calculation step 12 1, the shallow vertical stress continues to increase, with the maximum value of 8 MPa; the vertical stress in the deeper part from 17 to 35 m starts to decrease, with the stress value in the range from 29 to 35 m less than the initial value in calculation step 1-1. Calculation to the calculation step 13-1, the maximum value of vertical stress in the shallow part continues to increase to 12 MPa; 15–35 m vertical stress below the coal seam is less than the initial value, gradually upward expansion of the vertical stress reduction zone. However, since the vertical stress is always greater than 0, tension damage occurs in the deep part of the fault mainly under horizontal stress before crossing the fault. As shown in Fig. 4.33d, there is an abrupt change in the vertical stress of the fault after crossing the fault. Calculation to calculation step 14-1, the vertical stress of the fault in the shallow 7 m range is rapidly changed from compressive stress to tensile stress in calculation step 13-1. Calculation to the calculation step 14-20, the fault surrounding rock in the shallow 9 m range under the effect of vertical tensile stress occurs tensile damage, with a new tensile stress zone formed at 9–19 m. The vertical tensile stress of the fault continues to transfer to the depth as the calculation goes on. Calculation to calculation step 14-25 and calculation step 14-30 resulted in tensile stresses in the ranges of 19–27 m and 27 to 33 m, respectively, with damage occurring in the faults above them. Calculation to the calculation step 14-33, the vertical stress distribution range is basically unchanged, as the fault damage is no longer extended to the deeper part. From this analysis, after crossing the fault, the surrounding rock of the fault mainly undergoes tension damage from top to bottom under the effect of vertical tensile stress.

4. Analysis of seepage evolution of mining floor rock mass

From Fig. 4.34 Seepage distribution of mining basement rock, it is noticed that the surrounding rock of the fault is still in a stable state and the basement rock is almost free from seepage when calculating to Calculation Step 14-1. In the process of calculating to Calculation Step 14-20 and Calculation Step 14-30, with the gradual destruction of the fault, the seepage flow mainly concentrates at the destroyed fault and gradually extends to the depth, distributed as 10–25 m below the coal seam. At step 14-33 of the calculation, the distribution of seepage on the fault is basically unchanged, with no seepage distribution in the water barrier (aluminium mudstone) above the pressurised aquifer, indicating that the floor is not destabilised and is acting as a water barrier. As shown in Fig. 4.35, taking the vertical seepage volume of the unit at the intersection of the fault and the coal seam for analysis, the vertical seepage volume of the bottom plate at the fault starts to increase at step 14-1 of the calculation; the vertical seepage volume is 5.6 m^3/min at step 14-20; the vertical seepage volume is 7.6 m^3/min at step 14-25; the vertical seepage volume is 11.3 m^3/min at step 14-30; the vertical seepage volume is 11.6 m^3/min at step 14-33; the vertical seepage volume is 11.3 m^3/min at step 14-30; the vertical seepage volume is 11.3 m^3/min at step 14-33; the vertical seepage volume is 11.3 m^3/min at step 14-33; the vertical seepage volume is 11.3 m^3/min at step 14-33. The vertical seepage flow is basically stable at about 11 m^3/min, which is consistent with the fault damage and stress change rule. Besides, before the workface crosses the fault (calculation step 2-1 to calculation step 13-1), the seepage volume decreases gradually with the slope increasing as the workface approaches the fault. In the process of structural deformation of the upper disc before crossing the fault, extrusion is formed between the shallow upper disc rock layer and the lower disc rock layer, which reduces the permeability of the surrounding rock in the shallow part of the fault and reduces the seepage and the smaller the distance between the working face and the fault, the faster the rate of seepage reduction.

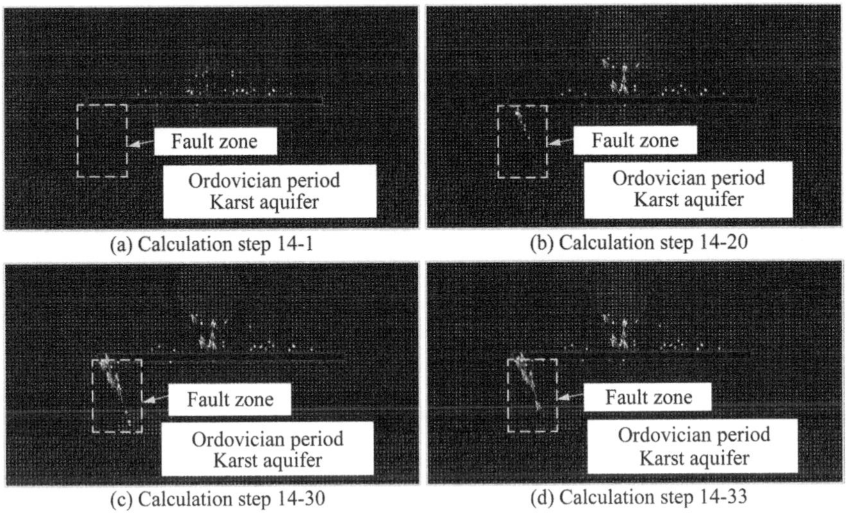

Fig. 4.34 Distribution of seepage in the rock mass at the mining floor

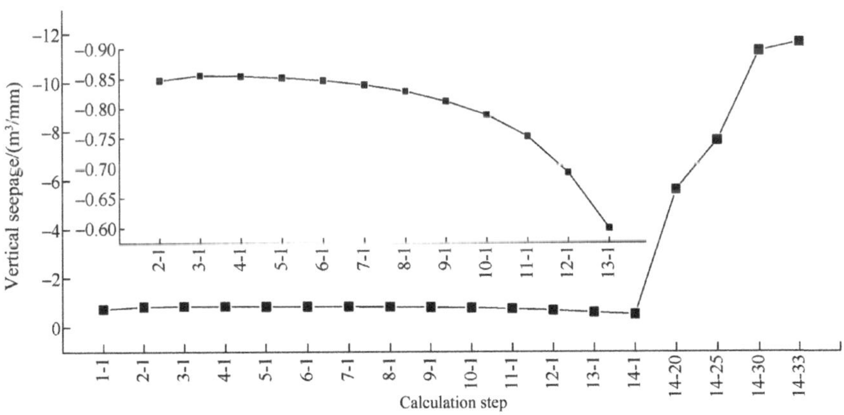

Fig. 4.35 Vertical seepage curve of the floor at the fault

4.7.5 Microseismic Activity Characteristics of Floor Fault Damage

Microseismic events on the floor of the 22517-working face during the period from 20 December 2015 to 31 March 2016 were selected and projected onto the working face strike section to analyse the activity pattern of microseismic events before and after crossing the fault, as shown in Fig. 4.36 (the domes in the microseismic event

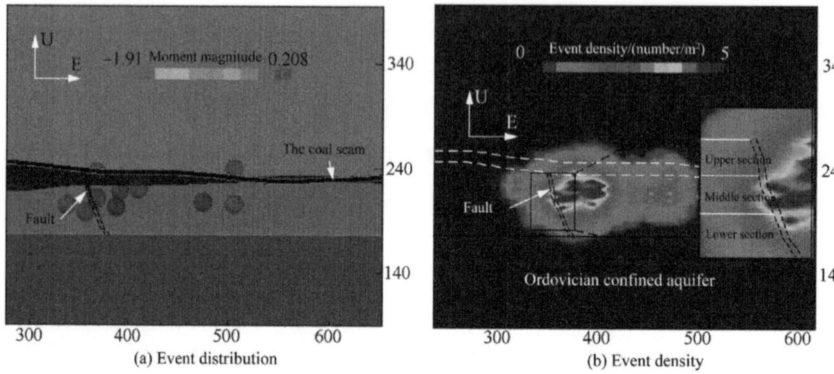

Fig. 4.36 Information on microseismicity of the floor plate before crossing the fault

distribution map represent microseismic events generated by the damage and destruction of the rock body, and the colors of the domes represent the moment magnitude of the microseismic events).

Combining Fig. 4.36a and b, it can be found that 8 microseismic events are collected in the floor fault area before the working face crosses the fault, with moment magnitude ranging from − 1.03 to − 0.02, while the microseismic events in the fault area are mainly clustered in the elevation of 10–25 m below the coal seam. The fault is divided into upper, middle and lower segments, in which the damage is mainly distributed in the middle segment, while the upper and lower segments are still in a stable state. After the workface crosses the fault, the microseismic activity law of the floor fault changes abruptly. As shown in Fig. 4.37, 24 floor microseismic events were collected, with moment magnitudes located from − 0.73 to 0.49, resulting in a significant increase in cumulative seismic moment of the microrupture. Microseismic events start to occur in the upper and lower sections of the fault, with a concentration of microruptures extending along the strike towards the open area up to 80 m. The area of damage to the surrounding rock in the fault is further expanded both horizontally and vertically.

Combined with the time distribution characteristics of microseismic events for analysis, the microseismic events before crossing the fault are mainly concentrated in December 2015. At this time, it is 75–85 m from the fault, which is basically the same as the distribution range of the over-head bearing pressure in the working face. The load at C of the beam structure is the largest. The upper disc floor structure has the largest bending moment under the combined effect of overhanging support pressure and bearing water hydraulic pressure, with the maximum stress on the fault surface. However, the surrounding rock is stable because the lower section of the fault is aluminous mudstone with large thickness. In contrast, the middle section is a rock formation with small thickness and low strength, in which the surrounding rock is the first to undergo micro-destruction. With the continued advancement of the work face to the fault, the work face overrunning support pressure zone gradually

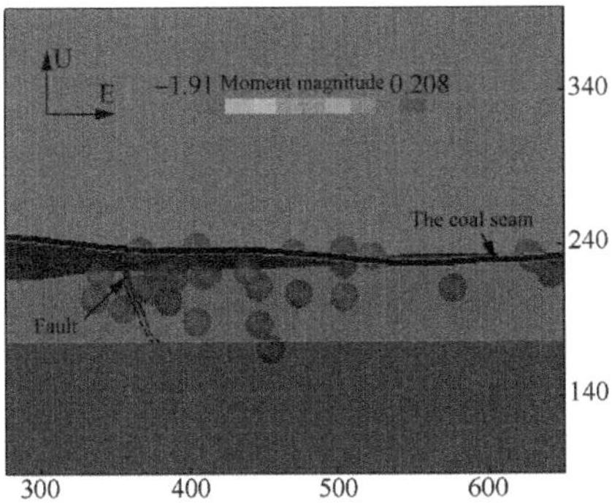

(a) Event distribution

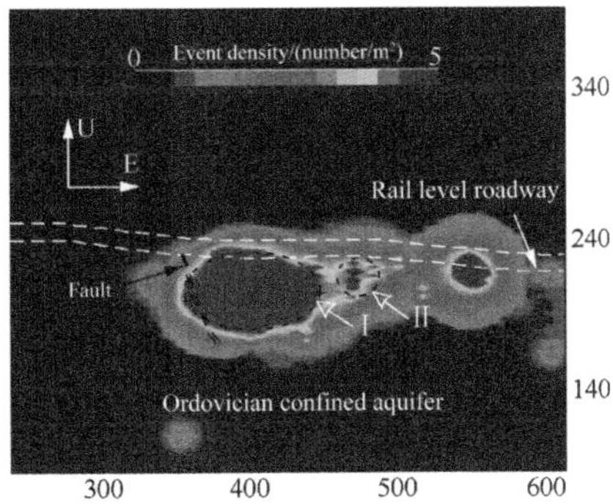

(b) Event density

Fig. 4.37 Information on microseismicity of the floor after crossing the fault

moved from the upper plate to the lower plate between 1 January 2016 and 20 February 2016, with the bending moment of the upper plate girder structure gradually decreasing, reducing the force on the fault face. The floor surrounding rock produces only sporadic microseismic events, and the distribution of areas of microrupture concentration in the vicinity of the fault remains essentially unchanged. After the workface crossed the fault, between 21 February 2016 and 31 March 2016, the zone of influence of the overbearing pressure moved completely to the lower plate, with the upper plate girder structure buckling upwards under the water pressure of the pressurised water and gradually separating from the lower plate, expanding the degree and extent of damage to the surrounding rock of the fault, leading to the acquisition of a large number of microseismic events.

As shown in Fig. 4.38a, the fault damage before the cross-fault is concentrated in the middle section of the fault and is located near the fault without extending further up the upper plate. As shown in Fig. 4.38b, after crossing the fault, the high energy zone extends to the upper disc mining area and the depth. It extends along the strike of the working face to within about 80 m of the upper goaf, extending in the depth direction to about 35 m below the coal seam. Based on the principle of energy dissipation, it can be seen that the occurrence of a rock microfracture represents a slight rock damage. With the occurrence of microseismic events, the surrounding rock internal energy is continuously released to cause a gradual decrease in the mechanical parameters of the surrounding rock, so that the surrounding rock damage area is defined by the microseismic energy release area. Comparing before and after the workface crosses the fault, the damage area of the floor plate expands from the vicinity of the fault to the mined-out area, indicating that the structural deformation mode of the floor plate has changed.

In the process of approaching the fault at the working face, the upper girder structure bends upwards. The surrounding rock above the neutral layer at the fault is under compressive stress with tensile stresses below, resulting in localised micro-damage in the middle section of the fault. As the fault is still below the solid coal

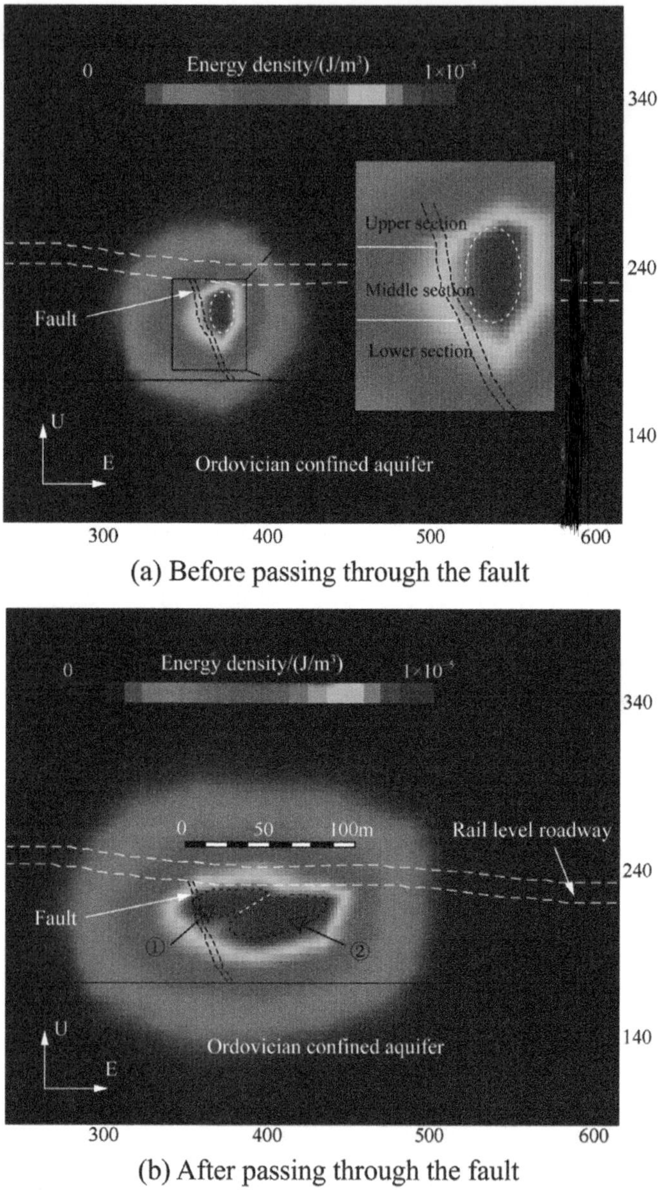

(a) Before passing through the fault

(b) After passing through the fault

Fig. 4.38 Energy density diagram

and the damage zone is located in the deep part, in which the surrounding rock deformation is constrained by the surrounding rock, with the high energy density zone concentrating near the fault and not expanding along the strike to the mining area. In Fig. 4.38b, the high-energy zone of the base plate can be divided into two sections along the white line, of which section ① is 0–50 m away from the fault and section ② is 50–100 m away from the fault. There is a significant difference in the distribution elevations of the high energy density zones in the ① and ② sections. After crossing the fault, the floor of section ① is completely unloaded, buckling upwards under the action of water pressure. The deformation range of the upper disc floor extends towards the direction of the mined-out area, with ② section of the floor damaged and destroyed, forming a high-energy density zone. The two floor segments have different deformation mechanisms.

Fault microseismic event aggregation is the result of mining disturbance and floor water pressure coupling, reflecting the fault stress change and its manifestation law in the process of working face advancement. The deformation mechanism of the floor structure before and after crossing the fault is different, with differences in the damage characteristics of the fault. In order to accurately grasp the microseismic aggregation characteristics under the domination of different factors before and after crossing the fault and the potential damage characteristics of the surrounding rocks of the fault, the aggregation area of microseismic events near the fault is taken as the object of the study, and the seismic source parameter E_S/E_P (the ratio of the energy of the shear wave to the energy of the compression wave) in the aggregation area is analysed. Seismologically, the ratio of the energy released by the P-wave to the S-wave, E_S/E_P is an important feature reflecting the earthquake source mechanism. Since the sensor site may be very close to the source, the source can be treated as a near-field, considering the seismic wave array as a hemispherical surface. According to the work of Gibowicz et al. [107], the P- and S-wave seismic energy is:

$$E = 4\pi \rho c r^2 \frac{J_c}{F_c^2} \tag{4.13}$$

where E denotes the energy released by a micro rupture event, J; ρ is the density of the rock mass, kg/m^3; c is the velocity of the elastic wave in the rock mass, m/s; r is the distance between the sensor and the source of the earthquake, m; J_c denotes the integral of the velocity of the motion of the mass, m; and F_c denotes the empirical coefficient of the type of radiation of the seismic wave.

For fault slip or shear-type induced seismic events, the S-wave energy is much higher than the P-wave energy, generally $E_S/E_P \geq 10$, while Gibowicz et al. [77] similarly found that excavation-induced shear-damage microseismic events are higher in E_S/E_P; whereas, for non-shear types such as strain-type rockbursts, tensile and volume-stress change-induced events, the P-wave energy is higher in general, with $E_S/E_P \leq 3$. Cai et al. [108] developed a tensile model to estimate fracture size and found that 78% of the 804 microseismic events had an E_S/E_P of less than 10, which is consistent with in-situ tension damage of the surrounding rock. Discriminating the damage nature of rock media by analysing the E_S/E_P values of microseismic

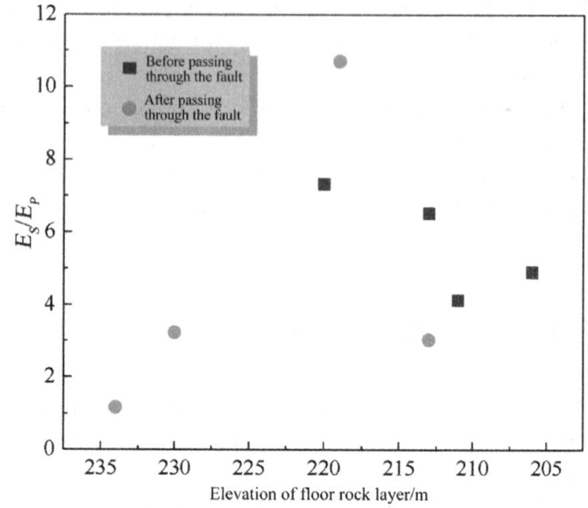

Fig. 4.39 E_S/E_P ratio for microseismic events in fault surrounding rocks

events is an important reference value for the study of fault damage mechanism. The E_S/E_P values and elevation distribution of microseismic events are shown in Fig. 4.39. Before the working face crossing the fault, the E_S/E_P values of the fault microseismic events are mostly in the range of 3–10, with the main distribution in the middle section of the fault, 10–25 m below the coal seam. From this analysis, the form of fault damage before crossing the fault is not shear damage, which can also be characterised as tensile damage according to the above research results of Cai et al. The reason for $E_S/E_P > 3$ is that the middle section of the fault is in the transition zone of tensile damage. According to the theoretical analysis in Sect. 4.7.3, the tensile damage zone is in the lower section of the fault, but the lower section of the aluminium mudstone has a high strength and a large thickness, without any damage. Based on the E_S/E_P value and elevation distribution characteristics, it can be found that the E_S/E_P value decreases with increasing depth, with the deeper part of the fault as a potential tension damage zone. After the workface crosses the fault, the E_S/E_P of fault microseismic event less than 3, with microseismic event elevation distribution in the upper section of the fault relative to that before the workface crosses the fault, indicating that tensile damage mainly occurs in the fault zone.

4.7.6 Summary

(1) Considering the influence of the deformation mode in floor structure on the fault damage and its destructive nature, the mechanical model of floor fault damage is established, determining the tensile stress distribution area of the floor fault before and after the working face passes over the fault, with the stress analytical formula of the fault deduced. Before the workface crosses the fault, the shallow

part of the fault is subjected to compressive stress, with the deep part subjected to tensile stress for tensile damage, and expanding upwards along the fault. After the workface crosses the fault, due to the difference in deformation of the floor structure, the fault damage mutates to a tensile damage by compressive stress in the deep part and tensile stress in the shallow part, extending downwards along the fault.

(2) The displacement, stress and seepage evolution of the fault are studied by $RFPA^{2D}$ numerical simulation. Before crossing the fault, the displacement direction of the upper disc in contact with the fault unit is approximately perpendicular to the fault, while after crossing the fault, the displacement mainly occurs in the vertical direction. The negative values of stress before and after crossing the fault are horizontal and vertical stresses, which extend upward and downward along the fault, respectively. The deformation of the floor plate before and after crossing the fault affects the permeability characteristics of the surrounding rock around the fault, with the seepage volume at the fault decreasing slightly during the process of the working face approaching the fault, increasing rapidly with the destruction of the fault after crossing the fault.

(3) Through the spatial and temporal distribution characteristics of microseismic activity, the damage area of the surrounding rock of the basement fault is circled, with the ratio of S-wave and P-wave energies discriminating the nature of micro-damage of the surrounding rock in the fault. Micro-damage occurs 10–25 m below the coal seam before the workface crosses the fault, which is located in the transition zone between tension damage and compression damage, with E_S/E_P values of micro-seismic events located at 4–8. After crossing the fault, the maximum micro-damage depth of the fault is 35 m, of which a penetration is formed within 20 m, with $E_S/E_P < 3$ in micro-seismic event, resulting in tensile damage to surrounding rock.

4.8 Conclusion

This chapter applies high-precision microseismic monitoring technology to the mining process of 22517-working face in Dongjiahe Mine. Through continuous and dynamic monitoring of the microseismic events and energy density in the floor area, combined with structural mechanics and elastic mechanics models to study the deformation and damage mechanism, identify and define the potential hazardous areas of the floor rock body, provide a theoretical basis for microseismic monitoring, and analyse the micro rupture events and mechanical characteristics of the hidden faults under the floor in combination with the specific geological conditions, and finally verify the theoretical analysis and results of microseismic monitoring by numerical simulation software RFPA2D, which provides a reference for the realisation of safe coal mining on bearing water. The main conclusions are as follows.

(1) By analysing the characteristics of the stress during the initial pressure along the working face of mining on bearing pressure water, we construct a mechanical model of the floor under the joint action of mining stress, rock stress and water pressure, deduce the formula of the floor stress under the action of water pressure, solve the analytical solution of the floor stress field by using the elasticity theory, and give the contour plot of the floor stress component by MATLAB programming.
(2) Combined with the contour map, the cause of horizontal tensile stress below the floor during the initial incoming pressure is qualitatively analysed and explained, from which it is concluded in combination with the Coulomb criterion of stage tension that: during the initial incoming pressure, tensile damage occurrs in the rock body within 4 m below the strike-direction goaf, with shear damage circles formed on both sides from 5 to 11 m below the floor, resulting in damage to the floor plate, which is at a depth of 11 m.
(3) The damage situation of the floor plate in the strike direction of the 22517-working face of Dongjiahe Mine is simulated by RFPA finite element software, and the damage form of the floor is "inverted saddle shape" through the acoustic emission map, indicating that the diagonal lower part of the working face is the danger zone of water surge, with the damage depth of 10.2 m, followed by the damage depth of 11.22 m from the micro-vibration monitoring, which is roughly in line with the theoretical calculation.
(4) The damage area of surrounding rock near the fault in the floor during the advancement of the working face is circled by the spatial and temporal distribution characteristics of the microseismic activity. Before crossing the fault, the depth of micro-damage of the floor near the fault reaches about 25 m, with micro-rupture having the characteristics of segmental local damage; after crossing the fault, the micro-rupture gradually extends downward, with the depth of micro-damage reaching 35 m, which can easily form a water-conducting channel at the fault, resulting in a water inrush accident.
(5) Through the combination of numerical simulation and microseismic monitoring, the expansion process of the passage of water inrush in the floor is divided into two stages: before crossing the fault and after crossing the fault. Before crossing the fault, the shear stress at the fault gradually decreases with downward slip tendency, the micro-rupture extends upward along the fault, mainly occurring pressure-shear damage to the surrounding rock; after crossing the fault, the micro-rupture extends from top to bottom, the surrounding rock undergoes tensile damage, with the risk of water inrush.
(6) With the deformation mode of the floor structure and the mechanical properties of the fault structure surface, the fault-containing floor is simplified as a structure with two cantilever beams fixed, and the mechanical models before and after the fault are established, with the internal force obtained through the force method. Combined with its deformation characteristics, the damage of the fault and its stress distribution are concluded to have segmental characteristics. Before the workface crosses the fault, the upper part of the fault is under pressure with the

lower part under tension; after the workface crosses the fault, the upper part of the fault is under tension with the lower part under pressure.

(7) We study the spatial and temporal distribution characteristics of microseismic activity during the monitoring period, and circle the damage area of surrounding rock of the basement fault. The nature of the micro-damage in the surrounding rock of the fault is identified by the ratio of S-wave and P-wave energies, which verifies the fault segmentation damage characteristics and deformation damage mechanism during the structural evolution of the base plate.

In this chapter, the elastic mechanics and structural mechanics models are combined with the results of on-site microseismic monitoring and numerical simulation to reproduce the damage process. From the on-site microseismic data, the damage depth and the distribution characteristics of the stress field of the floor during the first incoming pressure in the strike direction of the working face of the Dongjiahe Mine 22517 are investigated respectively. Besides, the study also includes the stress, displacement, and seepage at the fault before and after the working face crosses the fault, which enriches the study on the mechanism of the generation of the water inrush in the floor, providing a certain theoretical basis for the realization of the safe coal mining on the bearing pressure water in the defective geological structure of the floor.

References

1. China Energy Strategy Research Group (2006) Thematic study on China's sustainable energy development strategy. Science Press, Beijing
2. He YG, Ye XD, Wang Z (2015) Consideration on the 13th five-year plan of coal industry. Coal Econ Res 35(1):6–8
3. Hu WY (2005) Theory and methods of mine water hazard prevention and control. China Coal Industry Publishing House, Beijing
4. Wand LG, Song Y (2001) Nonlinear character and forecasting of water-inrush from coal floor. China Coal Industry Publishing House, Beijing
5. Zhang JC, Liu JQ (1993) Analysis and research on the influencing factors of water-inrush from coal floor. J Min Strata Control Eng 4:37–41
6. Peng SP, Wang JN (2001) Safe coal mining on pressurised water. China Coal Industry Publishing House, Beijing
7. Zhao YS, Hu YQ (2004) Theory and technology of coal mining over pressurised water. China Coal Industry Publishing House, Beijing
8. Shi LQ, Jin H (2004) Mechanism and prediction and forecasting of water-inrush from floor. China University of Mining and Technology Press, Xuzhou
9. Reibiec MC (1992) Hydrofracturing of rock as a method of water, mudmand gas inrush hazards in underground coal mining. 4th IMWA, Belgrale, Yugoslavia
10. Booth CJ (1984) A numerical model of groundwater flow associated with an underground coal mine in the Appalachian plateau. Pennsylvania State University, Pennsylvania
11. Sresalev B (1983) Conditions for safe coal mining in water. The development and practice of foreign mine water control technology. Anshan Ferrous Metallurgical Mining Design Institute, Ministry of Metallurgical Industry, Anshan
12. Dorchaninoff (1984) Tectonic stress and stability of shaft engineering (trans: Zhao DY). China Coal Industry Publishing House, Beijing

13. Mironenko V, Strelsky F (1993) Hydrogeomechanical problems in mining. Mine Water Environ 12(1):35–40
14. Sammarco O (1986) Spontaneous inrushes of water in underground mines. Int J Mine Water 5(3):29–41
15. Sammarco O (1988) Inrush prevention in an underground mine. Int J Mine Water 7(4):43–52
16. Motyka J, Bosch AP (1985) Karstic phenomena in calcareous-dolomitic rocks and their influence over the inrushes of water in lead-zinc mines in Olkusz region (South of Poland). Int J Mine Water 4:1–12
17. Kuznetsov SV, Trofimov VA (1993) Hydrodynamic effect of coal scam compression. J Min Sci 12:35–40
18. Shen GH, Li BY, Wu G (1992) Theory and practice of special mining in mines. China Coal Industry Publishing House, Beijing
19. Yang TH, Liu HL, Zhu WC et al (2011) Modification and application of critical water outburst coefficient in coalmine based on the concept of effective stress. Chin J Rock Mech Eng 30(Supp. 2):4011–4018
20. Shi L, Q, Qiu M, Wei WX et al (2014) Water inrush evaluation of coal seam floor by integrating the water inrush coefficient and the information of water abundance. Int J Min Sci Technol 24(5):677–681
21. Liu Q, Sun YJ, Xu ZM (2011) Application of modified water inrush coefficient method to evaluation of water inrush from mine floor. Coal Sci Technol 39(8):107–109
22. Li BY, Shen GH, Jing ZG et al (1988) Theory and practice of preventing water inrush from floor of mining face. Safety Coal Mines 5:47–48
23. Shi LQ, Han J (2005) Theory and practice of dividing coal mining area floor into four-zone. J China Univ Min Technol 34(1):16–23
24. Zhang JC, Liu TQ (1990) On depth of fissured zone in seam floor resulted from coal extraction and its distribution characteristics. J China Coal Soc 15(2):112–114
25. Zhang JC, Shen B (2004) Coal mining under aquifers in China: a case study. Int J Rock Mech Min Sci 41(4):629–639
26. Zhang JC (2005) Investigations of water inrushes from aquifers under coal seams. Int J Rock Mech Min Sci 42(3):350–360
27. Xu JL, Qian MG (2004) Study and application of mining-induced fracture distribution in green mining. J China Univ Min Technol 32(2):141–144
28. Li LJ, Qian MG, Li SG (1996) Mechanism of water-inrush through fault. J China Coal Soc 21(2):119–123
29. Wu Q, Zhang Z, Zhang S et al (2007) A new practical methodology of the coal floor water bursting evaluating II: the vulnerable index method. J China Coal Soc 2007(11):1121–1126
30. Wu Q, Liu Y, Liu D et al (2011) Prediction of floor water inrush: the application of GIS-based AHP vulnerable index method to donghuantuo coal mine, China. Rock Mech Rock Eng 44(5):591–600
31. Song ZQ, Hao J, Tang JQ et al (2013) Study on water inrush from fault's prevention and control theory. J China Coal Soc 9:1511–1515
32. Guo WJ, Zhang SC, Sun WB et al (2018) Experimental and analysis research on water inrush catastrophe mode from coal seam floor in deep mining. J China Coal Soc 2018(1):219–227
33. Lu HF, Shen D, Yao DX et al (2014) Analytical solution of critical water inrush pressure of mining floor affected by fault. J Min Safety Eng 6:888–895
34. Zhou Q, Herrera J, Hidalgo A (2018) The numerical analysis of fault-induced mine water inrush using the extended finite element method and fracture mechanics. Mine Water Environ 37(1):185–195
35. Hua X, Zhang WQ, Jiao DZ (2011) Assessment method of water-inrush risk induced by fault activation and its application research. Proc Eng 26:441–448
36. Hu X, Wang L, Lu Y et al (2014) Analysis of insidious fault activation and water inrush from the mining floor. Int J Min Sci Technol 24(4):477–483
37. Li LC, Yang TH, Liang ZZ et al (2011) Numerical investigation of groundwater outbursts near faults in underground coal mines. Int J Coal Geol 85(3):276–288

38. Li LC, Tang CA, LI Gen et al (2009) Damage evolution and delayed groundwater inrush from micro faults in coal seam floor. Chin J Geotech Eng 31(12):1838–1844
39. Liu S, Liu W, Yin D (2017) Numerical simulation of the lagging water inrush process from insidious fault in coal seam floor. Geotech Geol Eng 35(3):1013–1021
40. Cheng JL, Sun XY, Zheng G et al (2013) Numerical simulations of water-inrush induced by fault activation during deep coal mining based on fluid-solid coupling interaction. Disaster Adv 6(11):10–14
41. Liu WT, Wu Q (2008) Numerical simulations of water inrush of fault f0 in FanGeZhuang coal mine. Chin J Rock Mech Eng 27(Supp. 2):3604–3610
42. Yang TH, Chen SK, Zhu WC et al (2008) Water inrush mechanism in mines and nonlinear flow model for fractured rocks. Chin J Rock Mech Eng 27(7):1411–1416
43. Lu YL, Wang LG (2015) Numerical modeling of mining-induced fracturing and flow evolution in coal seam floor based on micro-crack growth. J Min Safety Eng 32(6):889–897
44. Li ZH, Zhai CZ, Li LF (2015) Experimental study on water inrush mechanism due to floor faults activation in mining above confined aquifer. J Central South Univ 46(5):1806–1811
45. Zhang SC, Guo WJ, Li Y et al (2017) Experimental simulation of fault water inrush channel evolution in a coal mine floor. Mine Water Environ 36(3):443–451
46. Zhang SC, Guo WJ, Sun WB et al (2015) Experimental research on extended activation and water inrush of concealed structure in deep mining. Rock Soil Mech 36(11):3111–3120
47. Sun WB, Zhang SC (2015) Development of floor water invasion of mining influence simulation testing system and its application. Chin J Rock Mech Eng 34(Supp. 1):3274–3280
48. Xu YC, Chen XM, Li JB et al (2013) Experimental research on floor heave and water inrush in the broken rock roadway under great depth and high water pressure. J China Coal Soc 38(Supp. 1):124–128
49. Cheng JL, Yu SJ, Song Y et al (1999) Detection of the failure depth of coal seam floor by acoustic wave computer tomography. J China Coal Soc 24(6):576–580
50. Zhang PS, Wu JW, Liu SD (2006) Study on dynamic observation of coal seam floors failure law. Chin J Rock Mech Eng 25(Supp. 1):3009–3013
51. Liu ZX, Wang MM (2015) Study on encircling face electromagnetic method for monitoring coal face floor inrush. J China Coal Soc 40(5):1117–1125
52. Liu SD, Wang B, Zhou GQ et al (2009) Experimental research on mine floor water hazard early warning based on response of geoelectric field in groundwater seepage. Chin J Rock Mech Eng 28(2):267–272
53. Huang Z, Jiang ZQ, Tang X et al (2016) In situ measurement of hydraulic properties of the fractured zone of coal mines. Rock Mech Rock Eng 49(2):603–609
54. Yang TH, Liu J, Zhu WC et al (2007) A coupled flow stress-damage model for groundwater outbursts from an underlying aquifer into mining excavations. Int J Rock Mech Min Sci 44(1):87–97
55. Li BY (1999) "Down Three Zones" in the prediction of the water inrush from coal bed floor aquifer theory, development and application. J Shandong Inst Min Technol 4:11–18
56. Gao YF, Li BY (1992) Investigation of rules of floor rockfailure in the workings with risk of ordovician confined water. J China Coal Soc 2:32–39
57. Liu WT, Mu DR, Xie XX et al (2018) Sensitivity analysis of the main factors controlling floor failure depth and a risk evaluation of floor water inrush for an inclined coal seam. Mine Water Environ 37(3):636–648
58. Lu HF, Liang XY, Shan N et al (2018) Study on the stability of the coal seam floor above a confined aquifer using the structural system reliability method. Geofluids 1–15
59. Lu YL, Wang LG (2015) Numerical simulation of mining-induced fracture evolution and water flow in coal seam floor above a confined aquifer. Comput Geotech 67:157–171
60. Wang LG, Wu Y, Sun J (2009) Three-dimensional numerical simulation on deformation and failure of deep stope floor. Proc Earth Planet Sci 1(1):577–584
61. Zhou WL, Zhang PS, Wu RX et al (2019) Dynamic monitoring the deformation and failure of extra-thick coal seam floor in deep mining. J Appl Geophys 163:132–138

62. Li H, Bai H, Wu JJ et al (2019) A set of methods to predict water inrush from an ordovician karst aquifer: a case study from the Chengzhuang mine, China. Mine Water Environ 38(1):39–48
63. Xu JP, Sui WH, Gui H et al (2009) Utilizing angular displacement to monitor failure of coal seam floor. Proc Earth Planet Sci 1(1):943–948
64. Yin HY, Wei JC, Lefticariu L et al (2016) Numerical simulation of water flow from the coal seam floor in a deep longwall mine in China. Mine Water Environ 35(2):243–252
65. Timoshenko SP, Goodier JN (1970) Theory of elasticity. Mc GRAW-HILL Book Company, New York
66. Guo WJ, Liu YX (1989) The concept of water-inrush coefficient of the floor and its application. Hebei Coal 2:56–60
67. Tang CA, Tham LG, Lee PKK et al (2002) Coupled analysis of flow, stress and damage (FSD) in rock failure. Int J Rock Mech Min Sci 39(4):477–489
68. Li TJ, Li LC, Tang CA et al (2019) A coupled hydraulic-mechanical-damage geotechnical model for simulation of fracture propagation in geological media during hydraulic fracturing. J Petrol Sci Eng 173:1390–1416
69. Yan FZ, Lin BQ, Zhu CJ et al (2015) A novel ECBM extraction technology based on the integration of hydraulic slotting and hydraulic fracturing. J Nat Gas Sci Eng 22:571–579
70. Tang CA, Tham LG, Wang SH et al (2007) A numerical study of the influence of heterogeneity on the strength characterization of rock under uniaxial tension. Mech Mater 39(4):326–339
71. Zhuang DY, Ma K, Tang CA et al (2019) Study on crack formation and propagation in the galleries of the Dagangshan high arch dam in Southwest China based on microseismic monitoring and numerical simulation. Int J Rock Mech Min Sci 115:157–172
72. Mendecki AJ (1997) Seismic monitoring in mine. Champman and Hall Press, London
73. Tang CA, Li LC, Xu NW et al (2015) Microscismic monitoring and numerical simulation on the stability of high-steep rock slopes in hydropower engineering. J Rock Mech Geotech Eng 7(5):493–508
74. Ma K, Tang CA, Wang LX et al (2016) Stability analysis of underground oil storage cavers by an integrated numerical and microseismic monitoring approach. Tunn Undergr Space Technol 54:81–91
75. Dai F, Li B, Xu NW et al (2016) Deformation forecasting and stability analysis of large-scale underground powerhouse caverns from microseismic monitoring. Int J Rock Mech Min Sci 86:269–281
76. Aki K, Richards PG (2002) Quantitative seismology, 2nd edn. University Science Books, California
77. Gibowicz SJ, Young RP, Talebi S et al (1991) Source parameters of seismic events at the underground research laboratory in Manitoba, Canada, scaling relations for events with moment magnitude smaller than −2. Bull Seismol Soc Am 81(4):1157–1182
78. Hudyma M, Potvin YH (2010) An engineering approach to seismic risk management in hardrock mines. Rock Mech Rock Eng 43(6):891–906
79. Xu NW, Dai F, Liang ZZ et al (2014) The dynamic evaluation of rock slope stability considering the effects of microseismic damage. Rock Mech Rock Eng 47(2):621–642
80. Lockner D (1993) The role of acoustic emission in the study of rock fracture. Int J Rock Mech Min Sci Geomech Abstr 30(7):883–899
81. Wu Q, Li B, Liu SQ et al (2013) Vulnerability assessment of coal floor groundwater bursting based on zoning variable weight model: a case study in the typical mining region of Kailuan. J China Coal Soc 9:1516–1521
82. Wu Q, Li B (2016) Determination of variable weight interval and adjust weight parameters in the variable weight assessment model of water-inrush from coal floor. J China Coal Soc 9:2143–2149
83. Shi LQ, Tan XP, Wang JA et al (2015) Risk assessment of water inrush based on PCA_Fuzzy_PSO_SVC. J China Coal Soc 1:167–171
84. Hu W, Xu DJ (2013) Application of finite element strength reduction method to risk assessment of groundwater inrush from coal seam floor. J China Coal Soc 38(1):27–32

85. He JH, Li WP, Qiao W (2018) P-H-g evaluation system for risk assessment of water inrush in underground mining in North China coal field, based on rock-breaking theory and water-pressure transmission theory. Geomat Nat Haz Risk 524–543
86. Zhang PS, Wei Y, Zhang WQ et al (2016) Mechanism of water inrush due to damage of floor and fault activation induced by mining coal seam with fault defects under fluid-solid coupling mode. Chin J Geotech Eng 38(5):877–889
87. Gao YB, Liu SQ, Lv B et al (2016) Mechanism study of floor water inrush around mining field based on micro-crack extension. J Min Safety Eng (4):624–629
88. Liu ZG, Cao AY, Guo XS et al (2018) Deep-hole water injection technology of strong impact tendency coal seam; a cate study in Tangkou coal mine. Arab J Geosci 11(2):1–9
89. He J, Dou LM, Gong SY et al (2017) Rock burst assessment and prediction by dynamic and static stress analysis based on micro-seismic monitoring. Int J Rock Mech Min Sci 93:46–53
90. Xin L, Wang ZT, Wang G et al (2017) Technological aspects for underground coal gasification in steeply inclined thin coal seams at Zhongliangshan coal mine in China. Fuel 191:486–494
91. Wang GF, Dou LM, Li ZL et al (2014) Space breeding mechanism of rock burst and its microseismic characteristics. J Min Safety Eng 1:41–48
92. Tang LZ, Pan CL, Yang CX et al (2006) Establishment and application of microseismicity monitoring system in Dongguashan copper mine. Metal Mine 10:41–44
93. Dong LJ, Sun DY, Li XB et al (2016) A statistical method to identify blasts and microseismic events and its engineering application. Chin J Rock Mech Eng 7:1423–1433
94. Zhang WD, Ma TH, Tang CA et al (2014) Research on characteristics of rockburst and rules of microseismic monitoring at diversion tunnels in Jinping II hydropower station. Chin J Rock Mech Eng 33(2):339–348
95. Tang CA, Wang J, Zhang J (2010) Preliminary engineering application of microseismic monitoring technique to rockburst prediction in tunneling of Jinping II project. J Rock Mech Geotech Eng 2(3):193–208
96. Xu NW, Li T, Dai F et al (2016) Stability analysis on surrounding rock mass in underground caverns based on discrete element simulation and microseismic monitoring. Adv Eng Sci 5:1–8
97. Zhang BH, Deng JH, Gao MZ et al (2012) Safety evaluation research based on microseismic monitoring in underground powerhouse of hydropower station. Chin J Rock Mech Eng 5:937–944
98. Ma K, Tang CA, Li LC et al (2013) Reinforcement effects of anti-shear gallery of dagangshan right bank slope based on microseismic monitoring and numerical simulations. Chin J Rock Mech Eng 32(6):1239–1247
99. Ma K, Tang CA, Liang ZZ et al (2016) Stability analysis of the surrounding rock of underground water-sealed oil storage caverns based on microseismic monitoring during construction. Chin J Rock Mech Eng 35(7):1353–1365
100. Lu CP, Dou LM, Zhang N et al (2014) Microseismic and acoustic emission effect on gas outburst hazard triggered by shock wave: a case study. Nat Haz 73(3):1715–1731
101. Li N, Wang EY, Ge MC (2017) Microseismic Monitoring technique and its applications at coal mines: present status and future prospects. J China Coal Soc 42(Supplement 1):83–96
102. Huang Z, Jiang ZQ, Zhu SY et al (2014) Characterizing the hydraulic conductivity of rock formations between deep coal and aquifers using injection tests. Int J Rock Mech Min Sci 7112–7118
103. Wu RX, Liu SD, Zhang PS et al (2010) Detection of limestone water-conducting channels in coal mine by parallel 3d electric method of surface boreholes. Chin J Rock Mech Eng (Supplement 2):3585–3589
104. Qian ZW, Huang Z, Song JG (2018) A case study of water inrush incident through fault zone in China and the corresponding treatment measures. Arab J Geosci 11(14):1–7
105. Liu WT, Liu SL, Liao SH et al (2015) Study on water inrush passage from floor affected by fault. Coal Eng 12:85–88

106. Chen ZH, Hu ZP, Li H et al (2011) Fracture mechanical model and criteria of insidious fault water inrush in coal mines. J China Univ Min Technol 5:673–677
107. Gibowicz SJ, Kijko A (1994) An introduction to mining seismology. Academic Press, New York
108. Cai M, Kaiser PK, Martin CD (1998) A tensile model for the interpretation of microseismic events near underground openings. Pure Appl Geophys 153:67–92

Chapter 5
Microseismic Monitoring and Stability Analysis of Underground Powerhouse in Pumped Storage Power Station

Abstract This chapter explores microseismic monitoring and stability analysis of underground powerhouses in pumped storage power stations, with a focus on the Huanggou Pumped Storage Power Station as a case study. It begins by discussing the research background and significance, followed by a review of existing stability analysis methods, including conventional monitoring techniques and microseismic approaches. The chapter highlights key challenges that require further investigation. The design and implementation of a microseismic monitoring system at the Huanggou project are then detailed, covering aspects such as system construction, positioning accuracy, and waveform recognition. By analyzing microseismic activity patterns, the study assesses the stability of surrounding rock masses during excavation, correlating observed activity with geological and structural conditions. Integrating monitoring data with stability evaluations, this work offers a comprehensive understanding of rock mass behavior and provides practical solutions for addressing the complexities of underground powerhouse construction. The findings contribute to advancing the application of microseismic monitoring in large-scale engineering projects and enhancing safety during underground excavation.

5.1 Research Background and Significance

Hydropower is a mature, flexible, clean, low-carbon, and renewable energy source with significant economic, social, and ecological benefits. At the end of the nineteenth century, with the completion of the first underground hydropower station, the Fénayaz Power Station in Switzerland, the layout of underground powerhouses became widely applied in hydropower engineering [1]. Hydropower projects utilize underground powerhouses to convert water energy into electricity, addressing energy shortages and achieving green and sustainable socio-economic development [2, 3]. Recently, China's hydropower construction has experienced vigorous growth, gradually becoming the country with the world's largest hydropower capacity. Some large hydropower stations have been put into operation, while others are currently under construction. In the future, a considerable number of hydropower stations are

K. Ma and C. Tang, *Microseismic Monitoring and Stability Analysis of Large Rock Mass Engineering*, https://doi.org/10.1007/978-981-96-6483-2_5

also planned for construction [4–6]. During the 12th Five-Year Plan period, China focused on building a safe, stable, economical, and clean modern energy industry system. The newly hydroelectric installed capacity reached 103.48 million kW. By the end of 2015, China's total hydroelectric installed capacity had reached 319.54 million kW, with pumped storage capacity accounting for only 23.03 million kW. With the continuous progress of economic development and social civilization, there has been a rapid increase in energy consumption. The rapid development of new energy and the transformation and upgrading of the energy structure require an accelerated construction of pumped storage projects. The "13th Five-Year Plan" for hydropower development explicitly states, "Scientifically and orderly develop large hydropower projects, strictly control small and medium-sized hydropower projects, and accelerate the construction of pumped storage power stations." It is anticipated that there will be an additional 60 million kW of pumped storage capacity by the year 2025 [7]. Although China started the construction of pumped storage power stations relatively late, with nearly an 80-year gap from the world's first pumped storage power station, the Nant de Drance Pumped Storage Power Station in Switzerland in 1882, China has benefited from the experience accumulated in the construction of large-scale conventional hydropower stations and the advancements in industrial technology. As a result, China's pumped storage power station development began from a high starting point, and the Guangzhou Pumped Storage Power Station, with the largest single-station installed capacity in the world, has already been successfully completed. With the increased emphasis on hydropower development and advancements in related engineering technology, the construction of underground powerhouses in Chinese hydropower stations has exhibited a trend towards large single-cavern dimensions, large-scale cavern complexes, and complex layouts. The scale of 'large span, high sidewalls, and multiple caverns' in terms of span and sidewall height is unparalleled worldwide. The deformation magnitude and the scale of damage in the cavern complexes of underground powerhouses in Chinese hydropower stations are rare on a global scale [8–10]. The underground powerhouse, deeply embedded in the mountainous terrain, is subject to significant ground stress, and the hydrogeological conditions are complex. During the excavation process, issues such as rock bursts, large deformations, and extensive collapses in the surrounding rock mass pose a major challenge to the construction of hydropower projects. Various engineering problems frequently arise during the construction of underground powerhouses in major hydropower stations [11–13]. For example, the main plant of the Dagangshan Hydropower Station experienced a collapse of the diabase dyke in the surrounding rock, with the scale of the collapse exceeding 2000 cubic meters. This was the first time such a large-scale collapse accident occurred in the underground plant of a hydropower station in China [14, 15]. During the excavation process of the underground powerhouse at Jinping I Hydropower Station, the surrounding rock of the main powerhouse experienced significant deformations that far exceeded expected values, with obvious time-dependent characteristics, and cracking and damage occurred in multiple locations [16, 17]. Compared with conventional hydropower stations, underground powerhouses of pumped storage power stations have their uniqueness. The characteristics of large capacity, high head, and special operating conditions make the requirements for the stability of surrounding rocks and seepage prevention during the construction and operation periods of pumped storage power stations' underground powerhouses even more stringent [18]. During the excavation construction period of the underground powerhouse at the Xilongchi Pumped Storage Power Station, water seepage issues were exposed, especially in the top arch of the main powerhouse, where the phenomenon of water seepage was obvious in areas with developed faults and fissures [19]. The underground powerhouse drainage gallery of the Pushihe Pumped Storage Power Station exhibited significant rock seepage. Research found that under the action of high-water head pressure, the fault fracture zones and weak structural planes such as micro-fractures and joints within the rock mass formed seepage channels, leading to water seepage in the surrounding rock [20]. The geological conditions at hydropower station sites are complex. There are many factors contributing to the deformation and failure of surrounding rocks, and it is not sufficient to rely solely on existing engineering experiences and techniques. Therefore, during the phase of accelerated development of pumped storage power stations in China, conducting targeted research on the stability issues of surrounding rocks in underground powerhouses of pumped storage power stations, which meets the safety needs of engineering construction, has significant engineering value and practical significance.

Underground powerhouses are mostly built inside mountains with favorable lithological conditions, featuring a complex array of chambers with stability significantly influenced by geo-stress, geological structures, and groundwater. These conditions exhibit typical engineering characteristics of large underground caverns. Weak areas in the chamber structures and inherent defects within the rock mass are prone to damage due to stress concentration and construction disturbances, leading to the deterioration of the mechanical properties of the surrounding rock mass and a decrease in its load-bearing capacity. This, in turn, affects the overall or partial stability of the underground powerhouse. The presence of rock mass damage has become a focal point of research on the stability of surrounding rocks among scholars both domestically and internationally [21–23]. The damage to surrounding rocks caused by excavation disturbances not only poses a threat to the stability of underground powerhouses but also easily forms seepage channels in fault zones, joints, and other fracture zones, presenting hidden dangers to the construction and operation of underground powerhouses [24–26]. Pumped storage power stations, designed to perform peak-shaving and valley-filling functions, subject their underground powerhouses to high water head and high external water pressure over long periods. This environment increases the likelihood of secondary damage to the surrounding rocks (damage occurring after initial excavation-induced damage), making the stability analysis during both the construction and operation phases particularly important. The deformation and instability phenomena of surrounding rocks are often the result of the initiation, development, and expansion of microcracks within the rock mass until they coalesce, leading to failure. Based on the principle of energy dissipation, the evolution of microcracks, accompanied by the release of energy, can effectively reveal the internal dynamic damage evolution process within the rock mass [27, 28]. In recent years, microseismic monitoring technology has been able to effectively capture micro-fracture signals occurring within rock masses. It has been widely applied in the monitoring and early warning of large-scale geotechnical disasters in coal mines [29], underground oil storage in aquifers [30], tunnels [31], slopes [32], and other areas, achieving very significant result. Dai et al. [33] and Li et al. [34] applied microseismic monitoring technology to monitor the stability of the surrounding rocks during the excavation of the underground powerhouses at the Baihetan and Wudongde Hydropower Stations. Based on the results of microseismic monitoring, they conducted related research on the identification of surrounding rock damage and deformation stability analysis during the excavation process of the underground powerhouses. Utilizing microseismic monitoring, a technology for three-dimensional spatial monitoring of micro-fractures in rock masses, to identify and analyze the damage and stability of surrounding rock masses will become an important means for ensuring the safe construction of underground powerhouses.

This chapter, based on the understanding that the degradation of surrounding rock mass under excavation disturbances induces micro-fractures in rocks, leading to damage to the surrounding rocks, takes the underground powerhouse of the Huanggou Pumped Storage Power Station in Heilongjiang as the research subject. It aims to overcome the limitations of traditional point and plane two-dimensional monitoring information. By employing microseismic monitoring and numerical simulation as research methods, the chapter studies the damage areas and the evolution pattern of damage in the surrounding rocks of the underground powerhouse under excavation disturbances. Through the spatio-temporal evolution pattern of micro-fractures, it establishes the intrinsic connection between construction dynamics, fault structures, and microseismic activity, identifying and delineating the potential danger areas of the surrounding rocks of the underground powerhouse. The research results can provide insights for the analysis of the stability of surrounding rocks during the excavation process of underground powerhouses at pumped storage power stations, and also offer some references for subsequent construction and support work.

5.2 Research Status and Analysis

5.2.1 Analysis of the Stability of Surrounding Rock in Underground Excavation

Ling [35] utilized FLAC3D finite difference software to simulate the excavation of underground cavern groups using three different excavation methods. The study analyzed the redistribution of stress displacement and the distribution of plastic failure zones caused by the excavation, and verified the superiority of adopting a layered excavation support system. Gehrels [36] analysed the displacement of surrounding rock and the plastic zone of underground excavation activities by importing ANSYS models into FLAC3D. Qi et al. [37] used the three-dimensional elasto-plastic finite element method to evaluate the stability of surrounding rock in underground excavation with different support methods and under different excavation sequences. Based on the numerical results, they provided the feasibility of on-site construction plans and anchoring measures. Zhang et al. [38] employed the three-dimensional finite difference method to analyze the stability of the surrounding rock of the Shuibuya underground cavern group. Through comparisons of various schemes, they demonstrated the rationality of local soft rock replacement, optimization of support types and support parameters. Zhang et al. [39] calculated the three-dimensional geo-stress field of the Dagangshan Hydropower Station underground plant area using FLAC3D based on the inversion analysis method. The results indicate that the geo-stress field of the plant area is influenced by a combination of tectonic stress, geological structure, and topography. Hao et al. [40] used the discrete element software UDEC to study the impact of faults in jointed rock masses on the stability of surrounding rock in underground caverns. They performed back-analysis calculations on fault parameters such as fault dip angle and shear strength, and successfully applied these findings to the analysis of the stability of surrounding rock in underground workshops. Wu et al. [41], taking the underground workshop of the Qingjiang Shuibuya Hydropower Station as an example, applied the discontinuous deformation analysis method to study the deformation and failure characteristics of the surrounding rock of the underground caverns from the geo-stress level, anchoring, rock mass structure and strength. With the deepening of rock mechanics issues and the rapid advancement of computer technology, numerical simulation methods have become an effective tool for evaluating the stability of surrounding rock in underground caverns. As summarized above, numerous scholars have studied the stability of surrounding rock during the excavation process of underground caverns from various perspectives such as construction, support measures, geological conditions, and stress states. This research provides a foundation and engineering experience for the analysis and early warning of the stability of the surrounding rock in caverns during the excavation process.

5.2.2 *Conventional Monitoring of the Stability of Underground Rock Mass*

The construction phase is a critical period that affects the stability of the surrounding rock mass in underground caverns. In China, the scale of underground construction is rapidly evolving towards larger spans, higher sidewalls, and multiple caverns. Excavation and construction of large underground caverns involve a dynamic process, with the deformation and stress information of the surrounding rock mass continuously changing as excavation progresses. Therefore, real-time data collection from on-site monitoring is particularly crucial for the assessment of rock mass stability [42]. Traditional on-site safety monitoring methods involve the installation of monitoring equipment, such as multipoint displacement meters, anchor stress meters, and anchor cable load cells, in areas with weak structural surfaces or at critical sections of engineering structures. By analyzing the monitoring data from these points, the stability status of the surrounding rock mass can be assessed [43–46]. Zhao et al. [47], based on deformation monitoring data of the underground rock mass at the Tai'an Power Station, introduced the theory of abrupt change and established a model for abrupt changes in the deformation rate of the surrounding rock mass. They conducted a study on the stability analysis of the underground factory rock mas. Wei [48] based on monitoring data, geophysical exploration, construction records, and geological information from the Jinping I Hydropower Station, conducted an analysis of the deformation and failure characteristics of underground rock mass during construction under high in-situ stress conditions. It is concluded that the time-dependent nature of rock mass deformation should be considered as a gradual process of rock mass failure expansion. Maejima et al. [49], based on the monitoring data during the construction process, provided recommendations for optimizing subsequent construction activities. They utilized conventional monitoring data such as stress, displacement, and strain data from the construction of a pumped storage power station to offer construction design and support. Tang et al. [50] initially apply total stations and sliding micrometers in the monitoring of surrounding rock during the construction phase of large-scale hydroelectric power plants in China. They provided guidance for the construction of underground powerhouses based on deformation monitoring data from the rock surface and deep-seated rock mass. Wang et al. [51] comprehensively evaluate stability of the surrounding rock in the underground powerhouse of the Xilongchi Power Station, based on displacement observation data from the surrounding rock and numerical analysis. Gan et al. [52] utilized monitoring data from both the construction and operational phases and studied on the deformation of the surrounding rock, the stress in the support anchors, the anchoring forces of anchor cables, and the stress in the grid beam reinforcement in the underground powerhouse of the Bailianhe Pumped Storage Power Station. They examined these change processes and the mutual influence patterns. Subsequently, a comprehensive assessment of the stability of the underground powerhouse's surrounding rock was carried out. Shen [53] investigated the monitoring and measurement scheme for the underground powerhouse of the Shiziping Hydroelectric Power Station, as well as the implementation of related work. They provided examples of monitoring data from typical cross-sections. A preliminary analysis and assessment of rock mass stability were conducted, and deformation patterns of the underground chamber roof were summarized. Most of the methods mentioned above are based on traditional conventional monitoring techniques such as stress, displacement, combined with geological, geophysical, and numerical methods for analyzing and assessing the stability of underground rock masses. Conventional monitoring methods can only reflect the deformation characteristics of specific locations or sections, and often require significant deformation or even macro-instability of the rock mass to be detected. They have significant limitations in providing real-time feedback or early warning during construction activities on-site. Conventional monitoring methods can only obtaine the deformation characteristics of specific locations or sections, and often require significant deformation or even macro-instability of the rock mass to be detected.

They have significant limitations in providing real-time response or early warning during construction activities on-site.

5.2.3 Microseismic Monitoring for the Stability of Underground Rock Mass

The deformation and instability problems in underground caverns are often responses of the initiation, development, and propagation of microcracks within the rock mass. The detaction of microcracks can effectively reveal the dynamic process of internal damage evolution in the rock mass. Recently, microseismic monitoring technology, as a novel three-dimensional spatial monitoring method, has proven to be effective in capturing signals of microcracking occurring within deep rock masses. It has been widely applied in rock engineering for monitoring and early warning of rock dynamic hazards, including coal mines [29], underground water-sealed oil depots [30], tunnels [54], dams [55], and other rock-related projects. In hydroelectric engineering, microseismic monitoring technology is increasingly being employed to the monitoring of underground caverns excavation processes. Zhang et al. [56] established an ISS (Integrated Seismic Source) microseismic monitoring system in the collapse zone of the underground powerhouse at the Dagangshan Hydropower Station. They investigated methods for overall stability assessment and safety evaluation prediction of the underground powerhouse based on microseismic source parameters. Xu et al. [57], based on microseismic monitoring, combined numerical simulation and conventional monitoring to analyze the relationship between microseismic activity and rock mass damage during the excavation process of the Monkey Rock Hydropower Station's underground powerhouse. They validated the effectiveness of microseismic activity. Dai et al. [58] established a microseismic monitoring system that covered the main powerhouse of the Baihetan Hydropower Station. They investigated the relationship between microseismic activity and construction dynamics, geological structures, and rock mass deformation. A deformation prediction method was proposed using apparent stress and cumulative apparent volume as indicators for rock mass deformation warnings.

5.3 Outstanding Problems

(1) Pumped storage hydropower plants have characteristics such as high head difference, significant submersion depth of the turbines, and low installation elevations. They also require high impermeability standards. During the excavation process, the surrounding rock of the underground pumped storage hydropower plants often incurs varying degrees of damage before macroscopic deformation and instability occur. Effectively monitoring the damage to the surrounding is

a crucial issue that controls the success or failure of the entire construction and operation. Traditional monitoring methods for underground powerhouses mainly focus on monitoring rock mass deformation and stress values. It is challenging to accurately and continuously capture the dynamic evolution of rock mass damage during construction. Furthermore, it is difficult to obtain the underlying nature of microcrack initiation, propagation, and coalescence in the surrounding rock, which eventually leads to macroscopic instability.

(2) Microseismic monitoring not only represent the extent and morphology of damaged areas in the surrounding rock from the perspective of microcracking, but also allow for the analysis of the microcracking mechanisms within the damaged regions using seismic source parameters. Excavation and unloading can lead to stress rebound and stress redistribution during construction, which are the primary causes of deformation and instability in the surrounding rock. However, microseismic monitoring cannot provide information regarding the changes in the stress field of the surrounding rock. Combining the microseismic monitoring in field, numerical simulation on the evolution of rock mass damage under the influence of various factors is of significant importance for a deeper understanding of the mechanical mechanisms within the damaged surrounding rock.

5.4 Research Contents

This chapter summarizes the extensive research conducted on the stability of the surrounding rock in underground plants. Based on the advantages of microseismic monitoring systems in the stability monitoring and early warning of large geotechnical projects, on September 30, 2017, the first microseismic monitoring system in China applied to an underground plant of a pumped-storage power station was attempted to be constructed at the Huanggou Pumped Storage Power Station in Heilongjiang. Using a combination of microseismic monitoring and numerical simulation, the gradual process of initiation, development, and expansion of microcracks in the surrounding rock of underground plants under excavation disturbance was reproduced. This revealed the laws and mechanisms of surrounding rock damage evolution, identified potential risk areas of the surrounding rock, and provided references and guidance for subsequent construction and support. Numerical simulations were further used to explore the influence of different lateral pressure coefficients and steeply dipping fault angles at different positions on the damage evolution of the surrounding rock in underground plants. The specific research content of this chapter is as follows.

(1) An optimized design for a three-dimensional spatial array network of sensors suitable for the Huanggou underground plant was developed to construct a microseismic monitoring system. Through targeted artificial tapping experiments, an equivalent P-wave velocity model was determined to control the accuracy of location errors. Methods such as threshold setting, spectral analysis,

and manual identification were used to eliminate noise signals, establishing a typical waveform library for microseismic monitoring in Huanggou.

(2) By analyzing the temporal and spatial evolution of microseismic events, the intrinsic relationship between construction dynamics, geological structures, and microseismic characteristics was established. Based on seismic source parameters such as microseismic energy, density, and the ratio of shear wave to compressional wave energy, the gradual evolution process of microcracking in the surrounding rock during excavation and its fracture mechanisms were revealed. This analysis also identified and delineated potential danger areas in the surrounding rock of the underground plant.

(3) The RFPA (Rock Failure Process Analysis) finite element method, which has the capability to simulate the real fracture process of rock, was utilized to study the damage evolution of the surrounding rock in underground plants. A simulation was conducted on the $0 + 60$ m cross-section of the underground plant, and the numerical results were compared and verified against microseismic monitoring information to reveal the laws and mechanisms of damage evolution in the surrounding rock. Based on this, the study simulated and analyzed the damage evolution of the surrounding rock under the influence of different lateral pressure coefficients and steeply dipping fault locations.

5.5 The Microseismic Monitoring System for Huanggou Pumped Storage Power Station

5.5.1 Overview of the Project and Construction

The Huanggou Pumped Storage Power Station is located within the territory of Sandaohezi Town, Mudanjiang City, Heilongjiang Province. It is constructed in a depression on the right bank of Sandaohezi, featuring an upper reservoir with a total capacity of 1.161×10^7 m^3. The built Lianhua Reservoir is utilized as the lower reservoir. With a maximum water height of 445.6 m and a total installed capacity of 1200 MW, the key structures of the project mainly include the main dam of the upper reservoir, the auxiliary dam at the tail of the reservoir, water conveyance tunnels, upstream and downstream surge shafts, and a central underground powerhouse, among others. The underground powerhouse system is located within the mountain above the middle part of the water conveyance tunnel, at a depth of about 310 m (Fig. 5.1).

The underground powerhouse system comprises three main caverns arranged in parallel from upstream to downstream: the main powerhouse, the main transformer cavern, and the tailrace cavern, with the axis of the caverns oriented at NW 311°. The dimensions of the main powerhouse are 143.00 m in length, 26.50 m in width, and 55.30 m in height. The auxiliary powerhouse is 19.50 m × 25.00 m × 45.60 m, with a total excavation length of 163.20 m and the crown elevation at 178.60 m. The main transformer cavern has excavation dimensions of 127.10 m × 21.20 m

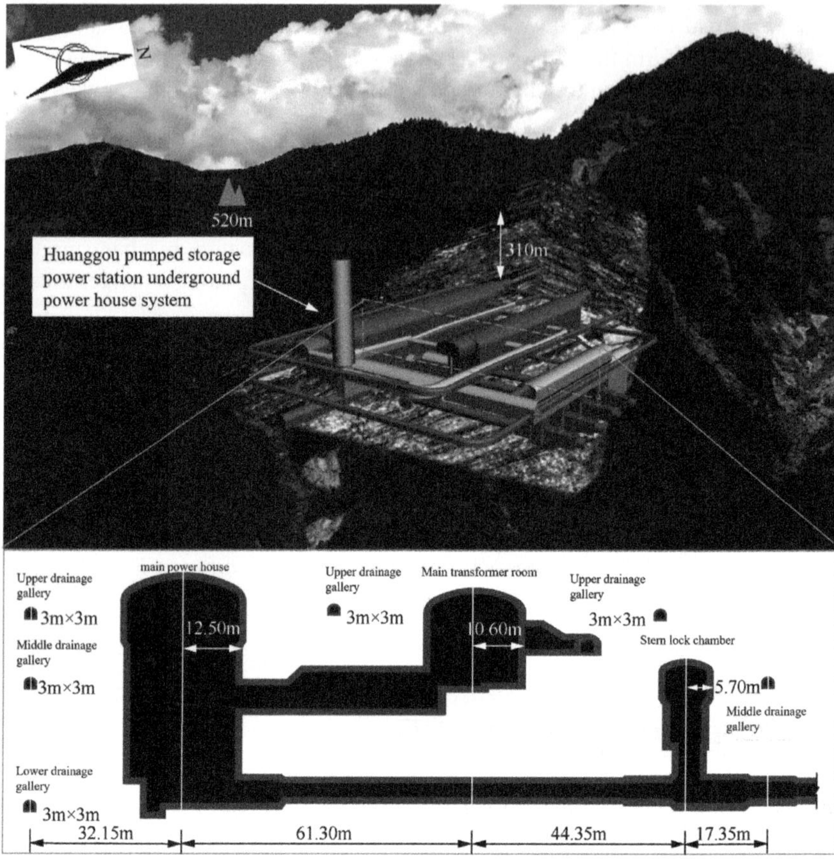

Fig. 5.1 Underground power house system and typical section diagram of Heilongjiang Huanggou Pumped Storage Power Station (along the axis of #1 water transfer tunnel)

× 28.10 m (length × width × height), with the crown elevation at 175.20 m. The tailrace cavern is excavated to dimensions of 127.10 m × 21.20 m × 28.10 m (length × width × height), with the crown elevation at 159.20 m. The main powerhouse is connected to the main transformer cavern by four bus ducts that are perpendicular to the longitudinal axis of the powerhouse. The rock mass thickness between these caverns is 37.45 m, and the thickness of the rock mass between the tailrace cavern and the main transformer cavern is 28.05 m. A three-level drainage gallery surrounds the three major caverns, forming a complex underground cavern group characterized by "high sidewalls, large spans, and multiple caverns" (Fig. 5.2).

1. In-site Geomorphology and Geology

The engineering site is located in the northeastern part of the Zhangguangcai Range, characterized by a terrain of medium to low mountainous and river valley areas within a small structural basin formed by tectonic erosion. The elevation of the

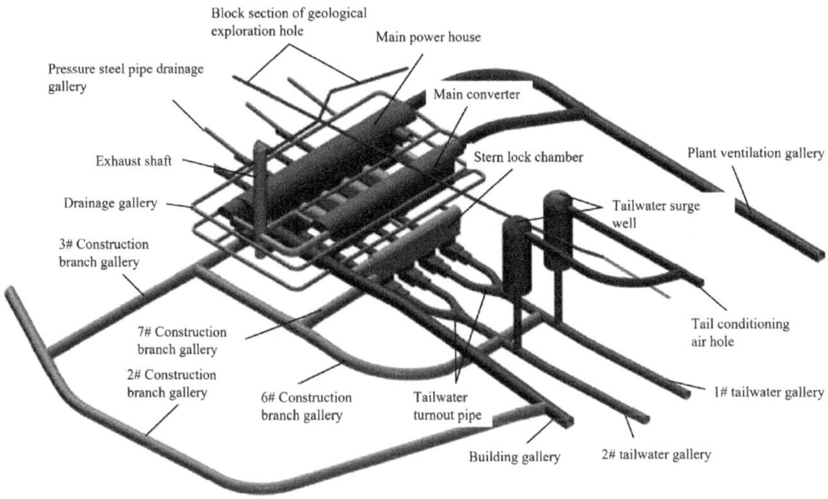

Fig. 5.2 Layout of underground powerhouse caverns for Heilongjiang Huanggou Pumped Storage Power Station

mountains generally ranges from 500 to 1184 m, with a relative height difference of about 300–500 m. The slopes of the mountains typically vary from 20° to 45°. Small structural basins are distributed between the valleys, and steep cliffs are commonly found along the rivers. The Mudanjiang River and its tributaries often have developed first and second level terraces; the first level terraces rise 2–8 m above the river water level, while the second level terraces are 8–28 m above the river water level. The local area features small-scale landslide deposits. The geological strata exposed in the engineering area primarily consist of metamorphic rocks from the Neoproterozoic era. There are scattered distributions of Mesozoic Jurassic volcanic rocks, along with extensive intrusions of Proterozoic mixed granite, late Hercynian Bai Gang granite, and Yanshanian period granite. The surface is covered with Quaternary loose accumulation layers, with the stratigraphic sequence from old to new being Neoproterozoic, Paleozoic, Mesozoic, and Cenozoic. The intrusive rocks are divided into three phases: Proterozoic intrusive rocks, late Hercynian intrusive rocks (which include granitic porphyry veins), and early Yanshanian intrusive rocks. Geotectonically, the engineering area is located in the uplift zone of the Zhangguangcai Ridge within the Jilin-Heilongjiang fold system of the Tianshan Xing'an geosyncline fold area. The regional structure is dominated by fault structures, with the earlier developments being nearly north–south oriented structures. Along with regional tectonic movements, the lower Neoproterozoic Mashan Group experienced intense metamorphism, leading to the formation of widespread migmatites and extensive granite magma intrusion. During the middle of the early Paleozoic, the crust was in a period of instability, leading to the formation of volcanic rock constructions and shallow marine sedimentation, followed by north–south folding and regional metamorphism. In the late Hercynian period, the engineering area underwent very strong tectonic movements, older structures were further activated, and there was extensive granite magma intrusion, with the north–south structures developed further since the end of the Proterozoic. During the Yanshanian orogeny, NNE-oriented faults were formed, along with nearly east–west, NE, and NW-oriented faults. The north–south structures showed obvious inherited activities, with local granite magma intrusion accompanied by extensive volcanic eruptions, forming the Middle Jurassic volcanic rock constructions. The engineering site is located in the eastern part of the Songliao Basin and its surrounding areas, where seismic activity is relatively weak. From January 1, 1918, to December 31, 2004, a total of 38 earthquakes with a moment magnitude (Ms) of 4.7 or higher were recorded in this area. Among these, 22 earthquakes were in the range of 5.0–5.9, and 3 earthquakes were of magnitude 6.0 or higher.

2. Engineering geological conditions

The power station's water conveyance and power generation system consist of #1 and #2 water conveyance tunnels, upstream and downstream surge shafts, pressure pipelines, and the power generation system. The underground powerhouse, comprising the main powerhouse, main transformer carvern and tailrace cavern, forms the core of the power generation system. The plant area is located in a low mountain terrain with shallow cuts, with mountain elevations ranging from 215 to 680 m and a relative height difference of 100 to 150 m. The terrain slope is between 20° to 40°, with some areas having steeper slopes of 50°–60°. The water conveyance tunnels are generally buried at depths of 60–430 m, with no shallow or near-mountain tunnel sections. The bedrock in the water conveyance and power generation area consists of late Hercynian Bai Gang granite, interspersed with a small amount of granitic porphyry veins. The width of these veins is generally less than 1 m, with the widest being about 20 m. Bai Gang granite is known for its hardness and relatively intact rock mass. The saturated compressive strength of fresh rock reaches 120 MPa. According to ultrasonic test, the longitudinal wave velocity of weakly weathered Bai Gang granite is between 3 to 4 km/s, for slightly weathered rock it is 4.5–5 km/s, and for fresh rock, it is 5–5.3 km/s. In the water conveyance and power generation area, 80 faults have been discovered (34 of which were exposed through exploration tunnels, most do not extend to the surface). These faults are primarily oriented nearly north–south and east–west, mostly being thrust faults, with a few being strike-slip or normal faults. The dip angles range from 55° to 85°, with widths generally between 0.5 to 1.5 m, though a few range from 2 to 5.6 m, and are mostly composed of breccia and rock debris mixed with mud. Additionally, from borehole drilling and exploration tunnel revelations, more than 20 low-angle faults have been discovered, with widths generally ranging from 0.1 to 0.5 m. A few of these faults are mainly composed of brecciated rock and rock debris mixed with mud. The joints primarily consist of two groups with steep dip angles, nearly north–south and east–west, and extend over long distances. The low-angle east–west joints are widely distributed and sometimes appear densely, but they usually do not extend far, showing discontinuous distribution. The spacing between joints is generally 1–2 m, with dense sections ranging from 5 to 25 cm, and extension lengths mostly between 4 to 7 m, with a few extending over 10 m. In exploration tunnels located in the area of the underground powerhouse caverns, stress relief method measurements of geostress were conducted at four points. The measured maximum principal stress direction is generally east–west, with stress magnitudes mostly between 5 to 10 MPa, which is consistent with the stress field characterized by the geological structure of the region. The direction of the maximum horizontal principal stress is approximately N71°W, with a magnitude of 12.2 MPa. The intermediate principal stress and the minimum principal stress are close in magnitude, at 6.5 MPa and 5.7 MPa, respectively, indicating a moderate level of geostress. The measurement results are presented in Table 5.1 [59]. The caverns of the powerhouse system are situated within the mountain at the middle to lower part of the water conveyance tunnel, with the maximum burial depth reaching 310 m. The surrounding rock is predominantly fresh Bai Gang granite, with the RQD (Rock Quality Designation) values reaching 90–100%. The rock quality is hard and intact, with strong resistance to weathering, and no adverse alteration phenomena have been found, classifying it as Class II surrounding rock. According to rock mass deformation tests in the underground powerhouse, the deformation modulus of fresh Bai Gang granite rock mass ranges from 23.7 to 30.1 GPa. Due to the deep burial of the underground powerhouse and the freshness and integrity of the rock mass, the measured maximum principal stress values range from 12.2 to 13.38 MPa. There might be a slight possibility of rockburst during excavation, which requires preventive measures.

3. Faults and joints

The geological exploration results for faults in the underground powerhouse are presented in Table 5.2. In the main powerhouse, two crack-like minor faults, f_{33} and f_{34}, were discovered, with widths of only 3–5 mm and short extensions. The main transformer carvern is intersected by three steeply dipping faults, f_{31}, f_{32}, and f_{33}. Among these, fault f_{31} has an EW direction, southward inclination, and a dip angle of 80°, with a width of 0.3–1.1 m; fault f_{32} has a direction of N15°W, northeastward inclination, and a dip angle of 70°, with a width of 0.05 m; fault f_{33} has a direction

Table 5.1 Results of ground stress measurement in underground power plant area

Monitoring point	Maximum principal stress			Intermediate principal stress			Minimum principal stress		
	Value/MPa	Direction	Dip angle/(°)	Value/MPa	Direction	Dip angle/(°)	Value/MPa	Direction	Dip angle/(°)
D_1	14.1	N60°W	− 7	8.4	N29°E	11	7.9	N63°E	− 77
D_2	14.7	N47°E	3	7.0	N32°W	− 74	5.1	N44°W	16
D_3	11.1	N70°E	− 39	8.7	N51°W	− 32	6.3	N3°E	34
D_4	15.7	N65°W	0	4.8	N26°E	− 67	1.8	N25°E	23
D_4'	13.7	N83°W	4	5.8	N6°E	− 11	4.7	N29°E	79
Comprehensive plant area	12.2	N71°W	3	6.5	N13°E	− 67	5.7	N20°E	23

of N21° to 30°W, northeastward inclination, and a dip angle of 80°, with a width of 0.005 m. Fault f_{33} is mainly composed of breccia and rock debris mixed with mud. The tailrace carvern is intersected by two steeply dipping faults, f_{31} and f_{32}. Notably, faults f_{32} and f_{33} exhibit distinct horizontal slickensides, indicating that the stress field in this area is significantly influenced by tectonic activity. The faults exposed within the ventilation and access tunnels are mostly steeply dipping and intersect the tunnel axis at a large angle, with most of these faults being of a smaller scale. In the rock mass, joints are not very well developed, but four main sets of joints can be identified: (1) Joints oriented N75°–85°W, dipping NE or SW with a dip angle of 60°–80°; (2) Joints oriented N45° to 50°E, primarily dipping NW and a few dipping SE, with a dip angle of 60°–70°. (3) Joints oriented N75° to 85°E, dipping NW with a dip angle of 50°–60°; (4) Joints oriented N75° to 85°E, dipping NW with a dip angle of 10°–20°. The spacing between joints generally ranges from 0.5 to 2.5 m, and they are mostly in a closed state. The first and second sets of joints are the most developed, while the third and fourth sets are less so. The first set of joints intersects the axis of the powerhouse at an angle of 26°–36°, which is relatively small, while the other three sets intersect at larger angles. Low-angle joints are only sporadically distributed and do not extend far. The main sidewall structure of the main powerhouse and the joint rose diagram of the underground powerhouse can be found in Fig. 5.3 [60].

4. Hydrologic condition

Groundwater in the area is primarily categorized into two types: bedrock fracture water and pore water in loose accumulation layers. Pour water is mainly found in river valley alluvial and colluvial layers, as well as in slope and landslide accumulation layers, with varying aquifer thicknesses. It is recharged by atmospheric precipitation and discharges into rivers. Bedrock fracture water is primarily contained within the fractures of bedrock, and its permeability depends on the degree of fracture development and the nature of the bedrock. In the engineering area, most of the mountains are composed of igneous rocks. Although fractures are relatively developed, they tend to close with increasing depth, generally resulting in weak permeability. According to the data from drilling and water pressure tests, the surrounding rock in the powerhouse area mostly belongs to weak or slightly fractured rock bodies. The groundwater level in the plant area is about 290 m above the crown of the tunnel. To reduce the adverse effects of external water pressure, it is necessary to implement drainage and pressure relief measures in the design. Additionally, proper seepage and drainage designs should be prepared for both the construction and operational phases, based on actual conditions.

5. Construction overview

The excavation of the various galleries and caverns in the underground powerhouse system primarily utilizes drill and blast construction methods. Full-face construction is mainly used for the drainage galleries, while the underground powerhouse, due to its large cross-section, is excavated in stages. The central parts are excavated using bench blasting, and the periphery employs smooth blasting or pre-split blasting

Table 5.2 Geological exploration results of fault in the factory area

No.	Occurrence			Width/m	Property	Characteristics of tectonic rocks
	Direction	Strike	Dip angle/(°)			
f_{31}	EW	S	80	0.3–1.1	Reverse fault	The composition is debris mixed with mud, affecting a width of 3.5 m, with dense joints and severe rust
f_{32}	N15°W	NE	70	0.05	Strike slip fault	The composition is debris mixed with mud, with an impact zone width of 4 m. The rock is in a block shape of 15 cm, with obvious horizontal scratches and steps
f_{33}	N21°–30°W	NE	80	0.005	Strike slip fault	The composition is debris mixed with mud, with severe rust on the fracture surface and horizontal scratches
f_{34}	N55°–67°W	SW	70	0.003	Strike slip fault	The composition is debris interbedded with mud, and the fault plane is relatively straight

techniques. The main powerhouse is excavated in seven levels, with the excavation heights from top to bottom being sequentially 10, 8, 8.1, 6.6, 7.4, 7.5, and 6.2 m. By September 30, 2017, when the microseismic monitoring system was completed, the construction progress of the underground powerhouse is as indicated in Fig. 5.4. The main transformer cavern and the tailrace cavern had been excavated to near the designed elevation. The middle and upper-level drainage galleries, along with the access and ventilation tunnels, were essentially completed. The busbar tunnel excavation had progressed to near the downstream sidewall of the main powerhouse. During the microseismic monitoring period, the main excavation work for the primary powerhouse was completed on Level III, excavating from an elevation of 154 m down to 152 m. The excavation proceeded from the installation space to the main machine area and then to the main and auxiliary powerhouse areas. The excavation width was 19 m, with a 3 m wide protective layer reserved on both sides to minimize the impact of construction on the powerhouse sidewalls and the concrete crane beams on the rock walls [61]. At the same time, excavation was also underway for the upper and lower drainage galleries of the pressure steel pipe. Site geological forecasting data

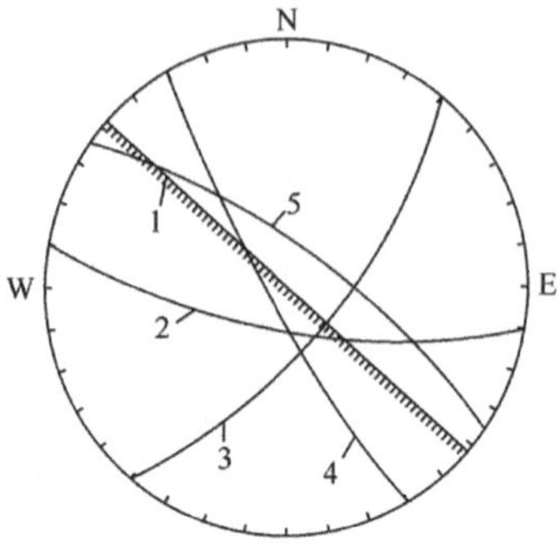

Serial number	Dip angle (°)	Tendency (°)	Type
1	90	41	Main powerhouse side wall
2	70	190	Joints
3	65	130	Joints
4	80	240	Fault f_{33}
5	70	35	Fault f_{34}

(a) The stereogram of main powerhouse side wall structural plane

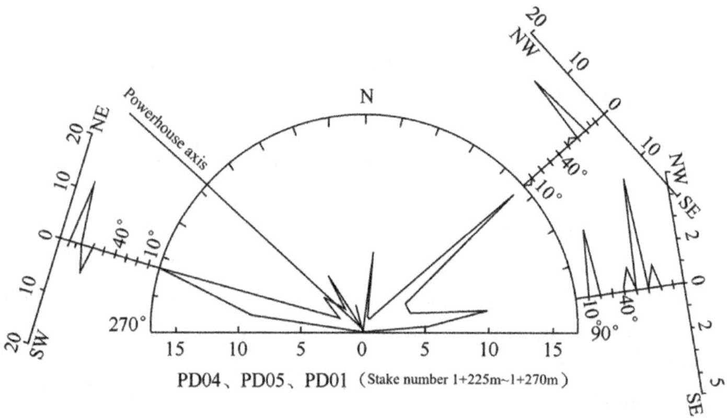

(b) The rose diagram of underground powerhouse joints

Fig. 5.3 Occurrence of faults and joints in underground power plants

indicated that during the excavation process of the underground powerhouse, several instances of local rock mass instability, such as slabbing and falling blocks, were observed, and there were even occurrences of slight rockbursts. These incidents not only posed threats to construction safety but also severely restricted the progress of the construction. To rock stability during the excavation and unloading process of the underground powerhouse, a microseismic monitoring system was successfully established in the underground powerhouse of the Heilongjiang Huanggou Pumped Storage Power Station on September 30, 2017. This system enables the real-time monitoring of micro-fractures in the surrounding rock of the underground powerhouse during the excavation and unloading process. During the on-site installation of the microseismic monitoring system, it was observed that in some areas of the already excavated middle and lower drainage galleries, there were serious occurrences of slabbing and falling blocks, and even slight rockbursts, characterized by obvious flaky spalling (Fig. 5.4).

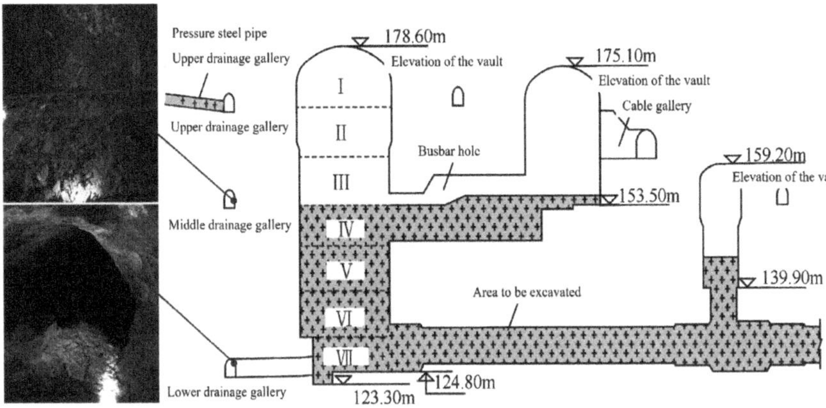

Fig. 5.4 Schematic diagram of underground power plant construction progress

5.5.2 The Construction of Microseismic System

Considering the engineering characteristics of the underground powerhouse, such as high sidewalls, large spans, multiple and complexly interconnected caverns, and numerous workfaces, and based on on-site inspections and subsequent excavation progress, in combination with the requirements of the Canadian ESG microseismic monitoring system, a microseismic monitoring system for the underground powerhouse was arranged as shown in Fig. 5.5. The microseismic monitoring system for the underground powerhouse primarily consists of three components: acceleration sensors (S01–S18), the Paladin signal acquisition system, and the Hyperion data processing system. Sensors placed within the surrounding rock of the underground powerhouse caverns capture the elastic wave signals released by micro-fractures in the rock. These signals are then transmitted via cable to the Paladin signal acquisition system, where the electrical signals are converted into digital signals and stored in the Hyperion data processing system for user processing and analysis. Through the network, real-time information exchange between the analysis center and decision-making departments is facilitated. The G1030 geophone sensor is made of stainless steel and serves as a single-component signal reception probe, with a diameter of 32 mm and a length of 146 mm. It has a sensitivity of 30 V/(m s) and a frequency response range of 50–5000 Hz, which makes it well-suited to adapt to the medium geostress characteristics of the Huanggou Pumped Storage Power Station. The Paladin signal acquisition system operates with a sampling frequency of 20,000 Hz and a 24-bit A/D conversion, utilizing threshold triggering. The microseismic monitoring system continuously monitors micro-fracture events within the surrounding rock of the underground powerhouse due to excavation unloading effects 24 h a day. By employing filtering, time–frequency analysis, and manual identification methods, noise events are filtered out. The collected microseismic signals

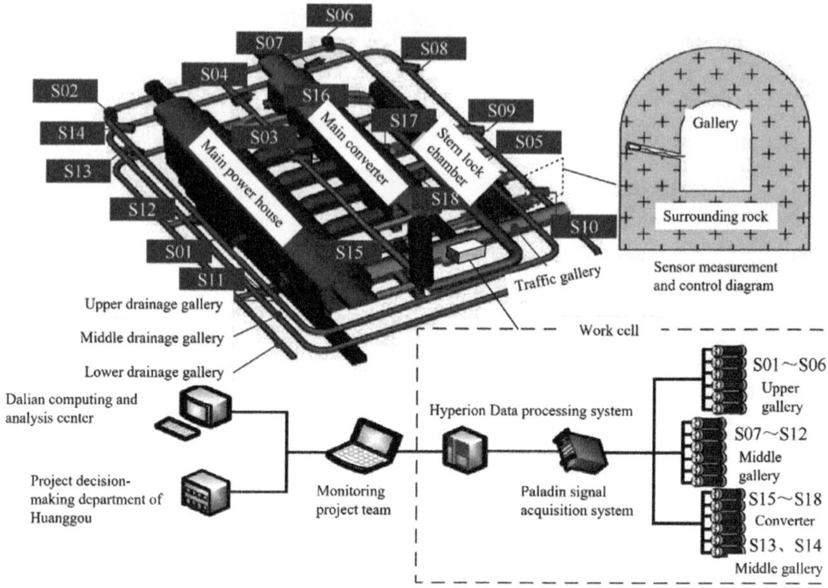

Fig. 5.5 Network topology diagram of microseismic monitoring system

are then classified, and those signals resulting from rock micro-fractures are effectively identified and processed. This involves analyzing their waveforms and time–frequency diagrams to invert and obtain detailed information on the micro-fractures' spatiotemporal distribution, magnitude, energy, and other seismic source parameters.

In line with the on-site construction progress and the requirements of the microseismic monitoring system, 18 acceleration sensors are installed on the sidewalls at different elevations of the upper and middle level drainage galleries and the main transformer carvern, which have been essentially excavated. A high-density sensor spatial array covers the three major caverns of the underground powerhouse. 6 sensors are installed in the upper-level drainage gallery, 8 sensors in the middle-level drainage gallery, and the remaining 4 sensors are positioned on the upstream and downstream sidewalls of the main transformer carvern as well as at the intersection of the main transformer carvern and the access tunnel, arranged in a three-dimensional spatial array distribution. To facilitate the reuse of sensors, a recyclable sensor installation scheme has been designed. The borehole diameter for sensor installation must be at least 32 mm. Considering the influence of surrounding rock deformation and construction disturbance, the designed borehole diameter for sensor installation is 40 mm, with a depth of approximately 3.5 m. The borehole angle is nearly horizontal, with the opening facing downwards, to ensure that the sensor installation environment remains dry. During installation, the sensor is connected in series with a disposable paper cup that has an opening at the bottom by tightening a bolt (not too tightly, to avoid affecting the sensor's later recovery). The prepared epoxy resin and hardener are filled into the paper cup. Using a sensor installation rod, the sensor and

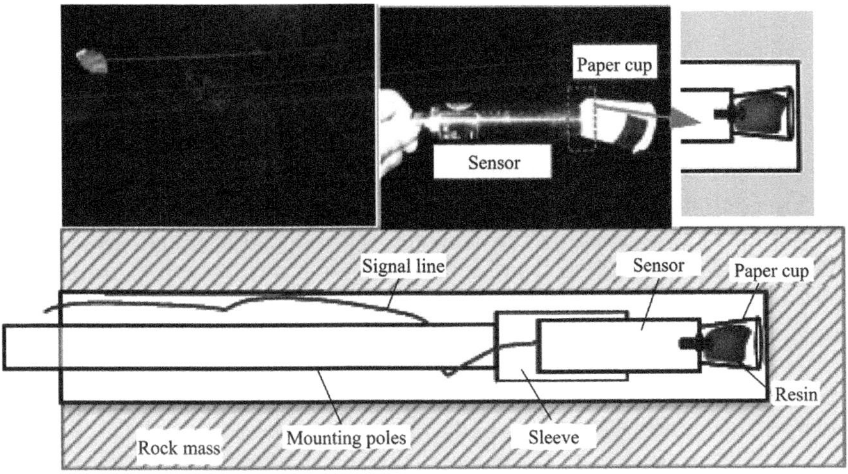

Fig. 5.6 Installation method of sensors

paper cup are quickly and steadily slid to the bottom of the borehole. The installation rod is pressed along the borehole to ensure the paper cup is in close contact with the bottom of the borehole (Fig. 5.6). The setup is left to solidify for 10 min, allowing the resin to react and harden thoroughly. Afterward, the installation rod is carefully removed, the sensor is connected in series with the cable, and the bias voltage is checked to be between 18 to 22 V to confirm the sensor circuit is operational. After the microseismic monitoring work was completed, the monitoring team successfully recovered the 18 sensors installed on-site. The project has demonstrated that this recyclable sensor installation scheme is suitable for microseismic monitoring during the excavation process of underground powerhouses, offering significant engineering and economic value.

5.5.3 *Positioning Error and Waveform Recognition of Microseismic Systems*

Microseismic monitoring technology is a geophysical real-time monitoring technique that evaluates the stability of rock masses in rock engineering by capturing micro-seismic signals generated from micro-fractures within the rock mass [62]. Accurate source location is fundamental to the effective application of microseismic monitoring systems in engineering surveillance. Its positioning accuracy is influenced by key factors such as the spatial coordinates of sensors, waveform arrival times, wave speed models, and the microseismic network array configuration. Among these, the wave speed model serves as a prerequisite for microseismic source location [63]. The geological conditions of large geotechnical projects are complex, and coupled with the heterogeneity of the rock and the effects of cavities created by excavation, it is challenging to accurately define and differentiate the true wave speeds within the rock mass of the monitored area [64, 65]. Currently, most microseismic monitoring systems in geotechnical projects within China adopt a certain simplified wave speed model, and their positioning errors can be controlled within 10 m [66, 67]. The underground powerhouse of a hydropower station features the typical structural characteristics of "high sidewalls, large spans, and multiple intercrossing chambers." The complexity of geological structures and lithological composition makes the propagation path and speed of elastic waves generated by micro-fractures in the surrounding rock under excavation unloading extremely complicated. Effectively determining the propagation speed of elastic waves within the rock mass in the monitored area is particularly important for the positioning accuracy and seismic source parameter calculations of the underground powerhouse microseismic monitoring system. The microseismic monitoring system of the underground powerhouse at the Huanggou Pumped Storage Power Station operates based on an overall simplified wave speed model. It assumes the propagation speed of microseismic signals within the monitored area to be equivalent to a series of integral wave speed models. Through artificial fixed-point percussion tests, the wave speed model of the rock mass within the monitored area is accurately determined. Referring to the geotechnical investigation report for the underground powerhouse of the Huanggou Pumped Storage Power Station, the longitudinal wave speed of the fresh Bai Gang granite in the plant area is 5.0–5.3 km/s. On September 28 and 29, 2017, nine points were selected within the spatial array of the underground powerhouse microseismic monitoring system's sensors for artificial knocking tests. The knocking times, locations, and captured waveforms were recorded. Table 5.3 includes the locations of the knocking points, the number and IDs of sensors triggered by each knock, and the number and IDs of sensors involved in the positioning. Based on the spectral characteristics of the waveform, 17 different wave speeds ranging from 4700 to 5500 m/s were designed for the system's P-wave velocity. The positioning errors of the knocking tests under different P-wave velocities were calculated (Fig. 5.7). When the system's P-wave velocity is set at 5150 m/s, the average system positioning error is 5.4 m (Table 5.4), which is controlled within 6 m. This level of accuracy meets the engineering requirements for microseismic monitoring positioning precision.

During the excavation of the underground powerhouse using the drill and blast method, construction disturbances such as blasting and air drilling pre-splitting generate a large amount of noise signals. By employing the threshold setting and time–frequency analysis technology of the ESG microseismic monitoring system, the microseismic signals collected during the excavation process are identified and classified into: artificial knocking waveforms (Fig. 5.8a), rock micro-fracture waveforms (Fig. 5.8b), air drilling waveforms (Fig. 5.8c), and excavation blasting waveforms (Fig. 5.8d).

Table 5.3 Manual tapping test positioning information table (September 28, 2017)

Time (hour:minute:second)	Strike point position	*X* (E) *Y* (N) *Z* (U)	Trigger sensor number	Number of participating positioning sensors
19:28:24	Upstream boundary of upper drainage gallery 0 + 34 m above the plant building, 0 + 40 m to the left of the plant building	553,772.7973 5,022,990.515 169.78	1, 2, 3, 4, 11, 12, 13, 14, 16, 17	1, 11, 2, 13, 3, 14, 4
19:34:53	Ventilation tunnel side boundary 0 + 34 m above the plant building, 0 + 40 m to the left of the plant building	553,736.9755 5,023,083.731 170.61	2, 3, 4, 6, 8, 13, 14, 16	4, 6, 8, 13, 14, 16
20:08:10	Upper drainage corridor side wall 0–97 m below the plant building, 0 + 62 m to the left of the plant building	553,841.0221 5,023,102.533 170.45	3, 4, 5, 6, 7, 8, 9, 10, 16, 17, 18	17, 8, 6, 3, 10, 18
21:20:36	Side walls of the installation room in the main plant building 0–11 m below the plant building, 0–13 m to the right of the plant building	5,022,988.876 553,841.5977 160.5	1, 3, 4, 10, 11, 12, 13, 15, 18	15, 12, 11, 10, 4, 13
21:22:33	Side walls of the installation room in the main plant building 0–11 m below the factory building, 0 + 7 m to the left of the plant building	5,023,001.997 553,826.5035 160.5	1, 3, 4, 10, 11, 12, 13, 15, 16, 18	4, 11, 13, 15, 16, 18, 10
21:48:25	Middle level drainage corridor side wall 0–48 m below the plant building, 0 + 119 m to the left of the plant building	5,023,103.739 553,766.5453 151.4	2, 3, 4, 6, 8, 12, 13, 14, 16, 17	16, 13, 2, 6, 12, 8, 17, 3

(continued)

Table 5.3 (continued)

Time (hour:minute:second)	Strike point position	*X* (E) *Y* (N) *Z* (U)	Trigger sensor number	Number of participating positioning sensors
21:57:07	Middle floor drainage corridor corner wall 0–120 m below the plant building, 0 + 116 m to the left of the plant building	5,023,155.205 553,815.2587 153.54	2, 4, 6, 7, 8, 9, 10, 13, 14, 16, 17	6, 8, 10, 4, 16, 17
21:59:50	Downstream boundary of middle level drainage gallery 0–120 m below the factory building, 0 + 50 m to the left of the plant building	5,023,111.905 553,865.0695 154.18	4, 5, 6, 7, 8, 9, 10, 17, 18	17, 10, 6, 7, 18, 4

Fig. 5.7 Relationship curve between impact point and positioning error under different P-wave velocities

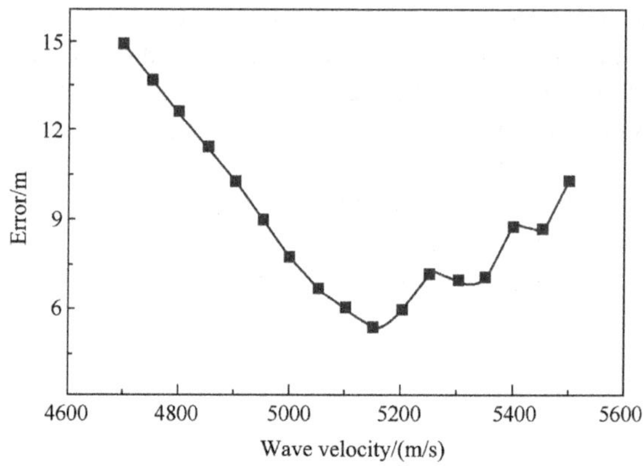

5.5.4 *Microseismic Activity Pattern*

The microseismic monitoring system for the underground powerhouse of the Huanggou Pumped Storage Power Station was successfully installed and commenced operations on September 30, 2017. Up until December 31, 2017, within the effective spatial range, a total of 339 microseismic (rock micro-fracture) events and 126 blasting events were detected. Under the strong excavation unloading effects, the microseismic events induced within the rock mass surrounding the underground powerhouse were mainly concentrated in two areas: One microseismic cluster area

Table 5.4 Table of manual knocking positioning error for P-wave velocity of 5150 m/s

Wave velocity/(m/s)	Knocking positioning error for P-wave velocity of 5150 m/s (/m)										Error mean value/m
	September 28, 2017								September 30, 2017		
	19:28:24	19:33:45	20:08:00	21:20:26	21:22:33	21:48:14	21:57:57	21:59:44	16:51:57	17:01:46	
4700	10.45	22.89	18.97	23.24	15.73	9.39	17.16	15.42	5.31	10.74	14.9
4750	8.66	20.99	18.01	21.63	14.18	8.93	15.88	14.40	4.89	9.59	13.7
4800	7.38	19.03	17.06	20.03	12.57	8.48	14.93	13.38	4.59	8.61	12.6
4850	6.23	16.93	16.13	18.42	10.67	8.03	13.65	12.36	4.35	7.65	11.4
4900	5.25	14.85	15.21	16.42	8.87	7.36	12.63	11.33	4.10	7.03	10.3
4950	3.91	12.74	12.44	14.82	6.96	6.80	11.55	10.31	4.00	6.63	9.0
5000	3.53	10.75	9.15	13.26	4.67	6.32	9.99	9.26	4.00	6.47	7.7
5050	3.67	8.97	8.22	11.29	2.68	5.74	8.25	8.19	4.00	6.53	6.8
5100	3.62	7.62	7.29	9.87	2.70	5.08	5.81	7.09	4.10	6.66	6.0
5150	4.31	6.67	6.39	8.60	4.19	5.25	1.03	5.99	4.37	6.92	5.4
5200	5.02	5.97	11.12	6.89	6.36	4.70	2.75	4.89	4.71	7.40	6.0
5250	5.50	5.63	10.29	5.96	21.27	4.06	4.18	2.07	5.04	7.80	7.2
5300	6.16	5.40	9.48	4.54	18.88	3.28	5.75	2.50	5.50	8.14	7.0
5350	6.75	5.34	8.68	4.03	17.66	3.08	5.27	5.01	5.94	8.82	7.1
5400	7.24	6.28	7.89	3.74	29.64	3.44	4.95	8.31	6.48	9.18	8.7
5450	5.04	6.74	7.09	3.94	29.87	3.16	4.72	9.45	7.00	9.45	8.6
5500	5.59	7.19	6.33	3.65	30.30	10.62	4.60	17.35	7.60	9.73	10.3

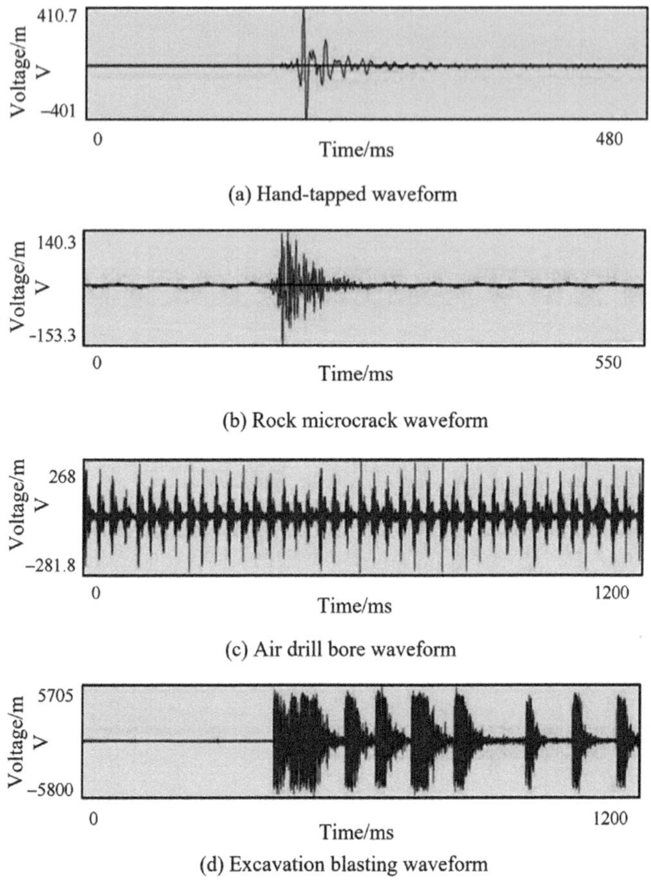

Fig. 5.8 Identification of typical event waveforms

(Cluster Area I) is located between the upstream sidewall of the main powerhouse and the downstream sidewall of the middle-level drainage gallery, forming a strip-like cluster from pile number plant left 0 + 20 m to plant left 0 + 80 m. Another microseismic cluster area (Cluster Area II) is located along the downstream sidewall of the main powerhouse, sporadically distributed along the lower part of the sidewall, and has a small concentrated distribution at the bottom of the intersection between the installation space and the access tunnel (Fig. 5.9a). Microseismic events are primarily concentrated within the surrounding rock of the upstream sidewall of the main powerhouse, showing the highest clustering density and evolving in a strip-like pattern along the sidewall (Fig. 5.9b and c). A combination of on-site field inspections and geological data analysis revealed that at pile number factory left 0 + 45 m of the middle-level drainage gallery, fault f_{34} is exposed, and joints are developed. There exists a steeply dipping fault f_{34} within the surrounding rock of the upstream sidewall of the main powerhouse. The strip-like distribution pattern

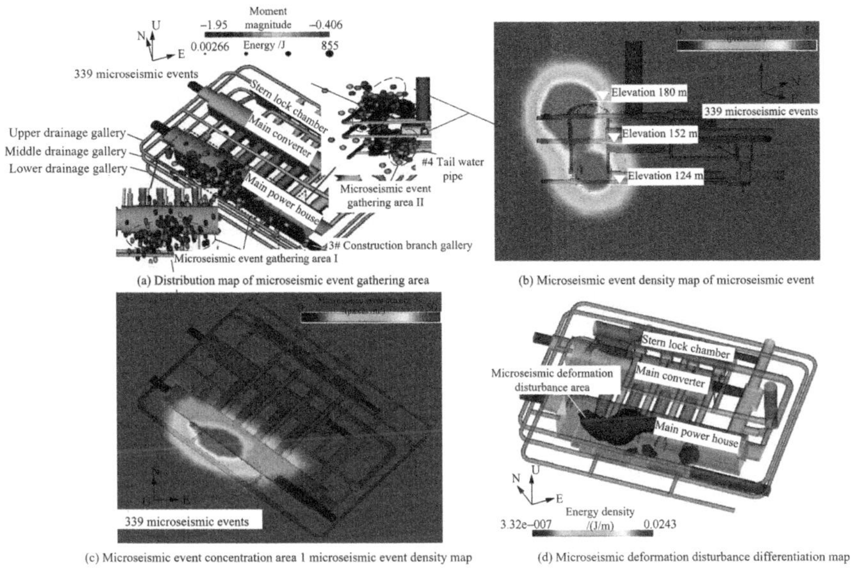

Fig. 5.9 Spatial distribution pattern of microseismic events

of the micro-fracture clustering area on the upstream side of the main powerhouse is consistent with the spatial orientation of this fault, indicating that rock mass near the fault fracture zone was damaged during the intense unloading excavation process (Fig. 5.9d).

Since the microseismic monitoring system of the Huanggou underground powerhouse commenced normal operations on September 30, 2017, up until December 31, 2017, it has achieved round-the-clock monitoring of the surrounding rock of the underground powerhouse during the excavation unloading process. A total of 465 effective events were detected, including 339 microseismic (rock micro-fracture) events and 126 blasting events (Fig. 5.10). From Fig. 5.10, it is observed that during October, the underground powerhouse was in a period of cessation, and micro-fracture events rarely occurred. In early November, construction resumed with the excavation of the central installation space of the main powerhouse, leading to an increase in micro-fracture events, with a frequency of less than 10 events per day. In early December, continuous excavation and blasting of the main powerhouse took place, and at the same time, excavation was also underway for the busbar tunnel and the #1 upper-level pressure steel pipe, resulting in a high intensity of construction activities. This led to a surge in microseismic event frequency, peaking at 24 events per day. In late December, as construction focused mainly on sidewall support and trimming, the frequency of microseismic events decreased. The cumulative number of microseismic events is positively correlated with the cumulative number of blasting events. The more continuous excavation and blasting events occur,

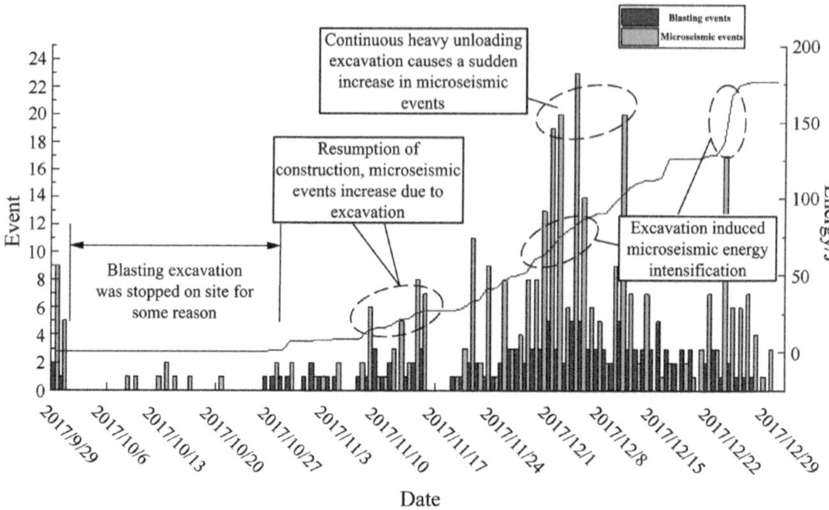

Fig. 5.10 Time distribution pattern of microseismic activity

the faster the trend of increase in the cumulative number of microseismic events, indicating frequent microseismic activity. This suggests that the microseismic activity of the surrounding rock during the excavation process of the Huanggou Pumped Storage Power Station's underground powerhouse is significantly affected by construction disturbances. Construction disturbances are one of the main factors influencing the occurrence of microseismic events within the surrounding rock of the underground powerhouse.

5.5.5 *Stability Evaluation of Surrounding Rock Based on Microseismic Activity*

Under the effect of excavation unloading, the stress within the surrounding rock adjusts, leading to stress redistribution and concentration, which in turn induces micro-fracture events in the rock. The evolution from the initiation and development of these micro-fractures to their eventual coalescence is the ultimate result in the macroscopic instability and failure of the surrounding rock [68]. The excavation of underground caverns requires the use of blasting construction methods. Construction disturbances have a significant impact on the change in stress state and the decline in mechanical properties of the surrounding rock. Furthermore, the geological conditions of the surrounding rock are complex, often containing faults, joints, and other geological defects, which play an important role in the generation of micro-fractures [69, 70]. Based on microseismic monitoring outcomes, analyzing the relationship between microseismic activity, construction disturbances, and geological structures allows for a detailed investigation and assessment of the local stability in geotechnical engineering. This approach enables a deeper insight into the mechanisms underlying the damage to the surrounding rock, enhancing the understanding of how these factors interact and impact the integrity and safety of geotechnical constructions [71, 72].

1. The relationship between construction steps and microseismic activity

Figure 5.11 is a spatial comparison chart of the construction dynamics of the underground powerhouse during different monitoring periods and the aggregation of microseismic events. Microseismic events frequently occur within the surrounding rock near the open faces close to construction blasting points, and their migration and evolution patterns are consistent with the advancement of the construction workface. A small number of microseismic events are concentrated in a limited area at the intersections of the #3 tailwater pipeline, the #4 tailwater pipeline, and the #3 construction adit with the main powerhouse. These are primarily caused by excavation disturbances, and the locations of the microseismic events coincide with the blasting construction sites (Fig. 5.11a). In mid-November 2017, excavation work at the main powerhouse resumed. From November 15 to 30, 2017, excavation was conducted in the section from pile number plant left 0 + 0 m to plant left 0 + 40 m. Microseismic events clustered around the arch shoulder and arch crown's open faces on the upstream sidewall of the main powerhouse within the same pile number range. Due to the excavation and blasting of the main powerhouse and the unloading effect, local surrounding rock stress continuously adjusted. Stress concentration occurred at the arch shoulder of the open face in the main powerhouse, a structurally weak part of the chamber, leading to the aggregation of micro-fracture events (Fig. 5.11b). From December 1 to 10th, the construction workface of the main powerhouse advanced forward to the section from pile number plant left 0 + 40 m to 0 + 80 m (Fig. 5.11c). From December 11 to 31st, 2017, the excavation of the main powerhouse extended to near the left end wall of the main powerhouse, with microseismic events distributed in a strip-like pattern in the section from pile number plant left 0 + 60 m to 0 + 80 m (Fig. 5.11d). From Fig. 5.11e, it can be observed that the characteristics of frequent excavation and unloading-induced microseismic activity have a certain temporal and spatial relationship with the on-site construction dynamics. The microseismic event density cloud map clearly reflects that the surrounding rock within the elevation range of 150–180 m on the upstream shoulder of the main plant is the area with the highest concentration of microseismic activity, indicating it as a potentially hazardous area. By comparing the blasting and excavation locations with the characteristics of microseismic activity in different time periods using Fig. 5.11, a mechanism can be established to understand the dynamic response relationship between construction activities and microseismic events. It can be observed that the microcracks induced by the upstream sidewall of the underground plant tend to cluster in a strip-like pattern, and this clustering has a dynamic response relationship with the advancement of the excavation workface in the main plant. The aggregation of microcracks exhibits a clear dynamic migration and evolution process. This indicates that the stress distribution of the surrounding rock of the underground plant has changed under the influence of construction disturbances. As a result, a significant number of microcracks have formed in the local surrounding rock, creating a high-energy release zone, which further exacerbates the local damage to the surrounding rock. The excavation of large cross-section structures like underground factories involves a systematic construction process that is carried out in layers and stages. During the continuous and intensive excavation and unloading, the stress in the surrounding rock of the caverns undergoes constant dynamic adjustments. Therefore, real-time monitoring of micro-fractures in the rock mass within the construction disturbance zone is essential. Researching the relationship between construction dynamics and micro-seismic activity can help clarify the impact of construction disturbances on the damage and stability of the surrounding rock. This information can provide valuable insights for subsequent construction and support measures.

2. The relationship between geological structure and microseismic activity

The construction scale of hydroelectric power stations is rapidly advancing towards larger and more complex projects. The areas where underground powerhouses are located often have complex geological conditions, and many of them are characterized by the presence of faults. The existence of these faults poses a potential threat to the stability of the excavation, construction, and long-term operation of large underground caverns. In seismology, the parameter "2" is commonly used as an important parameter for studying the rupture mechanism of seismic sources. Microseismic monitoring systems, which observe micro-fractures in rocks, also fall under the category of small-scale seismic sources. In this section, we attempt to utilize these microseismic source parameters for fault identification and the study of the damage mechanism of surrounding rock in underground factory caverns. The source rupture (microseismic events) is governed by a primary destructive mechanism. A thorough analysis of the ratio of horizontal-to-vertical wave energy (P-wave to S-wave energy ratio) can help determine whether the damage mechanism in the surrounding rock micro-fractures is due to shear failure or non-shear failure (volumetric stress changes). This is particularly important for a clear understanding of the damage mechanisms in the surrounding rock during the excavation process of underground factory chambers. Current research suggests that the criteria for revealing the mechanism of seismic source rupture are as follows: less than 3 indicates tensile rupture; between 3 and 10 indicates mixed tensile-shear rupture; greater than 10 indicates shear rupture. Combining field surveys and geological data analysis, it has been found that the migration and evolution of microseismic events on the upstream side of the main plant exhibit a zonal distribution. This feature is not only related to the dynamic construction activities on site but also has a certain connection with the geological structures within the rock mass of the sidewalls on the upstream side of the main plant. The establishment of a three-dimensional analysis model for the fault and microseismic activities of the underground plant shows that the spatial orientation of the microseismic event cluster area I, as seen in Fig. 5.12a, aligns with the direction of fault f_{34}. The strip-like clustering and evolutionary characteristics of microseismic events in this area are significantly influenced by fault f_{34}. Seismological and geological observations indicate that the mechanical and physical properties of fault zones are not uniform; there is a significant difference between the local stress field and

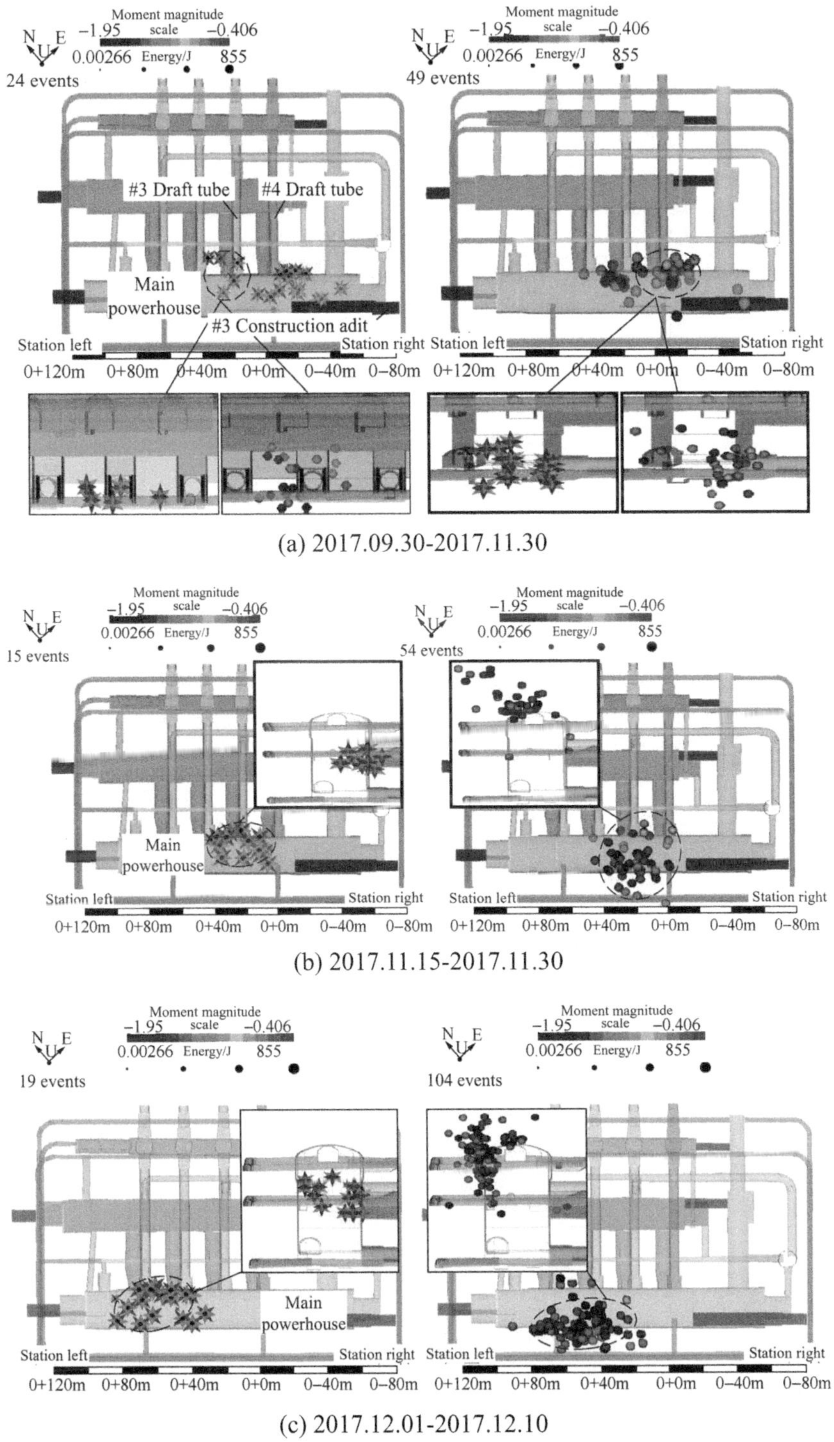

Fig. 5.11 Microseismic activity under the influence of construction disturbance

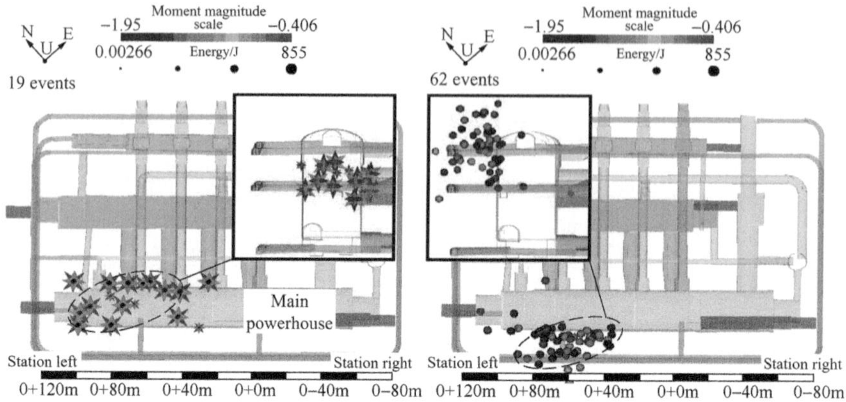

(d) 2017.12.11-2017.12.31

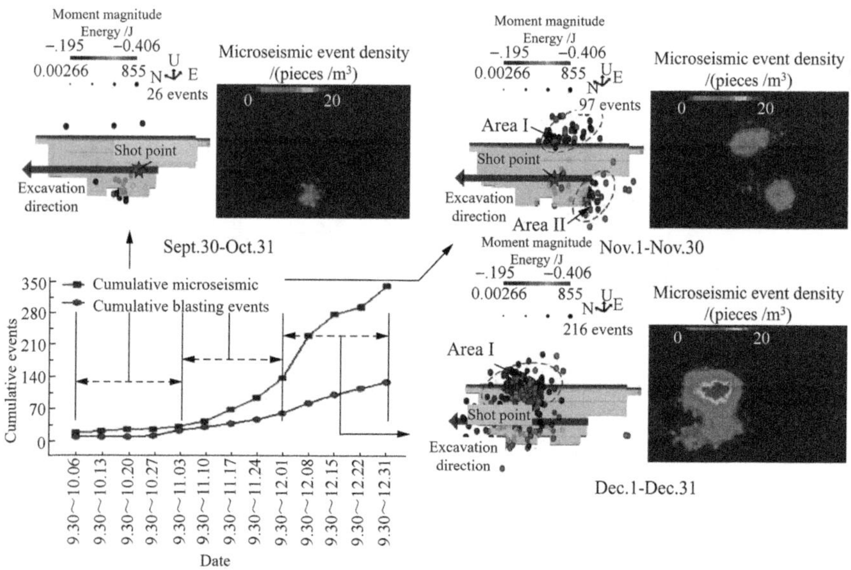

(e) The spatio-temporal evolution of microseismic

Fig. 5.11 (continued)

the regional stress field. Under stress, the fillings and fractured bodies on the fault zone are prone to sliding friction, leading to shear failure. Figure 5.12b presents the spatial distribution of microseismic events with values greater than 10 within the three-dimensional fault analysis model, as well as a fault distribution map at the plant area elevation of 172 m. It can be observed that the strip-like spatial distribution characteristics of rock micro-fracture events controlled by shear failure are consistent with the orientation of fault f_{34} as provided in the geological exploration report. The micro-fractures of the shear failure type are all distributed along the fault zone. This indicates that the pure shear failure micro-fractures in the surrounding rock induced by the excavation process of the Huangou underground plant are mainly controlled by faults. Microseismic events with values greater than 10 can effectively identify the development direction of faults within the rock mass of the engineering area. As the excavation area of the main plant is gradually excavated, the microseismic evolution pattern shown in Fig. 5.12 can provide a reference for the reinforcement and support of the surrounding rock with faults in the subsequent main plant area.

The temporal and spatial evolution pattern of microseismic events indicates that the excavation unloading induces a large number of microseismic events to cluster in a strip-like formation within the elevation range of 152–210 m in the main plant area. Figure 5.13 illustrates the distribution relationship of microseismic events with magnitudes less than 3 and those between 3 and 10. It is found that the microseismic events induced by blasting excavation are primarily characterized by tensile failures, with events less than 3 making up 60% of the total. These events cluster around the arch crown and shoulders at the pile number factory left 0 + 20 m of the main plant, forming a nucleus that extends horizontally in a strip-like fashion towards the factory left, following the direction of the sidewall. According to geological exploration data, the orientation of faults at the elevation of 172 m is depicted, with a small portion of tensile failure-type microseismic events occurring on fault f_{34}. Tensile failure-type microseismic events accumulate at the height of the exposed surface near the upstream side arch shoulder of the main plant and show a distinct V-shaped cluster. Compared to microseismic events dominated by tensile failure, the proportion of mixed tensile-shear failure type microseismic events (values between 3 and 10) is smaller, at only 29%. This indicates that the clustering and distribution pattern of microfractures resulting from mixed tensile-shear failure differs from those caused by tensile failure. Microfractures from mixed tensile-shear failure generally develop and distribute in a trend similar to the orientation of fault f_{34}, and they are located further away from the exposed surface near the arch shoulder on the upstream side of the main plant. Similarly, they form a V-shaped cluster at the arch shoulder as shown in Fig. 5.13b. Mixed tensile-shear failure microseismic events are distributed around the edge of the cluster of tensile failure events. The non-uniformity of the mechanical and physical properties of fault structures affects the adjustment and redistribution of geostress, resulting in significant differences between the local stress field around the fault and the cavern and the regional stress field. Hudson et al. [73] believe that due to the influence of fault structural surfaces, the maximum principal stress tends to align parallel to the structural planes, causing local stress concentration in certain areas. Frequent excavation unloading induces stress disturbances that disrupt the stress equilibrium of the surrounding rock in underground plant areas, leading to continuous adjustment and redistribution of the surrounding rock stress. Influenced by fault f_{34}, stress concentrates between the arch shoulder on the upstream side of the main plant and the fault, resulting in a large number of low-magnitude, low-energy tensile

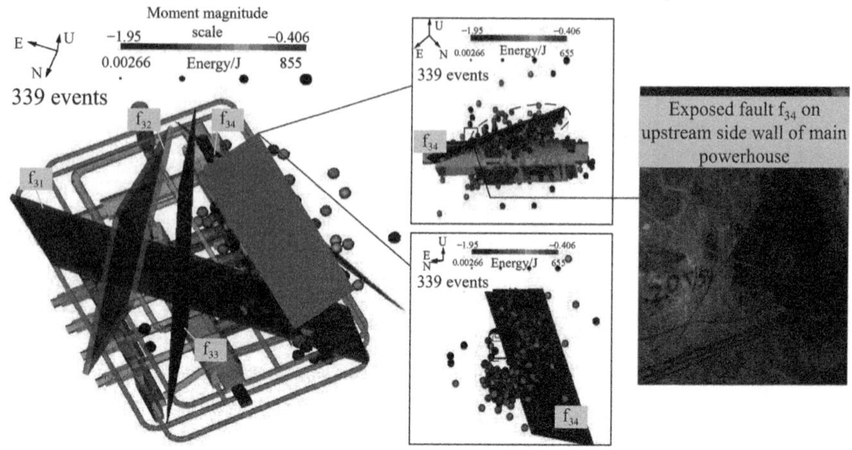

(a) Relationship between microseismic event distribution and fault location

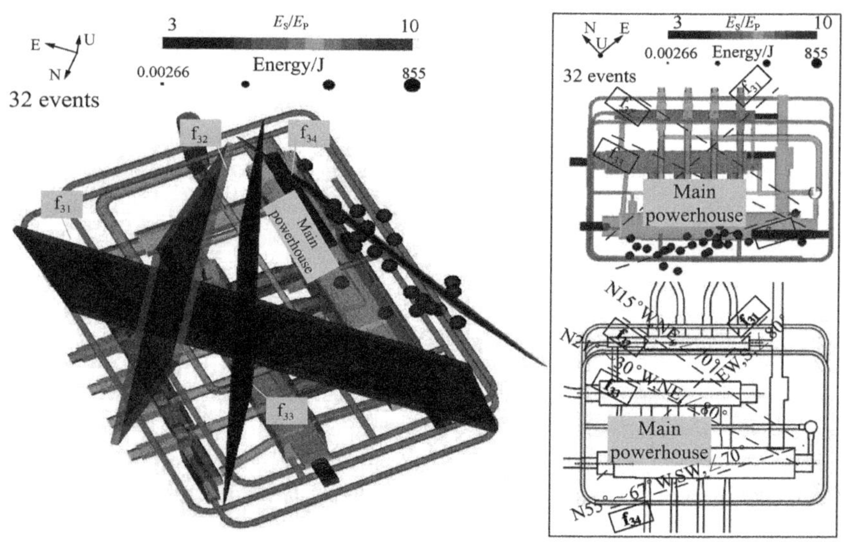

(b) E_s/E_p>10 Microseismic event distribution and fault location relationship

Fig. 5.12 Microseismic activity under the influence of faults

fractures concentrated at structurally weak points (such as the arch shoulder and crown) on the exposed surface. Tensile failure-dominated microfractures extend in a V-shape into the deeper rock mass near the upstream side arch shoulder, accompanied by a small number of mixed tensile-shear failures; whereas shear failures are mainly distributed along the fault zone. Microfracturing of the surrounding rock near the fault exhibits a complex fracture mechanism characterized primarily by tensile failure, with mixed tensile-shear and shear failures also present.

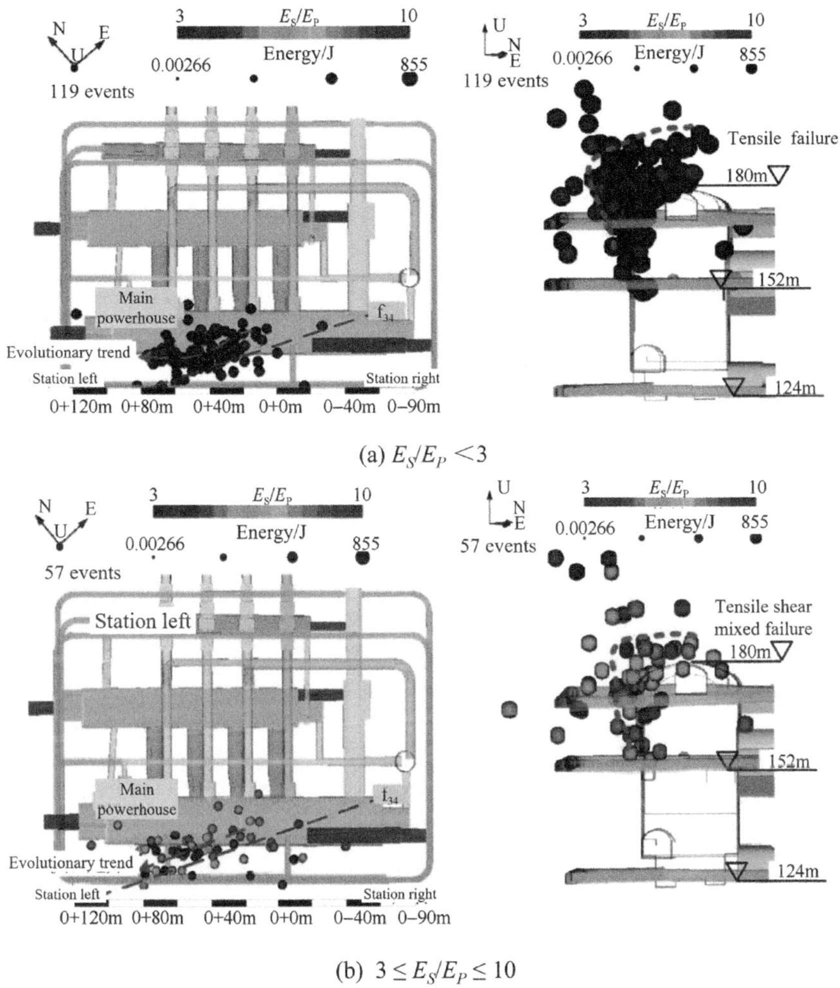

Fig. 5.13 Distribution relationship of microseismic events with different E_S/E_P at elevations of 152–210 m

3. Stability analysis of local surrounding rock in underground power plants

Based on the principle of energy dissipation [28], a rock microfracture accompanied by energy release represents a slight damage to the rock mass. As the evolution and clustering of rock microfracture events occur, the damage to the surrounding rock mass intensifies. The release of microseismic energy is a fundamental parameter in microseismic monitoring, and its magnitude can serve as an important basis for identifying damaged areas in the surrounding rock mass. By analyzing the microseismic energy density during the excavation process of the underground plant (Fig. 5.14), it is observed that microseismic events cluster and high energy is released in the section from the pile number factory left 0 + 20 m to factory left 0 + 80 m along the upstream sidewall of the main plant (i.e., microseismic cluster area I). Under the effect of excavation unloading, this section of the surrounding rock mass suffers the most severe damage, indicating a region of the main plant's surrounding rock where there is a localized risk of instability. During a field inspection in late December inside the drainage tunnel, it was discovered that in areas of high microseismic energy release, the inner sidewalls of the middle-layer drainage tunnel exhibited multiple instances of cracking, spalling, and falling blocks (Fig. 5.14a at pile number plant left 0 + 75 m, Fig. 5.14b at plant left 0 + 60 m, Fig. 5.14c) at plant left 0 + 45 m). The occurrences of surrounding rock instability within the middle-layer drainage tunnel were all located in areas of intense microseismic activity. The surrounding rock at locations of spalling and falling blocks showed well-developed jointing, with the rock experiencing obvious layered peeling along these joints. Moreover, at the location of plant left 0 + 45 m, fault f_{34} passes through.

Combining the previous analysis results, the blasting excavation construction of the main plant has caused stress accumulation in the surrounding rock mass of the upstream sidewall of the underground plant, thereby triggering microfracture events. The migration and evolution pattern of these microfractures are consistent with the direction of the blasting construction work face's advancement. Under the influence of fault structures, blasting construction causes the stress in the surrounding rock of the underground plant to concentrate towards the structurally weak areas (upstream side arch shoulders) of the rock mass on the upstream side of the main plant wall,

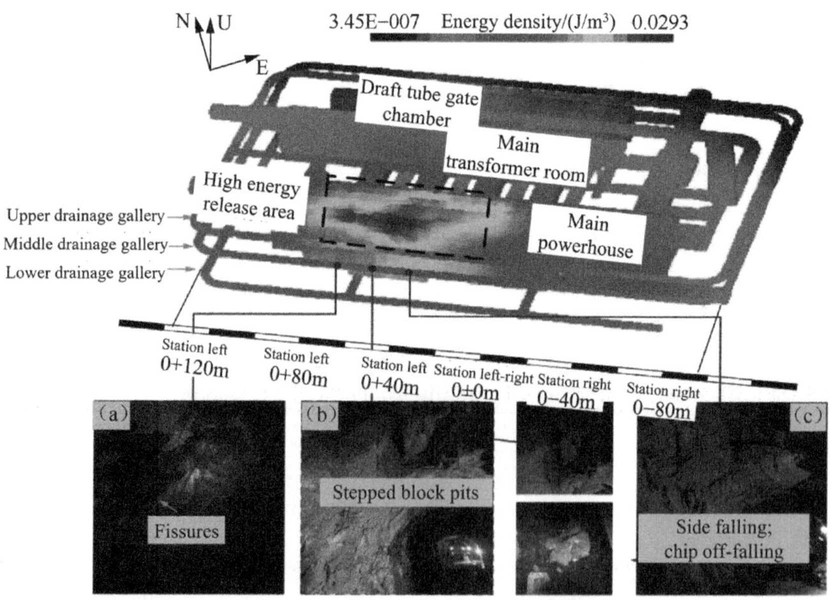

Fig. 5.14 Damage area identification based on microseismic energy density cloud map

inducing microfractures. Part of the energy is released through microfractures near the arch shoulder on the upstream sidewall (with microfractures occurring along the upstream sidewall and arch shoulder as the work face advances), while another part of the energy "migrates" along fault f_{34} into the interior of the surrounding rock, causing energy release through microfractures in weak areas of the rock mass (Fig. 5.14a, b, and c). It is evident that the intensity of microseismic activity is not only affected by construction disturbances but also related to the development of fault fracture zones and joints within the surrounding rock. Therefore, the segment of the surrounding rock from factory left 0 + 20 m to factory left 0 + 80 m along the upstream sidewall and arch shoulder of the main plant is designated as a potential damage area. This designation is essentially consistent with the potentially unstable block regions of the Huangou underground plant calculated by Yang Qing et al. [74] based on block theory, indicating a basic agreement in identifying areas of potential instability.

5.5.6 Brief Summary

This section provides the first attempt to apply a microseismic monitoring system to the stability monitoring of surrounding rock during the excavation process of an underground power plant in a pumped storage station. It involves real-time monitoring of micro-cracks within the internal rock mass of the underground power plant chambers, particularly under the conditions of strong excavation and stress release. Combining the on-site exploration during the excavation of the underground power plant at the Huanggou Power Station with microseismic monitoring, the following conclusions were drawn:

(1) By conducting manual point impact tests, the overall equivalent wave velocity of the rock mass in the monitoring area was determined to be 5,150 m/s, with a positioning error within 6 m, which meets the requirements of underground engineering. The reusable sensor installation plan was successfully implemented at the Huanggou Pumped Storage Power Station, and all sensors were successfully recovered.
(2) From the perspective of the spatiotemporal evolution of microseismic activity in the underground factory, it can be observed that within the effective monitoring area, microfracturing of the surrounding rocks is primarily concentrated in two areas: (a) Microseismic cluster area I, this area is located in the upstream portion of the main plant building at elevations ranging from 150 to 180 m. It extends from pile number 0 + 20 m to 0 + 80 m along the left side of the plant. (b) Microseismic cluster area II, this area is situated at the bottom end of the plant installation space, close to the side wall of the main plant building. In these regions, there is a higher concentration of microseismic events, indicating increased seismic activity and potential structural weaknesses, particularly in the underground plant's weak structural areas and along fault development zones.

(3) The microseismic activity in the underground factory is greatly influenced by construction disturbances, characterized by high-intensity continuous blasting and frequent microseismic activity. Microseismic Cluster Area I exhibits a strip-like distribution pattern that closely aligns with the orientation of fault f_{34}. Under the influence of excavation and unloading, the stress in the surrounding rocks of the main factory building adjusts, leading to the transfer and concentration of stress near the upstream side wall of the factory building in the vicinity of fault f_{34}. This concentration of high-energy release exacerbates damage to the surrounding rocks. Based on this information, it has been determined that the surrounding rocks in the section from pile number 0 + 20 m to pile number 0 + 80 m along the upstream side wall of the main plant building are at risk of instability. The microseismic monitoring results can serve as a valuable reference for subsequent excavation and support measures, as well as for preventing seepage in underground caverns.

5.6 Numerical Simulation of Underground Powerhouse Based on RFPA^{2D}-SRM

The underground powerhouse of the Huanggou pumped storage power station is located in Mudanjiang City, Heilongjiang Province. The underground powerhouse caverns are constructed within mountain body on the right bank of Sandaohezi, with a burial depth of around 310 m. The main powerhouse boasts an arch crown elevation of 180 m, and the surrounding rock of the cavern is primarily composed of pristine alaskite granite, displaying favorable lithology and classified as Class II surrounding rock. Four faults traverse the three main caverns in the plant area, creating an intricate cavern network (Fig. 5.15). A microseismic monitoring system was established for real-time continuous monitoring of the surrounding rock stability during the excavation of the underground powerhouse. Throughout the monitoring period, a substantial number of microseismic events clustered within the rock mass of the upstream sidewall of the main powerhouse. Field surveys revealed varying degrees of surrounding rock instability in the mid-level drainage gallery near the area of microseismic event clusters.

The surrounding rock of the underground powerhouse consists of pristine alaskite granite. The physical and mechanical parameters of the rock mass and faults in the plant area are detailed in Table 5.5, based on the geological survey and testing conducted in the region. According to the rock test results obtained from the Huanggou pumped storage power station, the average uniaxial compressive strength of multiple pristine alaskite granite samples is 147 MPa (dry) and 124 MPa (saturated).

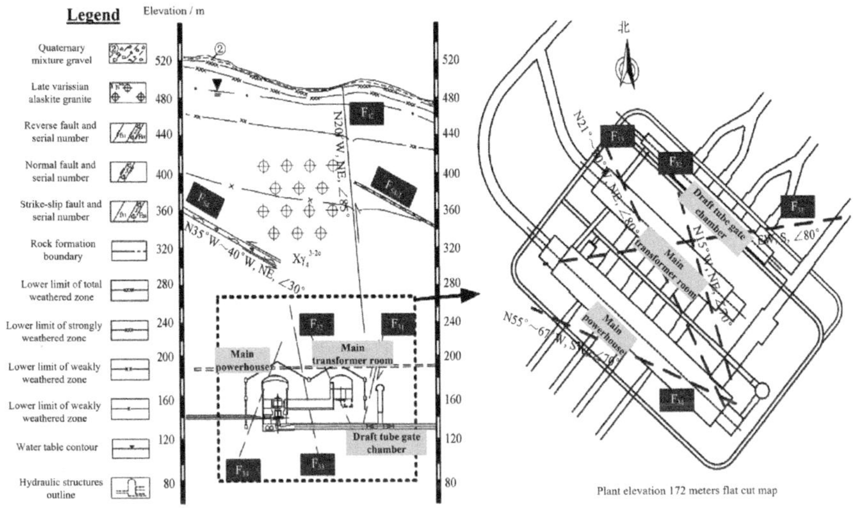

Fig. 5.15 Geological condition of faults in plant area

Table 5.5 Table of physical and mechanical parameters of plant area rock mass and fault zone

Rock	Elastic modulus E/GPa	Density (kg/m^3)	Poisson's ratio μ	Shear failure strength		Shear strength	
				f'	c'/MPa	f	c/MPa
Class II surrounding rock	40	2610	0.17	1.2~1.3	1.5	0.75	0
Fault	0.5~0.8	1700	0.30	0.30~0.35	0.02~0.05	0.30	0

5.6.1 Basic Principles of RFPA^{2D}-SRM

RFPA^{2D}-SRM (RFPA Strength Reduction Method) incorporates the fundamental principles of finite element strength reduction methods into the RFPA software. It gradually reduces the strength parameters (shear and tensile strength) throughout the finite element calculation process until they reach the failure state. RFPA-SRM employs the statistical analysis of failure occurrences at the microscale element as an instability criterion, fully considering the micro-scale non-uniform characteristics of rocks. It utilizes both shear and tensile failure criteria to achieve stability analysis of rock structures.

1. Microscale element assignment based on statistics

RFPA assumes that the mechanical properties of discretized microscale elements follow a certain statistical distribution pattern. It introduces the Weibull statistical distribution function to describe the mechanical strength parameters of rock microscale elements, aiming to reflect the non-uniform distribution of mechanical properties in rock microscale mechanics.

$$f(\alpha) = \frac{m}{\alpha_0} \times \left(\frac{\alpha}{\alpha_0}\right)^{m-1} \times \mathrm{e}^{-\left(\frac{\alpha}{\alpha_0}\right)^m} \tag{5.1}$$

In the equation, α represents the mechanical parameters (elastic modulus, strength, Poisson's ratio, self-weight, etc.) of the material (rock) medium element. α_0 is the average value of the mechanical parameters of the medium element. m is the distribution function parameter, whose physical meaning reflects the homogeneity of the material (rock), defined as the homogeneity coefficient of the material (rock), indicating the degree of material homogeneity. $f(\alpha)$ is the statistical distribution density of the α.

Using elastic modulus as an example to explain the assignment of model elemental mechanical parameters in RFPA. The average elastic modulus of all model elements is denoted as E_0, $\varphi(E)$ representing the distribution value of elements with a certain elastic modulus E. Based on Eq. (5.1), the integral of the elastic modulus Weibull distribution function is given by:

$$\varphi(E) = \int_0^e \phi(x)\mathrm{d}x = \int_0^e \left(\frac{m}{\alpha_0} \cdot (\frac{\alpha}{\alpha_0})^{m-1} \cdot \mathrm{e}^{-(\frac{\alpha}{\alpha_0})^m}\right)\mathrm{d}x = 1 - \mathrm{e}^{-(\frac{E}{E_0})^m} \tag{5.2}$$

With an increase in the homogeneity coefficient m, the mechanical properties of the elements concentrate within a narrow range, signifying a relatively uniform nature of the rock medium. Conversely, as the homogeneity coefficient m decreases, the distribution range of the mechanical properties of the elements widens, indicating a tendency toward non-uniformity in the rock properties (Fig. 5.16).

2. Strength criteria of RFPA-SRM

RFPA-SRM primarily introduces the Mohr–Coulomb strength criterion with tensile truncation as the criterion for microscale elemental failure. In the initial state of force on the microscale elements, it is assumed that the microscale elements are elastic, and their mechanical properties can be entirely expressed by their elastic modulus and Poisson's ratio. As the stress on the elements increases, the Mohr–Coulomb strength criterion and the maximum tensile stress criterion serve as thresholds for elemental failure. Firstly, considering tensile failure, if the maximum tensile stress of a microscale element reaches its designated threshold, the element initiates tensile failure. Secondly, when there is no occurrence of tensile failure, shear failure is assessed. If the stress state of the microscale element satisfies the Mohr–Coulomb strength criterion, the element experiences shear failure. Following failure,

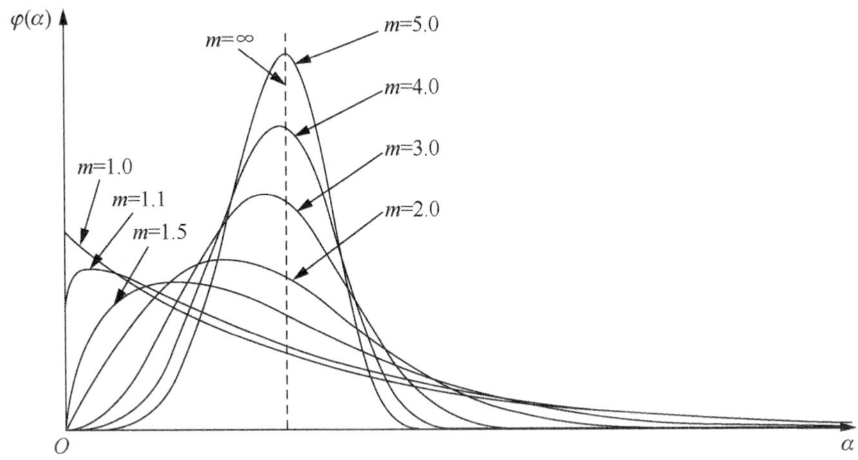

Fig. 5.16 Elastic modulus distribution values of rocks with different uniformity coefficients

the element can sustain a certain load based on the predetermined residual strength coefficient. The constitutive relationship between shear failure and tensile failure, as satisfied by the elements under ideal uniaxial stress conditions, is illustrated in Fig. 5.17.

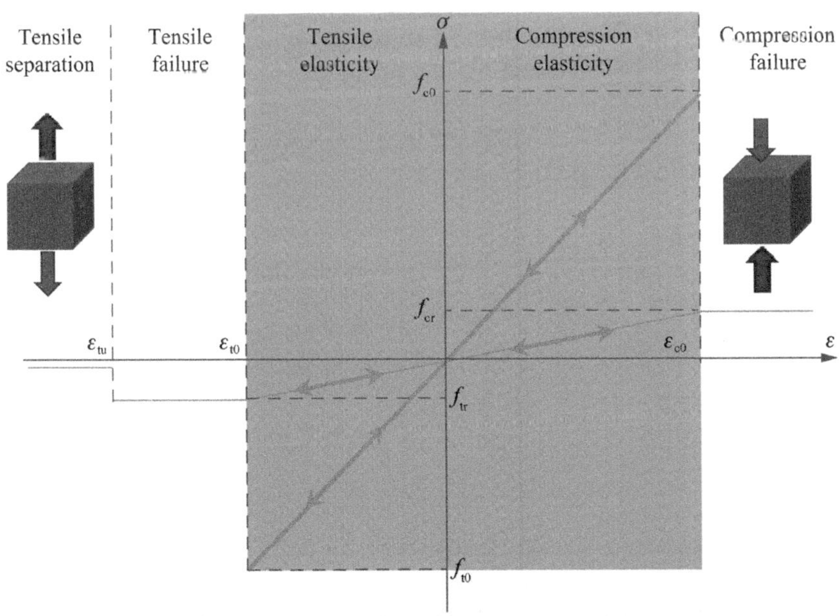

Fig. 5.17 Constitutive diagram of shear failure and tension failure satisfied by the elements under ideal uniaxial stress state

3. Stability criterion

RFPA-SRM gradually reduces the strength of micro-scale elements in a linear relationship, with a certain proportionate step size for each reduction. In each reduction step, the finite element calculation program iteratively strives to achieve equilibrium between external and internal forces, concurrently performing failure analysis, ultimately leading to the instability failure of the surrounding rock in tunnels. Owing to the utilization of full-load calculation, every calculation step produces convergent results. During the calculation analysis, the failure count of each reduction step for the microscale elements is automatically documented, enabling the maximum failure count to serve as the instability criterion. The safety reserve factor F_s, commonly referred to as the stability factor, constitutes a crucial concept in the analysis of stability for rock mass structures. Using the traditional example of slope stability limit analysis, when the material's shear strength parameters c and φ are substituted with their critical strength parameters c' and φ', the slope reaches a critical equilibrium state. In finite element calculations aimed at determining F_s, there is a continuous reduction of strength parameters in certain steps or proportions. Upon reaching an equilibrium state in the calculation, specifically when the slope experiences instability failure, the acquired F_s represents the strength reserve safety factor for the slope.

$$\begin{cases} c' = c/F_s \\ \tan\varphi' = \tan\varphi/F_s \end{cases} \tag{5.3}$$

Expanding on the previously mentioned strength reserve concept, RFPA-SRM proposes reduction criteria for the initial strength f_0.

$$f_0^{trial} = \frac{f_0}{F_s^{trial}} \tag{5.4}$$

In the formula, f_0^{trial} represents the tested strength, and F_s^{trial} denotes the safety factor for the test. When the number of element failures reaches the maximum, signifying instability in the rock mass structure, the test safety factor F_s^{trial} transforms into the ultimate safety factor F_s.

5.6.2 Typical Section Surrounding Rock Stability Analysis

Utilizing the geological data supplied for the underground powerhouse region of the Huanggou pumped storage power station, particularly focusing on the mechanical parameters of alaskite granite, numerical experiments in uniaxial compression were conducted using RFPA. The microscopic values within RFPA underwent validation through numerical experiments. The stress–strain curve derived from these experiments demonstrated a uniform uniaxial compressive strength consistent with the strength determined in rock tests conducted on-site (Table 5.6 and Fig. 5.18).

In this section, focusing on the impact of stress and faults, a numerical simulation analysis is performed to investigate the patterns of damage evolution and failure modes in the surrounding rock during the excavation process of the underground powerhouse at the Huanggou Pumped Storage Power Station. Comparisons and feedback are drawn with the microseismic monitoring results, unveiling the mechanisms of surrounding rock damage in the underground powerhouse of the Huanggou pumped storage power station under the influence of excavation unloading. Field reconnaissance and the examination of microseismic monitoring outcomes reveal that, while excavating and blasting at an elevation of 152 m in the Huanggou underground powerhouse, micro-fractures in the surrounding rock on the upstream side of the main powerhouse exhibit a banded distribution along the sidewalls and faults. This forms a V-shaped aggregation in a plane perpendicular to the powerhouse axis. Micro-fractures initiate at the station left 0 + 20 m and extend to the station left 0 +

Table 5.6 The parameters of numerical simulation

Mechanical parameters	Parameter values	Mechanical parameters	Parameter values
Elastic modulus/MPa	40,000	Uniaxial compressive strength/MPa	124
Input elastic modulus/MPa	45,721	Input uniaxial compressive strength/MPa	300
Poisson's ratio	0.3	Tension compression ratio	10
Frictional angle/(°)	36	Residual threshold coefficient	0.1

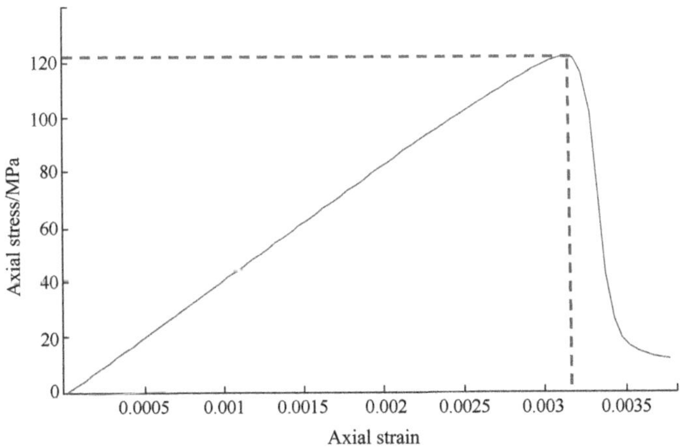

Fig. 5.18 Numerical simulation of stress–strain curve for uniaxial compression

80 m. Combining the research content from Sect. 5.4, the typical section at station left 0 + 60 m is selected, which exhibits the highest degree of microseismic aggregation. An RFPA numerical model is established for conducting numerical calculation studies. Stress boundary conditions are determined based on the measured in-situ stress results, with 12 MPa in the X-direction and 8 MPa in the Y-direction, and constraints at the model bottom. The model dimensions are 200 m × 120 m, with a grid division of 1000 × 600 = 600,000 elements, using a plane strain model for solving and a strength reduction factor of 0.01. The numerical model of the typical section at station left 0 + 60 m is shown in Fig. 5.19c. The established numerical model corresponds to the construction conditions of the underground powerhouse during the microseismic monitoring period. Specifically, the main powerhouse is excavated to an elevation of 152 m, while the main transformer room and tailrace gate chamber are excavated to elevations of 152 m and 139 m, respectively. Boundary stress conditions are applied in RFPA, and the excavation is completed step by step. By simulating the progressive damage process of the surrounding rock induced by the excavation and blasting at an elevation of 152 m during the actual construction of

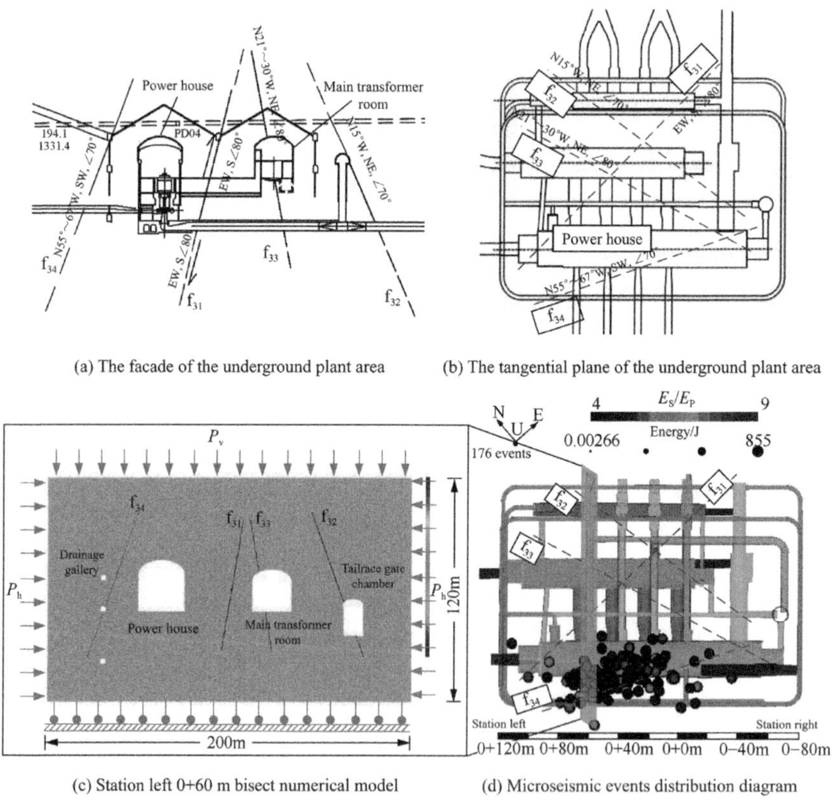

Fig. 5.19 Faults distribution diagram of the plant area

the main powerhouse, utilizing the strength reduction method, the potential failure mode and stability safety factors of the surrounding rock are determined.

1. Analysis of numerical calculation results

Figure 5.20 illustrates the shear stress diagrams derived using the strength reduction method for various excavation steps in the RFPA numerical model of the typical section at stake station left 0 + 60 m. Within the shear stress diagram, increased brightness correlates with elevated stress values. Analyzing the damage pattern of the surrounding rock in the main powerhouse through numerical simulation results, the rock stress undergoes adjustments, leading to the formation of an asymmetric stress concentration zone in the arch section of the main powerhouse. Stress concentrates notably on the upstream side of the arch shoulder rock (step 85-1 in Fig. 5.20). This phenomenon arises from the presence of a steeply dip angle fault f_{34} on the upstream side of the main powerhouse, causing an uneven adjustment in the stress field of the surrounding rock near the fault. The interaction between the arch shoulder and the fault creates a high-stress zone, leading to the occurrence of "stress bias phenomenon" in the stress distribution of the surrounding rock on the upstream side

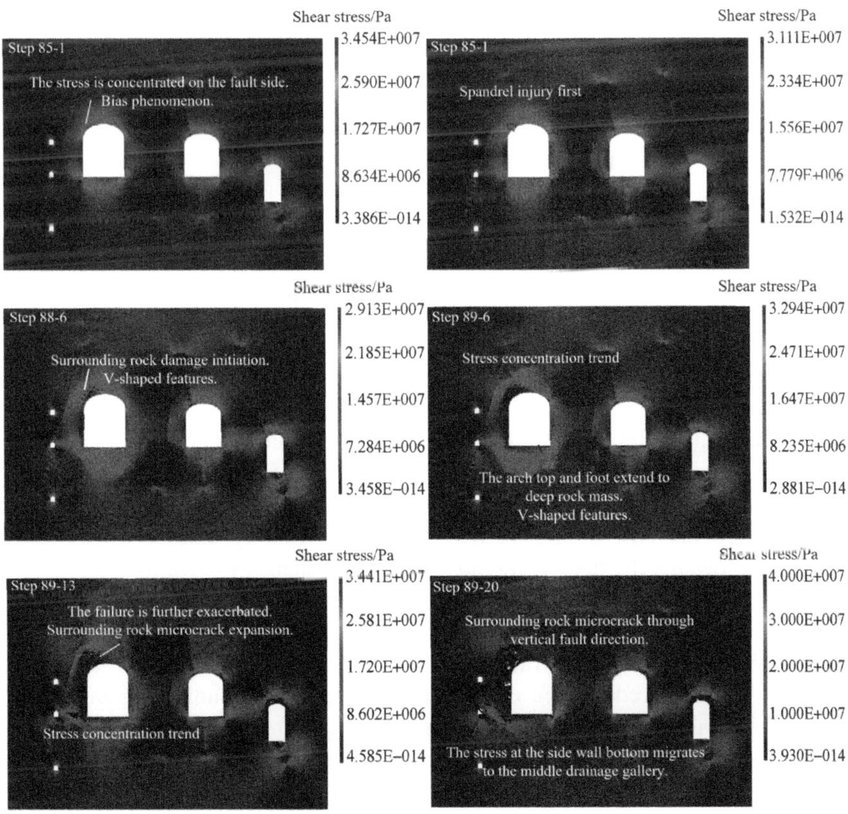

Fig. 5.20 Shear stress diagram of station left 0 + 60 m

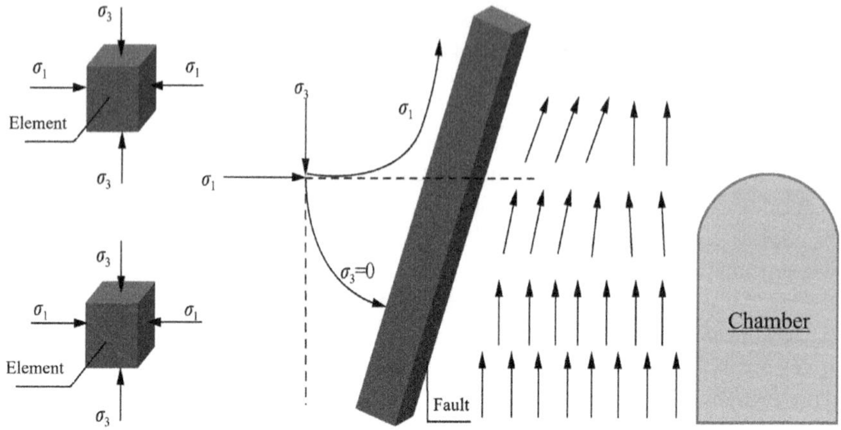

Fig. 5.21 Influence of fault on stress field of surrounding rock ($E_D << E$)

of the underground powerhouse (see Fig. 5.21) [73]. Tangential stress concentration occurs on the free surfaces of the arch shoulder on the upstream side of the powerhouse, where the surrounding rock initially experiences damage, resulting in the formation of microcracks (step 87-1 in Fig. 5.20). This provides a comprehensive explanation for the observed initiation and development of rock micro-fracture events on the upstream side of the arch shoulder in practical engineering. Based on the RFPA simulation results, the safety reserve coefficient of the surrounding rock in the underground powerhouse at this moment is 2.446, calculated using the step corresponding to the acoustic emission mutation.

2. Comparison between numerical simulation and microseismic monitoring

Considering the real-time monitoring outcomes of micro-fractures in the surrounding rock during the excavation process, we perform a comparative analysis between the numerical simulation results for the segment at station left 0 + 60 m and the concurrent monitoring results of 69 microseismic events spanning an elevation range from 152 to 210 m at station left 0 + 48 m to 0 + 72 m (Fig. 5.22). Illustrated in Fig. 5.22, under the impact of excavation disturbance, micro-fractures in the surrounding rock on the upstream side of the main powerhouse primarily exhibit tensile failure, extending towards the deep rock and forming a V-shaped excavation damage zone (Fig. 5.22a). The evolution pattern of surrounding rock damage, as derived from numerical simulation results, aligns entirely with the spatial distribution characteristics of observed rock micro-fracture events. Additionally, acoustic emission results indicate that the predominant damage mechanism in the numerical simulation is also tensile failure (Fig. 5.22b). Simultaneously, shear-type damage occurs alongside the generation of tensile fractures, corresponding to the destruction mechanism reflected by the hypocenter parameters. The observed data from microseismic events

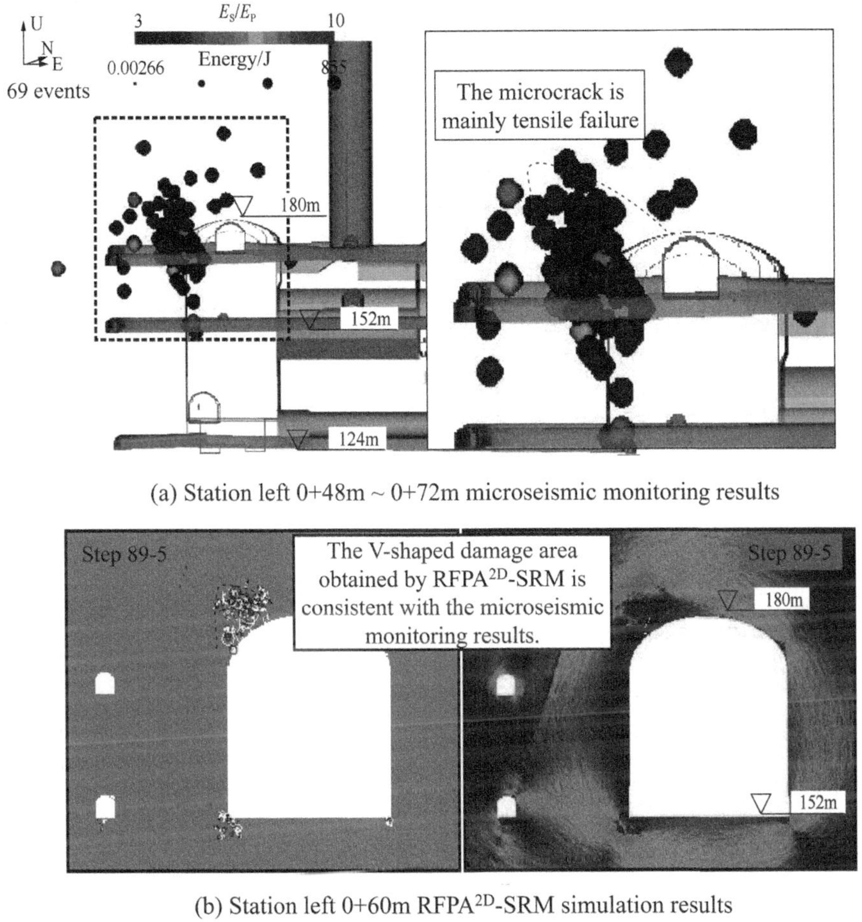

Fig. 5.22 Comparison of numerical simulation and microseismic monitoring results

corresponds with the results of the RFPA^{2D}-SRM numerical simulation, thereby validating the reliability and accuracy of the established numerical model for the section at station left 0 + 60 m and the chosen microscale elements physical–mechanical parameters. This revelation highlights the manifestation of damage evolution patterns and mechanisms in the surrounding rock, influenced by factors like faults and in-situ stress during the excavation disturbance process. The research findings suggest that stress disturbance induced by frequent blasting excavation disturb the stress equilibrium state of the surrounding rock in the underground powerhouse, triggering adjustments and redistribution of stress. The existence of a steeply dip angle fault intensifies the stress field between the fault and the arch shoulder rock of the underground powerhouse, resulting in noticeable stress concentration in the upstream side arch shoulder rock. Owing to the concentration of tangential stress, the free surfaces of the arch shoulder is prone to experiencing tensile failure. The predominant type of surrounding rock damage is tensile failure, characterized by tension fractures and shear slips occurring due to the movement of original defects, leading to the development of tensile-shear failure. The collective impact of tensile and shear failure causes cracks to propagate into the deep rock mass, ultimately giving rise to the V-shaped excavation damage zone. The evolution patterns and mechanisms of surrounding rock excavation damage in the underground powerhouse can serve as a valuable reference for subsequent excavation construction and reinforcement support design.

5.6.3 Influence of Fault Position on Surrounding Rock Stability

The numerical simulation results presented in the preceding section highlight the impact of the fault on the stress distribution in the surrounding rock of the underground cavern, resulting in a phenomenon known as "stress bias phenomenon". The deterioration of the stress field in proximity to the fault renders it prone to the formation of damaged areas. An analysis of the spatiotemporal evolution patterns of micro-fractures in the surrounding rock during the excavation process also indicates that micro-fractures in the section from station left 0 + 20 m to 0 + 80 m on the upstream side of the main powerhouse arch shoulder display a banded distribution. Micro-fractures initiate and aggregate at station 0 + 20 m (where fault f_{34} intersects with the main powerhouse sidewall at an elevation of 172 m), extending towards station 0 + 80 m (where fault f_{34} is 18 m away from the main powerhouse sidewall at an elevation of 172 m), with the maximum aggregation occurring near station 0 + 60 m. In order to deepen our comprehension of the impact of the steeply dip angle fault f_{34} at different locations on the patterns of damage evolution and stability of the surrounding rock in the underground powerhouse, RFPA numerical models are developed for analysis. The models explore fault f_{34} at various distances from the main powerhouse sidewall. Numerical models are formulated to represent practical engineering conditions at station left 0 + 20 m, 0 + 40 m, 0 + 60 m, and 0 + 80 m, corresponding to profile numbers 1, 2, 3, and 4, respectively. Distances between fault f_{34} and the sidewall are defined as d = 0 m, 6 m, 12 m, and 18 m (at an elevation of 172 m), as depicted in Fig. 5.23. Model parameters, boundary conditions, and the application of the strength reduction method for calculations remain consistent with the previous methodology.

Analogous to the Fig. 5.20, when the fault is positioned at varying distances from the upstream side of the main powerhouse arch foot, the arch shoulder rock in proximity to the fault side becomes a stress concentration area. With the degradation of rock strength, the surrounding rock experiences an evolution of damage, ultimately culminating in macroscopic instability and failure. The rock near the arch shoulder on the upstream side of the main powerhouse initiates damage on the exposed surface initially, gradually transforming into a V-shaped damaged area,

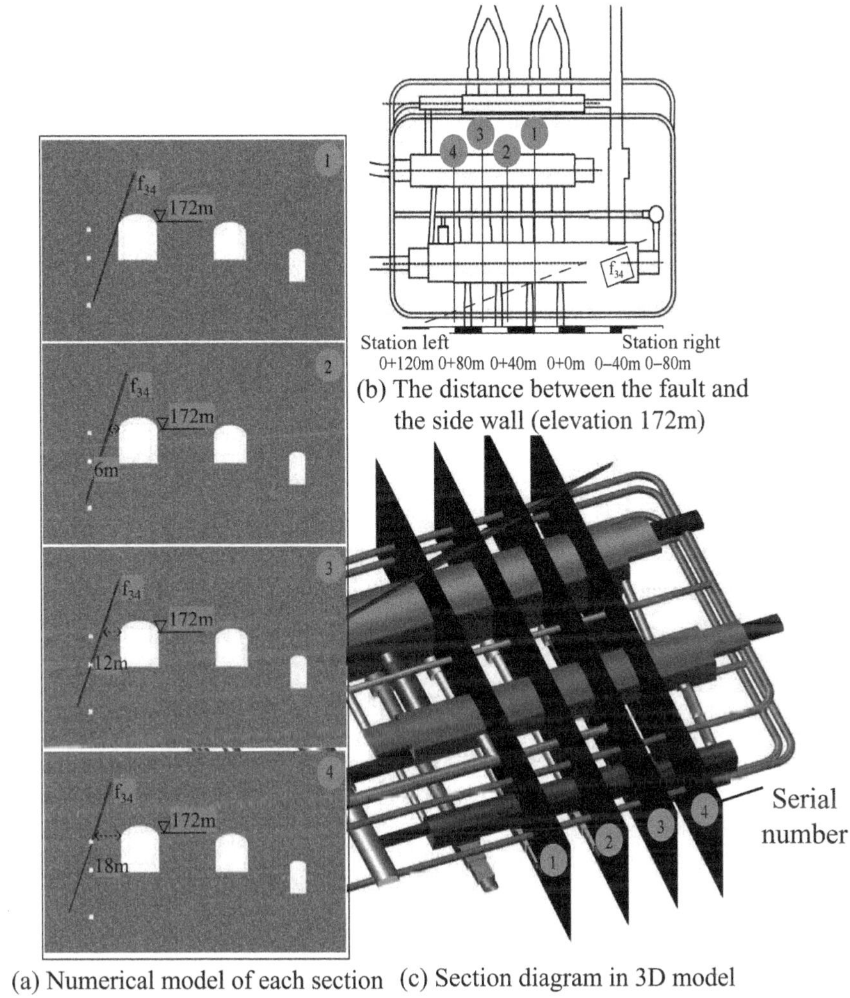

Fig. 5.23 Schematic diagram of the sectional numerical model at the same fault location

and eventually developing macroscopic cracks with a tendency toward the vertical fault. The stress concentration zone near the upstream side of the arch shoulder shifts in the direction of the vertical fault as the damage evolution in surrounding rock. In Fig. 5.24, the numerical simulation results of the elastic modulus of the surrounding rock for various station stake's profiles are compared with the practical engineering microseismic monitoring outcomes. It is evident that, at different distances between the fault and the main powerhouse sidewall, the configuration and position of the simulated damage areas in the arch shoulder rock, align well with the features of observed micro-fracture aggregation in the practical engineering. This suggests that RFPA, utilizing strength reduction modeling of construction disturbances, can depict the cumulative evolution patterns of surrounding rock damage and potential instability failure modes during construction disturbance in the underground powerhouse. The simulation outcomes offer guidance and serve as a reference for subsequent construction and support reinforcement measures in practical engineering. It is advisable to closely monitor the rock condition in regions where the fault is in proximity to the sidewall (0 + 20 m to 0 + 60 m) during the excavation process.

Consequently, safety storage coefficient for the stability of the surrounding rock at various distances between the fault and the powerhouse are computed by extracting the step numbers corresponding to acoustic emission mutations at different profiles, as detailed in Table 5.7. The stability of the surrounding rock is at its weakest when the

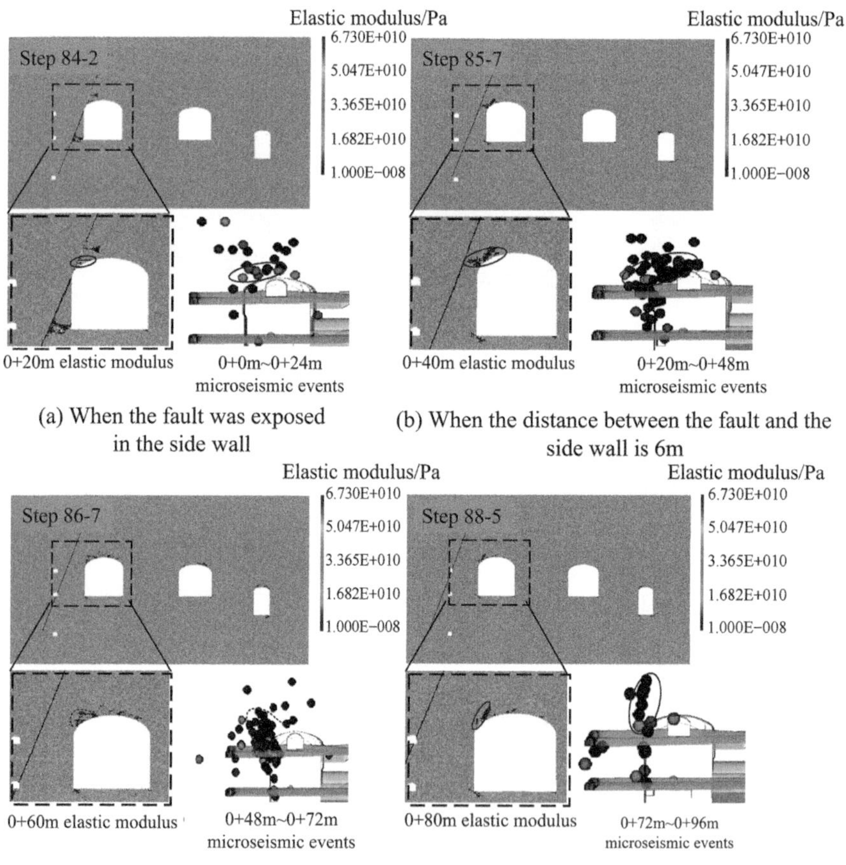

(c) When the distance between the fault and the side wall is 12m

(d) When the distance between the fault and the side wall is 18m

Fig. 5.24 Comparison of numerical simulation and microseismic monitoring results

Table 5.7 Safety coefficient of surrounding rock in different sections

Section number	Distance from the side wall/m	Instability step	Safety coefficient
1 (0 + 20 m)	0	84	2.326
2 (0 + 40 m)	6	85	2.350
3 (0 + 60 m)	12	86	2.373
4 (0 + 80 m)	18	88	2.424

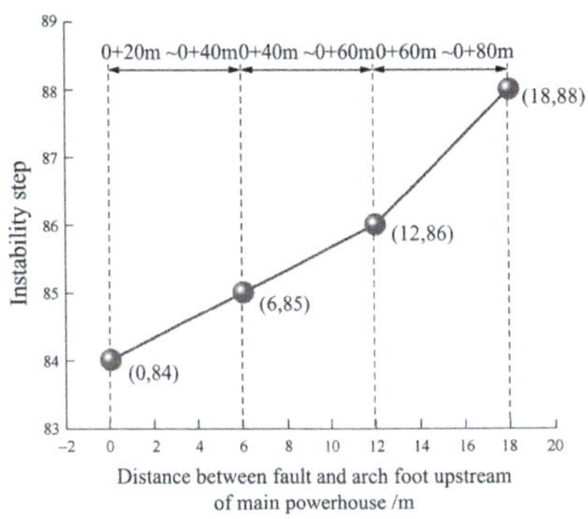

(a) The relationship between instability step and distance

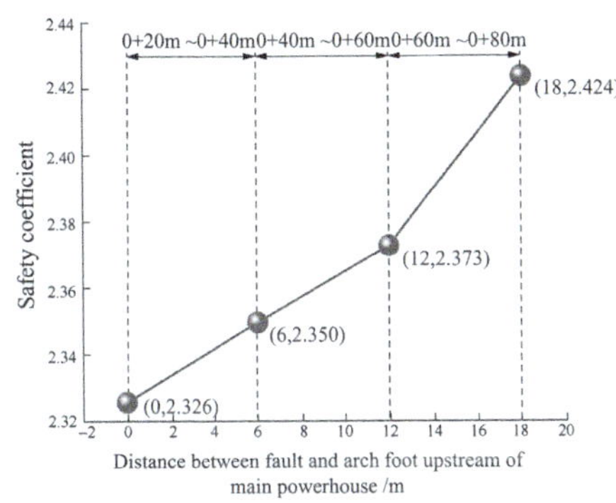

(b) The relationship between safety coefficient and distance

Fig. 5.25 Relation curve between surrounding rock stability and fault location

fault is exposed at the main powerhouse sidewall. However, as the distance between the fault and the sidewall increases, the influence on the stability of the surrounding rock diminishes. With the fault positioned 18 m away from the sidewall (stake station left 0 + 80 m), the safety storage coefficient measures 2.424, surpassing the calculated safety storage coefficient of 2.397 in the absence of the fault. It can be inferred that the presence of the fault at this distance has no discernible impact on the stability of the surrounding rock (Fig. 5.25). Consequently, the influence distance of the fault on the stability of the surrounding rock is established as 18 m (horizontal distance between the fault and the arch foot). This aligns with the region identified by microseismic monitoring results, acknowledging the damaged area in the surrounding rock near the upstream side arch shoulder at stake station left 0 + 20 m to 0 + 80 m.

5.6.4 Influence of Various Lateral Pressure Coefficients on the Stability of Surrounding Rock

Stress levels are important factors that affect the stability of the surrounding rock in underground caverns. The mechanical mechanism of instability in the surrounding rock during excavation primarily involves the adjustment and redistribution of the

Table 5.8 Horizontal deviant stress values corresponding to different lateral pressure coefficients when Y = 8 MPa (dead weight stress)

Lateral pressure coefficient K	Horizontal stress value X/MPa
0.5	4
1.0	8
1.25	10
1.5	12
2.0	16

initial ground stress. Whether the redistribution of stress exceeds the strength limit of the surrounding rock depends on the magnitude and direction of the ground stress, as well as the state of the rock mass [75]. In this section, the damage evolution law and failure mode of the horseshoe-shaped underground powerhouse are numerically simulated based on the excavation of the underground powerhouse in the Huanggou pumped storage power station.

The size, parameters, and boundary conditions of the numerical model RFPA in this section are consistent with those in Sect. 5.6.2. The dead weight stress of 8 MPa remains unchanged, while the horizontal stress load value in the X direction is varied to simulate different lateral pressure coefficients (Table 5.8). The damage evolution law and failure mode of the surrounding rock in the underground power house caisson group under different deviating stresses are determined. The numerical simulation model does not consider the influence of structural planes such as faults, and it is based on the actual construction conditions during microseismic monitoring.

Figure 5.26 depicts the shear stress diagram and acoustic emission diagram of the simulation results under different side pressure coefficients, illustrating the damage evolution process and potential instability failure mode of the surrounding rock of the underground powerhouse. As can be observed from Fig. 5.26, the stress field of the surrounding rock is symmetrically distributed along the axis of the cavern without considering the fault, thus confirming the uneven distribution of stress caused by the existence of a fault from the side. When K = 0.5, the horizontal stress is lower than the gravitational stress. Under the influence of the gravitational stress, damage initially occurs on both sides of the spandrel and the bottom foot of the side wall of the main powerhouse, and then progresses and expands from the bottom foot upwards, resulting in distinct “< >” shaped damage areas on both sides of the side wall and ultimately leading to the formation of macroscopic cracks. The stability of the partition rock mass between the caves is significantly affected by the gravitational stress. By comparing the simulation results of K = 1.0 and K = 1.25, it can be observed that the weak parts of the cavern structure, such as the spandrel and foot of the side wall, are the first to experience damage, and micro-cracks form at the upstream side arch foot of the main powerhouse. The difference lies in the fact that when the horizontal stress exceeds the dead weight stress, the stress does not concentrate on the partition rock mass between the two caverns, and the potential instability failure of the partition rock mass does not occur (K = 1.25). However, when K = 1.0, the rock mass of the cavern partition wall still experiences stress concentration, leading to damage evolution and eventual instability. When K = 1.5

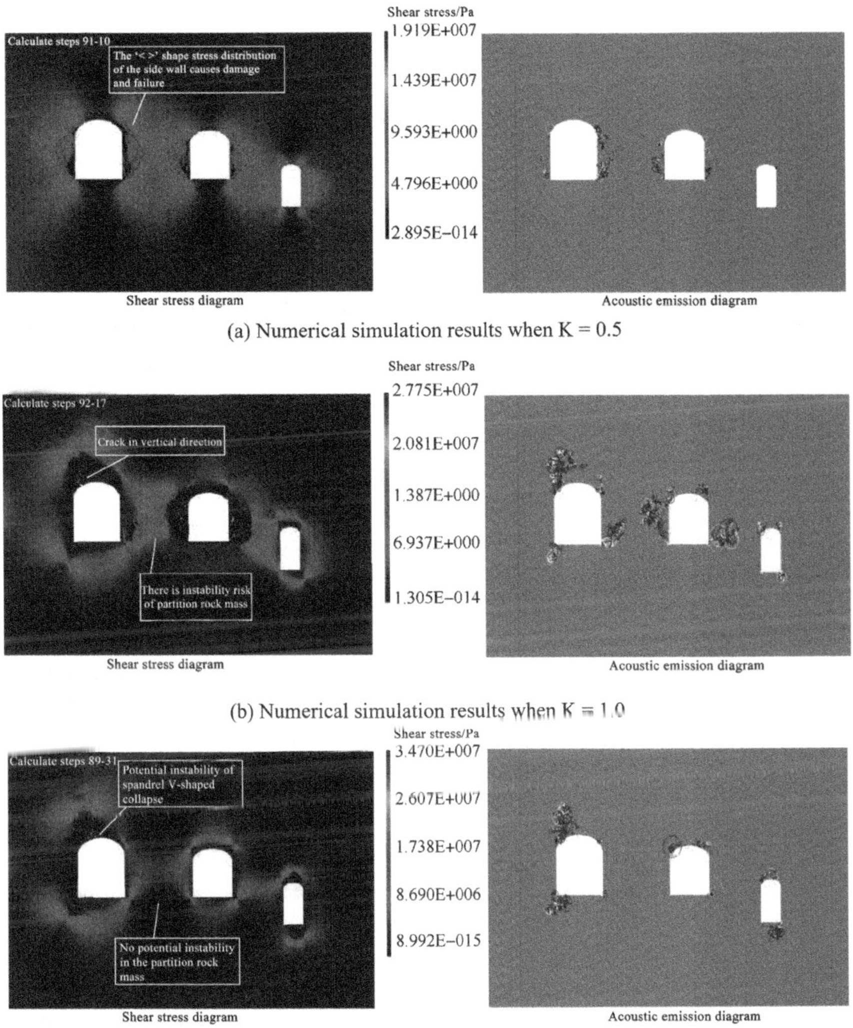

(a) Numerical simulation results when K = 0.5

(b) Numerical simulation results when K = 1.0

(c) Numerical simulation results when K = 1.25

Fig. 5.26 Numerical simulation results of different lateral pressure coefficients

and K = 2.0, the stress primarily concentrates in the surrounding rock of the arch and floor of the main powerhouse, and the damage evolution of the surrounding rock results in a V-shaped damage area. The initiation and expansion of micro-cracks in the arch and the subsequent breakthrough led to a potential instability failure mode of arch collapse, while the rock mass of the partition wall between the caverns remains stable. The acoustic emission results highlight that the damage mechanism on the open surface of the surrounding rock is mainly due to tension failure, accompanied by shear failure. The numerical simulation results indicate that the horizontal stress

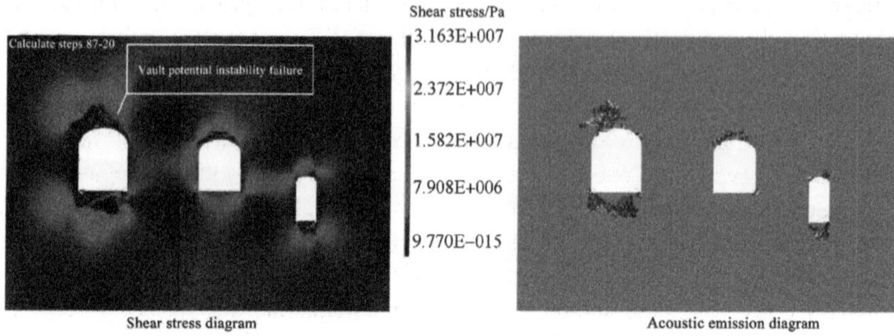

(d) Numerical simulation results when K = 1.5

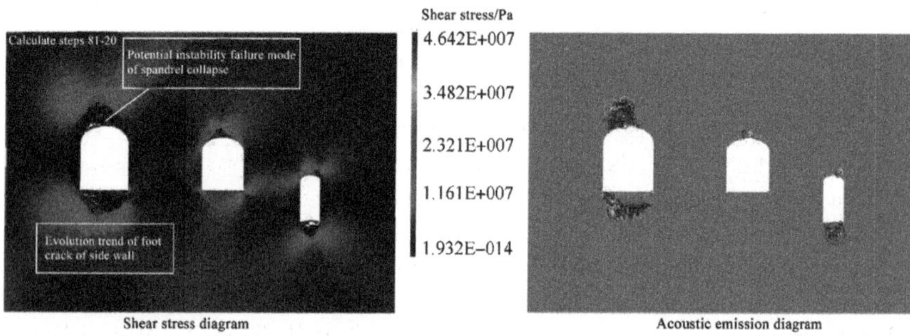

(e) Numerical simulation results when K = 2.0

Fig. 5.26 (continued)

influences the damage evolution pattern and potential instability failure mode of the surrounding rock in long-span underground caverns.

The calculation steps for determining the maximum number of unit failure times in the RFPA intensity reduction calculation process was extracted, which corresponds to the abrupt point of AE. Furthermore, the safety reserve coefficients of the underground plant were calculated for different side pressure coefficients, as presented in Table 5.9. When K equals 1.0, the maximum safety reserve factor is 2.521. This implies that the horizontal stress is comparable to the dead weight stress and contributes to the stability of the underground caverns. Conversely, as the lateral pressure coefficient increases, the stability of the underground powerhouse deteriorates (refer to Fig. 5.27).

Table 5.9 Safety reserve coefficients of underground powerhouse under different side pressure coefficients

Variable	Lateral pressure coefficient K				
	0.5	1.0	1.25	1.5	2.0
Number of instabilities	91	92	89	87	81
Safety reserve factor	2.496	2.521	2.466	2.397	2.257

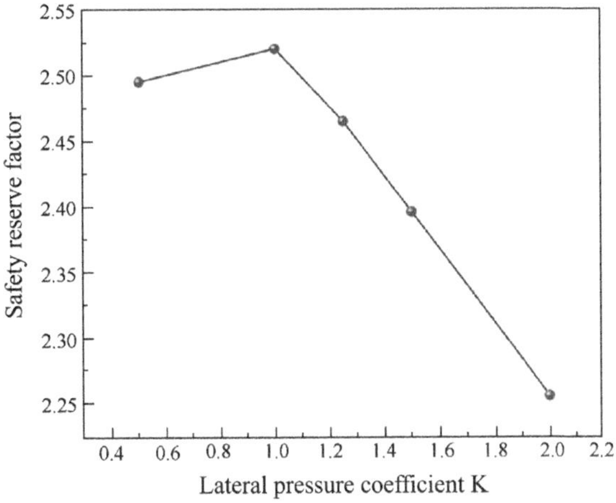

Fig. 5.27 Relationship between safety reserve factor and lateral pressure coefficient

5.6.5 Summary

The spatial and temporal evolution of microseismic events indicates that the damage to surrounding rock in underground chambers is mainly caused by stress disturbances generated by frequent blasting and excavation. Therefore, in this section, the strength reduction method of rock fracture process analysis (RFPA) is employed to simulate the progressive evolution of surrounding rock damage and potential instability failure modes resulting from the deterioration of rock strength under construction disturbances from the perspective of micro-fracturing of rocks. Furthermore, the different evolution processes of surrounding rock damage leading to instability under different fault locations and lateral pressure coefficient conditions are discussed. Combining with comparative analysis of microseismic monitoring results, the following main conclusions are drawn.

(1) The micromechanical parameters of surrounding rock physical–mechanical properties were validated through uniaxial compression numerical simulation tests. A numerical model of RFPA was established for the profile of the factory area at 0 + 60 m. The simulation results indicate that the damage to the upstream side of the main plant's surrounding rock originates from the arch foot and arch top, develops towards the deep rock mass of the arch shoulder, and eventually forms a V-shaped damage zone pointing towards the fault. The numerical simulation results are consistent with the microseismic clustering characteristics observed on-site, revealing the dominant tensile failure mechanism of the surrounding rock.
(2) The presence of fault f_{34} leads to an uneven distribution of stress in the surrounding rock, causing a "biased pressure" phenomenon in the arch shoulder. This phenomenon triggers the initiation and development of micro-fracturing, leading to progressive damage in the surrounding rock. The numerical simulation results explain the progressive evolution of microseismic events observed

in the upstream side of the arch shoulder of the main plant from the perspective of stress field changes in actual engineering. The numerical simulation results of the progressive damage process of the surrounding rock profiles at different positions of the fault match the results of microseismic monitoring. The closer the fault is to the sidewall, the poorer the stability of the surrounding rock. Based on the simulated safety reserve coefficient of surrounding rock stability, the distance affected by the fault is determined to be 18 m.

(3) When the gravity stress is constant, the magnitude of horizontal structural stress determines the evolution pattern and potential failure mode of the surrounding rock in underground chambers. When $K < 1$, the gravity stress dominates and forms a "< >" shaped damaged zone on the sidewalls of the chambers, eventually leading to macroscopic instability failure, with the rock mass of the partition wall between two chambers prone to through-going failure. When $K > 1$, damage under the dominance of horizontal stress mainly occurs in the tensile damage locations of the arch shoulder and base of the sidewall, extending towards the deep rock mass of the chamber to form a V-shaped damage zone. The progressive evolution of the arch shoulder damage ultimately leads to potential instability and collapse of the arch top. When $K = 1$, the vertical progressive damage evolution of the upstream side of the arch shoulder in the main plant occurs, with the highest safety reserve coefficient at this point, indicating the best stability of the chamber surrounding rock.

5.7 Conclusion

This chapter establishes a microseismic monitoring system for the underground plant of a pumped storage power station during the excavation process. Through the monitoring of microfracturing in the surrounding rocks under the influence of excavation and unloading, the spatiotemporal evolution patterns of microfracturing is studied. The chapter also investigates the relationship between construction dynamics, fault structures, and microseismic activity characteristics. It identifies the areas at risk of damage in the surrounding rocks of the underground plant. Furthermore, by using comparative analysis with numerical simulations, the chapter reveals the damage mechanisms in the surrounding rocks. The influences of fault locations and lateral pressure coefficients on the evolution of rock damage are investigated. These findings provide valuable insights for construction and support during the construction period. The main conclusions drawn from this study are as follows:

(1) By conducting manual point impact tests, the overall equivalent wave velocity of the rock mass in the monitoring area was determined to be 5,150 m/s, with a positioning error within 6 m, which meets the requirements of underground engineering. Threshold setting, spectral analysis, manual identification, on-site construction comparison, and other methods are used to filter out noise signals other than rock micro-fractures, ensuring the timeliness and reliability of microseismic monitoring result processing.

(2) Based on the spatiotemporal evolution patterns of microseismic events and seismic source parameters (E_S/E_P), this study investigates the relationship between microseismic activity characteristics and construction dynamics, as well as fault structures. The findings indicate that the frequent blasting induces stress perturbations, leading to the initiation and development of numerous small-magnitude tensile microcracks in the surrounding rock mass of the underground powerhouse. The construction dynamics involving excavation and blasting exhibit consistent spatiotemporal responses with microseismic activity.

(3) Based on the principle of energy dissipation and in conjunction with the microseismic energy loss density cloud map and on-site survey data, the potential hazard zone comprising the upstream side arch shoulder surrounding rock in the range of stationing from 0 + 20 m to 0 + 80 m and elevations from 152 to 200 m is identified and delineated.

(4) Numerical simulation on the cross-section at station 0 + 60 m of the underground powerhouse indicate that the deterioration of surrounding rock strength leads to a modification in the stress field of the surrounding rock. Stress is transferred and concentrated towards the rocks around the fault, resulting in the formation of a V-shaped damaged zone in the upstream side arch shoulder of the main powerhouse. Ultimately, this can lead to a potential destabilization mode with the development of vertical fault-penetrating cracks.

(5) When the steeply inclined fault is at different distances from the sidewall, there are distinct patterns of damage evolution in the surrounding rock of the underground powerhouse. Numerical results are consistent with actual microseismic monitoring results. Based on the safety reserve factor, it is determined that the influence distance of the fault on the stability of the surrounding rock is 18 m. When the self-weight remains constant, the magnitude of horizontal stress directly influences the damage evolution pattern of the surrounding rock and potential instability modes. Under the dominance of self-weight ($K < 1$), V-shaped damaged zones occur on both sides of the chamber walls. On the other hand, under the dominance of horizontal stress ($K > 1$), V-shaped damaged zones are more likely to form at the cavern roof and bottom. The stability of the intervening rock masses between caverns is significantly influenced by self-weight, and when $K = 1$, the maximum safety reserve factor for the surrounding rock is 2.521.

References

1. Yan Y (2015) Seismic response analysis of surrounding rock and structure for underground powerhouse of hydropower station. Wuhan University, Wuhan
2. Koutsoyiannis D (2011) Scale of water resources development and sustainability: small is beautiful, large is great. Hydrol Sci J 56(4):553–575
3. Zarfl C, Lumsdon AE, Berlekamp J et al (2015) A global boom in hydropower dam construction. Aquat Sci 77(1):161–170

4. Huang H, Yan Z (2009) Present situation and future prospect of hydropower in China. Renew Sustain Energy Rev 13(6–7):1652–1656
5. Barnes MJ (2009) Hydropower in Europe: current status, future opportunities. Hydro Rev Worldwide 17(3):24
6. Sternberg R (2010) Hydropower's future, the environment, and global electricity systems. Renew Sustain Energy Rev 14(2):713–723
7. The National People's Congress Finance and Economy Committee, National Development and Reform Commission (2016) 2016—2020 Outline of the thirteenth-year plan for national economy and social development in People's Republic of China. Explanatory material. China Plan Press, Beijing
8. Xu GL, Li ZP, Song SW et al (2016) Characteristics of hydropower underground caverns engineering in China. Bull Geol Sci Technol 35(2):203–208
9. Meng GT, Fan YL, Jiang YL et al (2016) Key rock mechanical problems and measures for huge caverns of Baihetan Hydropower Plant. Chin J Rock Mech Eng 35(12):2549–2560
10. Wang AL, Deng ZG (2015) Development of underground powerhouse of hydropower station in China. Water Power 41(6):65–68
11. Qian QH (2012) Challenges faced by underground projects construction safety and countermeasures. Chin J Rock Mech Eng 31(10):1945–1956
12. Ishida T, Kanagawa T, Uchita Y (2014) Acoustic emission induced by progressive excavation of an underground powerhouse. Int J Rock Mech Min Sci 71(S29):362–368
13. Dong JX, Xu GL, Li ZP et al (2014) Classification of failure modes and controlling measures for surrounding rock of large-scale caverns with high geostress. Chin J Rock Mech Eng 33(11):2161–2170
14. Cai B, Deng ZW, Wu GZ (2012) Stability analysis of surrounding rock in collapse area of underground powerhouse of Dagangshan Hydropower Station. Yangtze River. 43(22):33–35
15. Wei ZY, Xu GL, Shen YJ et al (2013) Characteristics and stability evaluations of diabase dikes group surrounding underground caverns of Dagangshan hydropower station. J Eng Geol 21(2):206–215
16. Li ZK, Zhou Z, Tang XF et al (2009) Stability analysis and considerations of underground powerhouse caverns group of Jinping I Hydropower Station. Chin J Rock Mech Eng 28(11):2167–2175
17. Lu B, Wang JM, Ding XL et al (2010) Study of deformation and cracking mechanism of surrounding rock of Jinping I underground powerhouse. Chin J Rock Mech Eng 29(12):2429–2441
18. Luo SJ (2013) Risk control of geological disaster in underground engineering of pumped storage power station. Collection of Papers on Pumped Storage Power Station Engineering Construction 013. CSHE
19. Liu YG (2010) Underground cavity seepage drainage problems of Xilongchi Pumped Storage Power Station. Water Power 36(1):42–44, 99
20. Jing K, Wang Q, Feng YM et al (2017) Analysis of water seepage in underground powerhouse of Pushihe pumped storage power station. Water Resour Power 35(1):108–110
21. Oliveira DAF, Indraratna B (2010) Comparison between models of rock discontinuity strength and deformation. J Geotech Geoenviron Eng 136(6):864–874
22. Li LC, Tang CA, Liang ZZ et al (2009) Influence of weak interlayer on the damage mode of rock mass surrounding deep underground opening. Chin J Undergr Space Eng 5(5):856, 859, 866
23. Zhu WS, Li XJ, Zhang QB et al (2010) A study on sidewall displacement prediction and stability evaluations for large underground power station caverns. Int J Rock Mech Min Sci 47(7):1055–1062
24. Ni SH, He SH, Wang XG et al (2012) Hydraulic properties of fractured rock mass. Chin J Rock Mech Eng 31(3):488–498
25. Wang M, Xiao M, Li LZ (2015) Stability analysis of surrounding rock under seepage action of fractured rock mass. Water Resour Power 33(4):130–134

26. Read RS (2004) 20 years of excavation response studies at AECL's underground research laboratory. Int J Rock Mech Min Sci 41(8):1251–1275
27. Tang CA, Wang JM, Zhang JJ (2010) Preliminary engineering application of microseismic monitoring technique to rockburst prediction in tunneling of Jinping II project. J Rock Mech Geotech Eng 2(3):193–208
28. Xie HP, Ju Y, Li LY (2005) Criteria for strength and structural failure of rocks based on energy dissipation and energy release principles. Chin J Rock Mech Eng 24(17):3003–3010
29. Dou LM, Jiang YD, Cao AY et al (2017) Monitoring and pre-warning of rockburst hazard with technology of stress field and wave field in underground coalmines. Chin J Rock Mech Eng 36(4):803–811
30. Ma K, Tang CA, Liang ZZ et al (2016) Stability analysis of the surrounding rock of underground water-sealed oil storage caverns based on microseismic monitoring during construction. Chin J Rock Mech Eng 35(7):1353–1365
31. Xiao YX, Feng XT, Li SJ et al (2016) Rock mass failure mechanisms during the evolution process of rockbursts in tunnels. Int J Rock Mech Min Sci 83:174–181
32. Xu NW, Dai F, Liang ZZ et al (2014) The dynamic evaluation of rock slope stability considering the effects of microseismic damage. Rock Mech Rock Eng 47(2):621–642
33. Dai F, Li B, Xu NW et al (2016) Deformation forecasting and stability analysis of large-scale underground powerhouse caverns from microseismic monitoring. Int J Rock Mech Min Sci 86:269–281
34. Li A, Dai F, Xu NW et al (2017) Failure mechanism and mode of surrounding rock of underground powerhouse at the right bank of Wudongde Hydropower Station subjected to excavation. Chin J Rock Mech Eng 36(4):781–793
35. Ling Y (2010) Back analysis of initial stress field and stabilization analysis of surrounding rock in underground power house. Dalian University of Technology, Dalian
36. Gehrels T (2010) Numerical analysis of groups of underground cavity powerhouse during excavation process. World Build Mater 64(2):229–252
37. Qi DQ, Xu LM, Li R (2009) The rock stability analysis of underground powerhouse excavation on at a power plant. Chin Q Mech 30(2):337–346
38. Zhang L, Ding XL, Fu J (2003) Three dimensional numerical analysis of stability of surrounding rocks of Shuibuya Underground Powerhouse. Rock Soil Mech S1:120–123
39. Zhang YH, Wei Q, Sheng Q et al (2011) Three-dimensional back analysis of geostress field in underground powerhouse zone of Dagangshan Hydropower Station. Rock Soil Mech 32(5):1523–1530
40. Hao YH, Azzam R (2005) The plastic zones and displacements around underground openings in rock masses containing a fault. Tunn Undergr Space Technol 20(1):49–61
41. Wu AQ, Ding XL, Chen SH et al (2006) Researches on deformation and failure characteristics of an underground powerhouse with complicated geological conditions by DDA method. Chin J Rock Mech Eng 1:1–8
42. Sa WQ, Zhang SR, Zhang LM (2014) Real-time safety evaluation and early warning of large-scale underground cavern group during construction period based on internet of things. Chin J Rock Mech Eng 33(11):2301–2313
43. Fan QX, Liu YY, Wang Y (2011) Construction and monitoring study of large underground powerhouse caverns of Xiangjiaba Hydropower Station. Chin J Rock Mech Eng 30(4):666–676
44. Li JH, Wu WF, Li JC (2013) Control and monitoring of rock mass engineering of super large underground powerhouse cavern groups at Xiluodu Hydropower Station. Chin J Rock Mech Eng 32(1):8–14
45. Cai DW, Wang DK, Chen XP (2014) Analysis and evaluation system of rapid monitoring of large-scale underground cavern group during construction period in Southwest China. Chin J Rock Mech Eng 33(11):2341–2350
46. Li SJ, Feng XT, Xu DP et al (2016) Study of the characteristics and mechanism of surrounding rock deformation during the first layer excavation in Baihetan hydropower station underground main powerhouses. Chin J Rock Mech Eng 35(S2):3947–3959

47. Zhao Y, Zhang JW, Yuan SF (2014) Safety evaluation for stability of underground powerhouse surrounding rock based on catastrophe theory. Chin J Rock Mech Eng 33(S2):3973–3978
48. Wei JB, Deng JH, Wang DK et al (2010) Characterization of deformation and fracture for rock mass in underground powerhouse of Jinping I Hydropower Station. Chin J Rock Mech Eng 29(6):1198–1205
49. Maejima T, Morioka H, Mori T et al (2003) Evaluation of loosened zones on excavation of a large underground rock cavern and application of observational construction techniques. Tunn Undergr Space Technol 18(2):223–232
50. Qin WM, Sun Y, Chen RF et al (2008) Application of total station instrument and sliding micrometer to monitoring Shuibuya Underground Powerhouse. Rock Soil Mech 2:557–561
51. Wang KZ, Cai MF (2005) In-situ monitoring and stability analysis of surrounding rock mass for Xilongchi underground caver. Chin J Rock Mech Eng S2:5956–5960
52. Gan XQ, Tan Y, Li DY (2012) Stability of surrounding rockmass in the underground powerhouse of Bailianhe Pumped Storage Station. J Yangtze River Sci Res Inst 9:107–110
53. Shen W (2010) The monitoring and numerical simulating of construction in underground powerhouse of Shiziping Hydropower Station. Wuhan University of Technology, Wuhan
54. Xiao YX (2013) Frequency fractal feature of microseismic events in the process of immediate rockburst in deep tunnel. The ISRM Commission on Design Methodology. Rock Characterisation, Modelling and Engineering Design Methods—Proceedings of the 3RD Isrm Sinorock Symposium. The ISRM Commission on Design Methodology
55. Ma K, Jin F, Tang CA et al (2017) Deformation mechanism of the Dam Heel of Dagangshan High Arch Dam based on microseismic monitoring during initial impoundment. Chin J Rock Mech Eng 36(5):1111–1121
56. Zhang BH, Deng JH, Gao MZ et al (2012) Safety evaluation research based on microseismic monitoring in underground powerhouse of hydropower station. Chin J Rock Mech Eng 31(5):937–944
57. Xu NW, Li TB, Dai F et al (2015) Microseismic monitoring and stability evaluation for the large-scale underground caverns at the Houziyan Hydropower Station in Southwest China. Eng Geol 188:48–67
58. Dai F, Li B, Xu MW et al (2016) Microseismic characteristic analysis of underground powerhouse at Baihetan Hydropower Station subjected to excavation. Chin J Rock Mech Eng 35(4):692–703
59. Wu ML, Liao CT (2001) The measurement and study on in situ stress in the central area of the Huanggou Accumulation Power Station, Heilongjang. J Geomech 1:61–68
60. Liu LJ, Tian ZY, Zheng YB et al (2016) Study of site selection of deep underground powerhouse in Huanggou Pumped Storage Power Station. Resour Environ Eng 30(1):96–99
61. Cai GZ, Yu CZ, Xiong L et al (2016) Schedule and excavation of underground powerhouse for hydropower station. Water Resour Hydropower Northeast China 34(11):1, 2, 7, 71
62. Zhuang DY, Ma K, Tang CA et al (2015) Application of microseism monitoring technology in underground water-sealed oil storage caverns. Oil Gas Stor Transp 34(8):799, 806, 862
63. Li N (2014) Research on mechanisms of key factors and reliability for microseismic source location. China University of Mining and Technology, Wuhan
64. Wyllic R, Gregory A, Gardner G (1985) An experimental investigation of factors affecting elastic wave veloicties in porous media. Geophysics 23(3):459–493
65. Nur A, Simmond G (1969) Stress-induced velocity anisotropy in rock: an experimental study. J Geophys Res 74(27):6667–6674
66. Li B, Dai F, Xu NW et al (2014) Microseismic monitoring system and its engineering applications of deep-buried underground powerhouse. Chin J Rock Mech Eng 33(S1):3375–3383
67. Jia BX, Jia ZB, Zhao P et al (2017) Microseism location in local scale region based on high-density array. Chin J Geotech Eng 39(4):705–712
68. Huang RQ (2008) Geodynamical process and stability control of high rock slope development. Chin J Rock Mech Eng 27(8):1525–1544
69. Zhao ZN, Feng XT, Xiao YX et al (2016) Microseismic characteristics and rockburst risk of deep tunnel constructed by different excavation methods. Chin J Geotech Eng 38(5):867–876

70. Li ZH (2013) The formation and evolution mechanism of excavation damaged zone under deep tunnel excavation. Northeast University, Shenyang
71. Zhang WD, Ma TH, Tang CA et al (2014) Research on characteristics of rockburst and rules of microseismic monitoring at diversion tunnels in Jinping II Hydropower Station. Chin J Rock Mech Eng 33(2):339–348
72. Chen BR, Feng XT, Zeng XH et al (2011) Real-time microseismic monitoring and its characteristic analysis during TBM tunneling in deep-buried tunnel. Chin J Rock Mech Eng 30(2):275–283
73. Hudson JA, Cooling CM (1988) In situ rock stresses and their measurement in the UK-Part I, the current state of knowledge. Int J Rock Mech Min Sci Geomech Abstr 25:363–370
74. Yang Q, Yang G, Wang ZC et al (2007) Application of block theory to surrounding rock stability of underground caverns in Huanggou Pumped Storage Station. Chin J Rock Mech Eng 8:1618–1624
75. Zhu YW, Huang KJ, Li W (2004) Study on effect of in-situ stresses on plastic region induced by excavations of underground openings. Chin J Rock Mech Eng 8:1344–1348

Postscript

Time flies and with a myriad of thoughts in mind. This book explores the application of microseismic monitoring technology across a variety of projects, from slope engineering, underground water-sealed oil caverns, to high arch dams, incidents of water bursting into coal mines, and pumped storage facilities located underground. In fact, it is penned down following the sequence in which I joined these engineering projects. Reflecting on life as I approach the age of forty, the scenes from my student days remain vividly etched in my memory. At this time, I am grateful for my relentless perseverance. During my doctoral studies, I faced moments of confusion about an uncertain future, worries over financial constraints, and the distress of having papers rejected. In the early stages of my postdoctoral work in Beijing, I felt depressed due to financial difficulties and repeated job search failures. It seemed as if the first half of my life was spent being denied, doubted, and selected against. Though life was as fragile as an ant, I aspired to have the ambition of a great swan; though fate seemed as thin as paper, I possessed an indomitable spirit. Life need not be filled with confusion or blind conformity; time will eventually bring the right people and opportunities to us! This book is dedicated to every dream chaser who is determined to move forward.

The sudden outbreak of the COVID-19 pandemic at the end of 2019 had a significant impact on all sectors. At the time, the trend of leaving jobs to embark on postgraduate or doctoral studies gained momentum. However, without the bravery to commit fully and the persistence to work tirelessly, attaining the sought-after results in research becomes challenging. Rashly pursuing advanced studies due to prevailing trends, driven by a state of restlessness and worry, might eventually result in outcomes that defy expectations. "Not braving the vast ocean for fear of stormy waves, one punch to calm the surging tides"—in this era of both risk and opportunity, many people face unemployment and layoffs, yet there are also industry leaders like Jack Ma, Yu Minhong, and Ren Zhengfei. These individuals, during their entrepreneurial and striving phases, have all endured hardships beyond the imagination. In these times of swift economic expansion, by maintaining our focus, holding steadfast to our convictions, and committing fully to our tasks, we are bound to achieve happiness

K. Ma and C. Tang, *Microseismic Monitoring and Stability Analysis of Large Rock Mass Engineering*, https://doi.org/10.1007/978-981-96-6483-2

in our lives, irrespective of our chosen fields. It is my wish that all readers achieve meaningful success in their endeavors.

Upon completing this book, I would like to extend my special thanks to Professor Tang Chun'an, who has guided me for many years, and to my fellow mentors and peers with whom I have long collaborated. The generation, verification, modification, and perfection of various new ideas have all benefited from explorations in engineering. During my postdoctoral period, Dr. Zhuang Duanyang visited me in Beijing several times. The scenes of us dining together at China Agricultural University (East Campus) are still vivid in my memory. Later, we collaborated on the petroleum cavern project. Dr. Liu Xingzong, another junior colleague, joined me for a research trip to the Dahuaqiao Hydropower Station in Yunnan during my postdoc. I am grateful to the senior brothers in Professor Tang Chun'an's research group, CRISR. Additionally, this book references a vast array of literature, all cited at the end of each chapter, for which I extend my thanks to the original authors.

Sincere thanks to Cao Zhiguo, Xing Zhengguo, Zhang Kai, Fang Jie, Liu Xinjie, Yang Yingming, and others from the State Key Laboratory of Coal Mining Water Resource Protection and Utilization of China National Energy Group.

I appreciated Academician Chen Zuyu and Professor Li Shihai from the Institute of Mechanics, Chinese Academy of Sciences, as well as the collaborative efforts of young scholars including Hong Chunhua, Wang Wei, Feng Chun, Li Jianxiang, Zhou Dong, Sun Hui, Qiao Jiyuan, Liu Xiaoyu, Chen Weimin, Wu Mengxi, Li Qing, Yao Bo, Meng Da, Fan Yongbo, and Yao Bo.

I would like to express my gratitude to my dear friends who have consistently understood and supported me over the years, listed in the order of our acquaintance: Jiang Song, Hu Xujun, Li Runji, Wang Bo, Yuan Hui, Lv Gang, Guan Shuang, Ma Xianda, Chi Yanfei, Li Junhao, Wang Chuantao, Wang Dongping, Guo Kefei, Wang Jinming, Tu Dan, Tang Dehong, Wang Lixiang, and others. I would also like to extend special thanks to the mentors and friends who have assisted me during my academic and professional journey; I won't list them all here, but their contributions are deeply appreciated and will be remembered always.

This book has received assistance from various collaborating entities, including Guodian Dadu River Basin Hydropower Development Co., Ltd., Guoneng Dadu River Dagangshan Power Generation Co., Ltd., Shanxi Coal and Chemical Industry Group Co., Ltd., China Petroleum Pipeline Engineering Co., Ltd., and others. We extend our sincere thanks to them for their support.

The author of this book has conducted systematic research with the support of various research projects, including: Outstanding Youth Science Fund Project "Rock Mechanics and Rock Engineering"(Project No. 42122052); National Natural Science Foundation of China "Premonitory Characteristics and Digital Twin Intelligent Early Warning Research of Engineering Disasters in High Seismic Zone High Arch Dams" (Project No. 52379098); Project "Premonitory Characteristics of Coal Mine Water Inrush Based on Microseismic Monitoring and Numerical Calculation" (Project No. 51774064). Project "Multi-Information Fusion and Premonitory Identification of Deep Coal Mine Floor Water Inrush in Cloud Computing Environment" (Project No. 51974055); National Natural Science Foundation of

China Youth Science Fund Project "Microseismic Damage Effects and Mechanical Parameter Feedback Research in Large Underground Water Barrier Oil Caverns" (Project No. 51504233); Central University Fundamental Research Project "Numerical Analysis of Underground Water Barrier Oil Caverns Based on Rock Micro-Fracture" (Project No. DUT20GJ216). China Petroleum Technology Innovation Fund Research Project "Research on Microseismic Interpretation Methods for Underground Water Barrier Oil Caverns Based on Artificial Intelligence" (Project No. 2020D-5007-0302); Shanxi Province Natural Science Basic Research Program—Shaanmei Joint Fund Project "Dynamic Characterization of Rock Fracture Processes and Fracture Degree Based on Microseismic Parameters" (Project No. 2021JLM-11). Coastal and Offshore Engineering National Key Laboratory Young Scholar Innovation Fund "Numerical Simulation Methods for Premonitory Analysis of Disaster Sources in Large Subsea Oil Storage Caverns" (Project No. LY-1805); Yunnan Provincial Basic Research Program Project "Stability Analysis of Underground Petroleum Water Barrier Caverns in Chuxiong, Yunnan" (Project No. 202001AT070150); Open Fund from the State Key Laboratory of Coal Mining Water Resource Protection and Utilization, affiliated with the National Energy Group, for the research on "Solid–Liquid Coupling Mechanisms in the Development of Shallow-Buried, Deep-Excavation-Induced Water-Conducting Fracture Zones" (Project No. SHJT-17-42.15).

The publication of this book has received support from the Dalian Municipal People's Government, and the author would like to express heartfelt gratitude for this assistance.

In a certain sense, the compilation of this book can be seen as a summary and review of research work on various rock engineering microseismic monitoring techniques. Due to the complex and variable geological conditions in large-scale rock engineering projects, microseismic monitoring technology is often researched from a specific aspect or even a single specific engineering case. Therefore, it is inevitable that there may be limitations in this book. Further in-depth research and discussion by experts in the field are needed. We sincerely invite expert readers to provide criticism and corrections to help improve and advance the knowledge and technology in this area.

Ke Ma
September 1, 2023
Bosi Park, Beijing

GPSR Compliance

The European Union's (EU) General Product Safety Regulation (GPSR) is a set of rules that requires consumer products to be safe and our obligations to ensure this.

If you have any concerns about our products, you can contact us on ProductSafety@springernature.com

In case Publisher is established outside the EU, the EU authorized representative is:

Springer Nature Customer Service Center GmbH
Europaplatz 3
69115 Heidelberg, Germany

Batch number: 10584686

Printed by Printforce, the Netherlands